Second-Generation HTS Conductors

Second-Generation HTS Conductors

Edited by

Amit Goyal
Oak Ridge National Laboratory
Oak Ridge, Tennessee, USA

KLUWER ACADEMIC PUBLISHERS
Boston / Dordrecht / New York / London

Distributors for North, Central and South America:
Kluwer Academic Publishers
101 Philip Drive
Assinippi Park
Norwell, Massachusetts 02061 USA
Telephone (781) 871-6600
Fax (781) 871-6528
E-Mail <kluwer@wkap.com>

Distributors for all other countries:
Kluwer Academic Publishers Group
Post Office Box 322
3300 AH Dordrecht, THE NETHERLANDS
Telephone 31 78 6576 000
Fax 31 78 6576 474
E-Mail <orderdept@wkap.nl>

Electronic Services <http://www.wkap.nl>

Library of Congress Cataloging-in-Publication

Second-generation HTS conductors / edited by Amit Goyal.
p. cm.
Includes bibliographical references and index.
ISBN 1-4020-8117-0– ISBN 1-4020-8118-9 (ebook)
1. High temperature superconductors. I. Goyal, (Amit)

TK454.4.S93S43 2004
621.3'5—dc22

2004053783

Printed on acid-free paper.

Printed in the United States of America

CONTRIBUTORS

Paul N. Arendt
Superconductivity Technology Center
Materials Science and Technology Division
Los Alamos National Laboratory
Los Alamos, NM 87545
USA

M. Azoulay
School of Physics and Astronomy
Raymond and Beverly Sackler Faculty of Exact Science
Tel Aviv University
69978 Tel Aviv
Israel

Markus Bauer
Technical University Munich
Physics Department E10
James-Franck Str. 1
85747 Garching
Germany

Hans M. Christen
Oak Ridge National Laboratory
Solid-State Division
Oak Ridge, TN 37831-6056
USA

Paul G. Clem
Materials Chemistry Department 01846
Sandia National Laboratories
Albuquerque, NM 87185-1411
USA

G. Deutscher
School of Physics and Astronomy
Raymond and Beverly Sackler Faculty of Exact Science
Tel Aviv University
69978 Tel Aviv
Israel

Y. Di
Jet Process Corporation
24 Science Park
New Haven, CT 06511
USA

K. Fujino
Superconductor R&D Department
Electric Power System Technology Research Labs
Sumitomo Electric Industries, LTD
1-1-3, Shimaya, Konohana-ku
Osaka 554-0024
Japan

Amit Goyal
Metals and Ceramics Division
Oak Ridge National Laboratory
Oak Ridge, TN 37831
USA

B.L. Halpern
Jet Process Corporation
24 Science Park
New Haven, CT 06511
USA

K.S. Harshavardhan
Neocera, Inc.
10000 Virginia Manor Road
Beltsville, MD 20705
USA

S. Honjo
Power Engineering R&D Center
Tokyo Electric Power Company
4-1, Egasaki-cho, Tsurumi-ku
Yokohama 230-8510
Japan

Alex Ignatiev
Space Vacuum Epitaxy Center and
Texas Center for Superconductivity
University of Houston
Houston, TX 77204-5507
USA

Teruo Izumi
Superconductivity Research Laboratory
International Superconductivity
Technology Center
10-13 Shinonome 1-chome
Koto-ku, Tokyo 135-0062
Japan

Donald M. Kroeger
Metals and Ceramics Division
Oak Ridge National Laboratory
Oak Ridge, TN 37831-6116
USA

R. Krupke
School of Physics and Astronomy
Raymond and Beverly Sackler Faculty
of Exact Science
Tel Aviv University
69978 Tel Aviv
Israel

J.Y. Lao
Department of Chemistry
State University of New York
Buffalo, NY 14260
USA

Dominic F. Lee
Metals and Ceramics Division
Oak Ridge National Laboratory
Oak Ridge, TN 37831-6116
USA

Keith J. Leonard
Metals and Ceramics Division
Oak Ridge National Laboratory
Oak Ridge, TN 37831-6116
USA

Fredrick A. List III
Metals and Ceramics Division
Oak Ridge National Laboratory
Oak Ridge, TN 37831-6116
USA

Song-Wei Lu
Metals and Ceramics Division
Oak Ridge National Laboratory
Oak Ridge, TN 37831-6116
USA

K. Ohmatsu
Superconductor R&D Department
Electric Power System Technology
Research Labs
Sumitomo Electric Industries, LTD
1-1-3, Shimaya, Konohana-ku
Osaka 554-0024
Japan

M. Parans Paranthaman
Chemical Sciences Division
Oak Ridge National Laboratory
Oak Ridge, TN 37831-6100
USA

Todd A. Polley
MicroCoating Technologies
5315 Peachtree Industrial Blvd.
Chamblee, GA 30341
USA

Werner Prusseit
THEVA Dünnschichttechnik GmbH
Hauptstr. 1b
85386 Eching-Dietersheim
Germany

Z.F. Ren
Department of Physics
Boston College
Chestnut Hill, MA 02460
USA

Y. Sato
Power Engineering R&D Center
Tokyo Electric Power Company
4-1, Egasaki-cho, Tsurumi-ku
Yokohama 230-8510
Japan

Yuh Shiohara
Superconductivity Research Laboratory
International Superconductivity
Technology Center
10-13 Shinonome 1-chome
Koto-ku, Tokyo 135-0062
Japan

Shara S. Shoup
MicroCoating Technologies
5315 Peachtree Industrial Blvd.
Chamblee, GA 30341
USA

V.F. Solovyov
Division of Materials and Chemical
Sciences
Energy Sciences and Technology
Department
Brookhaven National Laboratory
Upton, NY 11973-5000
USA

M. Strikovski
Neocera, Inc.
10000 Virginia Manor Road
Beltsville, MD 20705
USA

M. Suenaga
Division of Materials and Chemical
Sciences
Energy Sciences and Technology
Department
Brookhaven National Laboratory
Upton, NY 11973-5000
USA

Y. Takahashi
Power Engineering R&D Center
Tokyo Electric Power Company
4-1, Egasaki-cho, Tsurumi-ku
Yokohama 230-8510
Japan

T. Tamagawa
Jet Process Corporation
24 Science Park
New Haven, CT 06511
USA

D.Z. Wang
Department of Physics
Boston College
Chestnut Hill, MA 02460
USA

J.H. Wang
Department of Chemistry
State University of New York
Buffalo, NY 14260
USA

H.J. Wiesmann
Division of Materials and Chemical
Sciences
Energy Sciences and Technology
Department
Brookhaven National Laboratory
Upton, NY 11973-5000
USA

Judy Wu
Department of Physics and Astronomy
University of Kansas
Lawrence, KS 66045
USA

L. Wu
Division of Materials and Chemical
Sciences
Energy Sciences and Technology
Department
Brookhaven National Laboratory
Upton, NY 11973-5000
USA

S.X. Yang
Department of Physics
Boston College
Chestnut Hill, MA 02460
USA

Y. Zhu
Division of Materials and Chemical Sciences
Energy Sciences and Technology Department
Brookhaven National Laboratory
Upton, NY 11973-5000
USA

PREFACE

The discovery of high temperature superconductors (HTS) in 1986 by two IBM scientists led to an unprecedented explosion of research and development efforts world-wide because of the significant potential for practical applications offered by these materials. However, the early euphoria created by the exciting prospects was dampened by the daunting task of fabricating these materials into useful forms with acceptable superconducting properties. Progress towards this goal has been hindered by many intrinsic materials problems, such as weak-links, flux-creep, and poor mechanical properties.

The earliest studies of critical current density J_c in HTS materials revealed that for a polycrystalline material containing a distribution of grain boundaries is much lower than that for a single crystal. High angle grain boundaries act as Josephson coupled weak-links leading to a significant field-dependent suppression of the supercurrent across the boundary. For clean stoichiometric boundaries, the grain boundary critical current density depends primarily on the grain boundary misorientation. The dependence of J_c(gb) on misorientation angle has been determined $YBa_2O_3Cu_{7-\delta}$(YBCO) in boundary types which can be formed in epitaxial films on bicrystal substrates. These include [001] tilt, [100] tilt, and [100] twist boundaries. In each case high angle boundaries were found to be weak-linked. These experiments have also been extended to artificially fabricated [001] tilt bicrystals in $Tl_2Ba_2CaCu_2O_8$, $Tl_2Ba_2Ca_2Cu_3O_x$, $TlBa_2Ca_2Cu_3O_x$ (Tl-1223) and $Nd_{1.85}Ce_{0.15}CuO_4$. In each case it was found that, as in YBCO, J_c depends strongly on grain boundary misorientation angle. Data on current transmission across artificially fabricated grain boundaries in Bi-2212 also indicate that most large angle [001] tilt and [001] twist boundaries are weak links. It is likely that the variation in J_c with grain boundary misorientation is similar in all high-T_c superconductors. Hence, the low J_c observed in randomly oriented polycrystalline HTS can be understood on the basis that the population of low angle boundaries is small and that frequent high angle boundaries impede long-range current flow. Using conventional processing techniques, three HTS materials were successfully fabricated in polycrystalline form with modest J_c's. These are the Bi-2223 powder-in-tube conductors, the Tl-1223 spray-pyrolyzed films and the Bi-2212 melt-processed thick films. These three types of conductors comprised the ***First-Generation*** HTS conductors or wires. Since bicrystal studies using most HTS compounds show that high angle boundaries are weakly-linked, it was important to determine how the current flows in these materials in order to further increase the properties. In this case one must talk about the grain boundary misorientation distribution (GBMD) and its relation to the measured critical current density. In the last ten years, significant progress has been made to experimentally determine the distribution of misorientation angles in high-J_c

superconductors. Measurements of grain orientations in hundreds of contiguous grains in high-J_c Bi-2223 powder-in-tube, Tl-1223 thick films, and melt-processed Bi-2212 thick films performed using electron backscatter Kikuchi diffraction (BKD) indicate that percolative networks of low angle boundaries with fractions consistent with the active cross-sectional area of the conductor, exist in each of these materials. The nature of the percolative paths is peculiar to each of the compounds and the processing method used to fabricate them. The general view has therefore emerged that long-range conduction in polycrystalline superconductors utilizes connected networks of low angle boundaries. This of course suggests that significant improvement in the properties of the Bi- and Tl-based materials will be made by increasing the percolative options for current flow, i.e., increasing the number of small angle boundaries while decreasing the number of large angle boundaries. Production of biaxial texture may be the only practical way to achieve this goal. No standard metallurgical processing route to fabricate conductors using the YBCO compound has so far been successful. Most methods result in a conductor with primarily high-angle boundaries resulting in low critical current densities. These observations suggest that in order to fabricate high J_c conductors using any of the HTS compounds, production of long-range biaxial texture with a greatly reduced population of high-angle boundaries is necessary. Essentially, for optimal properties, a kilometer long, flexible, crystallographically single-crystal-like HTS wire is required. The first generation HTS wires not only had modest superconducting properties but required the use of significant amounts of silver. This increased their cost to a level wherein it was not possible to compete on a price/performance basis to copper wires. Furthermore, the first-generation HTS wires could not be made using the YBCO compound since in polycrystalline form it would exhibit a very poor critical current density. With respect to intrinsic properties, the YBCO compound is slated to have the best superconducting properties at higher temperatures of operation close to 77 K since it is the most three dimensional of all HTS materials. Ideally, for best properties over a broad temperature range for applications, what was required is a method to produce kilometer long lengths, of flexible YBCO wire which is essentially single-crystal-like crystallographically.

The above problems led to the development of the ***Second-Generation*** of HTS wires. Three methods were invented to produce flexible metallic substrates which were also crystallographically biaxially textured and resembled a long, mosaic single crystal. In each case, the surface of the flexible, metallic substrate consisted of a ceramic oxide upon which epitaxial growth of a thick YBCO layer was possible. The first method invented is the Ion-Beam-Assisted-Deposition (IBAD). The second method developed was the Inclined-Substrate-Deposition (ISD). The third method invented is called the Rolling-assisted-biaxially-textured-substrates (RABiTS). None of these methods use silver as the substrate. Moreover, in each case the superconducting properties such as the critical current density approach that of a YBCO single crystal. This book covers in detail the three methods to form biaxially textured substrates as well as various possible methods to deposit epitaxial YBCO and other HTS materials on these substrates. Since successful scale-up to achieve large-scale commercialization is primarily dictated by the price/performance ratio of the conductor, the particular method of film deposition and the choice of the substrate is crucial.

These Second-Generation HTS conductors, also referred to as "Coated conductors" represent one of the most exciting developments in HTS technology. HTS wires based on this technology have the potential to carry 100–1000 times the current without resistance losses of comparable copper wire. HTS power equipment based on these HTS conductors has a potential to be half the size of conventional alternatives with the same or higher power rating and less than one half the energy losses. Clearly, the prospective dollar and energy savings are enormous. Upgrading of the world-wide electric power transmission and distribution with HTS based devices can significantly help in meeting the growing demand for electricity world-wide. There is little question that superconducting technology based on the Next-Generation HTS Superconducting Wires will make a substantial impact on the way we generate, transmit, distribute and use electric power. Of course the question is—how soon? The chapters contained in this book pertain to various aspects of these second-generation conductors and address questions such as scale-up issues.

The book is divided into three sections. The first section discusses the three methods to fabricate biaxially textured substrates, upon which, epitaxial YBCO or other HTS materials can be deposited to realize a single-crystal-like HTS wire. The second section includes chapters on various methods of HTS deposition such as—pulsed laser ablation (PLD), thermal co-evaporation, sputtering, pulsed electron beam deposition, *ex-situ* BaF_2 by co-evaporation flowed by annealing, chemical solution based *ex-situ* processes, jet vapor deposition, metal organic chemical vapor deposition (MOCVD), and liquid phase epitaxy (LPE). The third section includes detailed chapters on other HTS materials such as the various Tl-based and Hg-based conductors.

CONTENTS

C. Deposition of Other HTS Materials

Section A

Methods to Produce Biaxially Textured Substrates

Chapter 1

IBAD TEMPLATE FILMS FOR HTS COATED CONDUCTORS

Paul N. Arendt

Superconductivity Technology Center
Materials Science and Technology Division
Los Alamos National Laboratory
Los Alamos, NM 87545
USA

1.1 INTRODUCTION

For second generation high-temperature superconducting (HTS) coated conductors, there are vigorous programs among various institutions and consortia to develop long lengths of high-quality superconducting tapes using cost efficient manufacturing techniques. These efforts are exciting in that there are a large variety of competing methodologies being investigated for fabricating these conductors in a cost-effective fashion. These variations include the methods used to apply the $YBa_2Cu_3O_{7-\delta}$ superconducting films as well as the methods used to form the crystallographically textured template platforms on which the superconductors are deposited. This review outlines the status of and future prospects for the use of ion-beam assist deposited (IBAD) films as a template for these conductors. To this end, we begin with a historical review of the use of ion bombardment to modify the crystal texture of materials. Next, a history of IBAD technology and performance results for HTS films fabricated on IBAD templates is presented. Finally, estimates are made of the costs to commercially manufacture IBAD templates using known fabrication techniques.

1.2 MODIFICATION OF CRYSTAL STRUCTURE IN METAL FILMS BY ION BOMBARDMENT

Before beginning a discussion of the history of IBAD film development, it is instructive to outline some of the early ion bombardment experiments, which contributed to the understanding of this technology. Ion bombardment causes changes in the properties and the crystal structure of thin films. The earliest studies of the modification of film properties by ion beam bombardment were performed at normal incidence. The

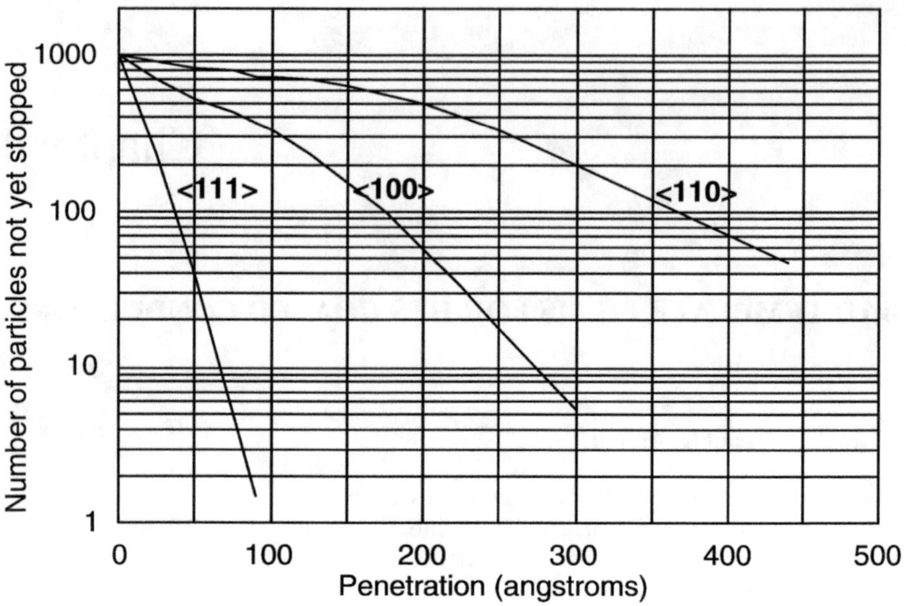

Figure 1.1. Calculated range curves for 5-keV Cu atoms slowing down in a static Cu lattice. The initial directions of the incident atoms are given on the curves. (Reprinted by permission of the authors.)

experiments were generally performed after the films were deposited, rather than during deposition. In many instances, the damage induced by random collisions of the ions with the lattice led to a decrease in the crystalline texture. For certain polycrystalline metal films, however, ion bombardment at medium energies (1–75 keV) caused preferred orientations to develop in the films. In one of the first such experiments, Trillat et al. (1956) demonstrated the conversion of randomly oriented gold films to a (110) orientation after bombardment by 12 keV Ar^+ ions. Next, Dobrev and Marinov (1973, 1975) reported initially (111) oriented films of silver and gold converting to a (110) orientation after bombardment by 10 keV Ar^+ ions. Similarly, van Wyk and Smith (1978) reported that randomly oriented copper films irradiated with 40 keV Cu^+ ions also exhibited a predominantly (110) orientation after bombardment. For all of these examples, the face centered cubic (fcc) metals realigned to a (110) orientation along the ion bombardment direction, after irradiation.

Experimental studies of the penetration of medium energy ions (15–75 keV) in fcc metals (gold and aluminum) (Nelson and Thompson, 1963; Andreen and Hines, 1966; Piercy et al., 1963) found the ion penetrations to increase sequentially along the $\langle 111\rangle$, $\langle 100\rangle$, and $\langle 110\rangle$ crystallographic directions. Figure 1.1 illustrates computer simulations using Born–Mayer potentials to describe the slowing of 5 keV Cu atoms in fcc Cu crystals as a function of their initial direction of motion (Robinson and Oen, 1963). The calculated range variations with crystallographic orientation corroborated the experimental distributions observed for gold and aluminum. It was concluded that channeling processes along the principal axes of the fcc crystal lattices were responsible for the qualitative resemblance between the experimental and calculated penetration range results (Andreen and Hines, 1966; Piercy et al., 1963).

Marinov and Dobrev (1977) found bombardment of hexagonal close packed (hcp) thin films of cadmium and cobalt with 10 keV Ar^+ ions resulted in the development of texture as well as phase changes. As deposited, the cadmium films had no preferred orientation but after bombardment the films had an $(11\bar{2}0)$ orientation. This direction was described as the most favorable for channeling and along which the energy loss for the bombarding ions was minimal. For non-$(11\bar{2}0)$-oriented crystallites, it was explained that the energy loss per unit length was much higher. This resulted in higher densities of thermal spikes and atoms being displaced into interstitial positions, initiating a recrystallization process. The $(11\bar{2}0)$ oriented crystallites served as recrystallization centers, leading to an increase in the number of crystallites with this orientation and to the final preferred orientation. As deposited, cobalt films exhibited a $(11\bar{1}0)$ orientation. After bombardment, they were transformed to a (110) oriented fcc phase. For cobalt films, this phase is normally stable only for very thin films heteroepitaxially deposited on fcc templates (Atrei et al., 1997). Marinov and Dobrev attributed the appearance of the fcc cobalt phase to local temperature increases leading to melting of the cobalt, followed by very rapid quenching. They then attributed the final (110) orientation of the cubic phase to recrystallization around centers that were favorably oriented for penetration of the incident ions.

Simultaneous vapor deposition with energetic particle bombardment normal to the surface of depositing metal films was found to affect their texture and morphology. In the absence of bombardment, gold deposited onto NaCl exhibits mixed (111) and (100) orientations with multiple twinning along the $\langle 111 \rangle$ plane (Lewis and Jordan, 1970). Under deposition with electron bombardment, the (111) orientation was less distinct and the twinning was restricted. One of the first reported investigations of simultaneous ion bombardment on the growth of thin films emphasized the surface coverage of silver films under different ion bombardment conditions (Marinov, 1977). The beam energy was varied from 1 to 10 keV Ar^+ and *in situ* electron diffraction patterns of the growing silver films were collected. Films grown without simultaneous ion bombardment had no preferred orientation. A preferred crystal orientation was observed for films grown during bombardment. However, no more specific description was made of these "preferred" orientations. A subsequent investigation of the texture of silver films deposited with and without simultaneous 10 keV Ar^+ ion bombardment found the films to be (110) and (111) oriented, respectively (Dobrev, 1982).

Van Wyk (1980) performed the first off-normal bombardment studies of texture modification using randomly oriented copper films irradiated by 40 keV Cu ions at a variety of beam incidence angles. After irradiation, the film crystallites tended to have the $\langle 110 \rangle$ directions aligned with the incident beam. Simultaneous off-normal bombardment with vapor deposition was first done by Yu et al. (1985, 1986), with low energy (200 eV) Ar^+ ions irradiating growing Nb films at an angle of 70° from the substrate normal. Pole figure analysis of the films showed that, without bombardment, a (110) fiber texture was obtained. With bombardment, a restricted fiber texture with non-uniform azimuthal distribution of the (110) reflections near the 60° circle were observed, with the central spot displaced by 5° toward the ion source. The investigators attributed the texture mechanism of the growing film to the difference in sputtering yields between grains oriented such that the ion beam was or was not channeled. Bradley et al. (1986) expanded upon this with a model that described how texture at the surface of the film evolved with increasing thickness. This texture evolution was a combination of homoepitaxial growth of grains aligned with the ion-assist beam and resputtering of grains whose crystallographic axes were poorly aligned with the ion-assist beam.

1.3 YBCO HTS FILMS ON METAL AND ON YSZ BUFFERED SUBSTRATES

Shortly after high-temperature superconducting (HTS) materials were discovered investigators attempted to deposit them in thin film form on polycrystalline flexible metal alloy substrates with the goal of forming flexible HTS conductors. The high temperatures required to deposit superconducting $YBa_2Cu_3O_{7-\delta}$ (YBCO) led to interfacial reactions of the YBCO with the metal substrates so that very low critical current density (J_c) values ($\leqslant 10^3$ A/cm^2, 64 K) were obtained (Witanachchi et al., 1990; Russo et al., 1990). Thin films of an intermediate, heteroepitaxially deposited, cubic yttria-stabilized zirconia (YSZ) buffer layer were used to ameliorate interfacial reactions on single crystal silicon and gallium arsenide substrates (Fork et al., 1990; Jia, 1990; Tiwari et al., 1990). Critical current density values greater than 10^6 A/cm^2 were reported for YBCO films heteroepitaxially grown on these intermediate YSZ buffers. YSZ buffers were then also deposited on polycrystalline metal substrates but the overcoated YBCO films still had poor J_c (77 K) values of less than 3×10^4 A/cm (Kumar et al., 1990; Narumi et al., 1991; Reade et al., 1991). In the former case, the initial YSZ films had a good lattice registry with their substrates and were well textured which in turn led to good *c*-axis and biaxial texture in the heteroepitaxial YBCO. In the latter case, the YSZ buffers on the metal substrates were either randomly oriented or fiber textured while the YBCO films were only *c*-axis textured. The above results corroborated earlier studies by Dimos et al. (1989), of intergranular J_c being greatly reduced across high angle grain boundaries.

1.4 ION BEAM ASSISTED DEPOSITION OF YSZ BUFFERS

Cubic oxide buffer layers, such as YSZ, could not be biaxially textured on polycrystalline metal substrates using standard heteroepitaxial deposition techniques so researchers began to investigate other methods to orient these films. Ion-beam assisted deposition (IBAD) of cubic zirconia films initially used near-normal ion assist experiments to modify fiber texture (Kao and Gorman, 1990). In the absence of the ion assist, the ion-beam-sputtered zirconia was amorphous. With a 100 eV Ar^+ ion assist, the room temperature deposited zirconia film was cubic with a (111) fiber orientation in the direction of the assist beam. Shortly afterward in a seminal work by Iijima et al. (1991), off-normal IBAD of ion-beam sputtered cubic yttria-stabilized zirconia (YSZ) was used to promote in-plane texture of the YSZ on polished, polycrystalline Hastelloy C-276 (Ni-based alloy) substrate platforms. The Ar^+ ion-assist angle was varied between 30 to 60 degrees from the substrate normal and the assist beam energy varied from 300 to 1000 eV. Pole figure analysis of the (111) reflection showed the YSZ to have in-plane texture with one of the $\langle 111 \rangle$ axes oriented in the direction of the assist beam. The [100] direction was perpendicular to the substrate. The YSZ films were then overcoated at 700°C with pulsed laser deposited (PLD) films of YBCO. Pole figure analysis of the (103) reflection showed the YBCO to be biaxially oriented on the YSZ template. A J_c (77 K, 0 T) value for a 0.1 μm thick YBCO films was 6.0×10^4 A/cm^2. This was the first reported example of the use of IBAD films as templates for heteroepitaxially deposited YBCO.

Because of the great interest in forming flexible superconducting tapes utilizing high quality YBCO, refinements of the IBAD YSZ techniques on polycrystalline metal substrates proceeded rapidly. Iijima et al. (1992) next used a fixed ion assist energy of

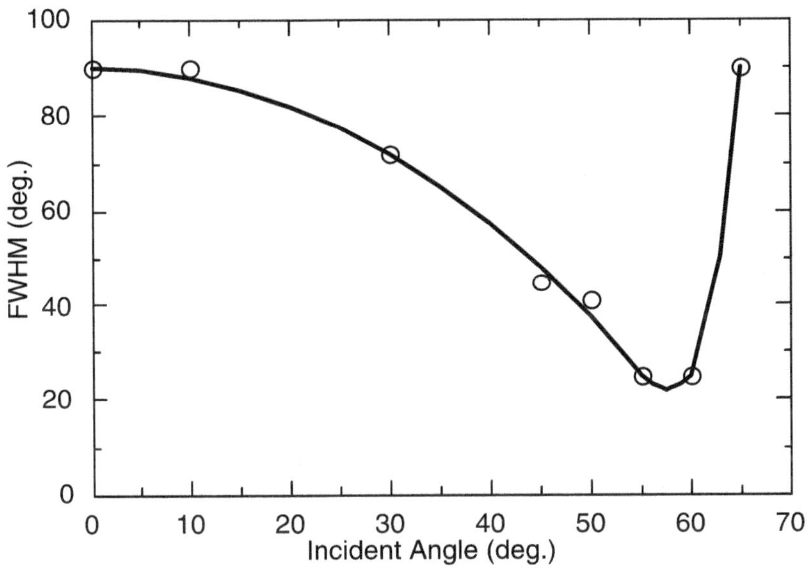

Figure 1.2. The FWHM of azimuthal distributions of (111) pole figures as a function of the beam incident angle. (Reprinted by permission of the authors.)

300 eV and a set bombardment angle of 45°. The in-plane texture of the YSZ {111} peak reflections ($\Delta\phi$) had an estimated best full-width half maximum (FWHM) value of 30°. Rocking curve measurement ((200) reflection) of the YSZ out-of-plane texture ($\Delta\omega$) resulted in a FWHM value of 5.3°. One micrometer thick YBCO films heteroepitaxially deposited on the YSZ templates had ($\Delta\phi$) ((103) reflection) values of 20° to 30°. The YBCO J_c (77 K, 0 T) had been improved to 2.5×10^5 A/cm^2. Next, Reade et al. (1992) used PLD to evaporate both the YSZ template as well as the final YBCO film. They reported that their best IBAD YSZ texture was obtained using an Ar^+ ion assist energy of 200 eV while the assist angle could be varied between 30 to 60 degrees from the substrate normal. It was noted that (100) texture in the IBAD YSZ might be enhanced if the ions were directed near an angle of 54.7°, corresponding to the [111] channeling direction in cubic YSZ. A best YBCO J_c (77 K, 0 T) of 6.0×10^6 A/cm^2 was reported. A $\Delta\omega$ of 1.3° was reported for their YBCO ((005) reflection); however, no $\Delta\phi$ values were reported for either their YSZ or YBCO films. Iijima et al. (1993a) reported a detailed experimental study of the effects of YSZ in-plane texture as a function of the assist ion bombardment angle. They found the optimum in-plane texture to occur near an angle of 55° from the substrate normal (Figure 1.2). Two other groups essentially verified Figure 1.2 with similar results several years later (Mao et al., 1998; Freyhardt et al., 1997). Iijima et al. (1993a) pointed out that optimum biaxial texture was obtained for an ion-beam-substrate geometry which corresponded to the angle of the $\langle 111 \rangle$ axes with the substrate normal for (100) oriented YSZ. They attributed this to channeling in favorably oriented crystallites and selective resputtering of unfavorably oriented crystallites. A YBCO $\Delta\phi$ value of 18° and a J_c (77 K, 0 T) of 4.3×10^5 A/cm^2 was reported for a 0.5 μm thick YBCO film deposited on their most highly aligned YSZ template, which had a $\Delta\phi$ value of 22°. Lower J_c values were reported for the YBCO films deposited on those less favorably aligned YSZ templates formed using ion beam assist angles other than 55°. Iijima et al. (1993b), then used plan view transmission electron microcroscopy to study the distribution of misorien-

tation angles of the YBCO grains. By mapping the in-plane ⟨110⟩ twin axes of a large number of YBCO grains, they found that 50% of the grains were misaligned within ±5° and 82% of the grains were misaligned within ±10° of the average grain alignment direction. They noted that to obtain J_c (77 K, 0 T) values above 1 MA/cm^2 the in-plane texturing of the IBAD YSZ layer had to be improved. Arendt et al. (1994), using a dual ion beam deposition system similar to that used by Iijima (1992), deposited thick (800 nm) IBAD YSZ films and reported IBAD YSZ $\Delta\phi = 14°$. The overcoated PLD YBCO was further improved to a $\Delta\phi$ of 10.5°. They reported a J_c (75 K, 0 T) of 8×10^5 A/cm^2 for a 280 nm thick YBCO film which had a 100 nm thick PLD CeO_2 buffer film deposited between the IBAD YSZ and the final YBCO film.

1.5 ALTERNATE TEXTURING MECHANISM BY LOW ENERGY IBAD

Other IBAD deposition studies by Sonnenberg et al. (1993) on amorphous substrates suggested that low energy (75–300 eV) ion channeling was not the dominant mechanism inducing biaxial texture in YSZ films. Pole figure studies of IBAD YSZ films showed the azimuthal symmetry of the crystal axes varied with ion-assist beam energy, ion to molecule arrival ratio (r value), ion-beam to substrate bombardment angle, and substrate temperature. In a subsequent study of (100) textured IBAD YSZ films by Ressler et al. (1997), field plots of two types of in-plane orientation (⟨111⟩ or ⟨110⟩ axes aligning with the ion-assist beam) as a function of ion bombardment angle and r value were generated. (These data were consistent with other IBAD YSZ experiments (Iijima et al., 1992, 1993a; Arendt et al., 1994).) Nuclear stopping cross section plots onto ⟨111⟩, ⟨110⟩, and ⟨100⟩ ZrO_2 planes were presented for 75 and 300 eV Ar^+ ion bombardment. The most open channels into these planes did not necessarily correspond to experimental data of which planes oriented toward the ion-assist beam. Energy dependent ion-etch rate experiments of IBAD YSZ films with ⟨111⟩ or ⟨110⟩ planes facing the ion-etch beam showed an anisotropy in the etch rates which corroborated the field plot data. Ion-induced damage anisotropy of different crystal planes was advanced by the authors to account for the types of biaxial texturing observed under the differing conditions of low energy ion-assisted deposition. A subsequent study by Iijima et al. (2001a), corroborated this concept of damage tolerance to low energy ions contributing to film texturing. This latter work correlated low energy IBAD deposition texturing of cubic oxide compounds with their lattice energy densities. Under similar deposition conditions, compounds with higher lattice energy densities generally exhibited lower mosaic spreads.

1.6 CRITICAL CURRENT MEASUREMENTS ON SHORT LENGTH YBCO/IBAD YSZ/Ni-ALLOY TAPES

For the YBCO/IBAD YSZ/Ni-alloy experiments described to this point, all of the reported critical current density values were measured using narrow bridges ($\leqslant$0.5 mm). Wu et al. (1994) reported the first total transport critical current (I_c) values for macroscopic superconductor film widths on Ni-based alloy platforms. Their YBCO film dimensions were 1 cm wide by 1 cm long. An I_c (75 K, 0 T) value of 23 A was reported for a 1.5 μm YBCO thick film (J_c (75 K, 0 T) $= 1.5 \times 10^5$ A/cm^2). They noted that the Ni-alloy substrate platform used was not very smooth ($R_a \sim 100$ nm) and that better I_c values could be expected with a better substrate surface finish. Using

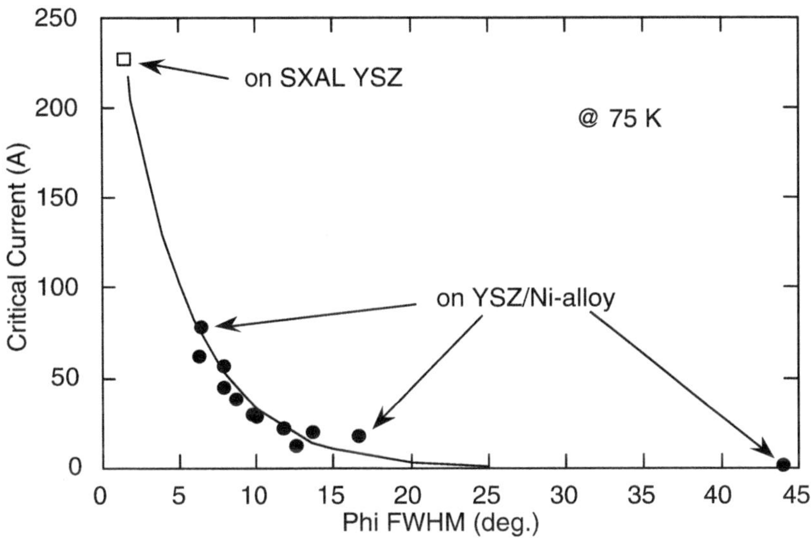

Figure 1.3. Transport critical currents at 75 K and zero field as a function of the FWHM of the YBCO (103) peaks in x-ray phi scan. All the YBCO samples were 1 μm thick and 1 cm wide. All the data except one from YBCO on (100) YSZ single crystal were from YBCO thick films on Ni alloys with an ion beam assisted deposited buffer layer.

a smoother, mechanically polished substrate ($R_a \sim 20$ nm), the same group (Foltyn, 1995) next reported an encouraging I_c (75 K, 0 T) = 120 A, for a 2.0 μm thick YBCO film (J_c (75 K, 0 T) = 6×10^5 A/cm^2). A second result reported in (Wu et al., 1994) was a J_c (75 K, 0 T) of 1.3×10^6 A/cm^2 for a 1 μm YBCO thick film (narrow bridge measurement). This was the first demonstration of $J_c \geqslant 1$ MA/cm^2 for YBCO films on polycrystalline metal substrates. Shortly afterward, Kohno et al. (1995) also reported a best I_c (77 K, 0 T) = 103 A on a 1 cm length of YBCO tape, with a film thickness of 2.0 μm (J_c (77 K, 0 T) = 5.2×10^5 A/cm^2). Another impressive result in this latter report was a 20 cm length of tape that had a total I_c (77 K, 0 T) of 27 A in a 1 μm thick YBCO film (J_c (77 K, 0 T) = 2.7×105 A/cm^2). The critical current improvements in this short time span (<1 year) culminated with a report by Peterson (1995) of 199 A (75 K, 0 T) for a short (1 cm) length of conductor.

Figure 1.3 illustrates a systematic study of I_c as a function of YBCO $\Delta\phi$ values, for 1 cm wide × 1 μm thick YBCO films on IBAD YSZ/Ni-alloy platforms and on a single crystal YSZ template, as reported by Wu et al. (1995). This data showed the importance of having good biaxial alignment ($\Delta\phi < 10°$) of the YBCO polycrystalline films (and the underlying IBAD YSZ template) in order to obtain superconducting films with high I_c values, at liquid nitrogen temperatures. Yang et al. (1995) reported a numerical study which calculated the mean value of J_c across large numbers of grains with differing in-plane grain alignment distributions, showing higher (normalized) J_c vs. texture values than those presented in Figure 1.3. Corroborating narrow bridge experimental J_c values were presented which fit the calculated J_c vs. texture dependence curve. There were other experimental differences between the two sets of data presented by Wu and Yang. First, the films used in the latter work were thinner (0.2–0.5 μm) than those in Figure 1.3. For the film thickness ranges used in the two studies, it had been shown for YBCO films on single crystal substrates that J_c decreases rapidly with increasing film thickness (Foltyn et al., 1993;

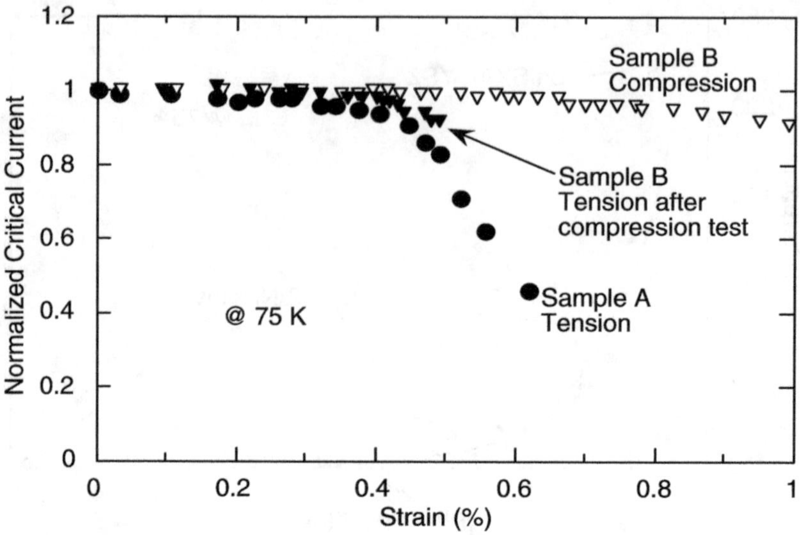

Figure 1.4. Normalized critical current as a function of a bending strain at 75 K and zero field. The tension test was done for the sample B after a compression test.

Miura et al., 1997). This may account for some of the differences between the two sets of experimental values. Second, Yang's YBCO deposition parameters may have been better optimized. Finally, Wu et al. (1995) also presented measurements of normalized I_c values as a function of the coated conductor bending strain (Figure 1.4). The data showed that, in compression, the critical current dropped less than 10% for strains of 1%. In tension, a critical bending strain of 0.4% was obtained. It was noted that this data illustrated the high flexibility of coated conductors on Ni-alloy substrates.

1.7 DEVELOPMENT OF LONGER LENGTH YBCO/IBAD YSZ/Ni-ALLOY COATED CONDUCTORS

Three institutions (Fujikura Electric Company, Los Alamos National Laboratory (LANL), and the combined groups at the University of Göttingen and the Zentrum für Funktionswerkstoffe gem. GmbH (ZFW)) have continued to develop PLD YBCO/IBAD YSZ/metal-alloy coated conductors with the concurrent goal of maintaining high I_c values while increasing tape lengths. A fourth group at Siemens has been making large area (10 cm × 10 cm) YBCO/IBAD YSZ film depositions on polished zirconia plates and etching lengths of spiral conductors on the final YBCO films (Nies et al., 2000). Table 1.1 is a chronological history of pertinent I_c and J_c values for different length conductors achieved by these groups and their collaborators. For the entries which are collaborations, the first group listed indicates where the IBAD template film was fabricated and the second group listed indicates where the YBCO film was fabricated. The quoted lengths of the conductors are the distances between voltage taps. All conductors are 1 cm wide except for those noted in the table, which are macroscopic 4–8 mm wide conductors. In-plane $\Delta\phi$ values are included but out-of-plane ($\Delta\omega$) values are not. On average, $\Delta\omega$□□□□□□ are 5–7° for the YSZ templates and 1.5–3° for the final YBCO films. (Iijima et al. (1998) showed that the final out-of-plane texture is achieved more readily and 4 to 5 times more rapidly than

Table 1.1. Transport critical current (I_c) progress for YBCO/IBAD YSZ/metal-alloy coated conductors*

Year	Group[1]	YBCO deposition method[2]	YSZ $\Delta\phi$/ YBCO $\Delta\phi$ (deg.)	J_c[3] (MA/cm^2)	I_c[3] (A/cm width)	Length of conductor (cm)	Reference
1994	LANL	PLD	14/10	0.12	23	1	(Ressler et al., 1997)
1995	LANL	PLD	12/7	0.6	120	1	(Iijima et al., 2001a)
1995	Fujikura	PLD	NA	0.52	104	1	(Wu et al., 1994)
				0.27	27	20	(Wu et al., 1994)
1995	LANL	PLD	12/5	0.99	199	1	(Peterson, 1995)
1995	Fujikura	PLD	NA/16	0.26	52	10	(Iijima et al., 1995)
1996	Fujikura	PLD	NA/19	0.12	7.7	76	(Iijima et al., 1996)
1996	Fujikura	MOCVD	NA	0.59	10.4	1	(Onabe et al., 1996)
1996	LANL	PLD	21/18	0.38	7	13	(Foltyn, 1996)
1997	Fujikura	PLD	23/NA	0.21	20.8	70	(Hosaka et al., 1997)
1997	LANL	PLD	13/8	0.41	70	12	(Foltyn, 1997; Arendt et al., 1998)
			23/22	0.02	4	95	(Foltyn, 1997; Arendt et al., 1998)
1998	LANL	PLD	17/12	0.39	47	98	(Foltyn, 1998)
1999	LANL	PLD	15/9	1.0	96	87	(Foltyn et al., 1999)
1999	LANL	PLD	12/6	0.87	122	97	(Foltyn, 1999; Foltyn et al., 2000)
1999	LANL/ORNL	e-beam	12/5	3.2	99[4]	1	(Feenstra, 1999)
2000	LANL/ORNL	e-beam	14/9	1.92	200[4]	1	(Feenstra et al., 2000)
2000	ANL/IGC	MOCVD	17/4	1.4	140[4]	1	(Balachandran, 2000)
2000	ANL/LANL/IGC	MOCVD	12/4	1.4	130[4]	1	(Selvamanickam et al.,
						1	2001)
2000	Fujikura	PLD	18/10	0.38	45	90	(Iijima et al., 2000c)
			18/10	0.24	25	210	(Iijima et al., 2000c)
2000	Siemens[5]	TCE	12/7	2.4	71[4]	67	(Nies et al., 2000)
2000	Fujikura	PLD	12/7	0.76	84	100	(Iijima et al., 2001b)
			12/11	0.25	35	400	(Iijima et al., 2001b)
2001	ZFW	PLD	NA/8.5	>0.6	141	180	(Dzick et al., 2001)
2001	Fujikura[6]	PLD	18/9	0.42	50	960	(Iijima et al., 2001c)
2001	ZFW/LANL	PLD	11/4	0.95	189	91	(Foltyn, 2001)

*NA = no FWHM data provided.
[1]ORNL = Oak Ridge National Laboratory, ANL = Argonne National Laboratory, IGC = Intermagnetics General Corporation.
[2]LANL films have a PLD CeO_2 buffer layer deposited between the YBCO and the YSZ layers, MOCVD = Metallorganic chemical vapor deposition, TCE = Thermal coevaporation, e-beam = *ex-situ* decomposition of fluoride based precursor deposited by electron beam evaporation.
[3]LANL measurements performed at 75 K, SF, all other measurements are at 77 K, SF.
[4]Conductor widths vary from 4 to 8 mm. Values quoted are for A/cm width.
[5]Siemens data is on polycrystalline ceramic substrate.
[6]Fujikura data for this entry uses $Gd_2Zr_2O_7$ for the template instead of YSZ.

the final in-plane texture.) For completeness, the 1 cm length values discussed in the previous section are included in Table 1.1. For tapes shorter than the longest dimension of the ion-assist gun, the IBAD YSZ templates were deposited in stationary mode. When the tape lengths exceeded the longest dimension of the ion assist gun, the films were deposited with the substrates translating through the deposition zone. As can be seen in Table 1.1 for the IBAD YSZ $\Delta\phi$ values for LANL (Iijima et al., 2001a; Foltyn, 1997), substrate translation initially resulted in a severe degradation of the biaxial texture. Consideration of the spatial variation of the ion and vapor fluences had to be taken into account to optimize the length of the shields in the deposition zone so that biaxial texture was improved for moving substrates (Arendt et al., 1998). Further optimization was also done on assist-ion gun divergence in order to obtain YSZ texture that approached that for films made in stationary mode (Iijima et al., 1998). Achieving equivalence of IBAD YSZ texture for moving and stationary tapes took several years (e.g., Table 1.1—LANL (Wu et al., 1994; Foltyn, 1999; Foltyn et al., 2000), Fujikura (Hosaka et al., 1997; Iijima et al., 2001b)). Both groups achieved good uniformity of the IBAD YSZ texture along the length of the tapes by passing the substrates through the deposition zone several times before the final thickness was attained (Arendt et al., 2000b; Iijima et al., 2000c). Note some latter entries in the table indicating both the Fujikura and the ZFW groups have scaled their processes to lengths greater than 1 meter (Dzick et al., 2001; Iijima et al., 2001c). (Iijima et al. (2000b) has stated that Fujikura intends to fabricate 100 m lengths of HTS coated conductors on IBAD templates by 2002.) Note also that the final Fujikura result uses a new template material ($Gd_2Zr_2O_7$) in place of YSZ. This compound is a pyrochlore, a structure related to the fluorite structure of YSZ. Using $Gd_2Zr_2O_7$, their final texture values on multi-meter lengths of tape ($\Delta\phi = 9.8°$ at 0.5 m/h) are achieved at a thickness approximately one-half that needed for YSZ. Studies of these compounds for their radiation damage tolerance suggest that $Gd_2Zr_2O_7$ is more sensitive to radiation damage than is YSZ (Sickafus et al., 2000). Whether this speeds renucleation of the grains and contributes to the faster rate of texture formation remains a matter of speculation.

1.8 OTHER IBAD INVESTIGATIONS

Many institutions have reported achieving biaxial texture for oxide, nitride, and metal films using IBAD techniques. These films were grown as templates for subsequent YBCO over coats as well as for fundamental growth/texture studies. Biaxially textured films have also been reported using IBAD deposition onto cylindrical and large area planar platforms by the ZFW group (Freyhardt et al., 1998). The ZFW group has also fabricated long lengths of YSZ templates by deposition onto metal alloy tapes wrapped around a rotating cylindrical mandrel (Garcia-Moreno et al., 1999). Table 1.2 lists these investigating institutions, the evaporation techniques employed, the materials investigated, and their best reported $\Delta\phi$ values. As the table illustrates, this is an active field of study with good progress being made for biaxial texturing of several types of cubic oxide materials.

1.9 IBAD MgO

For the biaxial texture values listed in Table 1.2, all but one of the oxides listed have IBAD film thicknesses varying from 0.5 to 2.0 μm thick. The exception is IBAD magnesia, which requires a film thickness of only 10 nm to achieve the single digit $\Delta\phi$ values listed for (Wang et al., 1997; Groves et al., 2000, 2001a). Thus for comparable IBAD film deposition rates, the time required to form the MgO template film is 50 to 200 hundred times faster than for the other cubic oxides. From industrial scale up and

Table 1.2. Biaxially textured materials fabricated using IBAD techniques*

Group	Evaporation technique	Material(s)	$\Delta\phi$ (deg.)	Reference
International Business Machines	Ion-beam sputter	Nb	NA*	(Yu et al., 1985)
International Business Machines	Magnetron sputter	Nb, Mo	NA	(Harper et al., 1997)
University of Michigan	e-beam	Nb	NA	(Ji, 1999)
Oak Ridge National Laboratory	PLD	CeO_2	28	(Zhu et al., 1994)
Chalmers University	PLD	CeO_2	10	(Xiong, 2000)
INFP[1]	Ion-beam sputter	CeO_2	12	(Zhu et al., 1994)
INFP	Ion-beam sputter	YSZ	19	(Knierim et al., 1997)
Argonne National Laboratory	e-beam	YSZ	17	(Chudzik et al., 1999)
Lawrence Berkeley Laboratory	PLD	YSZ, CeO_2	29, NA	(Reade et al., 1992; Reade, 1996)
Shanghai Inst. Metallurgy	Ion-beam sputter	YSZ, CeO_2	14, NA	(Mu et al., 2000; Mao et al., 1998)
CSIRO[2]	Biased magnetron sputtering	YSZ, CeO_2	10, 39	(Saavides et al., 2001; Gnanarajan, 1999)
Neocera Incorporated	PLD	YSZ	12	(Harshavardhan et al., 2001)
Massachusetts Inst. of Tech.	e-beam	YSZ, LCMO	NA	(Ressler et al., 1997)
Univ. of Göttingen	Ion-beam sputter	YSZ	7	(Weismann et al., 1998)
Inst. Metallic Materials, Dresden	PLD	YSZ, Pr_6O_{11}	9, 11	(Holzapfel et al., 1998)
Inst. Metallic Materials, Dresden	PLD	MgO	20	(Huhne et al., 2000)
Stanford University	e-beam	MgO	7	(Wang et al., 1997)
Los Alamos National Laboratory	e-beam	MgO	7	(Groves et al., 2000; 2001a)
Los Alamos National Laboratory	Ion-beam sputter	YSZ	11	(Jia et al., 2001)
Fujikura	Ion-beam sputter	HfO_2, CeO_2, Y_2O_3, $Zr_2Sm_2O_7$	21, 34, 26, 17	(Iijima et al., 2001a)
Fujikura	Ion-beam sputter	YSZ	12	(Iijima et al., 2001b)
Fujikura	Ion-beam sputter	$Gd_2Zr_2O_7$	9.8	(Iijima et al., 2001c)
Univ. Leipzig, Univ. Augsburg	e-beam	TiN	12	(Rauschenbach, 2000)

*NA = no FWHM data provided.
[1]INFP = Forschungszentrum Karlsruhe, Institut für Nukleare Festkörperphysik, Karlsruhe, Germany.
[2]CSIRO = Commonwealth Scientific and Industrial Research Organization—Australia.

cost related considerations for manufacturing of HTS coated conductors, this makes IBAD magnesia an attractive template. Do et al. (1995) first reported that this template is also different from the others listed in that it must be deposited on an amorphous surface (e.g., Si_3N_4) and that the optimum ion-assist beam energy is approximately three times greater (~750 eV) than that used for optimal biaxial texture formation of IBAD YSZ. Also, optimally textured IBAD MgO films are grown with the assist beam oriented at an angle of 45° to the substrate surface (while IBAD YSZ films are

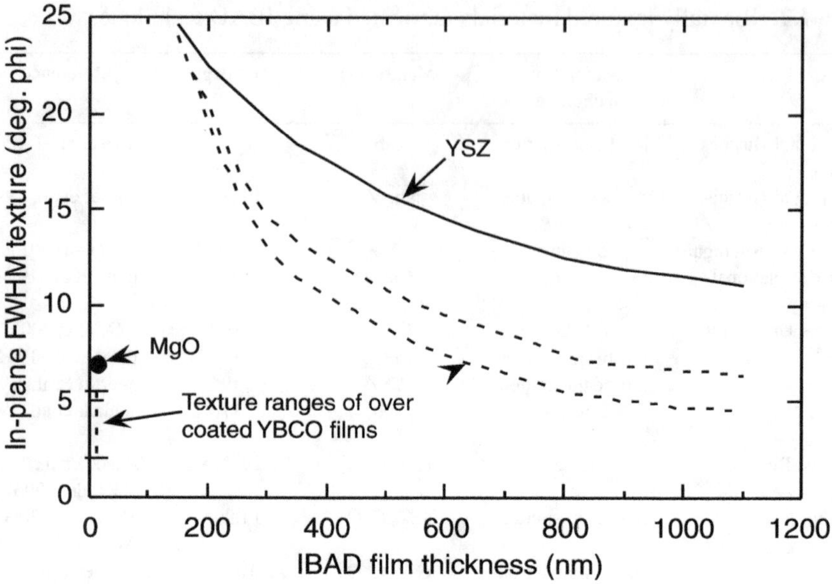

Figure 1.5. IBAD YSZ and MgO $\Delta\phi$ values vs. film thickness. The over coated YBCO texture values are for PLD films deposited at LANL.

optimum at 55°). The films nucleate with a ⟨001⟩ axis perpendicular to the substrate surface and a ⟨011⟩ axis parallel to the ion assist beam. As noted initially by Wang et al. (1997), using *in situ* reflection high-energy electron diffraction (RHEED) to monitor the growing film, these orientations are present during the nucleation phase of the films. The rapid texture formation and the higher beam energies employed suggest that IBAD MgO develops biaxial texture by channeling phenomena and not by the slower evolutionary texturing mechanism of anisotropic damage tolerance of different crystallographic planes, as suggested by Ressler et al. (1997), for IBAD YSZ.

Figure 1.5 illustrates the best in-plane $\Delta\phi$ values vs. thickness for IBAD MgO and IBAD YSZ templates fabricated at LANL. Several other groups have reported comparable data for IBAD YSZ texture evolution vs. thickness (Iijima et al., 2000b, 2001b; Chudzik et al., 1999). (Note that the best IBAD YSZ $\Delta\phi = 7°$ value reported in Table 1.2 by the Univ. of Göttingen group is for a film that was several micrometers thick.) Also included in Figure 1.5 are the ranges of $\Delta\phi$ values obtained for YBCO films over coated onto these two templates. The best YBCO texture improvement is several degrees for both templates. On the IBAD MgO template, the over coated YBCO $\Delta\phi$ value in the figure (2.1°) is the best texture reported in the literature (Groves et al., 2000) for YBCO films deposited on a polycrystalline template platform and approaches values reported for YBCO films deposited on single crystal platforms (Fork et al., 1990; Wu et al., 1995; Miura et al., 1997). The J_c (75 K, 0 T) reported for this film (3.1×10^5 A/cm^2) is also comparable to critical current density values reported in the literature for YBCO films of this thickness (0.45 μm) on single crystal substrates (Foltyn et al., 1993; Miura et al., 1997). Hammond and Matijasevic (1998) have also reported very low YBCO $\Delta\phi$ values (3.6°) and high J_c values (>1 MA/cm^2) for YBCO films deposited on IBAD MgO templates. The substrate platforms used by Groves and Hammond were highly polished ($R_a < 0.5$ nm) YSZ and Al_2O_3, respectively. The thinness of the MgO

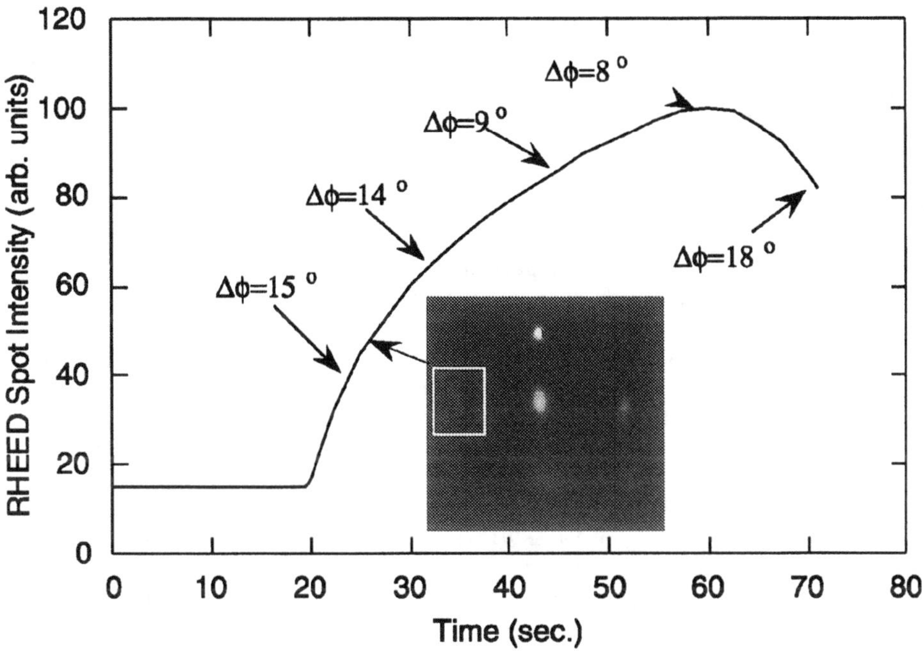

Figure 1.6. RHEED spot intensity trace vs. during growth of an IBAD MgO film. A sample RHEED diffraction pattern is shown in the inset. The box outlines the diffraction spot being monitored. In-plane textures obtained from different sampling points during the growth of the template are also indicated.

template demands a smoother platform than the thicker IBAD YSZ template material. That is, high quality (J_c values >1 MA/cm^2) YBCO films can be obtained on IBAD YSZ/Ni-alloy when the surface finish of the metal platforms has a $R_a = 2$–5 nm. Initial J_c investigations by Groves et al. (1999), for YBCO/IBAD MgO on Ni-alloy substrates with similar R_a values reported $J_c = 0.46$ MA/cm^2, for a 1.35 μm thick film. Arendt et al. (2000a) reported that improving the surface finish of the metal substrate to $R_a \leqslant 1$ nm resulted in critical current densities $\geqslant$1.1 MA/cm^2 for 1.6 μm thick YBCO films ($I_c = 180$ A/cm width).

In spite of previously listed speed/thickness advantages of the IBAD MgO template, several issues need addressing before one may begin to use it for long length coated conductors. First, $R_a \leqslant$1 nm surface finishes have not been demonstrated for long length metal substrate platforms. This is not a great limitation and is expected to be achieved in the near future. Second, nearly all of the other IBAD film deposition investigations used crystal monitor and Faraday cup probes to monitor vapor and ion fluences, respectively, in the deposition region. *Ex-situ* x-ray diffraction measurements of the films were then used as feedback to optimize the ion/atom ratios and the resultant template texture. In a more efficacious technique, the Stanford group first demonstrated use of *in situ* RHEED to monitor the texture evolution of the growing IBAD MgO films and refine their deposition parameters (Do et al., 1995). Until recently, the rapid growth of the MgO templates has required such a monitor because biaxial texture does not continue to improve asymptotically with increasing film thickness, as it does in Figure 1.5 with YSZ and for the other cubic oxides listed in Table 1.2. Figure 1.6 (Groves et al., 2001c) illustrates the intensity of one of the RHEED diffraction pattern spots vs. IBAD MgO film deposition time. Also included in the figure are MgO $\Delta\phi$ values measured for films whose depositions were interrupted at the points indicated on the plot. The biaxial texture improves up to the RHEED spot intensity maximum and then worsens as the IBAD film thickness continues to increase. As Wang et al. (2001) have shown, the coalescence of grains for the growing MgO films occurs near the RHEED maximum or when the film thickness is $\sim$10 nm. It is speculated that the texture degradation occurs because the assist ion beam causes increasing damage to the film once coalescence occurs. As IBAD film growth progresses beyond the RHEED maximum, tilting of the $\langle 001 \rangle$ axis away from the surface normal is observed (Wang, 1999; Groves et al., in press). The corresponding tilting of the $\langle 011 \rangle$ axis away from the ion assist beam initiates a non-channeling condition and is speculated to exacerbate damage to the films causing the observed texture degradation.

Matijasevic et al. (1998), have reported the successful use of RHEED to monitor and control YBCO film quality in a production environment where the substrates are moving. However, until RHEED has demonstrated feedback control for the fabrication of high quality MgO templates on moving tapes, the coated conductor community will hesitate to use this technique. To date, IBAD MgO films deposited without the use of RHEED have some what poorer biaxial texture values than those reported in stationary mode, ranging from $\Delta\phi = 20°$ (Huhne et al., 2000) to $\Delta\phi = 7.8°$ (Arendt, 2001) than the best texture values obtained using RHEED (Wang et al., 1997; Groves et al., 2001a). The latter value reported by Arendt was the average measured value on meter-length tapes deposited in a continuous mode. In lieu of using RHEED, the films were deposited with very close control of vapor and ion fluences ($\pm 2\%$).

1.10 IBAD TEMPLATE MICROSTRUCTURES

Microstructural investigations of IBAD YSZ template films have primarily concentrated on detailed analysis of the evolution of the IBAD YSZ films (Iijima et al., 1998; Dzick et al., 1999) using cross sectional transmission electron microscopy (TEM). Similar studies for IBAD MgO have investigated growth characteristics in cross sectional as well as plan view modes (Wang et al., 1997; Groves et al., 2001a, 2001b). Kung et al. (2001), discuss the differences of these two templates in plan view mode as well as the characteristics of YBCO films grown on each of them. Figure 1.7 is a dark field; plan view TEM of an IBAD YSZ and an IBAD MgO film (Kung et al., 2001). Both films were made using ion-beam sputtering with ion-beam assisted deposition. The bright contrast regions are areas where the individual grains are oriented within $\pm 7°$ of the diffracted beam. As the notes below each TEM image indicate, the grain sizes of the two templates vary by nearly an order of magnitude—the IBAD MgO being larger. On the other hand the regions where the individual grains have relatively good alignment (colonies) are much larger for the YSZ than for the MgO. Both films had comparable x-ray diffraction in-plane texture values (e.g., $\Delta\phi \sim 12°$).

The different domain sizes of the template layers replicate to the YBCO films that are deposited on them. Kung et al. (1999, 2001) cited plan view TEM examples of YBCO films on both types of templates illustrating the replication phenomenon.

There are notable microstructural differences observed for IBAD MgO templates prepared using different types of vapor sources. Figure 1.8 is a bright field plan view TEM of an IBAD MgO film deposited using electron beam evaporation for the vapor source. Note that the grain and colony sizes for this MgO film are much smaller than for the MgO film of Figure 1.7. These smaller grains and colonies of grains are speculated to be a result of lower admolecule mobility of the thermally evaporated MgO from the electron beam evaporation source.

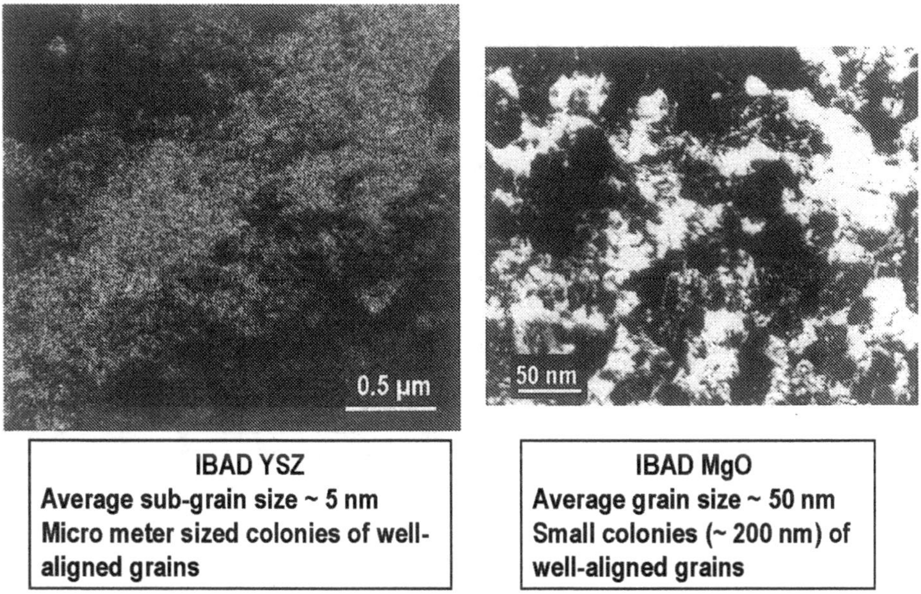

Figure 1.7. Dark field, plan view TEM images of IBAD YSZ and MgO templates.

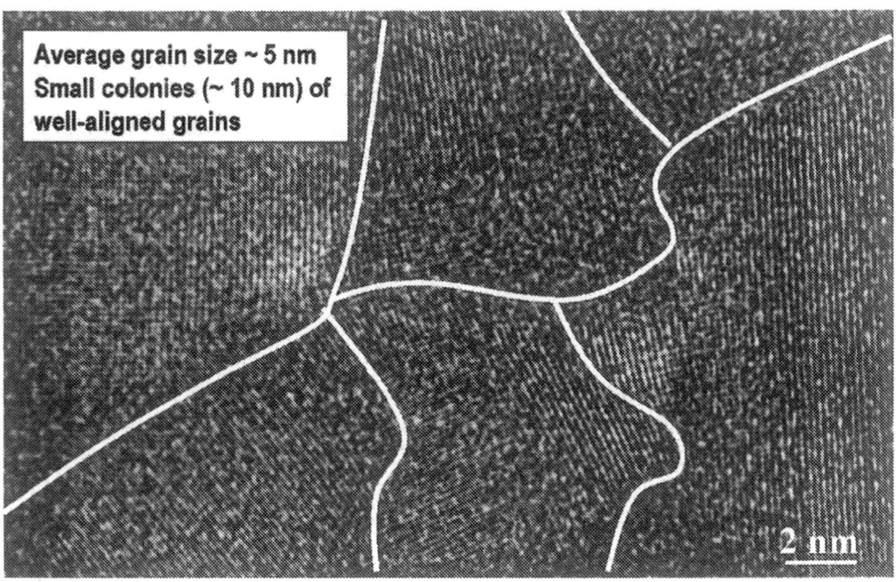

Figure 1.8. Bright field TEM plan view of an IBAD MgO film prepared using an electron beam evaporation source. The lines denote the nominal boundaries of colonies of well-aligned grains.

1.11 COST ESTIMATE OF IBAD MgO TEMPLATES

Scale-up issues and cost estimates of HTS coated conductor manufacturing facilities, which utilize IBAD MgO templates and YBCO superconducting films over coated using electron beam evaporation technology were discussed by Hammond

Table 1.3. IBAD template manufacturing parameters

Manufacturing assumptions	6,000 km/year	50,000 km/year
Tape process speed (cm/sec)	3	25
Process width (cm)	10	10
Number of 1 cm wide tapes after slitting	10	10
IBAD MgO deposition rate (Å/sec)	2	2
MgO film thickness (Å)	100	100
IBAD deposition zone length (meters)	1.5	12
Number of passes	1	1

Table 1.4. IBAD template manufacturing costs

	6,000 km/year	50,000 km/year
Manufacturing plant costs		
IBAD portion	6.3 M	38.0 M
Costs of tape ($/meter)		
Fixed		
Depreciation (20%)	0.21	0.15
Other fixed	(0.01)	(0.01)
Variable		
Material	0.20	0.20
Labor	(0.01)	(0.01)
Other variable	(0.06)	(0.06)
Total cost ($/meter)	0.49	0.43

(1995). This study assumed two different plants capable of producing 6,000 km/year or 50,000 km/year of 1 cm wide tape. In his analysis, Hammond assumed a demonstrated deposition rate of 2 Å/sec for the IBAD MgO (Wang et al., 1997; Groves et al., 2001a). The plants ran year round 24 hours a day, 7 days a week, at a 68% product acceptance rate.

Table 1.3 replicates the parameters that Hammond required for the two yields. In both cases, the processing width was assumed to be 10 cm. After processing, the tapes were slit into 1 cm widths. The lengths of the deposition zones are not unreasonable in that the dimensions of the largest linear Kaufman-type ion guns currently available are 67 cm × 6 cm. Such guns may be aligned in series along the desired length of the deposition zone.

As delineated in the previous section, high quality YBCO has been grown on these templates (Groves et al., 2001a). Also, the overall yield length may be greater than what was assumed, since high quality templates have been grown at a rate three times faster than the 2 Å/sec rate used in Table 1.3 (Arendt et al., 2000a).

Table 1.4 shows Hammond's estimates for the capital costs and the costs per meter for manufacturing IBAD templates on substrates by the process described above.

The costs of the manufacturing plants in the first entry line include land, architectural and engineering fees, the building, utility connections, and equipment. Table 1.4 assumes the price of the manufacturing plant is amortized over five years. The table does not include the cost of capital, which could vary widely. The costs delineated in parentheses are conservative in that the totals are for both IBAD and YBCO labor, as well as other variable costs. The material costs are due almost entirely to costs for the substrate and may be overestimates in that they are linear extrapolations based on the prices for small laboratory quantities. These calculations were made in 1995 so it would be prudent to increase them by a multiplier, which accounts for inflation since that time.

Hammond listed other technical issues related to IBAD processing:

1. The crystalline quality of the IBAD layer that can be achieved in manufacturing. This will determine the quality of the superconducting properties of the YBCO layer.
2. The possible requirement for a buffer layer between the substrate and the IBAD layer to promote template formation and/or to prevent diffusion of cations from the substrate.
3. The possible requirement for a buffer layer between the IBAD and YBCO layers to improve lattice matching.

1.12 BUFFER LAYERS

In the previous section, it was noted that the added costs of including the buffer layers of items 2 and 3 to the materials system would not significantly increase the total costs beyond those delineated in Table 1.4. Industrial manufacturers that coat continuously moving tapes with films using physical vapor deposition processes indicate that this is generally true. Adding buffer layer deposition stations to a continuous vapor deposition process with moving substrate tapes will not significantly increase the costs if the buffer layers are thin and do not require any complex deposition or monitoring techniques (D. O'Neill, private communication). The buffer layer used between the substrate platform and the IBAD MgO template satisfies these two requirements in that it is a 5 to 20 nm thick layer of amorphous Si_3N_4 and is deposited using electron beam evaporation at room temperature (Wang et al., 1997; Groves et al., 2001a). Buffer layers between the IBAD template and the final HTS YBCO film may or may not be required. For example, Hammond and Matijasevic (1998) demonstrated good superconducting properties for YBCO deposited directly on IBAD MgO/Si_3N_4/alumina. On Ni-alloy substrate platforms, good YBCO superconducting properties have been reported using thin (<250 nm) buffer layers of YSZ/Y_2O_3 (Groves et al., 1999) or YSZ/CeO_2 (Groves et al., 2001b) between the IBAD MgO and the final YBCO overcoat.

The buffer layers deposited between the IBAD MgO template and the final YBCO film have not been optimized to the extent that they have been for buffer layers on IBAD YSZ templates. Holesinger et al. (2001) performed extensive transmission electron micrograph (TEM) studies of various buffer layer architectures for the latter template and found that interfacial buffer layer/YBCO reactions occurred in all of the architectures studied. The differing interfacial reactions lead to secondary phases and intergrowths in the YBCO films. Ceria buffer layers performed best in minimizing the interfacial reactions that reduce the transport properties of the YBCO. Holesinger et al. (2000a) also noted that the positive volume change that occurs with the ceria-YBCO reaction products may act as a kinetic barrier limiting the extent of the reaction. However, the reactions that inevitably occur do not preclude obtaining high transport values for the YBCO. Figure 1.9 is a TEM cross section from a continuously processed coated conductor with a ceria buffer layer between the YBCO and IBAD

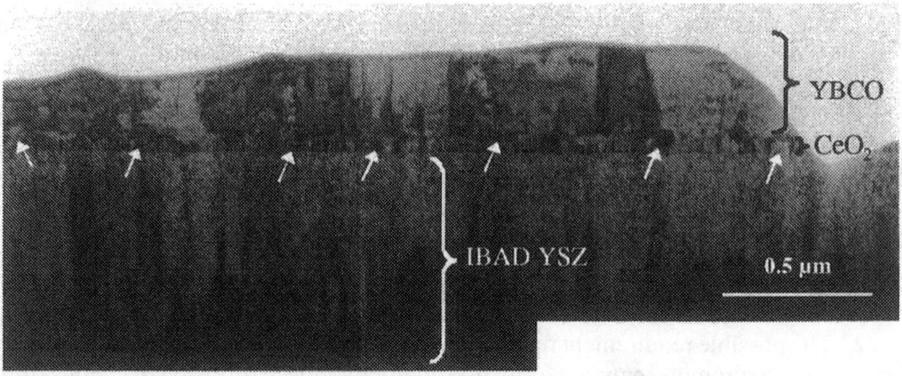

Figure 1.9. TEM cross-section of a continuously processed coated conductor employing a CeO_2 buffer layer between the YBCO and IBAD YSZ. The interfacial reactions (noted by arrows) were localized and spaced, on average, about 0.5 μm apart. The YBCO film was deposited to a thickness of 2 μm and had a J_c of 1 MA/cm^2. The measured I_c of this 1-cm wide sample was 200 A.

Figure 1.10. TEM micrograph showing the two-phase microstructure that results when YBCO reacts with CeO_2. In all cases, $YCuO_2$ was found above the $BaCeO_3$.

YSZ (Holesinger et al., 2000a). The I_c (75 K) of this particular section of conductor was 200 A across the 1-cm width of tape. The arrows indicate areas where interfacial reactions occurred along the interface. The reaction phases identified in the sample were $BaCeO_3$, $YCuO_2$, and CuO. Figure 1.10 is a magnified cross section of one of the reaction zones of Figure 1.9 illustrating two of the reaction phases. Figure 1.11 is a higher resolution TEM above another reaction zone illustrating intergrowths emanating from a YBCO grain boundary and intercalcating into the YBCO film (Holesinger et al., 2000a). Holesinger et al. (2000b) also did a study of the YBCO transport properties vs. ceria buffer thickness. They found that the optimum transport properties were obtained for ~10 nm thick ceria buffers. For IBAD YSZ templates, this certainly satisfies the manufacturer's criteria of thin buffer layers, which should not significantly add to the manufacturing costs.

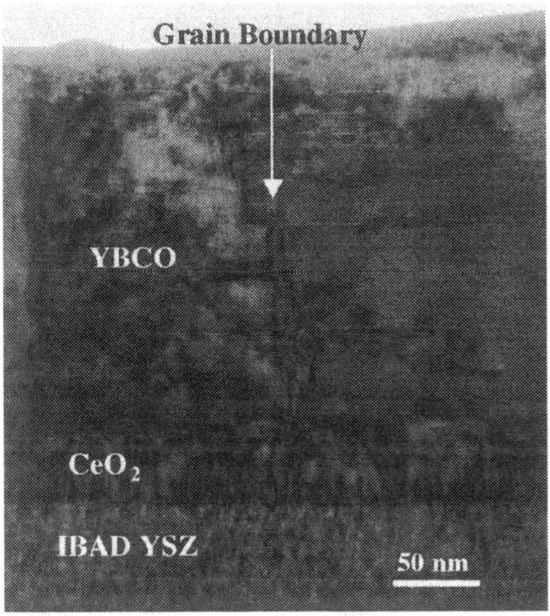

Figure 1.11. TEM micrograph showing the intergrowths emanating from a YBCO grain boundary. At the base of the grain boundary is an isolated $BaCeO_3$ grain resulting from the reaction between CeO_2 and YBCO. Excess yttrium and copper from this reaction diffuse along the grain boundary and intercalate into the YBCO film as intergrowths of the Y-247, Y-124, or Y-224 phases rather than forming the other secondary phases of $YCuO_2$ and CuO.

1.13 NON-IBAD TEMPLATES FOR COATED CONDUCTORS

Alternate template formation processes for coated conductors are being actively pursued on an international basis. These processes include inclined substrate deposition (ISD) and deformation texturing of fcc metals. In ISD the substrate is inclined to the vapor plume at a set deposition angle to obtain a textured template film. Sumitomo Electric Industries (Fujino et al., 1995) and the University of Cambridge (Quinton et al., 1999) have investigated texturing of YSZ by ISD. Sumitomo Electric forms this template on a 1-cm wide tape at a speed of 0.6 m/h. The University of Munich (Bauer et al., 1999) and Argonne National Laboratory (Balachandran and Gray, 2000) have investigated texturing of MgO by ISD.

Sumitomo Electric Industries (Ohmatsu et al., 2000) has demonstrated 10-meter lengths of YBCO coated conductor tape on their ISD YSZ template. The YBCO J_c (77 K) for this length of tape was $0.5–2 \times 10^5$ A/cm^2 with an individual 4.5 m section having a maximum I_c (77 K) $= 15$ A. Deformation texturing of silver substrates for coated conductors was first reported by Hitachi Ltd. (Doi et al., 1994). Oak Ridge National Laboratory (Goyal et al., 1996) has been preeminent in investigating deformation textured Ni and Ni-alloy substrates along with several of its industrial collaborators including American Superconductor Corp. (Rupich et al., 1999) and 3M Corp. (O'Neill, 2000). Furukawa Electric Co. Ltd. (Matsumoto et al., 1999) has also used deformation textured Ni to fabricate long lengths of template substrate materials. All of the institutions that are investigating deformation texturing of substrates have reported fabricating template tape lengths greater than fifty meters in length (O'Neill, 2000; Doi et al., 2000; Rupich, 2000; Goyal, 2000).

1.14 SUMMARY

Riley et al. (1999) point out that the three main factors which influence commercialization of HTS tape development are technical viability, reliability, and economic viability. Within the lengths of tape fabricated to date, IBAD templates have certainly satisfied the technical viability and reliability criteria necessary to fabricate high I_c second generation coated conductors using a variety of YBCO deposition techniques including: PLD (Foltyn, 1999; Iijima et al., 2001b), MOCVD (Balachandran, 2000; Selvamanickam et al., 2001), thermal coevaporation (Iijima et al., 2001b), and the *ex-situ* decomposition of fluoride based precursors (Feenstra et al., 2000; Shiohara, 2000). Whether these criteria continue to be met with progressively longer lengths of these templates will be a continuing challenge to the universities, companies, and national laboratories which are currently instituting IBAD scale-up efforts. These include 3M Corp. (Storer, 2000), IGC-SuperPower (Selvamanickam, 2000), LANL (Foltyn, 2000), University of Göttingen/ZFW (Dzick et al., 2000), and Fujikura Electric Company (Iijima et al., 2000a). Based on Hammond's IBAD scale up cost analysis presented in Section 1.10, the economic viability issue also appears to be satisfied. Riley et al. (1999), state that first generation HTS Bi-2223 wire is currently available at approximately \$300/kA m (77 K, SF) and that scale up of the current Bi-2223 technology will eventually result in a price of \$50/kA m. For second generation coated conductors, the United States Department of Energy Superconductivity for Electric System Program has stated that it would like its program to meet this latter cost-performance criterion in km lengths of tape by 2005 (*Coated Conductor Development Roadmapping Workshop*, St. Petersburg, FL, January 18–19, 2001). YBCO I_c values of 200 A/cm width (Peterson, 1995; Foltyn, 2000; Feenstra et al., 2001) to 500 A/cm width (Goyal et al., 1996) have been demonstrated on IBAD templates on metal substrates. If these values can be maintained in a manufacturing environment, a maximum of five or a minimum of two IBAD templates would be required per kA m of coated conductor. Using Table 1.4 this translates to template costs between \$1.00 and \$2.50 per kA m. Thus, IBAD templates promise to be competitive with the other template technologies being developed for HTS coated conductors.

REFERENCES

Andreen, C.J. and Hines, R.L., 1966, Channeling of D^+ and He^+ ions in a gold crystals, *Phys. Rev.*, 151:341.

Arendt, P., Foltyn, S., Wu, X., Townsend, J., Adams, C., Hawley, M., Tiwari, P., Maley, M., Willis, J., Mosley, D., and Coulter, Y., 1994, Fabrication of biaxially oriented yttria-stabilized zirconia on polycrystalline substrates, *Mater. Res. Soc. Symp. Proc.*, 341:209.

Arendt, P.N., Foltyn, S.R., Groves, J.R., DePaula, R.F., Dowden, P.C., Roper, J.M., and Coulter, J.Y., 1998, YBCO/YSZ coated conductors on flexible Ni-alloy substrates, *Appl. Supercond.*, 4:429.

Arendt, P., Foltyn, S., Groves, J., Holesinger, T., Kung, H., Jia, Q., Peterson, E., DePaula, R., Dowden, P., Coulter, Y., and Stan, L., 2000a, MgO templates for YBCO coated conductors, *Superconductivity Program for Electrical Systems, 2000 Annual Peer Review*, Washington, DC, July 17–19.

Arendt, P.N., Groves, J.R., Foltyn, S.R., Jia, Q.X., Peterson, E.J., DePaula, R.F., Dowden, P.C., Coulter, J.Y., and Ma, M., 2000b, Fabrication of high-quality ion-beam deposited cubic oxide template films on meter-length substrates, *Mater. Res. Soc. Symp. Proc.*, 585:67.

Arendt, P.N., 2001, IBAD MgO templates for coated conductors, *Superconductivity Program for Electrical Systems, 2001 Annual Peer Review*, Washington, DC, August 1–3.

Atrei, A., Roveda, G., Torrini, M., Bardi, U., Gleeson, M., and Barnes, C.J., 1997, The growth mechanism and structure of ultrathin cobalt films deposited on the Pd(111) surface, *Surf. Sci.*, 372:91.

Balachandran, B., 2000, Development of coated conductors—practical and fundamental issues, *Superconductivity Program for Electrical Systems, 2000 Annual Peer Review*, Washington, DC, July 17–19.

Balachandran, B. and Gray, K., 2000, Development of coated conductors—practical and fundamental issues, *Superconductivity Program for Electrical Systems, 2000 Annual Peer Review*, Washington, DC, July 17–19.

Bauer, M., Semerad, R., and Kinder, H., 1999, YBCO films on metal substrates with biaxially aligned MgO buffer layers, *IEEE Trans. Appl. Supercond.*, 9:1502.

Bradley, R.M., Harper, J.M.E., Cuomo, J.J., and Smith, D.A., 1986, Theory of thin-film orientation by ion bombardment during deposition, *J. Appl. Phys.*, 60:4160.

Chudzik, M.P., Erck, R., Lanlagan, M.T., and Kannewurf, C.R., 1999, Processing dependence of biaxial texture in yttria-stabilized zirconia by ion-beam-assisted deposition, *IEEE Trans. Appl. Supercond.*, 9:1490.

Dimos, D., Chaudhari, P., Mannhart, J., and LeGoues, J., 1989, Orientation dependence of grain-boundary critical currents in $YBa_2Cu_3O_{7-\delta}$ bicrystals, *Phys. Rev. Lett.*, 55:2442.

Do, K.B., Wang, C.P., Marshall, A., Geballe, T.H., Beasley, M.R., and Hammond, R.H., 1995, Control of biaxial texture of MgO thin films by ion-beam assisted deposition (IBAD), *Fall MRS Meeting*, Boston, MA, Nov. 27–Dec. 1.

Dobrev, D., 1982, Ion-beam-induced texture formation in vacuum-condensed thin metal films, *Thin Solid Films*, 92:41.

Dobrev, D. and Marinov, M., 1973, Effect of ion bombardment on crystallographic orientation in vacuum-condensed metal films, *C. R. Acad. Bulg. Sci.*, 26:231.

Dobrev, D. and Marinov, M., 1975, Effect of ion bombardment on the structure of vacuum-condensed metal films, *Bulg. Acad. Sci. Comm. Dept. Chem.*, 8:334.

Doi, T.J., Yuasa, T., Ozawa, T., and Higashiyama, K., 1994, A new type of superconducting wire; biaxially oriented $Tl_1(Ba_{0.8}SrO_{0.2})_2Ca_2Cu_3O_9$ on $\{100\}\langle 100\rangle$ textured silver tape, *Adv. Supercond. VII*, 6:817.

Doi, T., Hakuraku, Y., Higashiyama, K., Akata, H., Osumara, K., Kashima, N., and Nagaya, S., 2000, Biaxially oriented $NdBa_2Cu_3O_7$ films prepared on $\{100\}\langle 100\rangle$ textured Ag tapes without any buffer layer, *Appl. Supercond. Conf.*, Virginia Beach, VA, September 17–22.

Dzick, J., Wiesmann, J., Hoffmann, J., Heinemann, K., Garcia-Moreno, F., Isaev, A., Freyhardt, H.C., and Lechner, W., 1999, YSZ buffer layers on large technical substrates, *IEEE Trans. Appl. Supercond.*, 9:2248.

Dzick, J., Sievers, S., Hoffman, J., Kautschor, L.-O., and Freyhardt, H.C., 2000, Mechanisms of ion-beam assisted texturing of YSZ layers, *Fall MRS Meeting*, Boston, MA, Nov. 27–Dec. 1.

Dzick, J., Hoffmann, J., Sievers, S., Kautschor, L.O., and Freyhardt, H.C., 2001, Ion-beam-assisted texturing of YSZ layers, *5th European Conference on Applied Superconductivity*, Copenhagen, Denmark, August 26–30 (paper submitted to *Physica C*).

Feenstra, R., 1999, An *ex situ* processed YBCO coated conductor, *Superconductivity Program for Electrical Systems, 1999 Annual Peer Review*, Washington, DC, July 26–27.

Feenstra, R., Christen, D.K., Verebelyi, D.T., Lee, D.F., Paranthaman, M., Goyal, A., Kang, B.W., Lu, S.W., Martin, P.M., and Kroeger, D.M., 2000, Enhancing the current transport in *ex situ* processed YBCO coated conductors, *Fall MRS Meeting*, Nov. 27–Dec. 1.

Feenstra, R., Goyal, A., Paranthaman, M., Lee, D.F., Martin, P.M., Christen, D.K., Kroeger, D.M., Arendt, P.N., Groves, J.R., and DePaula, R.F., 2001, High-I_c YBCO coated conductors produced by an *ex situ* BaF_2 process, in: *Proc. Int. Workshop on Superconductivity*, Honolulu, HI, p. 146.

Foltyn, S.R., 1995, *Spring Materials Research Society Meeting*, San Francisco, CA, April 17.

Foltyn, S.R., 1996, *Superconductivity Program for Electrical Systems, 1996 Annual Peer Review*, Washington, DC, July 31–Aug. 1.

Foltyn, S.R., 1997, *Superconductivity Program for Electrical Systems, 1997 Annual Peer Review*, Washington, DC, July 21–23.

Foltyn, S.R., 1998, *Superconductivity Program for Electrical Systems, 1998 Annual Peer Review*, Washington, DC, July 20–22.

Foltyn, S.R., 1999, *Superconductivity Program for Electrical Systems, 1999 Annual Peer Review*, Washington, DC, July 26–27.

Foltyn, S., 2000, Toward a 1000 A coated conductor: Increasing the I_c of thick YBCO films, *Superconductivity Program for Electrical Systems, 2000 Annual Peer Review*, Washington, DC, July 17–19.

Foltyn, S.R., 2001, High current coated conductors based on IBAD YSZ and thick YBCO/Sm-123 multilayers, *Superconductivity Program for Electrical Systems, 2001 Annual Peer Review*, Washington, DC, August 1–3.

Foltyn, S.R., Tiwari, P., Dye, R.C., Le, M.Q., and Wu, X.D., 1993, Pulsed laser deposition of thick $YBa_2Cu_3O_{7-\delta}$ films with $J_c > 1$ MA/cm^2, *Appl. Phys. Lett.*, 63:1848.

Foltyn, S.R., Arendt, P.N., Dowden, P.C., DePaula, R.F., Groves, J.R., Coulter, J.Y., Jia, Q.X., Maley, M.P., and Peterson, D.E., 1999, High-T_c coated conductors—performance of meter-long YBCO/IBAD flexible tapes, *IEEE Trans. Appl. Supercond.*, 9:151.

Foltyn, S.R., Arendt, P.N., DePaula, R.F., Dowden, P.C., Coulter, J.Y., Groves, J.R., Haussamen, L.N., Winston, L.P., Jia, Q.X., and Maley, M.P., 2000, Development of meter-long YBCO coated conductors produced by ion beam assisted deposition and pulsed laser deposition, *Physica C*, 341–348:2305.

Fork, D.K., Fenner, D.B., Barton, R.W., Phillips, J.M., Connel, G.A.N., Boyce, J.B., and Geballe, T.H., 1990, Synthesis of superconducting $YBa_2Cu_3O_{7-\delta}$ thin films on nickel-based superalloy using *in situ* pulsed laser deposition, *Appl. Phys. Lett.*, 57:1161.

Freyhardt, H.C., Hoffman, J., Weismann, J., Dzick, J., Heinemann, K., Isaev, A., Garcia-Moreno, F., Sievers, S., and Usoskin, A., 1997, YBaCuO thick films on planar and curved technical substrates, *IEEE Trans. Appl. Supercond.*, 7:1426.

Freyhardt, H.C., Hoffmann, J., Wiesmann, J., Dzick, J., Heinemann, K., Issaev, A., Usoskin, A., and Garcia-Moreno, F., 1998, Y-123 films on technical substrates, *Appl. Supercond.*, 4:435.

Fujino, K., Hasegawa, K., Mukai, H., Sato, K., Hara, T., Ohkuma, T., Ishii, H., and Honjo, S., 1995, 1 meter long thin film tape with high J_c of 1.5×10^5 A/cm^2 fabricated by pulsed laser deposition, *Adv. Supercond. VIII*, 7:675.

Garcia-Moreno, F., Usoskin, A., Freyhardt, H.C., Issaev, A., Wiesmann, J., Hoffmann, J., Heinemann, K., Sievers, S., and Dzick, J., 1999, Laser deposition of YBCO on long-length technical substrates, *IEEE Trans. Appl. Supercond.*, 9:2260.

Gnanarajan, S. and Saavides, N., 1999, Evolution of texture of CeO_2 thin film buffer layers prepared by ion-assisted deposition, *Thin Solid Films*, 350:124.

Goyal, A., 2000, Fundamental issues for coated conductors, *Superconductivity Program for Electrical Systems, 2000 Annual Peer Review*, Washington, DC, July 17–19.

Goyal, A., Norton, D.P., Budai, J.D., Paranthaman, M., Specht, E.D., Kroeger, D.M., Christen, D.K., He, Q., Saffian, B., List, F., Lee, D.F., Martin, P.M., Klabunde, C.E., Hartfield, E., and Sikka, V.K., 1996, High critical current density superconducting tapes by epitaxial deposition of $YBa_2Cu_3O_x$ thick films on biaxially textured substrates, *Appl. Phys. Lett.*, 69:1795.

Groves, J.R., Arendt, P.N., Foltyn, S.R., DePaula, R.F., Peterson, E.J., Holesinger, T.G., Coulter, J.Y., Springer, R.W., Wang, C.P., and Hammond, R.H., 1999, Ion-beam assisted deposition of bi-axially aligned MgO template films for YBCO coated conductors, *IEEE Trans. Appl. Supercond.*, 9:1964.

Groves, J.R., Arendt, P.N., Jia, Q.X., Foltyn, S.R., DePaula, R.F., Kinder, L.R., Fan, Y., and Peterson, E.J., 2000, High critical current density PLD YBCO deposited on highly textured IBAD MgO buffer layers, *Ceram. Trans.*, 104:219.

Groves, J.R., Arendt, P.N., Kung, H., Foltyn, S.R., DePaula, R.F., Emmert, L.A., and Storer, J.G., 2001a, Texture development in IBAD MgO films as a function of deposition thickness and rate, *IEEE Trans. Appl. Supercond.*, 11:2822.

Groves, J.R., Arendt, P.N., Foltyn, S.R., Jia, Q.X., Holesinger, T.G., Kung, H., Peterson, E.J., DePaula, R.F., Dowden, P.C., Stan, L., and Emmert, L., 2001b, High critical current density $YBa_2Cu_3O_{7-\delta}$ thick films using IBAD MgO bi-axially oriented template layers on nickel-based superalloy substrates, *J. Mater. Res.*, 16:2175.

Groves, J.R., Yashar, P.C., Arendt, P.N., DePaula, R.F., Peterson, E.J., and Fitzsimmons, M.J., 2001c, Ultra-thin biaxially textured IBAD MgO template layers resolved by grazing incidence x-ray diffraction, *Physica C*, 355:293.

Groves, J.R., Arendt, P.N., Foltyn, S.R., Jia, Q.X., DePaula, R.F., Dowden, P.C., Kung, H., Holesinger, T.G., Stan, L., Emmert, L.A., and Peterson, E.J., in press, Ion-beam assisted deposition of MgO with *in situ* RHEED monitoring to control bi-axial texture, *Mater. Res. Soc. Symp. Proc.*

Hammond, R.H., 1995, Thick film YBCO for wires and tapes: Scale-up issues and cost estimates, *Adv. Supercond. VIII*, 7:1029.

Hammond, R.H. and Matijasevic, V., 1998, *DOE Grain Boundary Workshop*, Chantilly, VA, July 22–23.

Harper, J.M.E., Rodell, K.P., Colgan, E.G., and Hammond, R.H., 1997, Control of in-plane texture of body centered cubic metal thin films, *J. Appl. Phys.*, 82:4319.

Harshavardhan, K.S., Christen, H.M., Silliman, S.D., Talanov, V.V., Anlage, S.M., Rajeswari, M., and Claassen, J., 2001, Low-loss YBCO films on flexible polycrystalline yttria-stabilized zirconia tapes for cryoelectronic applications, *Appl. Phys. Lett.*, 78:1888.

Holesinger, T.G., Foltyn, S.R., Arendt, P.N., Kung, H., Jia, Q.X., Dickerson, R.M., Dowden, P.C., DePaula, R.F., Groves, J.R., and Coulter, J.Y., 2000a, The microstructure of continuously processed $YBa_2Cu_3O_y$ coated conductors with underlying CeO_2 and ion-beam assisted yttria-stabilized zirconia buffer layers, *J. Mater. Res.*, 15:1110.

Holesinger, T., Foltyn, S., Arendt, P., Kung, H., Jia, Q., Dickerson, R.M., Dowden, P.C., DePaula, R.F., Groves, J.R., and Coulter, J.Y., 2000b, The effect of the final buffer layer thickness on microstructure development and transport properties in YBCO coated conductors, *Superconductivity Program for Electrical Systems, 2000 Annual Peer Review*, Washington, DC, July 17–19.

Holesinger, T.G., Foltyn, S.R., Arendt, P.N., Jia, Q.X., Dowden, P.C., DePaula, R.F., and Groves, J.R., 2001, A comparison of buffer layer architectures on continuously processed YBCO coated conductors based on the IBAD YSZ process, *IEEE Trans. Appl. Supercond.*, 11:3359.

Holzapfel, B., Betz, V., Arranz, M.A., Reger, N., and Schultz, L., 1998, Laser ablated buffer layers for YBCO-tape conductors, *Mater. Res. Soc. Symp. Proc.*, 526:257.

Hosaka, M., Iijima, Y., Sadakata, N., Saitoh, T., Kohno, O., and Yoshitomi, J., 1997, Long and high J_c YBCO tapes using continuously deposited YSZ buffer layers by IBAD method, *Adv. Supercond. X*, 9:749.

Huang, M.Q., Geerk, J., Massing, S., Meyer, O., Reiner, H., and Linker, G., 1999, Textured CeO_2 buffer layers on amorphous substrates by ion beam assisted deposition, *Nucl. Inst. Meth. Phys. Res. B*, 148:793.

Huhne, R., Beyer, C., Holzapfel, B., Oertel, C.-G., Schultz, L., and Skrotzki, W., 2000, Growth of biaxial textured MgO layers by ion-beam assisted pulsed laser deposition, *Cryst. Res. Technol.*, 35:419.

Iijima, Y., Tanabe, N., Ikeno, Y., and Kohno, O., 1991, Biaxially aligned $YBa_2Cu_3O_{7-x}$ thin film tapes, *Physica C*, 185:1959.

Iijima, Y., Tanabe, N., Kohno, O., and Ikeno, Y., 1992, In-plane aligned $YBa_2Cu_3O_{7-x}$ thin films deposited on polycrystalline metallic substrates, *Appl. Phys. Lett.*, 60:769.

Iijima, Y., Onabe, K., Futaki, N., Tanabe, N., Sadakata, N., Kohno, O., and Ikeno, Y., 1993a, In-plane texturing control of Y–Ba–Cu–O thin films on polycrystalline substrates by ion-beam-modified intermediate buffer layers, *IEEE Trans. Appl. Supercond.*, 3:1510.

Iijima, Y., Onabe, K., Futaki, N., Tanabe, N., Sadakata, N., Kohno, O., and Ikeno, Y., 1993b, Structural and transport properties of biaxially aligned $YBa_2Cu_3O_{7-\delta}$ films on polycrystalline Ni-based alloy with ion-beam modified buffer layers, *J. Appl. Phys.*, 74:1905.

Iijima, Y., Onabe, K., Tanabe, N., Sadakata, N., Saitoh, T., Kohno, O., and Yoshitomi, J., 1995, High current laser-ablated YBCO tape with biaxially aligned structure, *Adv. Supercond. VIII*, 7:613.

Iijima, Y., Hosaka, M., Tanabe, N., Sadakata, N., Saitoh, T., Kohno, O., and Yoshitomi, J., 1996, Fabrication of high-J_c YBCO tapes using continuously deposited YSZ buffer layers by IBAD method, *Adv. Supercond. IX*, 8:659.

Iijima, Y., Hosaka, M., Tanabe, N., Sadakata, N., Kohno, O., and Takeda, K., 1998, Growth structure of yttria-stabilized-zirconia films during off-normal ion-beam assisted deposition, *J. Mater. Res.*, 13:3106.

Iijima, Y., Kakimoto, K., Kimura, M., Takeda, K., and Saitoh, T., 2000a, Real to real continuous formation of Y-123 coated concuctors by IBAD and PLD method, *Appl. Supercond. Conf.*, Virginia Beach, VA, September 17–22.

Iijima, Y., Kakimoto, K., Kimura, M., Takeda, K., and Saitoh, T., 2000b, Reel-to-reel continuous formation of Y-123 coated conductors by IBAD and PLD method, presented at *Applied Superconductivity Conference*, Virginia Beach, VA, Sept. 17–22.

Iijima, Y., Kimura, M., Saitoh, T., and Takeda, K., 2000c, Development of Y-123-coated conductors by IBAD process, *Physica C*, 335:15.

Iijima, Y., Kakimoto, K., and Takeda, K., 2001a, Ion-beam-assisted growth of fluorite type oxide template films for biaxially textured HTSC coated conductors, *IEEE Trans. Appl. Supercond.*, 11:3451.

Iijima, Y., Kakimoto, K., Kimura, M., Takeda, K., and Saitoh, T., 2001b, Reel to reel continuous formation of Y-123 coated conductors by IBAD and PLD method, *IEEE Trans. Appl. Supercond.*, 11:2816.

Iijima, Y., Kakimoto, K., Takeda, K., and Saitoh, T., 2001c, Long length IBAD process of fluorite type oxides for Y-123 coated conductors, in: *Proc. Int. Workshop on Superconductivity*, Honolulu, HI, p. 47.

Ji, H. and Was, G.H., 1999, Linkage between crystallographic texture and surface roughness in niobium films synthesized by ion bean assisted deposition, *Nucl. Inst. Meth. Phys. Res. B*, 148:880.

Jia, Q.X. and Anderson, W.A., 1990, Critical current density enhancement in $YBa_2Cu_3O_{7-\delta}$ films on buffered metallic substrates, *Appl. Phys. Lett.*, 57:304.

Jia, Q.X., Arendt, P.N., Foltyn, S.R., Holesinger, T.G., and DePaula, R.F., 2001, Superconducting YBCO films on polycrystalline yttrium-iron-garnet using IBAD-YSZ as a template, *IEEE Trans. Appl. Supercond.*, 11:3489.

Kao, A.S. and Gorman, G.L., 1990, Modification of zirconia film properties by low-energy ion bombardment during reactive ion-beam deposition, *J. Appl. Phys.*, 67:3826.

Knierim, A., Auer, R., Geerk, J., Li, Y., Linker, G., Meyer, O., Reiner, J., Schweiss, P., and Smithey, R., 1997, Influence of the process parameters on the growth of YSZ layers prepared by ion-beam assisted deposition (IBAD), *Nucl. Inst. Meth. Phys. Res. B*, 127:832.

Kohno, O., Iijima, Y., Onabe, K., Tanabe, N., Sadakata, N., Saito, T., Yoshitomi, J., and Nagaya, S., 1995, YBCO tape conductor by thin film deposition process, in: *Proc. Int. Workshop on Superconductivity*, Maui, HI, p. 210.

Kumar, A., Ganapathi, L., Kanetkar, S.M., and Narayan, J., 1990, Synthesis of superconducting $YBa_2Cu_3O_{7-\delta}$ thin films on nickel-based superalloy using in situ pulsed laser deposition, *Appl. Phys. Lett.*, 57:2594.

Kung, H., Foltyn, S.R., Arendt, P.N., and Maley, M.P., 1999, Characterization of the structure of Y–Ba–Cu–O coated conductors, *IEEE Trans. Appl. Supercond.*, 9:2034.

Kung, H., Foltyn, S.R., Arendt, P.N., Groves, J.R., Holesinger, T.G., Jia, Q.X., and Maley, M.P., 2001, Microstructural characterization of YBCO thick film coated conductors on IBAD-YSZ and MgO template layers, in: *Proc. Int. Workshop on Superconductivity*, Honolulu, HI, p. 241.

Lewis, B. and Jordan, M.R., 1970, Nucleation and epitaxy of gold deposits on sodium chloride substrates during electron bombardment, *Thin Solid Films*, 6:1.

Mao, Y.J., Jiang, B.Y., Ren, C.X., Liu, X.H., and Zou, S.C., 1998, Low energy ion beam assisted deposition of biaxially aligned yttria-stabilized zirconia films on polycrystalline Ni–Cr alloy, *Nucl. Inst. and Meth. in Phys. Res. B*, 135:492.

Mao, Y.J., Liu, X.H., Zhang, F., Ren, C.X., and Zou, S.C., 1998, Ion beam assisted deposition of biaxially textured cerium dioxide films on polycrystalline Ni-based alloy, *Surf. Ctngs. Tech.*, 104:78.

Marinov, M., 1977, Effect of ion bombardment on the initial stages of thin film growth, *Thin Solid Films*, 46:267.

Marinov, M. and Dobrev, D., 1977, The change in the structure of vacuum condensed hexagonal close-packed metal films on ion bombardment, *Thin Solid Films*, 42:265.

Matijasevic, V.C., Lu, Z., Von Dessoneck, K., Taylor, C., and Bartlett, D., 1998, RHEED monitoring of rotating samples during large area homogeneous deposition of oxides, *Mater. Res. Soc. Symp. Proc.*, 526:249.

Matsumoto, K., Kim, S.-B., Wen, J.-G., Hirabayashi, I., Watanbe, T., Uno, N., and Ikeda, M., 1999, Fabrication of in -plane aligned YBCO films on polycrystalline Ni tapes buffered with surface-oxidized NiO layers, *IEEE Trans. Appl. Supercond.*, 9:1539.

Miura, S., Hashimoto, K., Wang, F., Enomoto, Y., and Morishita, T., 1997, Structural and electrical properties of liquid phase epitaxially grown $YBa_2Cu_3O_x$ films, *Physica C*, 278:201.

Mu, H., Ren, C., Jiang, B., Yu, Y., Luo, E.Z., and Wilson, Z.H., 2000, Low energy ion beam assisted deposition of biaxially aligned YSZ and CeO_2/YSZ films on r-plane sapphire, *Surf. Ctngs. Tech.*, 124:61.

Narumi, E., Song, L.W., Yang, F., Patel, S., Kao, Y.H., and Shaw, D.T., 1991, Critical current density enhancement in $YBa_2Cu_3O_{6.8}$ films on buffered metallic substrates, *Appl. Phys. Lett.*, 58:1202.

Nelson, R.S. and Thompson, M.W., 1963, The penetration of energetic ions through the open channels in a crystal lattice, *Phil. Mag.*, 8:1677.

Nies, R., Blanchard, S., Gromoll, B., Fischer, S., Kraemer, H.P., Schilling, W., Schmidt, W., Seebacher, R., Utz, B., Volkmar, R.R., v. Lochneysen, H., and Neumueller, H.W., 2000, Large area deposition of biaxially textured YSZ buffer layers and superconducting YBCO on polycrystalline substrates, in: *Proc. Int. Workshop on Superconductivity*, Matsue-shi, Japan, p. 269.

Ohmatsu, K., Muranaka, K., Taneda, T., Takei, H., Sato, K., Matsuo, K., and Takahashi, Y., 2000, Development of in-plane algned YBCO tapes fabricated by inclined substrate deposition, *13th International Symposium on Superconductivity*, Tokyo, Japan, October 14–16.

Onabe, K., Nagaya, S., Shimonosono, T., Iijima, Y., Sadakata, N., and Saitoh, T., 1996, Fabrication of YBCO superconducting tape by continuous MOCVD technique, *Adv. Supercond. IX*, 8:671.

O'Neill, D., 2000, Progress towards lengths at 3M, *Superconductivity Program for Electrical Systems, 2000 Annual Peer Review*, Washington, DC, July 17–19.

Peterson, D., 1995, *Superconductivity Program for Electrical Systems, 1995 Annual Peer Review*, Alexandria, VA, Aug. 1–2.

Piercy, G.R., Brown, F., Davies, J.A., and McCargo, M., 1963, Experimental evidence for the increase of heavy ion ranges by channeling in crystalline structure, *Phys. Rev. Lett.*, 10:399.

Quinton, W.A.J., Liang, W.Y., Baudenbacher, F., Foord, D.T., Bramley, A.P., and Newcomb, S.B., 1999, The microstructure of $YBa_2Cu_3O_{7-\delta}$ films and yttria-stabilized zirconia buffer layers deposited on inclined hastelloy substrates, *IEEE Trans. Appl. Supercond.*, 9:1498.

Rauschenbach, B. and Gerlach, J.W., 2000, Texture development in titanium nitride films grown by low-energy ion assisted deposition, *Cryst. Res. Technol.*, 35:675.

Reade, R.P., Mao, X.L., and Russo, R.E., 1991, Characterization of Y–Ba–Cu–O thin films and yttria-stabilized zirconia intermediate layers on metal alloys grown by pulsed laser deposition, *Appl. Phys. Lett.*, 59:739.

Reade, R.P., Berdahl, P., Russo, R.E., and Garrison, S.M., 1992, Laser deposition of biaxially textured yttria-stabilized zirconia buffer layers on polycrystalline metallic alloys for high critical current Y–Ba–Cu–O thin films, *Appl. Phys. Lett.*, 61:2231.

Reade, R.P. and Russo, R.E., 1996, Ion-assisted pulsed-laser deposition for the fabrication of Y–Ba–Cu–O multilayer structures using oriented intermediate layers of YSZ and CeO_2, *Appl. Surf. Sci.*, 96:726.

Ressler, K.G., Sonnenberg, N., and Cima, M.J., 1997, Mechanism of biaxial alignment of oxide thin films during ion-beam-assisted deposition, *J. Amer. Ceram. Soc.*, 80:2637.

Riley, G.N., Li, Q., and Fritzmeier, L.G., 1999, High temperature superconducting composite wires, *Current Opinion in Solid State & Mat. Sci.*, 4:473.

Robinson, M.T. and Oen, O.S., 1963, The channeling of energetic atoms in crystal lattices, *Appl. Phys. Lett.*, 2:30.

Rupich, M., 2000, Current and future status of HTS conductor manufacturing at ASC, *Superconductivity Program for Electrical Systems, 2000 Annual Peer Review*, Washington, DC, July 17–19.

Rupich, M.W., Palm, W., Zhang, W., Siegal, E., Annavarapu, S., Fritzmeier, L., Teplitsky, M.D., Thieme, C., and Paranthaman, M., 1999, Growth and characterization of oxide buffer layers for YBCO coated conductors, *IEEE Trans. Appl. Supercond.*, 9:1527.

Russo, R.E., Reade, R.P., McMillian, J.M., and Olsen, B.L., 1990, Metal buffer layers and Y–Ba–Cu–O thin films on Pt and stainless steel using pulsed laser deposition, *J. Appl. Phys.*, 68:1354.

Saavides, N., Thorley, A., Gnanarajan, S., Herrmann, J., Katsaros, A., and Molodyk, A., 2001, YBCO coated tapes fabricated by IBAD and magnetron sputtering techniques, *Physica C*, 341:2491.

Selvamanickam, V., 2000, YBCO coated conductor fabrication by MOCVD, *Superconductivity Program for Electrical Systems, 2000 Annual Peer Review*, Washington, DC, July 17–19.

Selvamanickam, V., Carota, G., Funk, M., Vo, N., Haldar, P., Baluchandran, U., Chudzik, M., Arendt, P., Groves, J.R., DePaula, R., and Newnam, B., 2001, High current Y–Ba–Cu–O coated conductor using metal organic chemical vapor deposition and ion-beam assisted deposition, *IEEE Trans. Appl. Supercond.*, 11:3379.

Shiohara, Y., 2000, Recent progress of coated conductors in Japan, *Fall MRS Meeting*, Boston, MA, Nov. 27–Dec. 1.

Sickafus, K.E., Minivini, L., Grimes, G.W., Valdez, J.A., Ishimaru, M., McClellan, K.J., and Hartmann, T., 2000, *Science*, 239:748.

Sonnenberg, N., Longo, A.S., Cima, M.J., Chang, B.P., Ressler, K.G., McIntyre, P.C., and Liu, Y.P., 1993, Preparation of biaxially aligned cubic zirconia films on pyrex glass substrates using ion-beam assisted deposition, *J. Appl. Phys.*, 74:1027.

Storer, J., 2000, IBAD substrate development at 3M, *Superconductivity Program for Electrical Systems, 2000 Annual Peer Review*, Washington, DC, July 17–19.

Tiwari, P., Kanetkar, S.M., Sharan, S., and Narayan, J., 1990, *In situ* single chamber laser processing of $YBa_2Cu_3O_{7-\delta}$ superconducting thin-films on Si(100) with yttria-stabilized zirconia buffer layers, *Appl. Phys. Lett.*, 57:1578.

Trillat, J.J., Terao, N., Tertain, L., and Gervais, H., 1956, Application de la methode de decapage ionique en diffraction electronique, *J. Phys. Soc. Jpn.*, 4:406.

van Wyk, G.N., 1980, The dependence of ion bombardment induced preferential orientation on the direction of the ion beam, *Rad. Eff. Lett.*, 57:45.

van Wyk, G.N. and Smith, H.J., 1978, Ion bombardment induced preferential orientation in polycrystalline Cu targets, *Rad. Eff.*, 38:245.

Wang, C.P., 1999, Ion-beam-induced texturing in oxide thin films and its applications, Ph.D. thesis, Stanford University.

Wang, C.P., Do, K.B., Beasley, M.R., Geballe, T.H., and Hammond, R.H., 1997, Deposition of in-plane textured MgO on amorphous Si_3N_4 substrates by ion-beam-assisted deposition and comparisons with ion-beam-assisted deposited yttria-stabilized-zirconia, *Appl. Phys. Lett.*, 71:2955.

Weismann, J., Dzick, J., Hoffmann, J., Heinemann, K., and Freyhardt, H.C., 1998, Growth mechanism of biaxially textured YSZ films deposited by ion-beam-assisted deposition, *J. Mater. Res.*, 13:3149.

Witanachchi, S., Patel, S., Zhu, Y.Z., Kwok, H.S., and Shaw, D.T., 1990, Flexible stainless steel foil as a substrate for superconducting Y–Ba–Cu–O films, *J. Mater. Res.*, 5:717.

Wu, X.D., Foltyn, S.R., Arendt, P., Townsend, J., Adams, C., Campbell, I.H., Tiwari, P., Coulter, Y., and Peterson, D.E., 1994, High current $YBa_2Cu_3O_{7-\delta}$ thick films on flexible nickel substraes with textured buffer layers, *Appl. Phys. Lett.*, 65:1961.

Wu, X.D., Foltyn, S.R., Arendt, P.N., Blumenthal, W.R., Campbell, I.H., Cotton, J.D., Coulter, J.Y., Maley, M.P., Safar, H.F., and Smith, J.L., 1995, Properties of $YBa_2Cu_3O_{7-\delta}$ thick films on flexible buffered metallic substrates, *Appl. Phys. Lett.*, 67:2397.

Xiong, X. and Winkler, D., 2000, Rapid deposition of biaxially-textured CeO_2 buffer layers on polycrystalline nickel alloy for superconducting tapes by ion assisted pulsed laser deposition, *Physica C*, 336:70.

Yang, F., Narumi, E., Patel, S., and Shaw, D.T., 1995, In-plane texturing and its effects on critical current densities of $YBa_2Cu_3O_{7-x}$ thin films grown on polycrystalline substrates, *Physica C*, 244:299.

Yu, L.S., Harper, J.M.E., Cuomo, J.J., and Smith, D.A., 1985, Alignment of thin films by glancing angle ion bombardment during deposition, *Appl. Phys. Lett.*, 47:932.

Yu, L.S., Harper, J.M.E., Cuomo, J.J., and Smith, D.A., 1986, Control of thin film orientation by glancing angle ion bombardment during growth, *J. Vac. Sci. Tech. A*, 4:443.

Zhu, S., Lowndes, D.H., Budai, J.D., and Norton, D.P,, 1994, In-plane aligned CeO_2 films grown on amorphous SiO_2 substrates by ion-beam assisted pulsed laser deposition, *Appl. Phys. Lett.*, 65:2012.

Chapter 2

EPITAXIAL SUPERCONDUCTORS ON ROLLING-ASSISTED-BIAXIALLY-TEXTURED-SUBSTRATES (RABiTS)

Amit Goyal

Metals and Ceramics Division
Oak Ridge National Laboratory
Oak Ridge, TN 37831
USA

2.1 INTRODUCTION

An overview of the fabrication of epitaxial, high-J_c, biaxially aligned YBCO thick films on Rolling-assisted-biaxially-textured-substrates (RABiTS) is provided. The RABiTS technique utilizes standard thermomechanical processing to obtain long lengths of flexible, biaxially oriented substrates with smooth surfaces (rms ~20 nm). The strong biaxial texture of the metal (in-plane 5–7° FWHM) is conferred to the superconductor by deposition of intermediate metal and/or oxide layers which serve both as a chemical as well as a structural buffer. Epitaxial YBCO films have been grown using a variety of techniques on RABiTS^{TM} with critical current densities exceeding 10^6 A/cm^2 at 77 K in self-field and have field dependences similar to that of epitaxial films on single crystal ceramic substrates. The texture of the base metal has been achieved in kilometer lengths and scaleable techniques are being pursued to deposit the epitaxial multilayers. Deposited conductors made using this technique offer a potential route for the fabrication of long lengths of high J_c wire capable of carrying high currents in high magnetic fields and at elevated temperatures.

2.2 BIAXIALLY TEXTURED METAL TEMPLATES

The formation of preferred orientations in metals as a result of cold work has been studied since the early 1920's and several books and reviews have been written on this subject (Barrett and Massalski, 1996). Advances within the last decade in understanding and representing orientations in three dimensions have allowed for significant understanding of deformation and annealing textures. Texture representation in

Euler (Bunge and Esling, 1982; Wenk and Kocks, 1987) and Frank Rodriguous (Frank, 1988) space is now commonly used to understand, predict and develop preferred orientation in metals and intermetallics. A recent overview article focusing on the advances within the last decade summarizes work to date on deformation and annealing textures in metals and intermetallics (Kad and Goyal, 1996). In the RABiTS process a metal or an alloy is first biaxially textured in long lengths by the processes of rolling and annealing (Goyal et al., 1996a, 1997; US Patents 5,739,086; 5,741,377; 5,846,912; 5,898,020). Substrates of interest for this application include cubic metals and alloys such as face-centered cubic (FCC) and body-centered cubic (BCC) materials. While BCC metals such as iron-based alloys could potentially be used, FCC metals and alloys are of primary interest for this application such as Ni and Cu based alloys. The only annealing texture in FCC metals that can be obtained in a single component as well as made to be very well-developed or sharp, is the cube texture. This texture corresponding to the orientation $\{100\}\langle 100\rangle$, consists of a cube plane parallel to the plane of the sheet and a cube edge parallel to the rolling direction. This is a unique texture, since it can be developed into an exteremely sharp texture, unlike any other primary recrystallization texture in metals. A fully developed cube texture resembles a single crystal with a mosiac or subgrains. Silver as a metal to form biaxially textured templates has also been studied extensively because of its potential chemical compatibility with high temperature superconductors. However, due to the unique deformation characteristics of Ag at room temperature, it does not exhibit the standard Cu-type rolling texture upon heavy reduction via rolling. Modification of the stacking fault energy in Ag by rolling at higher temperatures close to 200°C permits the formation of the cube texture. Nevertheless, it is very difficult to obtain a "clean" texture in Ag because of the high tendency to form annealing twins. Even on the $\{100\}\langle 100\rangle$, cube textured Ag grains, two epitaxial orientations of most HTS materials are favored. Hence, a buffer layer is required between the Ag substrate and the superconductor layer. This defeats the purpose of using an expensive Ag substrate as the template. Moreover, at the commonly used temperatures for HTS deposition and buffer layer deposition, Ag has a very high vapor pressure. Hence, depositing on the Ag template is akin to depositing on high "mobile" surface. Most oxide films deposited on such substrates are therefore very rough and have textures which have a full-width-half-maximum (FWHM) substantially larger than the Ag substrate itself. Most of the work in this area has therefore focused on Ni and Cu based alloys.

Representative rolling textures of Ni-based substrates formed by consecutive rolling of a polycrystalline, randomly oriented bar to total deformations greater than 99%, followed by recrystallization are shown in Figure 2.1. Figure 2.1(a) and (b) show the (111), and (200) pole figures for an as-rolled Ni sample. This texture is referred to as the "Cu-Type" rolling texture. The localization of intensities along the $\{112\}\langle 111\rangle$ and the $\{123\}\langle 634\rangle$ orientations along the β-fiber or the skeleton line is evident when viewing the data in Euler space (Goyal et al., 1996b). By controlling the surface condition of the work rolls, it is possible to obtain substrates with surfaces as smooth as those obtained by mechanical and chemical polishing. Average line scans in 50 μm $\times$ 50 μm region indicate a rms roughness of $\sim$10 nm (Goyal et al., 1996b). The surface condition of a substrate can greatly affect epitaxy and integrity of buffer layers and hence the J_c of the superconducting film. Obtaining substrates with surfaces adequate for film growth without the need for a cumbersome polishing step is important for scale up to long lengths. Subsequent annealing of the substrates in a wide temperature range results in the formation of a sharp $\{100\}\langle 100\rangle$ cube texture. Figure 2.1(c) and (d) show the (111) and (200) pole figures for Ni recrystallized at 1000°C for 4 hrs in a vacuum of

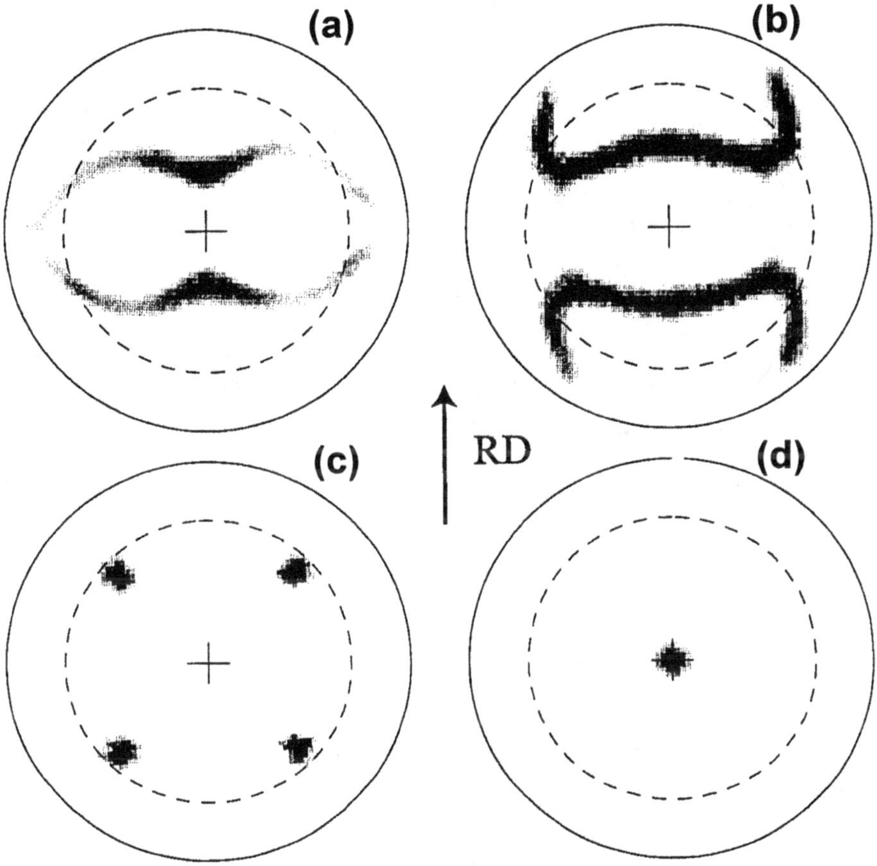

Figure 2.1. (a) Ni (111) pole figure and (b) Ni (200) pole figure of as-rolled substrate deformed greater than 99%. (c) Ni (111) pole figure and (d) Ni (200) pole figure of fully annealed substrate corresponding to figures (a) and (b).

$\sim 10^{-6}$ torr. The presence of a sharp, well-developed, single component cube texture is evident. Typical samples have X-ray ω- and ϕ-scans with full-width-half-maximum (FWHM) of 6° and 7° respectively. Depending upon the starting material, the impurity level and the deformation schedule, the cube texture can be made to be stable up to the melting point of Ni (Specht et al., 1998).

Figure 2.2 shows an orientation image micrograph of the Ni substrate. The micrograph was obtained using backscatter Kikuchi diffraction (BKD). Gray level shading on the micrograph is a reflection of the pattern quality or intensity of the Kikuchi bands observed at each point. Grain boundaries give rise to multiple diffraction patterns and hence have a poor pattern. Similarly, poor patterns are observed from any other crystallographic defect or strained region. BKD patterns were obtained on a hexagonal grid with a spacing of 3 μm. The total number of patterns obtained in the 0.5 μm × 0.5 μm region was close to 30,000. Indexing of the pattern at each location gave a unique measure of the orientation at that point. A hypothetical hexagonal lattice with a grain size of 3 μm was superimposed at each point from which a pattern was obtained. Grain boundary misorientations were then calculated for all the resulting boundaries using standard techniques. The micrograph was then regenerated with certain grain

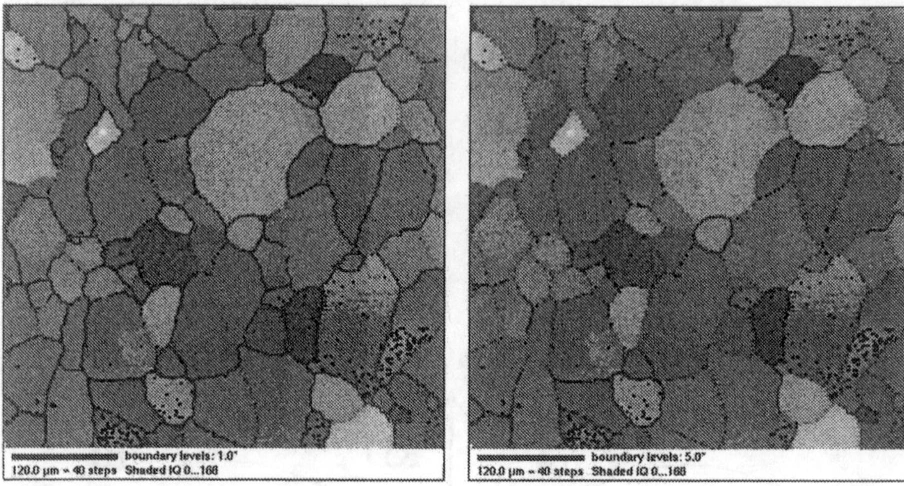

Figure 2.2. Electron backscatter Kikuchi diffraction created image of a macroscopic region of a cube textured Ni substrate. In Figure 2.2(a) and (b) all grain boundaries greater than 1° and 5° respectively are drawn.

(a) (b)

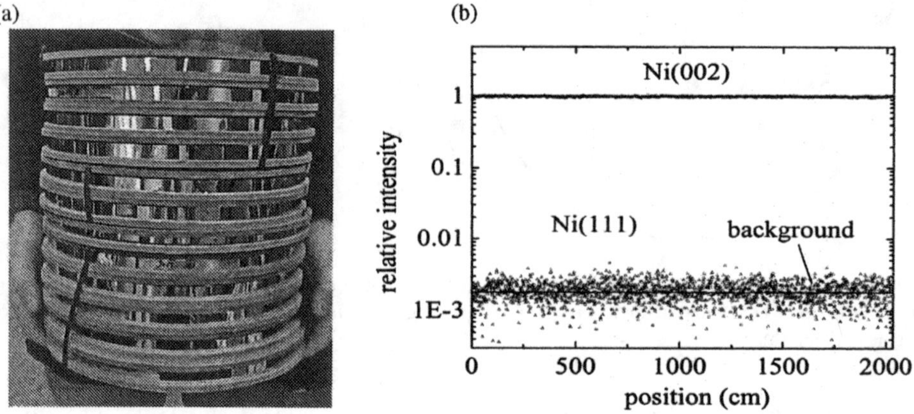

Figure 2.3. (a) Long lengths of heavily rolled Ni, 1 cm wide and 50 μm thick Ni substrate which upon annealing at 1100°C for 1 h yields a sharp cube texture. (b) (002) and (111) X-ray peak intensities measured continuously along the length of a 20 m long cube-textured Ni substrate, 1 cm wide and 50 μm thick. As can be seen, the (002) peak intensity is essentially constant along the length. The (111) peak intensity at all points is below the average background intensity.

boundary criteria. Figure 2.2(a) shows all grain boundaries greater than 1° and Figure 2.2(b) shows all grain boundaries greater than 5°. Substrates are typically textured in 4″ widths but widths greater than this are easily possible. Substrates have been fabricated in long lengths with piece lengths greater than 100 m. No technical limit on the length exists. Figure 2.3(a) shows 2 kilometers of textured Ni tape.

Figure 2.3(b) shows the (002) and (111) X-ray peak intensities measured continuously along the length of a 20 m long annealed Ni substrate, 1 cm wide and 50 μm thick. A sharp cube texture and constant cube texture is evident from the high and constant peak intensity of the (200) peak. The (111) peak at all locations is in the background.

2.3 MACROSCOPIC TEXTURE CHARACTERIZATION OF BIAXIALLY TEXTURED SUBSTRATES

The degree of macroscopic texturing in a substrate is typically measured by X-ray diffraction. A pole figure gives a nice representation of the texture in the substrates. For example a Ni (111) pole figure is a representation of the orientation of the [111] direction of every Ni grain in the substrate. Figure 2.4(a) and (b) shows typical, background corrected, Ni (111) pole figures in linear scale. Only intensity close to the cube orientation, {100}⟨100⟩ is observed in these pole figures. In order to observe the texture quality more closely, Figure 2.4(c) and (d) show the corresponding background corrected, log-scale pole figures of the linear pole figures shown in Figure 2.4(a) and (b). It can clearly be observed that while the substrate corresponding to Figures 2.4(b) and (d) has a "clean" cube texture, the substrate corresponding to the pole figures shown in Figures 2.4(a) and (c) has other orientations present. Quantification of the % cube texture in the substrate is done by summing the intensities at the cube orientation locations divided by the total integrated intensity in the log-scale pole figure. For the substrate corresponding to Figure 2.4(b) and (d), the % cube texture is ~100%, while

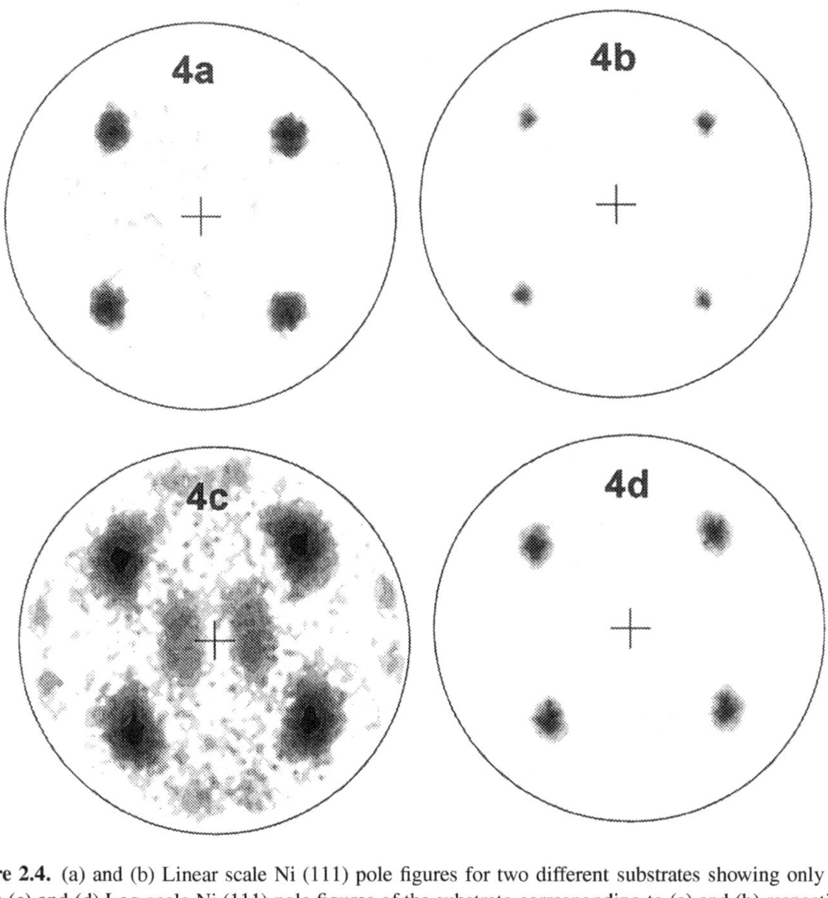

Figure 2.4. (a) and (b) Linear scale Ni (111) pole figures for two different substrates showing only cube poles; (c) and (d) Log-scale Ni (111) pole figures of the substrate corresponding to (a) and (b) respectively. It can clearly be seen that the substrate corresponding to figures (a) and (c) contains some retained rolling texture as well as twins.

that corresponding to Figure 2.4(a) and (c) is ∼94%. When the % cube texture is not close to 100%, numerous high angle grain boundaries can result in the substrate.

For determining the degree of in-plane and out-of-plane texture, omega-scans and phi-scans are done. The full-width-half-maximum (FWHM) of the out-of-plane texture is different for rocking about the rolling direction or in the direction. Typically the omega-scan is significantly broader for rocking about the rolling direction than for rocking in the rolling direction. Figures 2.5(a) and (b) show typical rocking curves in and about the rolling direction. The FWHM is determined by fitting a Guassian to the experimentally measured data to avoid erroneous determinations due to outliers from few large grains in the sample. A typical phi-scan is shown in Figure 2.5(c). The FWHM is determined by averaging the FWHM of the four peaks shown in Figure 2.5(c).

Depending upon the starting grain size of the material before rolling, the degree of randomness of the starting texture and the rolling parameters, it possible to obtain a substrate which shows no undesirable secondary recrystallization and maintains the primary recrystallization texture, i.e. the cube texture, till the melting point of the substrate. In such a case, significant sharpening of the out-of-plane and in-plane texture is possible by annealing at higher temperatures. Annealing at higher temperatures also reduces the twin density in the material. The grain size of the cube textured material continues to grow with annealing temperature but saturates when the substrate thickness is reached. Figure 2.6(a), (b) and (c) show the out-of-plane and in-plane texture, the corresponding twin density and grain size respectively of such a substrate as a function of the annealing temperature (Specht et al., 1998).

2.4 DEPOSITION OF THE SEED LAYER

Deposition of epitaxial buffer layers on textured metal or alloy substrates is a complex process because it involves the interaction between two surfaces with very different chemical properties. Under a wide range of deposition conditions for the oxide buffer layer, formation of NiO is favored. Typically this results in mixture of (111) NiO and (200) NiO. Even if a predominant (200) NiO is formed, the surface is quite rough. Hence, the deposition of oxide layers on metals and alloys of Ni an Cu is done under conditions where the formation of NiO or CuO is thermodynamically unfavorable. This is typically done by providing a background of 4%H_2 in argon or forming gas (Goyal et al., 1996a, 1996b, 1997; US Patents 5,739,086; 5,741,377; 5,846,912; 5,898,020; Norton et al., 1996; Paranthaman et al., 1997; He et al., 1997). If the deposition of the oxide layer is from a metal target such as during e-beam evaporation, a background of water of $\sim 1 \times 10^{-5}$ Torr is used to oxidize the material being deposited such as Ce (Paranthaman et al., 1997; He et al., 1997). Under these conditions, formation of a native metal oxide on the substrate, such as, NiO is thermodynamically unfavorable (Paranthaman et al., 1997; He et al., 1997).

Once oxidation of the Ni surface is prevented by choosing a seed layer that is thermodynamically more stable than NiO (e.g., CeO_2 or Y_2O_3-stabilized ZrO_2) and depositing it in reducing conditions, the lattice match between metal and oxide buffer layer is the main factor influencing epitaxial growth. This picture, however, is incomplete, since even under these conditions on many occasions, two in-plane epitaxial orientations of the seed oxide layer are observed. It is found that the presence of a fully developed c(2 × 2) sulfur superstructure on the surface of the Ni-alloy is essential for

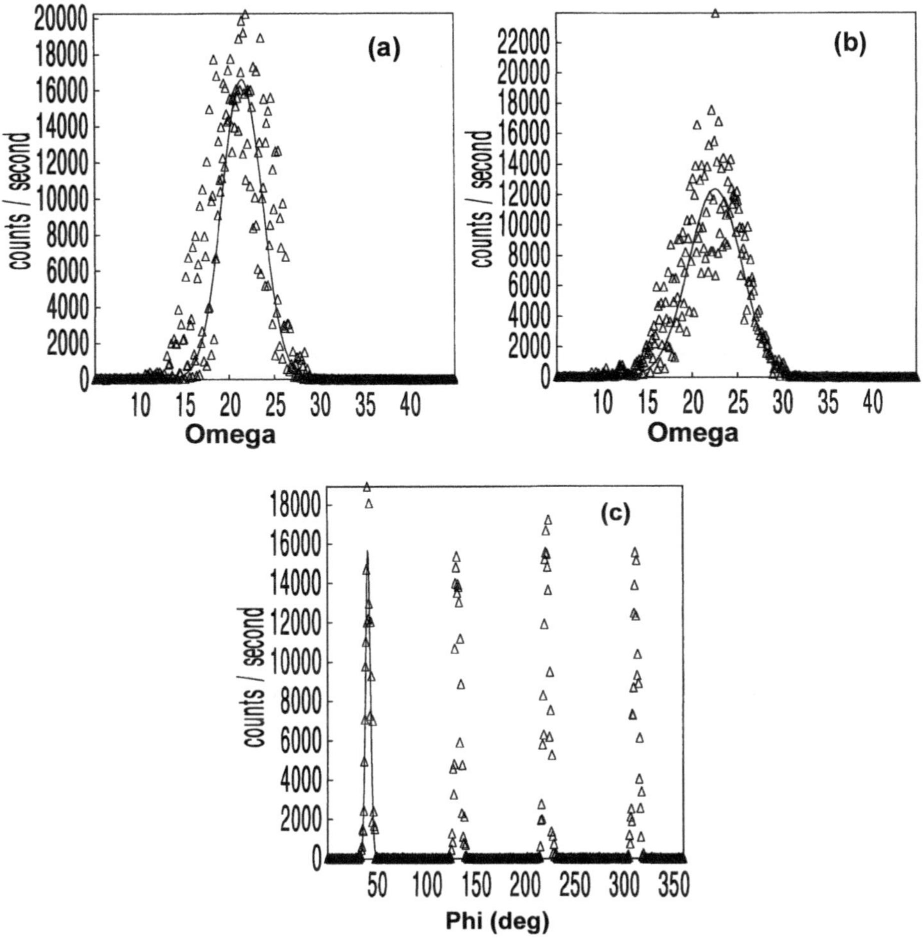

Figure 2.5. (a) Ni (200) Omega-scan or rocking curve when rocking *in—the* rolling direction. The fitted Guassian to the data is used to obtain the full-width-half-maximum (FWHM) of the out-of-plane texture in the rolling direction; (b) Ni (200) Omega-scan or rocking curve when rocking *about—the* rolling direction. The fitted Guassian to the data is used to obtain the full-width-half-maximum (FWHM) of the out-of-plane texture about the rolling direction, and (c) Ni (111) Phi-scan. The average FWHM obtained by fitting each of the four peaks by a Guassian to is used to obtain the full-width-half-maximum (FWHM) of the in-plane texture.

obtaining the correct epitaxial orientation of the seed layer (Cantoni et al., 2001, 2002, 2003). Figure 2.7(a) and (b) show *in-situ* reflection high energy electron diffraction patterns (RHEED) from the surface of a Ni with a c(2 $\times$ 2) superstructure and a Ni surface which has no c(2 $\times$ 2) superstructure (Cantoni et al., 2001, 2003).

Figure 2.8 shows the X-ray θ–2θ patterns and relative pole figures acquired from the YSZ films grown on textured Ni with and without superstructure, respectively (Cantoni et al., 2001). The YSZ films were grown under the same conditions and with the same procedure, while monitoring the process with RHEED. The YSZ films grown on the c(2 $\times$ 2) surface showed single (002) orientation, with a (111) pole figure indicating the same degree of grain alignment as the substrate. The resulting YSZ unit cell was rotated 45° in plane with respect to the Ni cell, consistent with the ex-

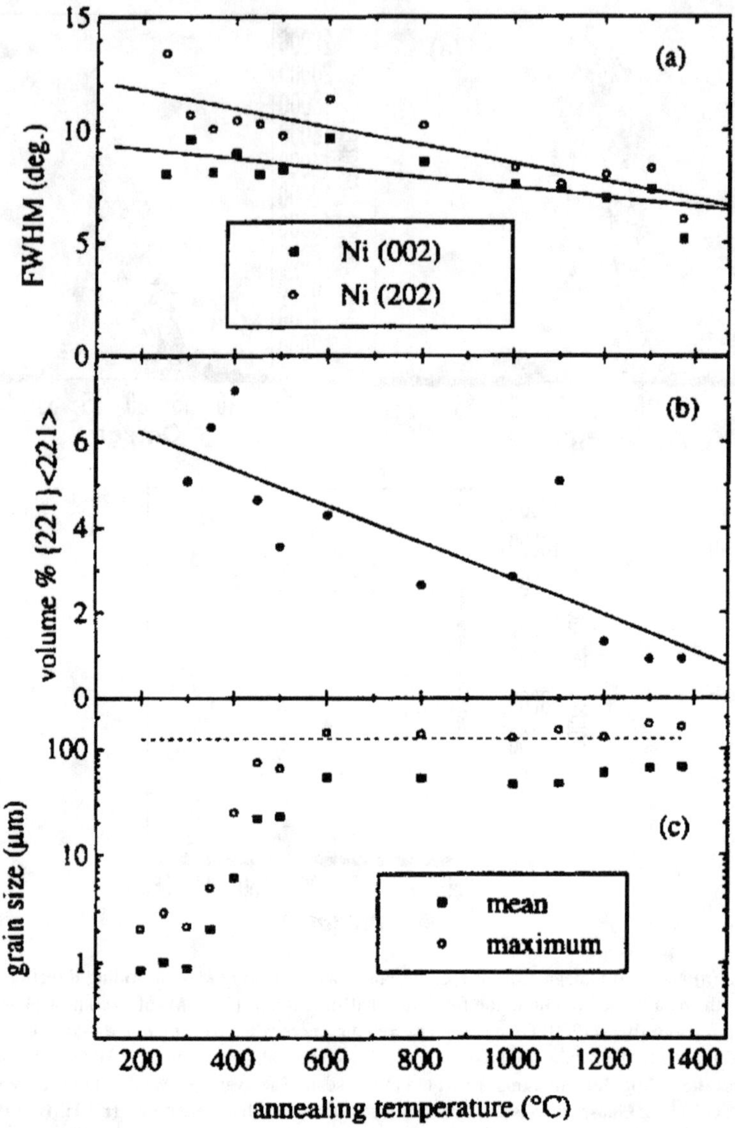

Figure 2.6. (a) FWHM of the omega and phi scan showing the out-of-plane and in-plane texture as a function of temperature; (b) volume % of twins in the substrate, and (c) grain size as a function of annealing temperature. The time at the annealing temperature was 10 mins for all temperatures studied.

pected cube-on-cube epitaxial orientation. In contrast, the YSZ films grown on the superstructure-free Ni overlayer showed only the (111) peak in the θ–2θ scan. In this case, the pole figure of the (200) reflection showed 4 different in-plane domains rotated 30° with respect to each other. This epitaxial relation is expected for the nucleation of a threefold symmetric lattice on a square symmetric lattice.

From the above it is clear that the presence of the c(2 × 2) sulfur superstructure is essential for obtaining cube-on-cube epitaxy of oxide seed layers. The above analysis holds true for deposition of oxides with a fluorite structure (CeO_2, YSZ, etc.),

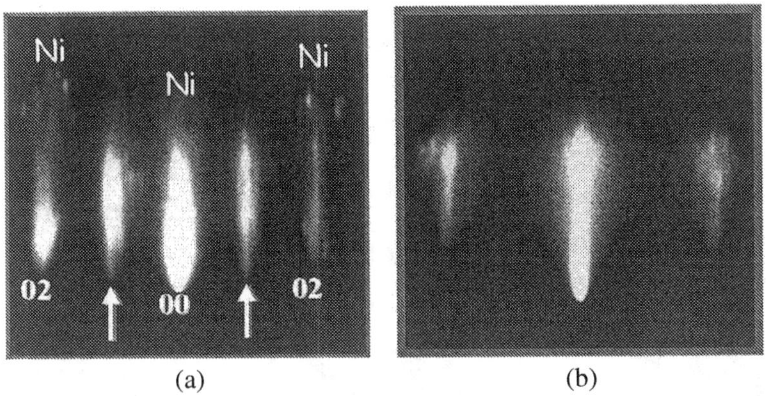

Figure 2.7. RHEED patterns obtained with the incident electron beam along ⟨100⟩ for: (a) RABITS Ni and (b) Ni overlayer deposited by PLD on RABiTS Ni. The arrows in (a) indicate the existence of a c(2 × 2) superstructure that is absent in (b) (from Cantoni et al., 2001).

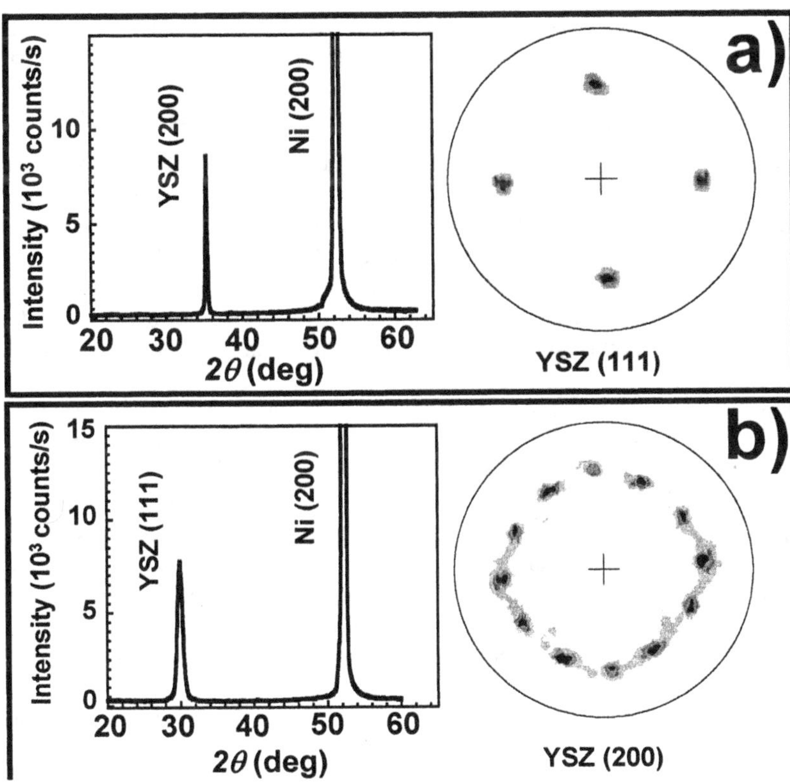

Figure 2.8. X-ray θ–2θ scan and pole figure for: (a) the YSZ film grown on the c(2 × 2) superstructure present on RABiTS Ni and (b) the YSZ film grown on the Ni overlayer epitaxially deposited on RABiTS Ni without the sulfur superstructure (from Cantoni et al., 2001).

perovskite structure (such as $SrTiO_3$, etc.), pyrochlore structure ($La_2Zr_2O_7$, etc.) and structure (Y_2O_3, etc.) on Ni and Ni-alloy substrates. The effect of the sulfur super-

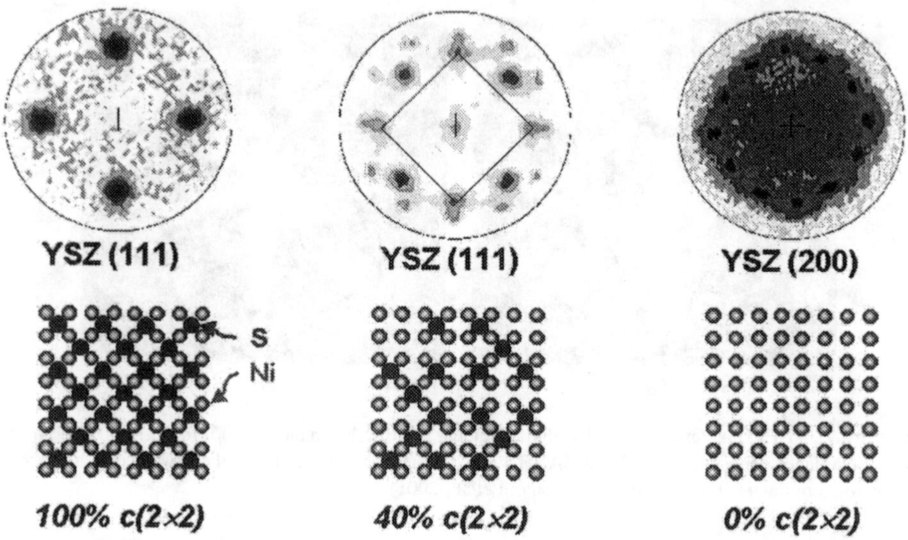

Figure 2.9. Comparison between pole figures of 3 different STO films grown on {100}⟨100⟩ Ni substrates having different sulfur coverage: 100% of a c(2 × 2) layer (0.5 ML) left; 40% middle; 0% right (from Cantoni et al., 2003).

structure has been explained on the basis of structural and chemical considerations. The S layer behaves like a template that well matches and mimics the arrangement of the oxygen atoms, in particular, the (001) sub-lattice planes for the seed layers mentioned above (Cantoni et al., 2001, 2002, 2003). Sulfur is chemically similar to oxygen and often exhibits the same electronic valence. The presence of such an ordered S template, thus, facilitates the bonding of the oxide cations in specific sites promoting the (001) epitaxial nucleation of the oxide film (Cantoni et al., 2001, 2002, 2003).

The c(2 × 2) sulfur superstructure can be formed on the surface of the Ni or Ni-alloy merely by surface segregation during the high temperature texture anneal (1000–1300°C) since sulfur is a common and almost unavoidable impurity in bulk Ni and Ni-alloys. However, depending upon the initial sulfur concentration in the metal or alloy as well as the specific texture annealing conditions, the sulfur superstructure can exhibit different coverage. The complete c(2 × 2) sulfur superstructure can be reproducibly formed on the surface of the metal/alloy tape by annealing the substrate at low temperatures (600–800°C) in very small amounts of H_2S (partial pressure of 1×10^{-7} to 1×10^{-6} Torr) for a few minutes (Cantoni et al., 2002, 2003). Previous surface studies have shown that H_2S molecules dissociate at the Ni surface and S atoms chemisorb forming a c(2 × 2) superstructure with a coverage that saturates to 0.5 ML corresponding to one complete atomic layer of the c(2 × 2)-S (Perdereau and Oudar, 1970; Andersson, 1976; Papageorgopoulos and Kamaratos, 1995). Incomplete c(2 × 2) superstructure results in mixed orientations of (200) and (111) textures of the oxide layer. Figure 2.8 shows comparisons between pole figures of YSZ films grown on cube textured Ni substrates having different sulfur coverages as determined by Auger Spectroscopy. In the first case, the substrate shows a complete S superstructure and the (111) logarithmic pole figure of the YSZ seed shows perfect cube texture. In the second case, the substrate superstructure exhibit a partial coverage (~40%) that translates to multiple orientation for the YSZ film. In the third case, the

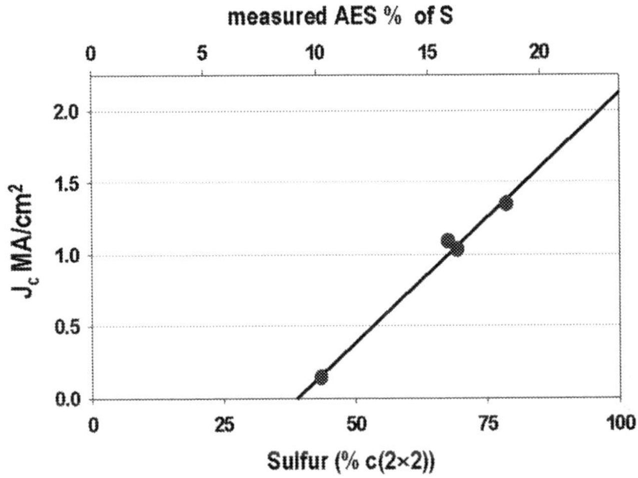

Figure 2.10. Critical current density versus initial substrate S concentration for four YBCO/CeO_2/YSZ/CeO_2/Ni RABiTS samples (from Cantoni et al., 2003).

surface of the substrate does not contain any S and the YSZ film shows only a (111) texture. In cases in which the substrate's S coverage was between 40 and 80% of a full c(2 × 2) layer, a cube texture with XRD peaks broader than those observed in seed layers grown on substrates with 100% c(2 × 2) were observed. The improvement in the texture of the seed layer determined by the S superstructure corresponds to an enhancement in J_c of the YBCO film subsequently deposited on the completed (seed plus buffer layer) RABiTS substrate.

This relation is illustrated in Figure 2.10, which plots the critical current density of 4 YBCO/CeO_2/YSZ/CeO_2/Ni samples versus the S surface content of the Ni substrate prior to seed layer deposition. The most commonly used seed layer at present is Y_2O_3. It is preferred over CeO_2 because it has a reduced tendency for cracking as well as reduced oxygen non-stoichiometry compared to CeO_2. Typical thickness of the seed layers in the range of 30–80 nm.

2.5 BARRIER AND CAP LAYER DEPOSITION

Once a high quality epitaxial seed oxide layer has been deposited, deposition of the barrier and cap layers is relatively straightforward. The most commonly used barrier layer for metal diffusion from the substrate to the superconductor layer is YSZ. Various techniques have been used to deposit the barrier layer. The important desired characteristic of the barrier layer besides texture is density. A dense layer is more effective in preventing metal diffusion to the superconductor layer. Pulsed laser ablation, rf-sputtering as well as reactive sputtering have been shown to result in dense YSZ layers. Typical thickness of the barrier layer is in the range of 150–300 nm depending on the time that will be required for deposition and/or formation of the superconductor layer during subsequent processing steps.

The most commonly used cap layer is CeO_2. Its purpose is to provide a good lattice match to YBCO. For growth of YBCO using the *ex-situ* BaF_2 process, CeO_2 is the only layer which results in good epitaxial growth of YBCO.

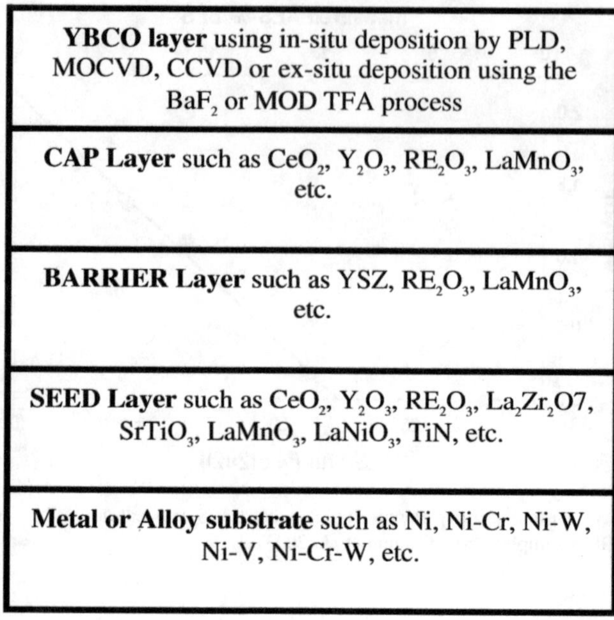

Figure 2.11. Various options for metal substrate, seed layer, barrier layer and cap layer that have been tried. It should be noted that for deposition of YBCO using the *ex-situ* BaF_2 process or the MOD TFA process, only the CeO_2 cap layer can be used.

Figure 2.11 shows a schematic of the buffer stack with the various options that have been tried for each layer in the stack.

Process parameters for deposition of barrier and cap layers via rf-sputtering have been published (List et al., 1988) and rf-sputtering is now routinely used to deposit these buffer layers in reel-to-reel configurations.

2.6 YBCO SUPERCONDUCTOR DEPOSITION

YBCO has been deposited on RABiTS substrates using pulsed laser ablation (PLD), metallorganic chemical vapor deposition (MOCVD) and the *ex-situ* BaF_2 process using e-beam co-evaporated precursors as well as using the metallorganic deposition (MOD) using the triflouroacetate (TFA) process. In each case, high J_c's have been reported exceeding 1 MA/cm^2 at 77 K, self-field. For short samples with YBCO deposition via PLD, J_c's approaching 3 MA/cm^2 at 77 K, self-field has been obtained (Mathis et al., 1988).

In long lengths, fully buffered tapes coated with a single layer of YBCO precursor by a commercial web-coating process with MOD using a trifluoroacetate (TFA)-based precursor has been demonstrated (Verebelyi et al., 2003). The organic components were decomposed in a humid, oxygen atmosphere up to a temperature of 400°C, to form a BaF_2-based precursor film with stoichiometric Cu and Y oxides for YBCO. This precursor was continuously converted to the epitaxial superconducting phase in a tube furnace in a humid, low oxygen partial pressure environment (Verebelyi et al., 2003). The resulting film thickness was 1.0 μm measured by SEM cross-section analysis. The substrate used was a 75 μm thick NiW alloy with a buffer sequence of

$Ni/Y_2O_3/YSZ/CeO_2$. The Ni layer was 2 μm thick, the Y_2O_3 layer deposited by e-beam evaporation was 50 nm, the YSZ layer deposited by rf-sputtering was 300 nm thick and the CeO_2 also deposited by rf-sputtering was 30 nm thick. Figure 2.12 shows I_c at 77 K, self field, for two wires, A (7.5 m) and B (8.0 m), measured at 50 cm intervals using the standard 1 $\mu V\,cm^{-1}$ criterion. End-to-end I_c performance, also determined at a 1 $\mu V\,cm^{-1}$ criterion, was 132 and 127 $A\,cm^{-1}$ width for A and B, respectively. The inset in Figure 2.12 provides I_c statistical data for both wires. The 3.0 $A\,cm^{-1}$ width standard deviation (σ) of I_c, measured for wire A, is the highest uniformity yet reported on any second generation wire of this length. Wire B is a replicate of A with nearly the same I_c and only slightly higher variability. The higher variability translates to a smaller '*n* value,' where *n* is the exponent of a power law fit to the end-to-end I–V curve near I_c, with wire B giving an exponent of 16 compared to 23 for wire A.

The chapter by Dominic et al. in this book covers details about *ex-situ* conversion of BaF_2 films.

Thick YBCO films have been grown on RABiTS substrates using PLD. Films of thickness 0.2–6.4 μm have been grown by PLD. Figure 2.13 shows the thickness dependence of J_c and I_c for YBCO films on two kinds of RABiTS substrates—Ni and Ni-3 at%W compared to the thickness dependence for YBCO on $SrTiO_3$ substrates by PLD (Kang et al., 2003).

The two kinds of RABiTS substrates correspond to $Ni/CeO_2/YSZ/CeO_2$ and Ni-3 at%W/Y_2O_3/YSZ/CeO_2. The Ni-3 at%W substrates were better significantly textured than Ni substrates. The NiW substrates are also significantly stronger than the Ni substrates and hence would have resulted in minimal damage during sample handling. Lastly, the presence of W, results in the formation of a $NiWO_3$ layer at the interface between NiW and NiO which restricts uncontrolled formation of NiO, thereby eliminating disruptions in the buffer stack (Leonard et al., 2003). The improved NiW substrates result in significantly better performance as indicated by the filled squares in Figure 2.13. The J_c is ~0.8 MA/cm^2 at 77 K, self-field, even for a 4.3 μm thick YBCO film. Also included in this study was a 6.4 μm thick YBCO film on the substrate Ni-3 at%W/Y_2O_3/YSZ/CeO_2 which could not be characterized in transport due to the contacts opening up during the measurement (Kang et al., 2003; Leonard et al., 2003). Detailed examination of the texture by X-ray diffraction showed that texture did not degrade with thickness of the YBCO film. Transmission electron microscopy (TEM) showed that no dead layer of YBCO formed even for the 6.4 μm thick YBCO film. Figure 2.14 shows a collage of cross-section TEM images for the 6.4 μm thick film. As can be seen from the figure, no dead layer was formed. The YBCO film contains numerous defects such as stacking faults, dislocations and second phase particles of CuO and Y_2O_3 but uninterrupted *c*-axis growth can be seen throughout the thickness of the film (Leonard et al., 2003).

2.7 FABRICATION OF ALLOY SUBSTRATES

Various Ni-alloy substrates have been successfully cube textured via rolling and annealing (Goyal et al., 1999, 2000). As has been specified in these references, cube-textured alloys can be designed and fabricated such that their stacking fault frequency parameter, alpha, is less than 0.01 at deformations greater than 50%. The alpha can be determined by X-ray diffraction as outlined in reference 25 and 26. If alpha is greater than 0.01, then the composition can be modified such that alpha is less than 0.01. Once

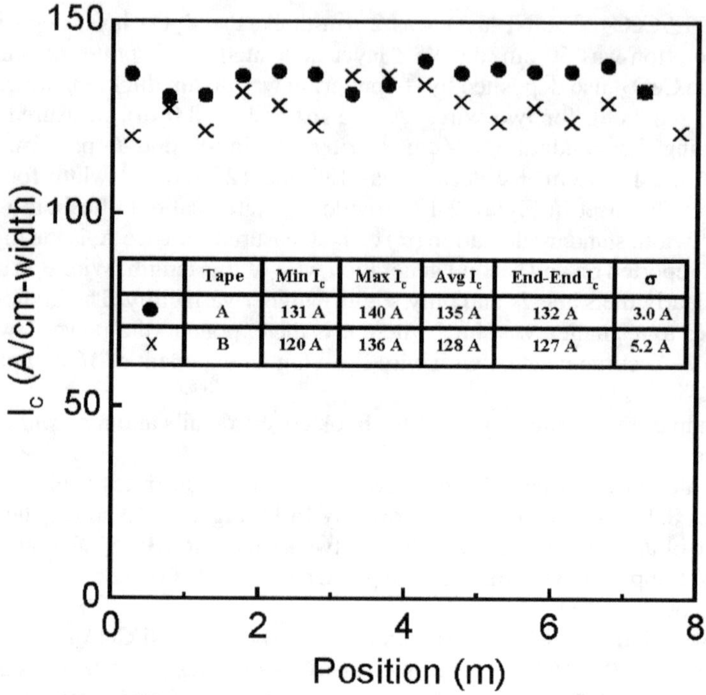

Figure 2.12. Critical current as a function of length, measured at 50 cm intervals, for two tapes, 7.5 m and 8 m long respectively at 77 K, self-field (from Verebelyi et al., 2003).

that is achieved the alloy can be rolled under conditions detailed below. If it is not possible to achieve alpha less than 0.01 at room temperature by compositional variation, then the rolling temperature can be increased to achieve it. The rolling then must be performed at this temperature. The rolling conditions in all cases are such that the total deformation is greater than 95%, the direction of rolling is preferably reversed at each pass and the annealing temperature is higher than the primary recrystallization temperature of the alloy. Using this method, single orientation, cube texture has been obtained in Ni–W alloys, Ni–Cr alloys, Ni–Mo alloys, Ni–V alloys, Ni–Cr–Al alloys and Ni–Cr–W alloys. Since cube textured NiW alloys were successfully developed (Goyal et al., 2002a), significant work has been performed in scaling these substrates up because of the ease of buffer layer deposition on these substrates. Tables 2.1 and 2.2 shows various binary and ternary substrates that have been cube-textured, indicating their %cube fraction, their yield strength in the annealed and textured state with the stress applied along the [100] crystallographic direction and their measured Curie temperatures. As can be seen from the figure, about 9 at%W and 13 at%Cr in binary alloys is required to get a completely non-magnetic substrate to minimize AC losses in application. Nevertheless, even lower alloying content substrates exhibit significantly reduced AC losses compared to a pure Ni substrate. As was shown in a study of Ni–Cr substrates with varying Cr contents, even small amounts of alloying element additions such as 7 at%Cr, result in significantly lower losses than that of pure Ni (Thompson et al., 2001). It is clear that there is significant world-wide interest in the fabrication of cube-textured, RABiTS alloy substrates (De Boer et al., 2001; Eickemeyer et al., 2001). For alloy substrates, in particular Ni-alloy substrates containing Cr, a trick is used to deposit high quality epitaxial buffer layers. Since deposition

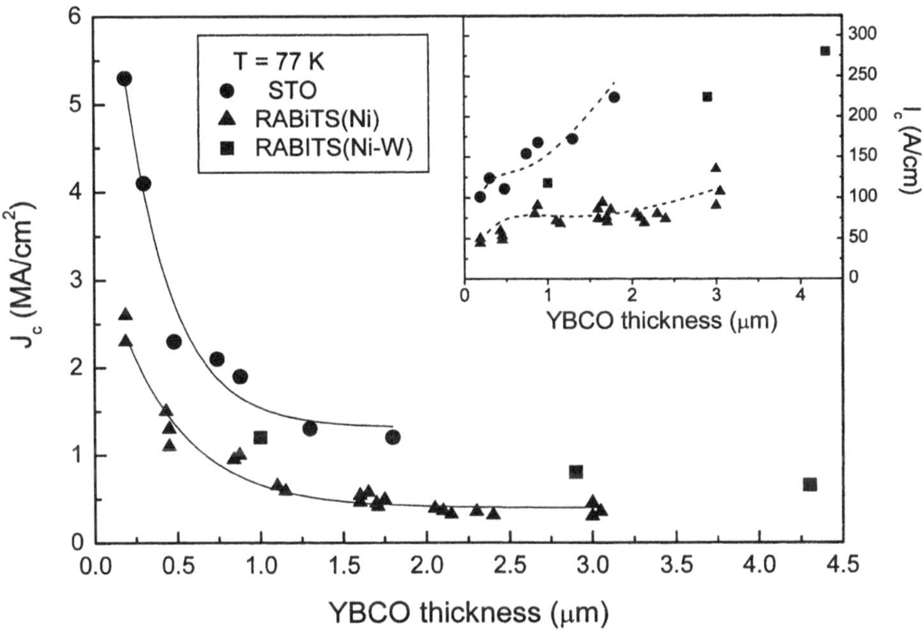

Figure 2.13. Critical current density as a function of YBCO film thickness on two kinds of RABiTS and STO substrates using PLD. Inset shows the critical current (A/cm-width).

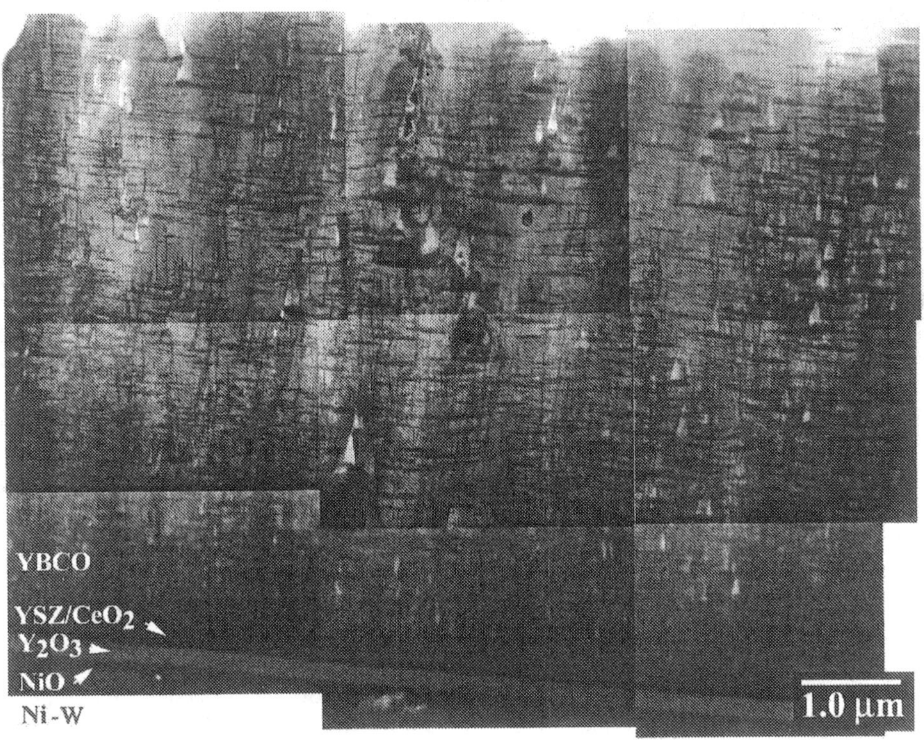

Figure 2.14. Composite cross-section image of the 6.4 μm thick YBCO film showing c-axis oriented growth throughout the full film thickness.

Table 2.1. Binary alloys of significant interest for use in RABiTS

Alloy	YS (0.2%)	Curie temp.
Ni	34 MPa	627 K
Ni-7 at%Cr	64 MPa	250 K
Ni-9 at%Cr	87 MPa	124 K
Ni-11 at%Cr	102 MPa	20 K
Ni-13 at%Cr	164 MPa	Non-magnetic
Ni-3 at%W	150 MPa	>400 K
Ni-5 at%W	165 MPa	335 K
Ni-6 at%W	197 MPa	?
Ni-9 at%W	270 MPa	~25 K

Table 2.2. Ternary alloys of significant interest for use in RABiTS

Ni-13 at%Cr-4 %Al	228 MPa	Non-magnetic
Ni-10 at%Cr-2 %W	150 MPa	Non-magnetic
Ni-8 at%Cr-4 at%W	202 MPa	Non-magnetic

of epitaxial oxides directly on the Cr containing alloy results in the formation of undesirable Cr_2O_3, a Ni overlayer or a Ni–W overlayer is first deposited epitaxially on the Ni–Cr alloy substrate (Goyal et al., 2002c). Using this process, J_c's over 1 MA/cm^2 in self-field, 77 K have been demonstrated (Goyal et al., 2002c).

Significant interest is currently in making alloy substrates using powder metallurgy as a route to making the starting coil (Goyal et al., 2001, 2002b, 2003). Powder metallurgy allows fabrication of alloys with homogeneous compositions without the detrimental effects of compositional segregation commonly encountered when using vacuum melting or casting to make alloys. Furthermore, powder metallurgy allows easy control of the grain size of the starting alloy body. Moreover, powder metallurgy allows a fine and homogeneous grain size to be achieved. Most importantly, using powder metallurgy, starting coils can be directly obtained at the desired thickness using well developed industry procedures without the need for complicated hot-rolling to homogenize the material with respect to grain size as is required for vacuum melted ingots. Using powder metallurgy, long lengths, over 2 km of Ni-3 at%W substrates have been fabricated by ORNL (Goyal et al., 2002a). Efforts are currently underway to fabricate Ni-5 and 9 at%W substrates via the powder metallurgy route.

An important point to mention is that AC losses from a substrate which is magnetic depends on the defect density in the substrate. Losses are an extrinsic feature and the Curie temperature is an intrinsic property of the substrate, much like the critical current density is an extrinsic property of the superconductor and depends upon defects while the superconducting transition temperature is an intrinsic property. Hence, depending on how a low alloying element substrate which is magnetic will contribute to losses depends on how the substrate has been handled after the final high temperature texture recyrstallization anneal.

Two additional methods exist to fabricate highly textured alloy substrates which reduced magnetism and have high mechanical strength. Both approaches are composite approaches to make substrates. In the first approach (Goyal et al., 2002) a powder-in-tube method using a Ni or Ni-alloy tube filled with non-magnetic and mechanically strong alloy powder or rod is heavily deformed followed by annealing. The result is cube-textured exterior and an un-textured interior. Alternatively, a rod of a hard, non-magnetic but deformable alloy is coated with Ni or a Ni-alloy which can result in the formation of a cube texture, is heavily deformed and annealed to form a textured exterior and an un-textured interior. In the second approach, a rolled substrate of a composition which can be cube textured is laminated in the final step with a mechanically strong and non-magnetic substrate (Goyal et al., 2000).

2.8 SUMMARY

The rolling-assisted-biaxially-textured-substrates (RABiTS) as shown schematically in Figure 2.15 appears to be a promising and industrially scalable method to fabricate long-lengths of high performance superconductors. Using this method, critical current densities exceeding 3 MA/cm^2 have been demonstrated for YBCO films. I_c/width approaching 300 A/cm width in thick films have also been demonstrated. Most importantly, reel-to-reel, continuously fabricated epitaxial YBCO/RABiTS have J_c's over 1 MA/cm^2 and I_c's of $\sim$130 A in 8 meter long lengths, showing the viability of this process. Efforts are presently underway world-wide to to achieve large-scale manufacturing of high performance and low-cost conductors using this process.

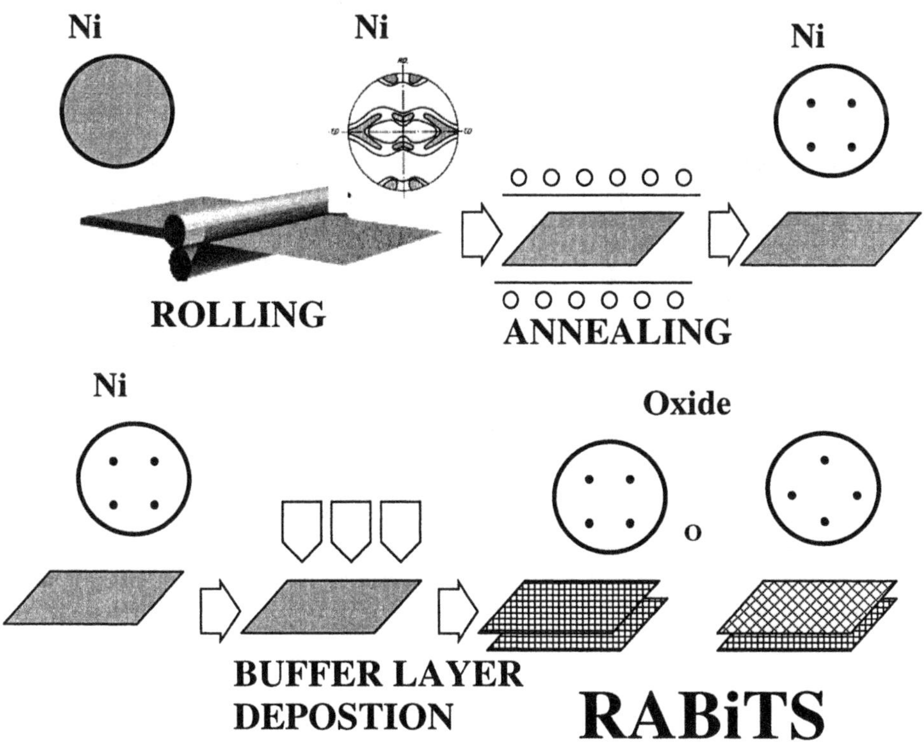

Figure 2.15. Schematic of the RABiTS process. First an untextured metal alloy is rolled to produce a particular desired rolling texture which upon annealing results in a sharp cube texture. Epitaxial buffer layers comprising generally of a seed layer, a barrier layer and a cap layer are then epitaxially deposited on the cube textured metal/alloy substrate. This comprises the **R**olling-**A**ssisted-**Bi**axially-**T**extured-**S**ubstrates (RABiTS) process. Epitaxial superconductors such as YBCO is then epitaxial deposited or formed on the substrate.

REFERENCES

Andersson, S., 1976, *J. Phys. C.*, 9:2721.

Barrett, C.S. and Massalski, T.B., 1996, *Structure of Metals*, McGraw-Hill, New York.

Bunge, H.J. and Esling, C., 1982, *Quantative Texture Analysis*, Dgm, Oberursel.

Cantoni, C., Christen, D.K., Feenstra, R., Norton, D.P., Goyal, A., Ownbyand, G.W., and Zehener, D.M., 2001, *Appl. Phys. Lett.*, 79:3077.

Cantoni, C., Christen, D.K., Heatherly, L., Kowalewski, M.M., List, F.A., Goyal, A., Ownby, G.W., Zehner, D.M., Kang, B.W., and Kroeger, D.M., 2002, *J. Mat. Res.*, 17:2549–2554.

Cantoni, C., Christen, D.K., Goyal, A., Heatherly, L., List, F.A., Ownby, G.W., Zehner, D.M., Christen, H.M., and Rouleau, C.M., 2003, *IEEE Trans. Appl. Supercond.*, 13:2646–2650.

De Boer, B., Eickemeyer, J., Reger, N., Fernandez, G.R., Ritcher, J., Holzapfel, B., Schultz, L., Prussiet, W., and Berberich, P., 2001, *Acta Met.*, 49:1421.

Eickemeyer, J., Selbmann, D., Opitz, R., De Boer, B., Holzapfel, B., Schultz, L., and Miller, U., 2001, *Supercond. Sci Technol.*, 14:152.

Frank, F.C., 1988, *Met. Trans.*, 19A:403.

Goyal, A., 2001, US Patent 6,180,570, January 30.

Goyal, A., 2002, US Patent 6,375,768, April 23.

Goyal, A., Norton, D.P., Budai, J.D. et al., 1996a, *Appl. Phys. Lett.*, 69(12):1795.

Goyal, A. et al., 1996b, *Appl. Supercond.*, 4:403–427.

Goyal, A. et al., 1997, *J. Mater. Res.*, 12:2924–2940.

Goyal, A. et al., 1999, US Patent 5,964,966, October 12.

Goyal, A. et al., 2000, US Patent 6,106,615, August 22.

Goyal, A. et al., 2001, US Patent 6,331,199, December 18.

Goyal, A. et al., 2002a, *Physica C*, 382:251.

Goyal, A. et al., 2002b, US Patent 6,447,714, September 10.

Goyal, A. et al., 2002c, US Patent 6,451,450, September 17.

Goyal, A. et al., 2003, US Patents 6,599,346, July 29; 6,602,313, August 5; 6,602,313, August 5; 6,607,838, August 19; 6,607,839, August 19; 6,610,413, August 26; 6,610,414, August 26; 6,635,097, October 21.

He, Q., Christen, D.K. et al., 1997, *Physica C*, 275:155.

Kad, B. and Goyal, A., 1996, Metals Information Analysis Center (MAIC), Report No. 9.

Kang, S., Goyal, A., Leonard, K.J., Rutter, N.A., and Kroeger, D.M., 2003, *J. American Ceramic Society* (submitted for publication).

Leonard, K.J., Kang, S., Goyal, A., Thomas, K.A., and Kroeger, D.M., 2003, *J. Mater. Res.*, 18:1723–1732.

List, F.A., Goyal, A., Paranthaman, M., Norton, D.P., Specht, E.D., Lee, D.F., and Kroeger, D.M., 1988, *Physica C*, 302:87–92.

Mathis, J.E., Goyal, A., Lee, D.F., List, F.A., Paranthaman, M., Christen, D.K., Specht, E.D., Kroeger, D.M., and Martin, P.M., 1988, *Jpn. J. Appl. Phys., Part 2—Letters*, 11B:L1379–L1382.

Norton, D.P., Goyal, A., Budai, J.D. et al., 1996, *Science*, 274:755.

Papageorgopoulos, C.A. and Kamaratos, M., 1995, *Surf. Sci.*, 338:77.

Paranthaman, M., Goyal, A., List, F.A. et al., 1997, *Physica C*, 231:266.

Perdereau, M. and Oudar, J., 1970, *Surf. Sci.*, 20:80.

Specht, E.D., Goyal, A., Lee, D.F., List, F.A., Kroeger, D.M., Paranthaman, M., Williams, R.K., and Christen, D.K., 1998, *Supercond. Sci. Technol.*, 11:945–949.

Thompson, J.R., Goyal, A., Christen, D.K., and Kroeger, D.M., 2001, *Physica C*, 370:169.

Verebelyi, D.T., Schoop, U., Thieme, C., Li, X., Zhang, W., Kodenkandath, T., Malozemoff, A.P., Nguyen, N., Siegal, E., Buczek, D., Lynch, J., Scudiere, J., Rupich, M., Goyal, A., Specht, E.D., Martin, P., and Paranthaman, M., 2003, *Supercond. Sci. Technol.*, 16:L19–L22.

Wenk, H.R. and Kocks, U.F., 1987, *Met. Trans.*, 18A:1083.

Chapter 3

INCLINED SUBSTRATE DEPOSITION

K. Fujino [1], K. Ohmatsu [1], Y. Sato [2], S. Honjo [2], and Y. Takahashi [2]

[1] Superconductor R&D Department
Electric Power System Technology Research Labs
Sumitomo Electric Industries, LTD
1-1-3, Shimaya, Konohana-ku
Osaka 554-0024
Japan
[2] Power Engineering R&D Center
Tokyo Electric Power Company
4-1, Egasaki-cho, Tsurumi-ku
Yokohama 230-8510
Japan

3.1 INTRODUCTION

High temperature superconducting thin films generally formed epitaxially on single-crystal substrates using physical vapor deposition have high-quality crystallinity. Also, their critical current densities (J_c) show the order of 10^6 A/cm^2 in a liquid nitrogen, which is two digits higher than the J_c of Bi2223 silver sheathed wire. In order to form a wire using physical vapor deposition, it is necessary to form the thin film on a substrate shaped like a long strip of tape. It is impossible to obtain a single crystal substrate in the shape of a strip of tape, and metal tape or another kind of tape with superior handling characteristics must be used for the substrate in power device applications such as a power transmission cable. Metal tape, however, is polycrystalline and it is difficult to epitaxially grow a thin film on metal tape. As a result, the superconducting property J_c is low at approximately 10^4 A/cm^2.

In order to solve these problems, the authors have developed an inclined substrate deposition (ISD) method using the pulsed laser deposition (PLD) method, which is a technology for forming thin film with bi-axially orientation on polycrystalline substrates.

In this report, all evaluations of the electric properties, magnetic field dependence and mechanical properties were carried out in a liquid nitrogen (77.3 K).

3.2 ISD METHOD

Figure 3.1 shows the structure of a high-temperature superconducting thin film made using the ISD method. In order to prevent mutual diffusion between the substrate metal and the superconductor, a buffer layer of yttria stabilized zirconia (YSZ) is formed on the metal substrate and the $YBa_2Cu_3O_{7-X}$ (YBCO) superconducting thin film is formed on the top of the buffer layer. A stabilizing layer of silver is then formed to protect the surface of the superconducting thin film. Since the current path in the high-temperature superconductor is in the a–b plane, the a and b axes must be aligned in the substrate plane in order to obtain a high J_c value. Therefore, as shown in Figure 3.1, a tape structure was adopted wherein the buffer layer is first made with a structure that is aligned in the substrate plane and the superconducting layer is then grown on the buffer layer epitaxially. The technology used to form such a buffer layer with a biaxially textured structure on a polycrystalline substrate is the ISD method.

Figure 3.2 is schematic diagram showing how a film is deposited using the ISD method. In the normal PLD method, an excimer laser is irradiated onto a sintered target of the buffer or superconducting material in a vacuum; a plasma is formed and a film is deposited onto the substrate, which is parallel to the target. In the ISD method, however, the substrate in inclined at a certain angle with respect to the target, as shown in the figure. With this geometric arrangement, the film is formed with a crystal structure that is biaxially aligned, having a crystal axis tending to be aligned perpendicularly to the substrate plane and a crystal axis tending to be aligned in the direction that the plasma travels. Thus, although the film is polycrystalline, it can be formed with a structure that not only has a crystal axis perpendicular to the substrate plane but also has a favorite orientation within the substrate plane (Fujino et al., 1996). The ISD method is unique in that it allows the formation of a thin film with a biaxially textured structure by merely changing the geometric orientation of the substrate; the equipment is simple and very advantageous in view of mass production processes. Deposition rates of the normal PLD method are also several 10 to several 100 times faster than sputtering, molecular beam epitaxy, and other methods, which also makes ISD a well-suited method for mass production. Deposition onto the tape substrate is conducted while feeding the tape substrate in a direction perpendicular to the plane of the paper in Figure 3.2.

3.3 TAPE PROPERTIES

YBCO (103) X-ray pole figure measurements were conducted to evaluate the substrate-plane alignment of the superconducting layer in tape made using the ISD

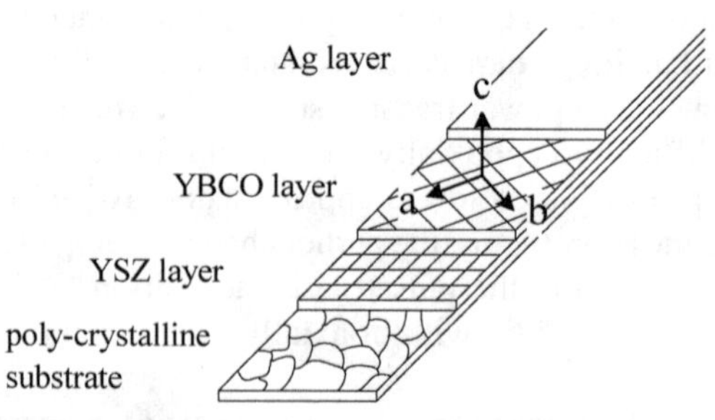

Figure 3.1. Structure of ISD tape.

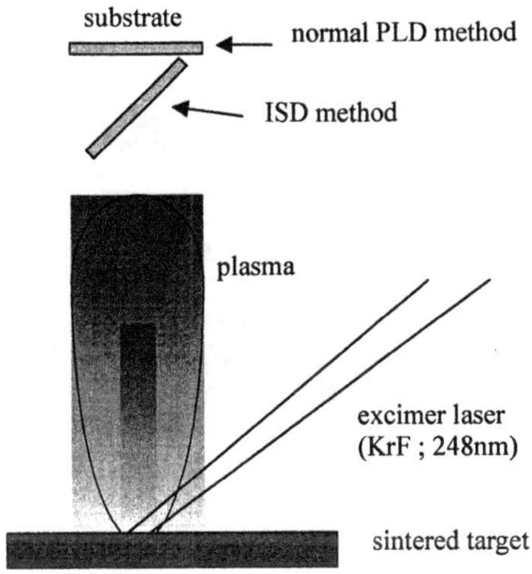

Figure 3.2. Schematic diagram of experimental set-up.

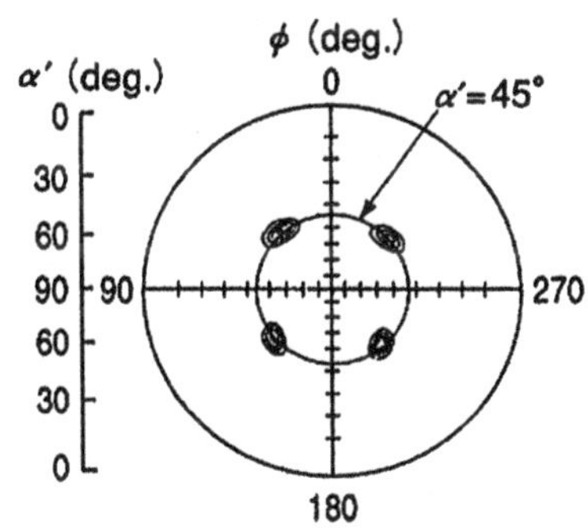

Figure 3.3. X-ray (103) pole figure of ISD tape.

method. The result is shown in Figure 3.3. The map for the ISD tape has four peaks that indicate the crystal axes are aligned in the substrate plane. When the *c*-axis of the YBCO is aligned perpendicularly with respect to the substrate plane, the diffraction peaks in the YBCO (103) plane are expected to appear at $\alpha' = 45^\circ$ in on the circumference. For the in-plane aligned film made using the ISD method, however, the position of the peaks is off by several degrees because the *c*-axis of the YBCO is slightly tilted with respect to the substrate plane. Figure 3.4 is a cross section SEM image showing how the YSZ buffer layer made using the ISD method has grown in columns that are slightly tilted with respect to the substrate plane. This is thought to result from a self-shadowing effect (Dirks and Lemy, 1997) caused by inclining the substrate (Hasegawa et al., 1998).

Figure 3.5 shows the relationship between deposition rate and in-plane orientation. There is no degradation of in-plane orientation up to the deposition rate of 0.7 μm/minute. We are considering the buffer materials other than YSZ in order to increase the speed of the ISD method and have confirmed that CeO_2 can be deposited 1 μm/minute faster than YSZ while still obtaining a similar orientation.

The J_c value and critical current (I_c) value of tapes manufactured using this ISD method is shown in Table 3.1. The superconducting current flows in the Cu–O plane

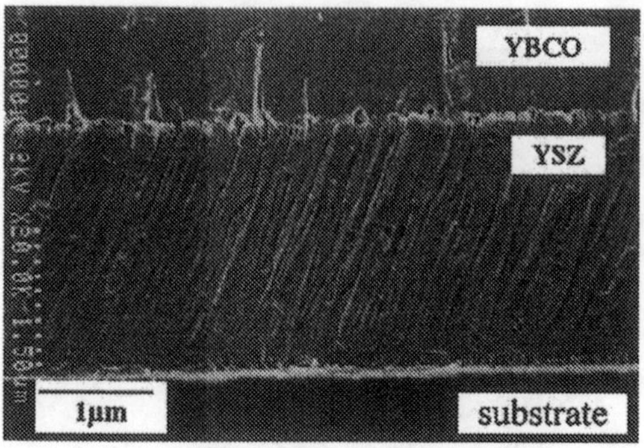

Figure 3.4. Cross section SEM image of biaxially aligned YSZ and YBCO.

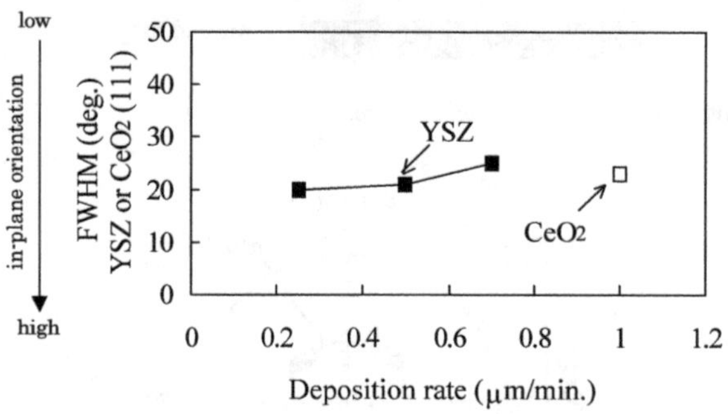

Figure 3.5. In-plane orientation versus deposition rate for ISD buffer layer.

parallel to the *a* and *b* axes of the YBCO superconductor. If there is no in-plane alignment in the substrate plane, the current path is obstructed and J_c will be low. This is why J_c values on the order of 10^4 A/cm^2 are all that can be obtained using the normal PLD method. Meanwhile, ISD tapes are structured as shown Figure 3.1 and show J_c values on the order of 10^5 A/cm^2, which is ten times greater than for PLD tapes.

In addition to J_c another effective property requirement for power device applications is high I_c achieved by increased film thickness. It is known that, in general, increasing the thickness of a thin film will result in decreased crystallinity caused by lattice defects and the like; the film properties will be closer to bulk and J_c will be reduced. We have also been investigating thicker films since the development of the ISD method and, as shown in Figure 3.6, in 1997 we succeeded in producing a 4 μm film with an I_c of 62.4 A while still maintaining a J_c value on the order of 10^5 A/cm^2.

Figure 3.7 shows the dependence of J_c on magnetic field for an ISD film along with the properties of a film with the normal PLD method without using the ISD method for comparison. Measurements were made with the magnetic field applied in two different directions: perpendicular and parallel to the tape plane. In either case, the degree of reduction in J_c caused by the magnetic field is smaller for the ISD tape. The fact that the ISD tape has an in-plane alignment is believed to result in less penetration of the magnetic field at the grain boundaries in the ISD tape than the normal PLD tape, for which there is no in-plane alignment.

In such power device related applications as power transmission cables and magnets, mechanical strength is important because the tape will experience bending, tensile, and other stresses. Figure 3.8 shows the dependence of I_c on bending strain for

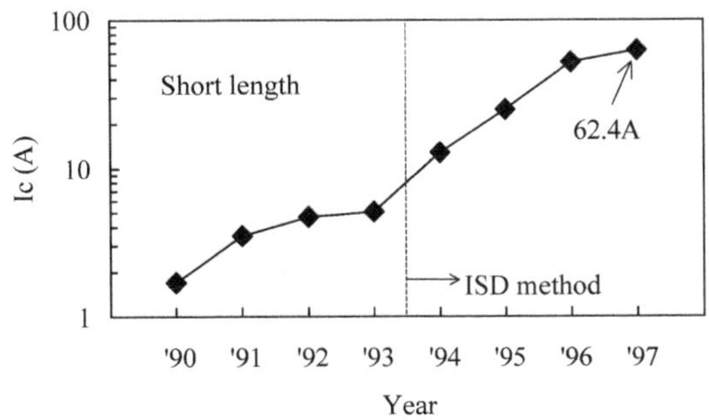

Figure 3.6. I_c value (77.3 K) versus year for short length tapes.

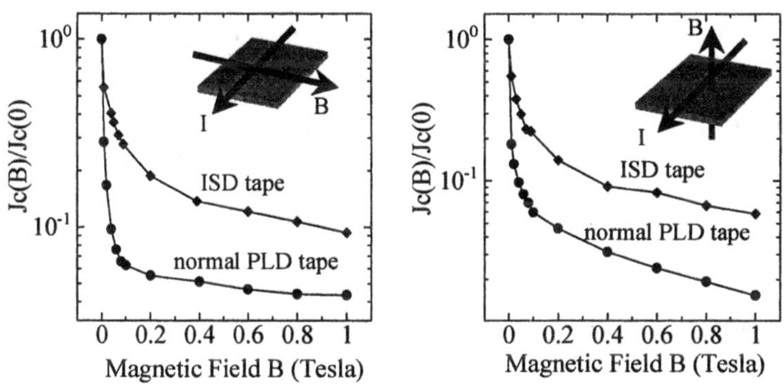

Figure 3.7. J_c dependence on magnetic field B at 77.3 K.

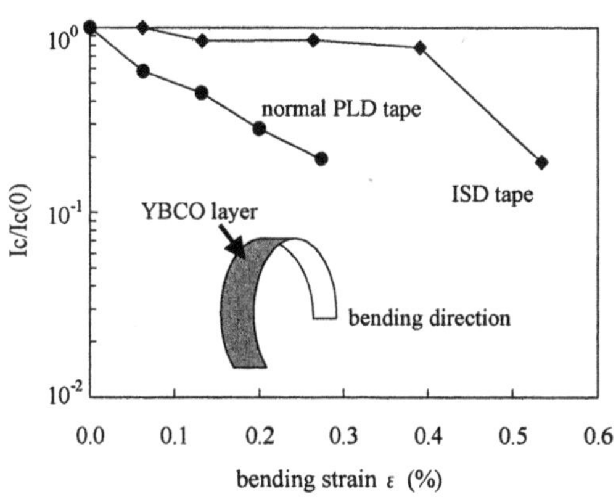

Figure 3.8. Bent test results of YBCO tapes at 77.3 K.

normal PLD tape and ISD tape. The measurements were made while bending the tape with the thin film surface on the outside so that a tensile strain developed in the YBCO superconductor. The strain rate ε (%) was estimated according to $t/2R \times 100$, where t is the thickness of the metal substrate and R is the radius of curvature of the substrate during the test. In the case of the PLD tape, I_c steadily declines as bending stress increase and the I_c value at a strain of 0.3% is only 10% of the initial I_c value. This is the result of the tensile stress causing cracks to form in the YBCO superconductor.

Conversely, the ISD tape drops slightly but holds at 80% of the initial I_c until the stress exceeds 0.4%. A bending strain of 0.4% corresponds to a radius of curvature of approximately 20 mm. The good mechanical characteristics of the ISD tape are thought to occur because of strong bonds between grain boundaries resulting from improved crystallinity. It has also been confirmed in a compressive strain test, wherein the tape was bent with the thin film on the inside, that both types of tape are free of deterioration in properties all the way up to the measurement limit (the elastic limit of the metal substrate).

3.4 SUMMARY

The development of the ISD method has made it possible to deposit a film with in-plane alignment on a polycrystalline metal substrate and achieve a J_c value ten times higher than normal PLD tape. We also confirmed that the resulting mechanical properties and other practical properties are improved. Since the ISD method can be used with metal substrates, which are inexpensive and easily made into long tape material, it is believed that it will be a very powerful method for future mass production.

REFERENCES

Dirks, A.G. and Lemy, H.J., 1997, *Thin Solid Films*, 47:219.

Fujino, K., Hasegawa, K., Mukai, H., Sato, K., Hara, T., Ohkuma, T., Ishii, H., and Honjo, S., 1996, in: *Advances in Superconductivity VIII*, Springer, p. 675.

Hasegawa, K., Fujino, K., Mukai, H., Konishi, M., Hayashi, K., Sato, K., Honjo, S., Sato, Y., Ishii, H., and Iwata, Y., 1998, Biaxially aligned YBCO film tapes fabricated by all pulsed laser deposition, *Appl. Supercond.*, 4:487.

Chapter 4

ISD BY THERMAL EVAPORATION

Markus Bauer

Technical University Munich
Physics Department E10
James-Franck Str. 1
85747 Garching
Germany

For the deposition of YBCO thin films with high critical current density on metal tape, a biaxially textured buffer layer can be used. Bauer (1999) developed an inclined substrate deposition technique using e-gun evaporation for the buffer layers. In contrast to the ISD process using pulsed laser deposition of Fujino (1995) and Quinton (1997) the buffer layer material is not YSZ but MgO. The setup for the buffer layer deposition is shown in Figure 4.1. MgO is either evaporated directly or by reactive thermal evaporation of magnesium. The substrate (Hastelloy C276) is inclined so that the vapor hits the substrate surface by the angle α. As high evaporation rates up to 500 nm/min can be used, this process is suitable also for the deposition onto long tapes.

The MgO buffer layers are highly textured and due to the inclined evaporation, the MgO [001]-axis is not oriented parallel to the substrate normal but tilted by a texture angle of typically 25° towards the deposition direction. The texture angle and the in-plane alignment depend on the deposition parameters of the MgO film. High deposition rates and film thicknesses above 1 μm are necessary for well-textured films, cf. Bauer (2000). The film exhibits a columnar structure with nearly perpendicular columns (Figure 4.1(b)) in contrast to ISD-YSZ where about 20° inclined columns were observed by Hasegawa (1998).

In order to explain the mechanism of texturing, Bauer (2000) performed Monte Carlo simulations of the film growth. By taking into account self shadowing of the growing columns, cf. Leamy (1980), and limited surface diffusion we were able to reproduce the columnar structure in the computational simulation. Furthermore, the observed crystalline orientation can be identified as the one giving the highest growth columns. The texture develops by an evolutionary shading process where only these columns survive. A similar process was suggested for the inclined substrate pulsed laser deposition of YSZ by Hasegawa (1998).

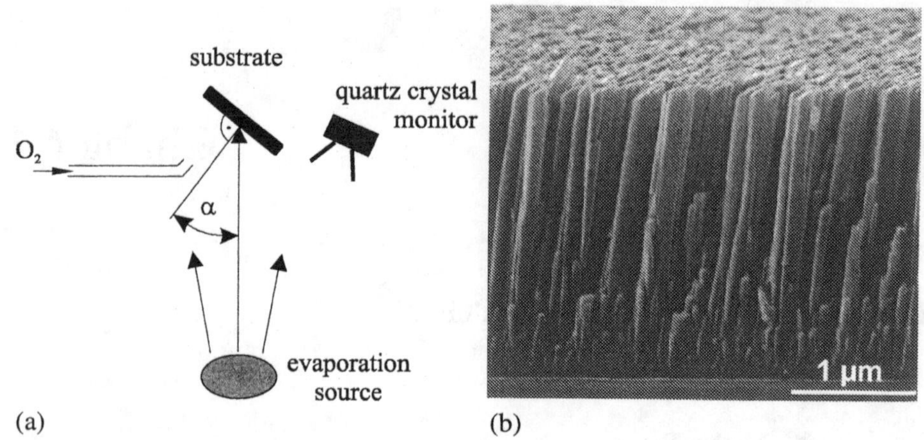

Figure 4.1. Experimental setup for inclined substrate deposition: (a) the evaporation source is an e-gun with MgO or Mg target; (b) scanning electron micrograph of a fracture cross section of ISD MgO.

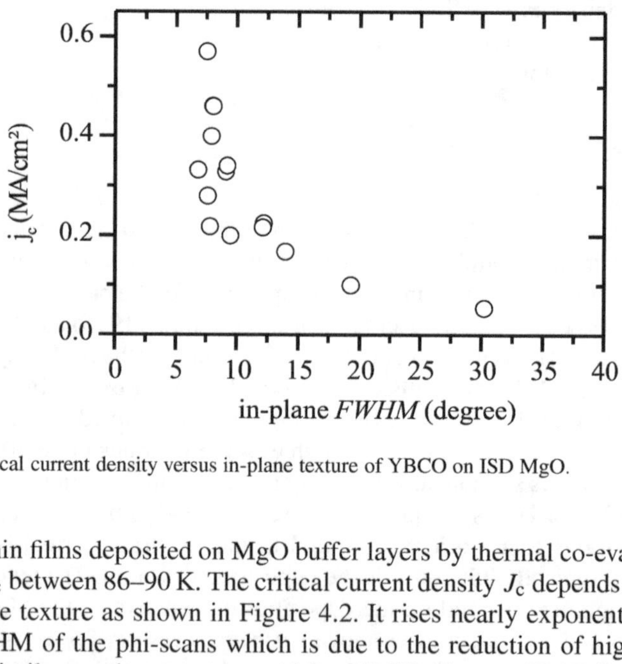

Figure 4.2. Critical current density versus in-plane texture of YBCO on ISD MgO.

YBCO thin films deposited on MgO buffer layers by thermal co-evaporation usually exhibit T_c between 86–90 K. The critical current density J_c depends on the quality of the in-plane texture as shown in Figure 4.2. It rises nearly exponentially with decreasing FWHM of the phi-scans which is due to the reduction of high angle grain boundaries. Similar results were reported for YBCO films on IBAD buffer layers by Iijima (1998).

The pole figure of a highly textured YBCO film with in-plane FWHM of 7° is depicted in Figure 4.3. The YBCO *c*-axis is inclined due to the tilted MgO buffer layer. The inclination leads to an anisotropic critical current density being lower in the tilt direction which can be attributed to the intrinsic J_c-anisotropy of YBCO, cf. Blatter (1994). The ratio between the high J_c value perpendicular and the low J_c value parallel to the tilt direction can range up to a factor of two. Maximum J_c values of 8×10^5 MA/cm^2 were measured in the high J_c-direction at 77 K. In order to obtain the highest possible J_c along the tape the substrate has to be inclined perpendicular to the tape axis during the MgO deposition.

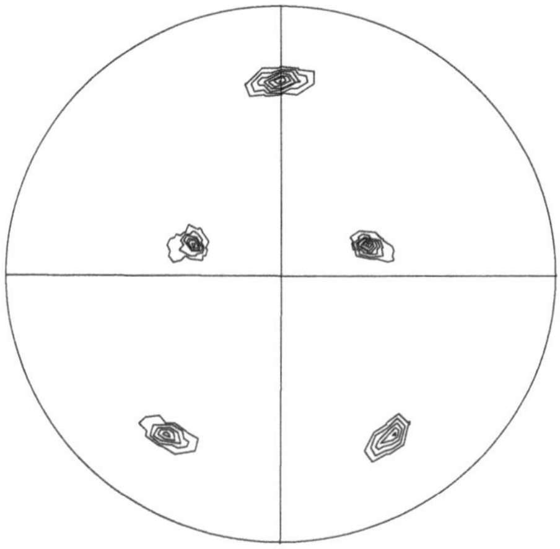

Figure 4.3. (103) pole figure of YBCO on a ISD-buffer layer of MgO.

REFERENCES

Bauer, M., 2000, Inclined substrate deposition by evaporation of magnesium oxide for coated conductors, in: *Proceedings of the Fall99 MRS Meeting in Boston*, submitted.

Bauer, M., Semerad, R., Kinder, H., 1999, YBCO films on metal substrates with biaxially aligned MgO buffer layers, *IEEE Trans. Appl. Supercond.*, 9:1502.

Blatter, G. et al., 1994, Vortices in high-temperature superconductors, *Rev. Mod. Phys.*, 66:1125.

Fujino, K. et al., 1995, One meter long thin film tape with J_c more than 10^5 A/cm^2 fabricated by pulsed laser deposition, in: *Advances in Superconductivity VII*, Springer, Tokyo, p. 629.

Hasegawa, K. et al., 1998, Biaxially aligned YBCO film tapes fabricated by inclined substrate pulsed laser deposition, in: *Advances in Cryogenic Engineering*, Vol. 44, Balachandran, ed., Plenum Press, New York.

Hasegawa, K., Fujino, K., Mukai, H., Konishi, M., Hayashi, K., Sato, K., Honjo, S., Sato, Y., Ishii, H., and Iwata, Y., 1998, *Appl. Supercond.*, 4:487.

Iijima, Y., Hosaka, M., Tanabe, N., Sadakata, N., Saitoh, T., Kohno, O., and Takeda, K., 1998, Processing and transport characteristics of YBCO tape conductors formed by IBAD method, *Appl. Supercond.*, 4.

Leamy, H.J., Gilmer, G.H., and Dirks, A.G., 1980, The microstructure of vapor deposited thin films, in: *Current Topics in Materials Science*, Vol. 6, E. Kaldis, ed., North-Holland, Amsterdam, p. 311.

Quinton, W.A.J. et al., 1997, Deposition of biaxially aligned YSZ films on inclined polycrystalline metallic substrates for $YBa_2Cu_3O_{7-\delta}$ tapes, *Physica C*, 292:243.

Section B

Methods of $YBa_2Cu_3O_{7-\delta}$ Deposition and Related Issues

Chapter 5

PULSED LASER DEPOSITION OF $YBa_2Cu_3O_{7-\delta}$ FOR COATED CONDUCTOR APPLICATIONS: CURRENT STATUS AND COST ISSUES

Hans M. Christen

Oak Ridge National Laboratory
Solid-State Division
Oak Ridge, TN 37831-6056
USA

5.1 INTRODUCTION

Amongst the numerous techniques currently being tested for the fabrication of coated conductors (i.e. high-temperature superconducting oxides deposited onto metallic tapes), pulsed laser deposition (PLD) plays a prominent role. Recent results from groups in the United States (e.g., Los Alamos National Laboratory), Europe (e.g., University of Göttingen), and Japan (e.g., Fujikura Ltd.) are most promising, yielding record numbers for J_c and I_c.

As a method for depositing films of complex materials, such as the high-temperature superconductors (HTS) relevant to this chapter, PLD is well established and conceptually simple. Issues that have kept PLD from becoming a successful technique for devices and optical materials, namely particulate formation and thickness non-uniformities, are much less of a concern in the fabrication of HTS tapes. Nevertheless, the approach still presents ongoing challenges in the fundamental understanding of the laser-target interaction and growth from the resulting energetic plasma plume. Numerous issues related to the scale-up, control, reproducibility, and economic feasibility of PLD remain under investigation.

It is clearly beyond the scope of this short chapter to give a complete treatise of such a complex subject—fortunately, several excellent reviews have already been published (Chrisey and Hubler, 1994; Lowndes et al., 1996; Willmott and Huber, 2000). It is thus the intent of the author to introduce PLD only briefly and with a strong focus on HTS deposition, summarizing the most important developments and referring the reader to the numerous cited works. A strong emphasis is placed on issues related to scale-up.

Finally, it is the goal of this chapter to introduce a preliminary cost-analysis of PLD for coated conductor fabrication. While an attempt to calculate the total cost of fabricating a length of superconducting tape would be beyond the scope of this chapter (and beyond our current knowledge of numerous issues), we can present a rather detailed analysis of the cost of laser operation. The point, of course, is not to show that PLD is either cheaper or more expensive than other techniques—in fact, for most other techniques, which have not been successfully applied to long length fabrication, such a calculation cannot yet be made. It is illustrative, for example, to consider that in an industrial setting, more meters of coated tape will have to be produced *per day* than some other methods have produced in their entire history, and that therefore many reliability issues (including mean-time between failure of equipment) have not been addressed for many of the approaches described in this book. It is the maturity of industrial laser technology combined with the maturity of the PLD process that allows us to estimate these costs for this technique.

5.2 BASIC PRINCIPLES OF PLD

History and Fundamental Mechanisms. The use of a pulsed laser to induce the stoichiometric transfer of a material from a solid source to a substrate, simulating the flash evaporation methods that have previously been successful, is reported in the literature as early as 1965 (Smith and Turner, 1965), where films of semiconductors and dielectrics were grown using a ruby laser. Pulsed laser evaporation for film growth from powders of $SrTiO_3$ and $BaTiO_3$ was achieved in 1969 (Schwarz and Tourtellotte, 1969). Six years later, stoichiometric intermetallic materials (including Ni_3Mn and low-T_c superconducting films of $ReBe_{22}$) were produced using a pulsed laser beam (Desserre and Floy, 1975). In 1983, Zaitsev-Zotov and co-workers for the first time reported superconductivity in pulsed laser evaporated oxide superconductor films after heat-treatment (Zaitsev-Zotov et al., 1983).

The real breakthrough for PLD, however, was its successful application to the *in-situ* growth of epitaxial high-temperature superconductor films in 1987 at Bell Communications Research (Dijkkamp et al., 1987).

Since then, PLD has been used extensively in the growth of high-temperature cuprates and numerous other complex oxides, including materials that cannot be obtained in an equilibrium process.

Conceptually, the process of PLD is extremely simple and illustrated schematically in Figure 5.1. A pulsed laser beam leads to a rapid removal of material from a solid target and to the formation of an energetic plasma plume, which then condenses onto a substrate. In reality, however, the individual steps—ablation and plasma formation, plume propagation, and nucleation and growth—are rather complex.

Ablation and Plasma Formation. Ablation has been studied extensively, not only in connection to PLD, but also because of its importance in laser machining. The mechanisms that convert the electromagnetic energy of the coherent light beam first into electronic excitations and then into thermal, chemical, and mechanical energy are complex (Kelly and Miotello, 1994; Miotello and Kelly, 1999) and still not fully understood. Heating rates as high as 10^{11} K^{-1} and instantaneous gas pressures of 10–500 atm. are observed at the target surface (Geohegan, 1994). The laser-solid interaction mechanisms depend strongly on the laser wavelength; in fact, significant changes in the energetics of species in a plume resulting from ablation of carbon using KrF (248 nm) and

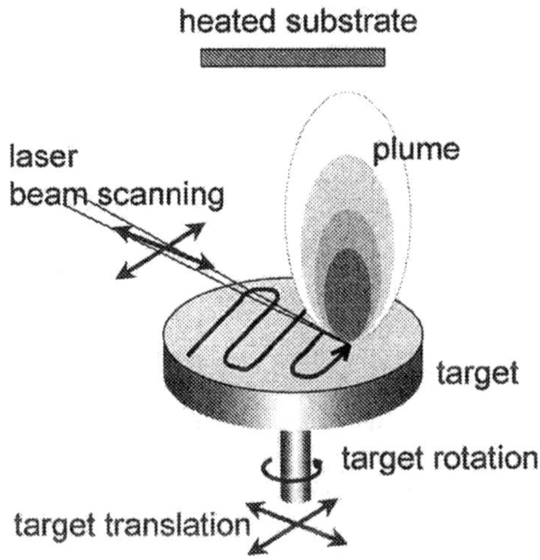

Figure 5.1. Schematic drawing of the basic components inside a vacuum system as used in pulsed laser deposition. The laser beam is scanned across a target, which may be a cylindrical disc, but could have other shapes, such as a cylindrical rod with ablation from the side.

ArF (193 nm) excimer lasers are observed (Puretzky et al., 1996), having a large effect on the growth of diamond-like carbon films. Quite surprisingly, some ablated surfaces show topographic features oriented along the polarization axis of the incident laser beam rather than the beam direction after ablation using a femtosecond laser (Henyk et al., 1999), illustrating that for very short pulses, thermal effects are not sufficient to describe the process.

For relatively long pulse durations, such as the tens of nanoseconds typical for excimer lasers, there is a strong interaction between the forming plume and the incident beam, leading to a further heating of the species. This may explain experiments of $YBa_2Cu_3O_{7-\delta}$ film growth where, for a given laser energy density at the target surface, ablation using a KrF excimer laser (248 nm, ≈30 ns pulse duration) resulted in far superior films than ablation using a frequency-quadrupled Nd : YAG (266 nm, ≈5 ns pulse duration) (Knauss, L.A., Christen, H.M., and Harshavardhan, K.S., Comparison of Nd : YAG and KrF excimer lasers for the pulsed laser deposition of $YBa_2Cu_3O_{7-\delta}$ at Neocera, Inc., unpublished). Similarly, certain aspects of a dual-laser approach (Witanachchi et al., 1995), where a CO_2 laser pulse with a 500 ns duration is allowed to interact with the plume formed by the ablation using a KrF excimer laser, have been attributed to increased laser heating of the plasma.

Plume Propagation. Extensive experiments have been performed to study the plume propagation using optical absorption and emission spectroscopy combined with ion probe measurements (Geohegan, 1994; Geohegan and Puretzky, 1996). Neutral atoms, ions, and electrons travel at different velocities, and strong interactions between the species of the plasma and the background gas are observed. In fact, it is generally assumed that some degree of thermalization needs to occur in order to obtain good film growth and to avoid resputtering of the growing film by the most energetic ions in the plume (Hau et al., 1995). Assuming that most of the species in the plume should be fully thermalized at precisely the time they reach the substrate (i.e. having equal lateral and forward velocities), a simple model predicts that the optimal growth rate should be close to 1 Å per pulse (Strikovski and Miller, 1998; Strikovski et al., 2000).

This is rather close to the actually observed values (typically ranging from 0.1 Å to 1 Å per pulse).

In excimer laser ablation experiments of $YBa_2Cu_3O_{7-\delta}$, the formation of nanoparticles in the plume has been observed at oxygen pressures above 175 mTorr and at room temperature (Geohegan et al., 1999). However, for a heated substrate, the temperature gradient moves nanoparticles away from the heater surface, such that in typical $YBa_2Cu_3O_{7-\delta}$ growth experiments, nanoparticle incorporation into the growing films appears not to be an issue.

Nucleation and Growth. Detailed descriptions of the growth modes observed in PLD have been published (Horwitz and Sprague, 1994; Kim et al., 2000; Blank et al., 2000; Norton, 1998). Island growth occurs most frequently, but layer-by-layer growth can occur at sufficiently low growth rates and sufficiently elevated temperatures. Recent experiments indicate that in general, crystallization is fast (as observed using time-resolved surface x-ray diffraction (Eres et al., 2001)), but changes at the surface are observed for times ranging from factions of a second to several seconds depending on the growth conditions (as observed both in RHEED (Rijnders et al., 1997) and in surface x-ray diffraction (Eres et al., 2001)). At very high temperature and low pressure, step flow growth was observed on some oxide materials (Nakagawa et al., 2000). However, for the fabrication of long lengths of HTS tapes, these slower growth modes are neither achievable nor necessary.

The term Laser-MBE has been used to describe a PLD system in which layer-by-layer growth is achieved and monitored by RHEED. The terminology, of course, is somewhat inaccurate, as a laser plume always contains a combination of ions, electrons, and neutral particles and is thus not a molecular beam. Nevertheless, "laser-MBE" has been successfully used to sequentially deposit single layers of SrO and BaO (Koinuma et al., 1998) and to intercalate SrO layers in manganites to form artificial crystalline structures (Tanaka and Kawai, 2000). The term has further been used when ablation occurred from a complex target (Ohashi et al., 1999; Chen et al., 1999; Yang et al., 2001), thus "laser-MBE" is often synonymous with "layer-by-layer growth by PLD."

Laser Requirements. The key parameters for a laser to be used in PLD are its wavelength, pulse duration, and energy per pulse. A sufficiently small wavelength assures that most of the energy is absorbed in a very shallow layer near the surface of the target; otherwise subsurface boiling can occur, leading to a large number of particulates at the film surface. The absorption of photons by oxygen molecules and optical elements in the beam path determines a lower practical wavelength limit of about 200 nm. The pulse duration must be short enough such as not to lead to a significant heating of the bulk of the target (again to avoid boiling, particulate formation, and changes in stoichiometry at the surface), but long enough to transfer some energy into the plasma. Finally, the laser energy at the surface of the target has to exceed a certain threshold, typically 1–3 J/cm^2 for a 30 ns pulse. Therefore, the laser energy per pulse will determine the spot size over which the laser must be focused, and thus the amount of material that can be ablated per pulse. In addition, it is generally observed that a small laser spot results in a broad angular distribution of the ablated species, so that it is impossible to change the energy per pulse without significant other modifications in the growth apparatus.

Excimer lasers appear to satisfy all of these requirements (see (Basting, 2001) for an excellent introduction to excimer lasers). Lasers using KrF excimers (248 nm, typically 20–35 ns pulse duration) have been used most often in PLD. Successful $YBa_2Cu_3O_{7-\delta}$ film growth has also been achieved using ArF (193 nm) (Mukaida et al., 1999; Kinoshita et al., 1994; Riabkina-Fishman and Zahavi, 1993) and XeCl (308 nm) (Wang et al., 2000; Schey et al., 1998; Boffa et al., 1997; Duhalde et al., 1998; Muenchausen et al., 1990) excimers.

For reasons mentioned above, Nd : YAG lasers have only rarely been used for the growth of $YBa_2Cu_3O_{7-\delta}$ (Kusumori and Muto, 2000).

The growth of less complex materials has been possible using a variety of lasers, including hybrid dye/excimer lasers (248 nm, 500 fs) for amorphous carbon nitride (Szorenyi et al., 1999), femtosecond Ti-sapphire lasers for ZnO (Millon et al., 2000; Okoshi et al., 2000), SnO_2 (Dominguez et al., 2001), carbon (Qian et al., 1999; Shirk and Molian, 2001) and AlN (Hirayama et al., 2001). A method of "ultrafast ablation" has been proposed and applied to the growth of amorphous carbon (Rode et al., 1999). Here, either a 10 kHz, 120 ns Q-switched Nd : YAG laser or a 76 MHz, 60 ps mode-locked Nd : YAG laser are used, resulting in very smooth films at higher growth rates than in conventional PLD, but again the approach has not successfully been applied to $YBa_2Cu_3O_{7-\delta}$.

5.3 PULSED LASER DEPOSITION OF $YBa_2Cu_3O_{7-\delta}$

Despite the fact that the first $YBa_2Cu_3O_{7-\delta}$ films were grown almost 15 years ago, new achievements and results are published continuously. For example, very recent publications indicate that it is now possible to grow double-sided $YBa_2Cu_3O_{7-\delta}$ reproducibly on large numbers of 3″ diameter substrates (for filter applications) (Lorenz et al., 2001), that $YBa_2Cu_3O_{7-\delta}$ can be grown onto flexible yttria-stabilized zirconia (YSZ) tapes for cryoelectronic applications (Harshavardhan et al., 2001), and that silver doping can result in higher J_c values (Xu et al., 2000).

Pulsed laser deposition has been used extensively in research efforts related to coated conductors. Numerous authors have reported critical current densities in excess of 10^6 A/cm^2 (Wang et al., 2000; Goyal et al., 1997, 1999; Park et al., 1998; Lee et al., 1999; Aytug et al., 2000b; Feldmann et al., 2000) PLD-grown films have contributed to studies on the effect of texture in coated conductors (Feldmann et al., 2000; Reeves et al., 2001), ac loss experiments (Kerchner et al., 1997), bend strain tolerance (Park et al., 1998), and, in many studies, as a means to quantify the quality of buffer layers grown by a variety of methods (Aytug et al., 2000b; List et al., 1998; Park et al., 1999; Huang et al., 2000; Xiong and Winkler, 2000; Boffa et al., 2000; Aytug et al., 2000a).

Recent results for long lengths of coated conductors are promising.* Reel-to-reel deposition of $YBa_2Cu_3O_{7-\delta}$ with end-to-end $I_c > 10$ A for a 4.5 m tape and $I_c = 62$ A (1 cm tape width) for shorter lengths and a film thickness of 4 μm were reported by Sato et al. (2001). Y. Iijima's group (Fujikura Ltd.) reported on a 9.6 m long carrying an $I_c = 50$ A (1 cm tape width), corresponding to a current density $J_c = 0.42$ MA/cm^2. On a shorter length of 8 cm (for which the buffer layer was deposited at a slower rate), $I_c = 140$ A (1 cm tape width, corresponding to $J_c = 1.2$ MA/cm^2) was measured (Iijima et al., 2001).

S. Foltyn and co-workers (Los Alamos National Laboratory) (Foltyn et al., 2001) have produced a 1 m long tape carrying $I_c = 189$ A (1 cm tape width) at 75 K, corresponding to $J_c \approx 1$ MA/cm^2. Researchers at the University of Göttingen (H. Freyhardt

* Unless otherwise noted, all data quoted in this section are obtained on $YBa_2Cu_3O_{7-\delta}$ on IBAD at 77 K and in self-field.

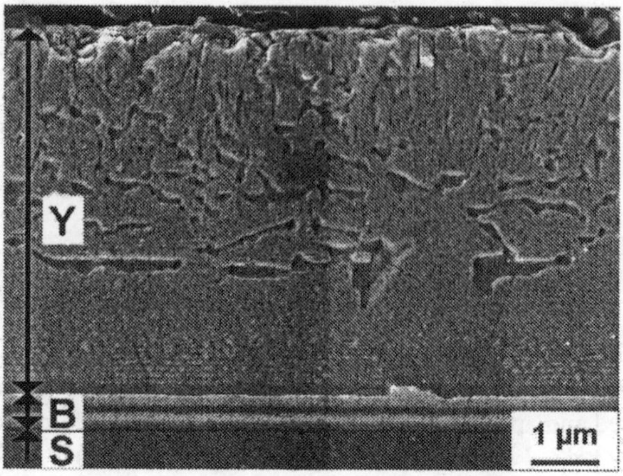

Figure 5.2. Cross-section SEM image of a PLD-grown $YBa_2Cu_3O_{7-\delta}$ film (labeled "Y") on an IBAD buffer layer ("B") on an Inconel 625 substrate ("S"). Porosity is clearly observed except for the initial $\approx$2 μm of the material. Micrograph courtesy of T. Holesinger, Los Alamos National Laboratory.

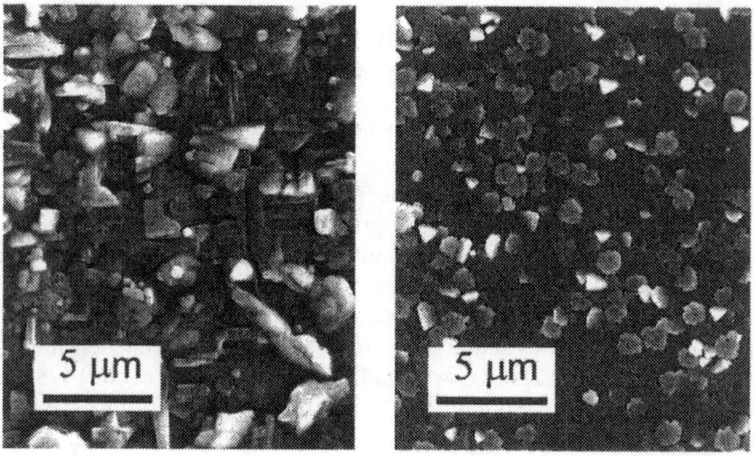

Figure 5.3. Plan-view SEM images of a 3.0 μm thick $YBa_2Cu_3O_{7-\delta}$ film (left) and a 3.7 μm thick $YBa_2Cu_3O_{7-\delta}$/$SmBa_2Cu_3O_{7-\delta}$ multilayer (right), showing the improved surface morphology of the tri-layers. Micrographs courtesy J.F. Smith, Los Alamos National Laboratory.

et al.) report $I_c = 142$ A (0.92 cm tape width) on a 1.9 m long tape, corresponding to $J_c = 1.23$ MA/cm^2 (Usoskin et al., 2001b).

To date, PLD appears to be the most mature technique for the growth of $YBa_2Cu_3O_{7-\delta}$ films with thicknesses above 2 μm. Results from Los Alamos National Laboratory have indicated that porous growth occurs beyond a critical thickness of about 2 μm, see Figure 5.2. This agrees well with the observation of decreased J_c for thicker layers. In fact, ion mill removal of the topmost portion of 3–6 μm thick films results in an appreciable decrease of the total critical current only when the total remaining thickness becomes less than 1.5 μm (Foltyn et al., 1999b).

To circumvent the problem of deteriorating microstructure in films above 2 μm thickness, Foltyn and co-workers have introduced an approach in which 0.2 μm thick layers of $SmBa_2Cu_3O_{7-\delta}$ are intercalated between separate 1 μm thick $YBa_2Cu_3O_{7-\delta}$ layers. The resulting films show a much better surface morphology (Figure 5.3) than

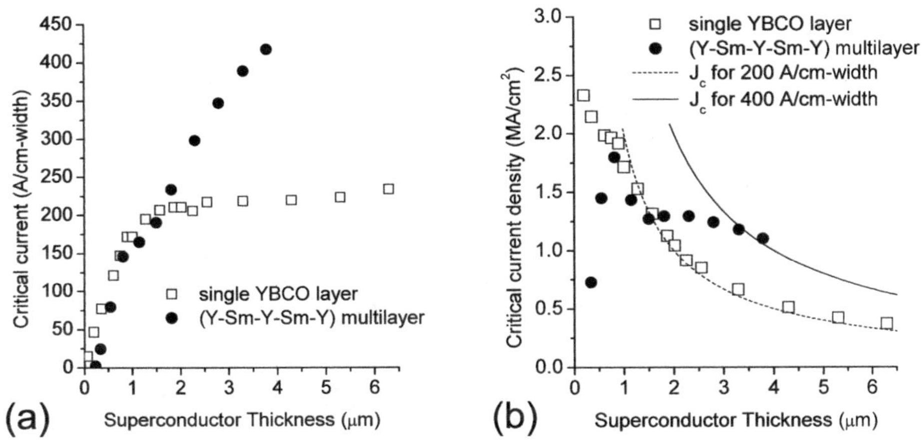

Figure 5.4. (a) Critical current (at 75 K, extrapolated from measurements on 200 μm wide bridges) as a function of superconductor thickness, both for single $YBa_2Cu_3O_{7-\delta}$ films and multilayers containing two $SmBa_2Cu_3O_{7-\delta}$ films (data from Foltyn et al., 2001). (b) Data from (a) plotted in terms of critical current density as function of superconductor thickness. Also shown are lines corresponding to the critical current densities required to achieve critical currents of 200 A/cm-width and 400 A/cm-width.

standard single layers of comparable thickness, and from the achieved current densities ($J_c = 1.1$ MA/cm^2) at a film thickness of 3.7 μm, a critical current in excess of 400 A/cm-width can be extrapolated. The effect of the intercalated $SmBa_2Cu_3O_{7-\delta}$ layers is most clearly visible when J_c data are plotted as a function of total superconductor thickness, as in Figure 5.4: for the single $YBa_2Cu_3O_{7-\delta}$ layer, the critical current does not increase beyond about 2 μm, whereas for the multilayer structures only a weak decrease of J_c with thickness is observed.

5.4 COMMERCIALLY AVAILABLE EQUIPMENT

There are numerous companies worldwide selling PLD equipment, including vendors that have been focusing on PLD for many years (for example, Neocera, Inc. (www.neocera.com), Surface (www.surface-tec.com), PVD Products, Inc. (www.pvdproducts.com), and others). (Neocera, Inc., for example, has sold over 55 systems since 1992.) Other companies offer PLD systems and components as part of a broader product line (for example, Johnsen Ultravac (www.ultrahivac.com), Meca2000 (www.meca-2000.com), Twente Solid State Technology B.V. (www.tsst.nl), JSQ GmbH (www.jsquid.com), BESTEC GmbH (www.bestec.com), Thermionics vacuum products (www.thermionics.com), and many more). Most of these systems are research-scale tools, allowing for the coating of wafers up to a few inches in diameter. It is worthwhile to note that numerous patents have been issued on the process of pulsed laser deposition (Venkatesan and Wu, 1991; Cheung, 1991; Zander et al., 1992; Noda et al., 1992; Roas et al., 1993; Hayashi and Yoshida, 1995) and on various improvements thereof, including approaches for scale-up, beam scanning, *in-situ* monitoring, etc., some of which are referenced elsewhere in this chapter. To the best of the author's knowledge, reel-to-reel PLD systems have not been offered commercially; however, at least two companies are currently working on the development of such tools, namely PVD Products (design includes an *in-situ* sputter source and *in-situ* an-

Figure 5.5. Photons at a bargain price and in a compact package: Lambda Physik's LAMBDA STEEL™ 1000 excimer laser delivers a peak optical power equivalent to that of 10 billion laser pointers. It measures 2.5 m × 2.2 m × 0.85 m (l × h × w). Photo courtesy of Lambda Physik.

nealing capability) and Neocera, Inc. (developing systems for both pulsed laser and pulsed electron deposition).

Excimer laser applications such as micromachining, UV lithography, thin film transistor annealing, laser marking, fabrication of fiber Bragg gratings, etc., have created a continuous demand for reliable, industrial lasers. In addition, medical applications (including, for example, refractive laser surgery) also rely on high-quality lasers.

Modern excimer lasers, such as Lambda Physik's STEEL 1000 (Figure 5.5) are designed to produce 300 W of optical power (300 Hz repetition rate and 1 J/pulse). Designed for firing 20 million pulses a day—more than the usage of many lasers in research laboratories over the time of several months—these instruments are capable of providing the optical power required to grow $YBa_2Cu_3O_{7-\delta}$ on long lengths of tape.

5.5 ISSUES RELATED TO SCALE-UP

5.5.1 Large-Area Deposition

Numerous approaches to scale the process of PLD from the original sample size of a few cm^2 to that of wafers with diameters of several inches have been described in the literature. Careful positioning of the laser plume with respect to the center of a rotating substrate, or scanning of the laser beam across a large target (Greer and Tabat, 1995) have been successful in uniformly coating wafers of 3″ diameter reproducibly (Lorenz et al., 2001). As an alternative to the scanning of a beam across a target, two laser beams can be used to form two plumes such as to create a region of sufficiently uniform particle flux (Dietsch et al., 1998). Finally, the use of cylindrical targets, a long line-

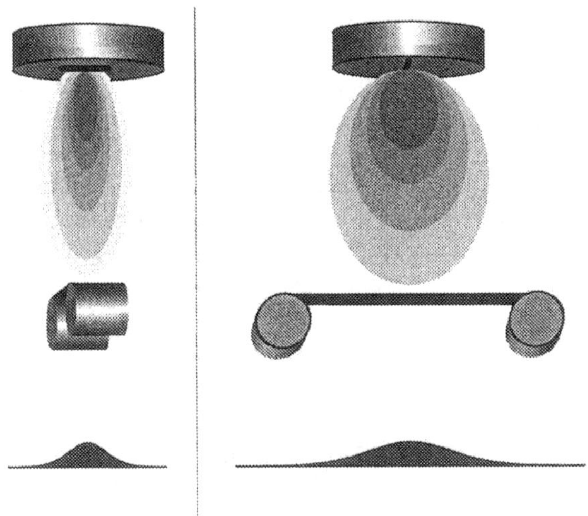

Figure 5.6. View from two orthogonal directions of a PLD plume resulting from a line-shaped laser spot, with a schematic representation of the growth rate profile as a function of position on the tape. The plume expansion is largest in the direction in which the laser spot dimension is smallest.

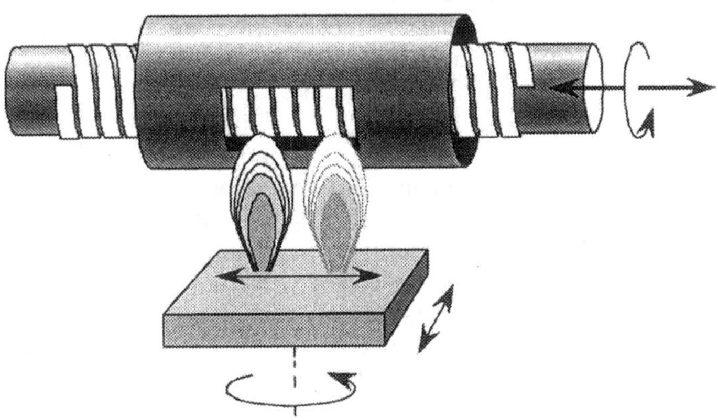

Figure 5.7. Schematic of the deposition apparatus used in (Usoskin et al., 2001a, 2001b). Target rastering and rotation is combined with beam scanning. The tape is wrapped in a helical fashion around a support rod which in turn is translated and rotated inside a heated cavity. Deposition occurs through a small opening in this heater.

shaped laser spot, and a plume direction that is not normal to the substrate, combined with a substrate translation inside a radiative "pocket" heater, has been successful in depositing uniform coatings over areas as large as 7 cm × 28 cm (Schey et al., 1998).

For the deposition of thin film coatings on tapes, however, the conditions are quite different from the situation encountered in the fabrication of films on large wafer substrates. The naturally occurring ellipsis-shaped deposition rate profile (typically measuring several centimeters along the long axis, but only 1–2 cm in the short direction depending on the target-substrate distance) is in fact quite appropriate for the coating of narrow tapes, as illustrated in Figure 5.6.

Alternatively, the tape can be wound around a support rod in a helical shape, and this assembly can be heated by placing it inside a cylindrical black-body heater having only a relatively small opening through which deposition occurs (see Figure 5.7)

(Usoskin et al., 2001b). For sufficiently rapid motion of the tape support assembly, the temperature will drop only insignificantly at the exposed deposition area. A further advantage of this approach (labeled "High-Rate Pulsed-Laser-Deposition" by the authors (Usoskin et al., 2001a)) is the large effective deposition area: the integral deposition rate (deposited volume per unit time) can be large while the local growth rate (thickness per unit time) is kept comparatively small. This may result in increased film quality under certain conditions, but the approach may not be applicable to the growth of tapes beyond a few meters in length. A reel-to-reel system utilizing this "quasi-equilibrium heating" is currently under development (H.C. Freyhardt, private communication).

5.5.2 Control of the Laser Spot on the Target

Because the laser energy density at the target strongly influences the ablation mechanisms, the plume propagation, and the generation of particulates, a laser spot with a uniform spatial energy distribution is usually desired. In the ideal case of a perfectly parallel laser beam, a single focusing lens would be adequate to achieve this goal reasonably, considering that a sufficiently uniform fraction of the laser beam can be selected using an aperture. In reality, however, the laser beam is divergent, and the divergences in the horizontal and vertical directions are different from each other.

Using a long beam path and imaging an aperture placed nearest to the laser can result in very uniform and reproducible spots on the target. Unfortunately, the length of the required beam path for such a configuration can be quite long. For example, assume that an energy density at the target of 2 J/cm^2 is needed and that an energy at the laser of 800 mJ over an area of 1 cm $\times$ 3 cm (portion of the beam passing an aperture nearest to the laser) is available. Considering losses of 8% at the imaging lens and the laser port to the chamber, the remaining 677 mJ need to be focused onto a spot of about 1 mm $\times$ 3 mm, i.e. requiring a linear reduction of a factor of 10. If the distance between the target and the laser port is 50 cm, a $f = 50$ cm focal length lens can be used (resulting in a distance of the lens to the target $l_2 = (1+1/10)f = 55$ cm). The distance between the lens and the aperture then has to be $l_1 = (1+10)f = 5.5$ m.

To gain even better spatial energy uniformity, a beam homogenizer can be used. Typically, beam homogenizers utilize an array of small (millimeter-sized) lenses to split the beam into an array of "beamlets," which are then superimposed at the focal plane directly at the target or imaged again using an additional lens (and requiring an additional length of beam path). Uniform growth has been achieved by using this approach (Schey et al., 1998), and improved target wear has been observed (Wagner et al., 1998).

5.5.3 Particulate Reduction and Target Wear

A number of approaches have been proposed (Chen, 1994; Schenck et al., 1998) to reduce particulates in PLD films for devices and optical applications where perfectly smooth surfaces are often required.

The simplest approaches are arrangements in which there is no line-of-sight between the substrate surface and the laser spot on the target, resulting in deposition only via gas phase collisions while "ballistic" particulates can't reach the substrate surface. Off-axis PLD (Wang et al., 2000; Greer and Tabat, 1995; Silliman, S.D., Christen, H.M., and Harshavardhan, K.S., Aluminum oxide films for gate insulator applications grown by pulsed laser deposition, to be published) is probably the simplest of these approaches. Alternatively, an "eclipse method" (Kinoshita et al., 1994) and variations thereof (Pique et al., 1995) have been proposed, in which a small shield is placed between the target and the substrate, and film growth can again occur only due to vapor transport.

In a method called "crossed-beam pulsed laser deposition," (Gorbunov et al., 1996; Tselev et al., 2001) two laser beams are fired onto two targets that are placed at angles in such a way that the two plumes cross each other about halfway between the targets and the substrate. Beyond this crossing point, gas phase collisions lead to a plume propagation predominantly in the direction parallel to the symmetry plane of the configuration, whereas the majority of the particles does not get deflected. This approach should not be confused with a different type of "crossed beam" configuration where a pulsed gas jet is used to promote reaction in the plume (Willmott and Antoni, 1998; Park and Moon, 2001), and which also leads to a reduction in particles for appropriate timing between the valve and the laser (Chen, 1994).

A different type of dual-laser approach can be used to reduce particulate *formation* (rather than to eliminate particulates once they're formed) (Witanachchi et al., 1995). Here, a CO_2 laser (500 ns pulse duration) is used to first locally melt the target, which is then ablated using a KrF excimer laser pulse. Note that this method, which was first demonstrated for simple oxides, does not result in decreased deposition rates and has more recently been applied to compound semiconductors (Mukherjee et al., 1998), but results for $YBa_2Cu_3O_{7-\delta}$ are not available. Alternatively, a second laser can also be used to vaporize particulates in the plume (Koren et al., 1990).

Unfortunately, all of these approaches lead to a reduced total deposition rate or an increased requirement of laser power (thus to an decreased deposition rate per unit of laser power).

Closely related to the issue of particulates is that of target wear. Non-uniform target erosion not only leads to changes in deposition rates and inefficient use of the target material. Most importantly, it has long been recognized that the formation of cones at the target surface (Foltyn, 1994) is a significant contribution to the particulates observed on PLD-grown films. The formation of such cones, which eventually break off and get transported onto the substrate, is most pronounced if the laser incidence angle remains constant for a large number of pulses. This can be avoided simply by a combination of target rotation with either beam scanning or target translation.

5.5.4 Window Coating

A significant fraction of the laser energy can be absorbed by material accidentally deposited onto the laser entry port of the PLD system, resulting in changes in process conditions or—when properly accounted for—in increased laser energy consumption. Therefore, for large-scale depositions, care must be taken to minimize this accidental coating. Combining careful positioning of baffles within the chamber with approaches such as "window purging" (i.e. the introduction of the process gas near the laser port (Greer, 1994)) or additional differential pumping of the area nearest to the window (C.M. Rouleau, private communication) may be sufficient. Alternatively, active devices such as PVD Products' "Intelligent Window" have been introduced, which periodically replace the coated area of the window with a clean portion.

5.5.5 *In-situ* Monitoring and Diagnostics

Determining the composition, energetics, and dynamics of the plasma plume using a variety of techniques, including optical emission and absorption spectroscopy, time-resolved imaging, and ion probes, has been the subject of a number of studies (see Geohegan, 1994 for a review).

Of particular importance to the fabrication of coated conductors are methods that allow real-time monitoring of certain process parameters, in particular to assure a uniform film thickness over long lengths. It is clear from the above remarks regarding window coating that a feedback mechanism is required that controls either the laser parameters (energy, repetition rate) or the tape travel speed during a long deposition.

One possible approach is to rely on an automated periodic measurement of the laser energy inside the deposition chamber to control both the laser energy and the operation of an "Intelligent Window" (as manufactured by PVD Products).

In order to determine changes both in laser energy and in the target surface condition, measurements on the growing film or on the laser plume are required. The thickness of the deposited film can be determined using ellipsometric techniques (Samano et al., 1998; Weissmantel et al., 1999), Raman measurements (Maguire et al., 200), or optical spectroscopy (Gottmann and Kreutz, 1999), and the amount of material deposited onto a reference point can be determined using a quartz microbalance system (Laube and Stark, 1996). Possible techniques to monitor the laser plume directly include magnetic probes (Kabashin et al., 1996), ion probes (Geohegan, 1994), and measurements of the optical emission from the plume (Li et al., 1995; Laube and Voevodin, 1998) combined with a fully automated window actuation and laser energy control program (Fujino et al., 1996, 1997).

5.6 SIMPLIFIED COST MODEL

5.6.1 Introduction

It is beyond the scope of this chapter to develop a complete cost model for the PLD fabrication of coated conductors. However, PLD has gained a reputation as a somewhat costly method due to the high price of excimer lasers and the high price of the gases required. A simple calculation, however, shows quickly that neither the capital cost of an excimer laser nor the excimer gases are the dominant contribution to the cost of depositing YBCO—the cost of replacing laser tubes and optics, however, plays a significant role.

Of interest here are considerations that compare the cost of YBCO deposition with PLD to other techniques. We can therefore ignore all aspects related to tape handling, buffer layer deposition, cap layer deposition, etc., and any other aspect that is not specific to the laser operation. We further ignore the cost of ceramic targets, which are also needed in sputtering and pulsed electron deposition approaches. These ceramics may be more expensive than the materials used in evaporation; however, the forward-oriented nature of PLD results in a more efficient use of the material.

It is important to note that the numbers presented here are estimates. The high estimate will be the cost under currently established procedures, and the lowest estimate will be the best-case scenario beyond which incremental improvements of equipment and processes will not be sufficient. One must keep in mind that not all of our assumptions may be applicable to the fabrication of thousands of kilometers of tape, but also that there may be future quantum-leap improvements that we cannot predict at this time.

5.6.2 Laser Cost per kWh of Optical Output

We start our calculation with a determination of the cost to produce a kWh of optical laser power, assuming the use of a Lambda Physik LAMBDA STEEL™ 1000 excimer laser. We further assume that we operate the laser 20 hours per day, 7 days per week, and 50 weeks per year. This corresponds to 21.6 million pulses per day, or $7.56 \cdot 10^9$ pulses per year at 1 J, i.e. 2100 kWh of optical energy per year.

Lambda Physik has kindly provided the author with a cost calculation spreadsheet, on which we will base the following considerations. All calculations are made for the KrF line at 248 nm (the XeCl line at 308 nm results in operation costs that are about 2% higher due to the higher price of the gases).

A complete cost model (cost of ownership and cost of operation) takes into consideration both fixed costs (capital, floor space, static life-time of components) and variable costs (consumables, dynamic life-time of components) and considers yield and equipment duty cycle. (Definitions of cost of ownership are given and established by SEMATECH (Dance and Jimenez, 1994) and have been applied previously to excimer lasers in lithography (Watson and Rowan, 1996).)

Current industrial lasers are very reliable, and in an early phase of tape production they are very unlikely to be a significant contribution to the overall downtime. Similarly, while there may be numerous parameters that are likely to contribute to yield issues (target changes, heating variations), the laser can be considered to be one of the more constant factors. Thus, for the costs related to the laser, yield and downtime issues may safely be ignored.

Fixed Costs. For the purpose of the current cost comparison, we have to ignore those costs that are similar for all techniques of coated conductor fabrication. This holds in particular for facilities' cost and floor space (the foot-print of the LAMBDA STEEL™ 1000 is 2.5 m × .85 m, requiring a floor space of 5 m^2 for access and maintenance).

Other fixed costs such as static lifetime of laser gases are negligible when the laser is operated at a high duty-cycle.

Therefore, the only fixed cost that enters into our calculation is the cost of capital. According to information provided by Lambda Physik, the expected laser life is likely to be longer than the technology cycle. In other words, while it is likely that a currently available excimer laser, such as the LAMBDA STEEL™, will be replaced by newer models in a few years, the actual life expectancy of the laser is much higher (10 to 20 years or more) if properly maintained (Jim Maclin, private communication). Note that in our considerations for variable costs, we will consider items such as laser tubes and optics, which have a life expectancy that is measured in laser pulses. However, other components such as electrical power supplies, fan motors, cooling components, etc., are not considered.

Assuming, therefore, a laser life of 5 to 20 years, and an estimated laser price of $800k, the capital cost per laser is between about $50k (optimistically assuming 3% annual interest and a 20 year depreciation) and $200k (pessimistically assuming 8% interest and a 5 year depreciation). Thus, the capital cost per kWh is about $20–100.

Variable Costs. According to the information provided by Lambda Physik, at the above-mentioned duty cycle, 484 gas fills are required per year, the window sets need to be exchanged 75 times, the thyrotron modules 10 times, and the laser tube between 3 and 4 times annually. Estimated costs for the window sets are about $128k per year and for the laser tubes $280k per year. In comparison, the laser gases cost only about

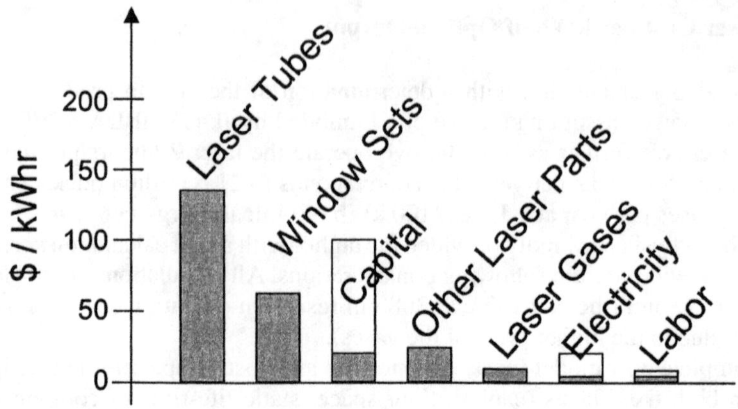

Figure 5.8. Fixed and variable costs considered for a LAMBDA STEEL™ 1000, resulting in a cost of \$250–350 per kWh of optical output.

\$19k per year, 75% of which is for Neon. In total, Lambda estimates a cost of \$480k per year for replacement parts and consumables, i.e. \$228/kWh.

Other variable costs are utilities and labor. Lambda estimates 212 hours of required labor per year. At a fully burdened technician cost of \$80k to \$160k, this adds about \$4–8/kWh of laser output.

The cost of the required electricity is harder to estimate. First we note that the electrical power consumption of the laser is on the order of 30 kVA, and that 20 kW of cooling capacity (using cooling water at 10°C) are required. Furthermore, an air flow of 1000 m^3/h is needed. While it is beyond the scope of this chapter to estimate the efficiency of the water cooling or the price of electricity 5 years from now, assuming 40–60 kW of power consumption at \$0.03 to \$.1/kWh (according to a June 15 press release of the Consumers Union (www.consumersunion.org), the wholesale electricity price in California in February 2000 was \$.03/kWh and rose to \$.36/kWh in February 2001) adds \$8k–42k annually or \$4–20 per kWh of laser output.

Total Cost of Laser Energy. Figure 5.8 summarizes the various fixed and variable cost components as describe above. It can therefore be estimated that the total cost of the optical energy falls into the range of \$250–350/kWh. It is impossible to predict if these costs will decrease significantly in the foreseeable future. Past cost savings have, for example, resulted from the introduction of Lambda Physik's NovaTube™ technology. A similar large change in laser technology would, however, be required for future appreciable price reductions (Jim Maclin, private communication).

5.6.3 Optical Energy Required to Grow a Meter of Tape

It is more difficult to estimate how much optical energy is required to grow a certain length of superconducting tape. However, there are a number of published reports that indicate the current status. Foltyn et al. reported the growth of 1 μm thick $YBa_2Cu_3O_{7-\delta}$ at a rate of 2.5 cm/min. using 14 W of excimer laser power (Foltyn et al., 1999a), corresponding to 0.0093 kWh/m. Using a 200 W laser, the same authors later reported a growth rate of 17 m/h, corresponding to 0.011 kWh/m (Foltyn et al., 2000), indicating in fact that the process scales almost perfectly with higher laser power and repetition rate. Iijima et al. predict that the growth of 3–4 m per hour with a thickness in excess of 1 μm should be possible with less than 100 W of optical power (Y. Iijima, private communication), corresponding to less than 0.025 kWh/m.

In a very different configuration, the researchers at the university of Göttingen report a growth rate of 45 nm m^2/h, using a LAMBDA 3308 excimer laser at 300 Hz and 0.4 J/pulse. These numbers correspond to 0.027 kWh/m for a 1 cm wide, 1 μm thick tape.

Several factors have to be considered when estimating the minimum laser power required for film growth. First, current implementations of PLD are typically not optimized for the lowest loss in laser energy. For the values from Los Alamos cited above, for example, it is estimated (S.R. Foltyn, private communication), that at least 25% in the laser energy on the target could be gained by using optimized optical elements and a shorter beam path.

A further significant factor is the shape of the laser spot and thus the shape of the plume, and the amount of material that gets deposited onto the tape rather than outside of the tape. While PLD is extremely forward-oriented when compared to some other methods, it is obvious that for cost-efficient manufacturing, some of the currently achieved film thickness uniformity can be traded in for an increased integrated collection efficiency. According to S. Foltyn (private communication), one can expect to gain up to a factor of two in the deposition rate at given laser power by optimizing beam path, optical elements, laser spot, and growth geometry.

Therefore, we can assume that the energy required to deposit one meter of tape (with a thickness of 1 μm and a width of 1 cm) lies between 0.005 kWh (assuming a two-fold increase in overall efficiency due to increased beam path and collection efficiencies) and 0.008 kWh (assuming only a 25% increase in the beam path efficiency).

5.6.4 Total Estimated Laser Cost per Length of Tape

To summarize the numbers calculated above, we find that

- 1 kWh of optical energy costs about \$250–350.
- The growth of 1 m of superconducting tape with a width of 1 cm and film thickness of 1 μm requires about 0.005–0.008 kWh (with currently achieved numbers sometimes significantly higher due to non-optimized beam paths and deposition geometries).

Using these numbers, the laser cost per meter of tape (1 cm wide, 1 μm thick) can be calculated to fall into the range of \$1.2–2.8. This corresponds to a laser operation cost of \$12–28 per kA m if a critical current density of 1 MA/cm^2 is assumed. Costs for other values of J_c and energy consumption can be determined from the plots in Figure 5.9.

To summarize, even with existing technology, the laser cost is less than \$30/kA m, but any currently discussed incremental improvements of equipment and process is very unlikely to result in a laser cost of less than \$10/kA m for $J_c = 1$ MA/cm^2 (this cost, obviously, varies as J_c^{-1}).

5.7 SUMMARY AND CONCLUSIONS

Pulsed laser deposition has been shown to be a powerful method for the fabrication of coated conductor tapes. Some of the highest values of I_c reported have been obtained by PLD, as this technique has been most successful in the growth of $YBa_2Cu_3O_{7-\delta}$ films with thicknesses exceeding 1–2 μm.

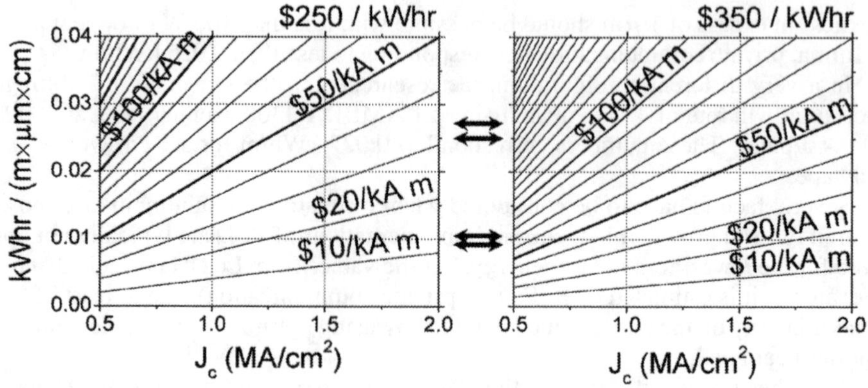

Figure 5.9. Laser cost contour plots for an assumed cost of \$250 (left) and \$350 (right) per kWh of optical energy. Both the expected J_c (x-axis) and the expected laser energy consumption per length of tape (y-axis) must be considered to determine the actual cost per kA m. The double arrows between the plots indicate the experimental values quoted in the text.

What are often described as the main disadvantages of PLD, namely the non-uniform deposition profile and the formation of particulates, appear to be much less of a concern in the fabrication of coated conductors than in thin films for device applications. In addition, many of the critical aspects for scale-up, such as laser port coating, target wear, and *in-situ* monitoring of deposition rates, have already been addressed successfully.

Unfortunately, excimer lasers are relatively expensive to operate. Using the current state of PLD, the cost of the required photons alone may contribute up to \$30/kA m. Incremental improvement of the process (optimized laser beam paths and collection efficiencies) may result in a decrease of the cost to just above \$10/kA m, assuming a J_c of 1 MA/cm^2. A significant further decrease of the cost would require quantum-leap changes in laser technology, deposition method, or achievable J_c.

ACKNOWLEDGMENTS

It is my great pleasure to acknowledge numerous fruitful conversations with S.R. Foltyn and V. Matijasevic (Los Alamos National Laboratory), Y. Iijima (Fujikura Ltd.), H.C. Freyhardt (University of Göttingen), K.S. Harshavardhan and G. Doman (Neocera, Inc.), J.A. Greer (PVD Products), J. Maclin (Lambda Physik), D.H.A. Blank (University of Twente), P.K. Schenck (NIST Gaithersburg), and D.B. Geohegan, G. Eres, C.M. Rouleau, A. Puretzky, and D.H. Lowndes (Oak Ridge National Laboratory). This work was sponsored by the U.S. Department of Energy under contract DE-AC05-00OR22725 with the Oak Ridge National Laboratory, managed by UT-Battelle, LLC, and by the DOE Office of Energy Efficiency and Renewable Energy, Office of Power Technologies—Superconductivity Program.

REFERENCES

Aytug, T., Wu, J.Z., Cantoni, C., Verebelyi, D.T., Specht, E.D., Paranthaman, M., Norton, D.P., Christen, D.K., Ericson, R.E., and Thomas, C.L., 2000a, Growth and superconducting properties of $YB_2Cu_3O_{7-\delta}$ films on conductive $SrRuO_3$ and $LaNiO_3$ multilayers for coated conductor applications, *Appl. Phys. Lett.*, 76:760.

Aytug, T., Wu, J.Z., Kang, B.W., Verebelyi, D.T., Cantoni, C., Specht, E.D., Goyal, A., Paranthaman, M., and Christen, D.K., 2000b, An all-sputtered buffer layer architecture for high-J_c $YB_2Cu_3O_{7-\delta}$ coated conductors, *Physica C*, 340:33.

Basting, D., ed., 2001, *Excimer Laser Technology: Laser Sources, Optics, Systems and Applications*, Lambda Physik, Göttingen.

Blank, D.H.A., Koster, G., Rijnders, G.A.J.H.M., van Setten, E., Slycke, P., and Rogalla, H., 2000, Epitaxial growth of oxides with pulsed laser interval deposition, *J. Cryst. Growth*, 211:98.

Boffa, V., Petrisor, T., Ciontea, L., Gambardella, U., and Barbanera, S., 1997, High-quality surface YBCO thin films prepared by off-axis pulsed laser deposition technique, *Physica C*, 276:218.

Boffa, V., Annino, C., Bettlinelli, D., Ceresara, C., Cioneta, L., Fabbri, F., Galluzzi, V., Gambardella, U., Celentano, G., Grimaldi, G., Mancini, A., Petrisor, T., and Scardi, P., 2000, Epitaxial growth of heterostructures on biaxially textured metallic substrates for $YBa_2Cu_3O_{7-x}$ tape fabrication, *Phil. Mag.*, 80:979.

Chen, L.-C., 1994, Particulates generated by pulsed laser ablation, in: *Pulsed Laser Deposition of Thin Films*, D.H. Chrisey and G.K. Hubler, eds., John Wiley & Sons, Inc., New York.

Chen, P., Xu, S.Y., Lin, J., Ong, C.K., and Cui, D.F., 1999, Angle-resolved X-ray photoelectron spectroscopy of topmost surface for $LaNiO_3$ thin film grown on $SrTiO_3$ substrate by laser molecular beam epitaxy, *Appl. Surf. Sci.*, 137:98.

Cheung, J.T., 1991, Method of thin film deposition using laser ablation, US Patent 5,049,405 (Sept. 17).

Cheung, J.T., 1995, Method of laser ablation for uniform thin film deposition, US Patent 5,411,772 (May 2).

Chrisey, D.B. and Hubler, G.K., eds., 1994, *Pulsed Laser Deposition of Thin Films*, John Wiley & Sons, New York.

Dance, D.L. and Jimenez, D.W., 1994, Applications of cost-of-ownership, *Semicond. Int.* (September).

Desserre, J. and Floy, J.F., 1975, Interaction of laser light beam pulses with complex targets: Application to production of compound thin films, *Thin Solid Films*, 29:29.

Dietsch, R., Holz, Th., Mai, H., Meyer, C.-F., Scholz, R., and Wehner, B., 1998, High precision large area PLD of X-ray optical multilayers, *Appl. Surf. Sci.*, 127–129:451.

Dijkkamp, D., Venkatesan, T., Wu, X.D., Shaheen, S.A., Jisrawi, N., Min-Lee, Y.H., McLean, W.L., and Croft, M., 1987, Preparation of Y–Ba–Cu oxide superconductor thin films using pulsed laser evaporation from high-T_c bulk material, *Appl. Phys. Lett.*, 51:619.

Dominguez, J.E., Fu, L., and Pan, X.Q., 2001, Epitaxial nanocrystalline tin dioxide thin films grown on (0001) sapphire by femtosecond pulsed laser deposition, *Appl. Phys. Lett.*, 79:614.

Duhalde, S., Lamagna, A., Villafuerte, M., Schwartzman, A., Correra, L., and Quintana, G., 1998, Influence of the deposition parameters on the structural and transport properties of YbaCuO thin films prepared by pulsed laser deposition, *Appl. Surf. Sci.*, 127–129:520.

Eres, G., Tischler, J.Z., Yoon, M., Larson, B.C., Rouleau, C.M., Lowndes, D.H., Zschack, P., and Chiang, T.C., 2001, Time resolved study of $SrTiO_3$ homoepitaxial pulsed laser deposition using surface x-ray diffraction, to be published.

Feldmann, D.M., Reeves, J.L., Polyanskii, A.A., Kozlowski, G., Biggers, R.R., Nekkanti, R.M., Maartense, I., Topmsic, M., Barnes, P., Oberly, C.E., Peterson, T.L., Babcock, S.E., and Larbalestier, D.C., 2000, Influence of nickel substrate grain structure on $YBa_2Cu_3O_{7-x}$ supercurrent connectivity in deformation-textured coated conductors, *Appl. Phys. Lett.*, 77:2906.

Foltyn, S.R., 1994, Surface modification of materials by cumulative laser irradiation, in: *Pulsed Laser Deposition of Thin Films*, D.H. Chrisey and G.K. Hubler, eds., John Wiley & Sons, Inc., New York.

Foltyn, S.R., Arendt, P.N., Dowden, P.C., DePaula, R.F., Groves, J.R., Coulter, J.Y., Jia, Q., Maley, M.P., and Peterson, D.E., 1999a, High-T_c coated conductors—performance of meter-long YBCO/IBAD flexible tapes, *IEEE Trans. Appl. Supercond.*, 9:1519.

Foltyn, S.R., Jia, Q.X., Arendt, P.N., Kinder, L., Fan, Y., and Smith, J.F., 1999b, Relationship between film thickness and the critical current of $YBa_2Cu_3O_{7-\delta}$-coated conductors, *Appl. Phys. Lett.*, 75:3691.

Foltyn, S.R., Dowden, P.C., Arendt, P.N., and Coulter, J.Y., 2000, Extremely high rate pulsed laser deposition of YBCO, presentation at the *Annual Peer Review* of the Department of Energy's *Superconductivity Program for Electric Systems*, Washington, DC, July 17–19.

Foltyn, S.R., Jia, Q.X., Dowden, P.C., Arendt, P.N., Smith, J.F., Holesinger, T.G., Kung, H., Coulter, J.Y., Gibbons, B.J., Jan, D.B., DePaula, R.F., Stan, L., and Groves, J.R., 2001, High current coated conductors based on IBAD YSZ and thick YBCO/Sm-123 multilayers, presentation at the *Annual Peer Review* of the Department of Energy's *Superconductivity Program for Electric Systems*, Washington, DC, August 1–3.

Fujino, K., Takano, S., Yoshida, N., Hara, T., and Ishii, H., 1996, 1997, Oxide superconducting film manufacturing apparatus, US Patent 5,489,338 (Feb. 6) and 5,601,649 (Feb. 11).

Geohegan, D.B., 1994, Diagnostics and characteristics of laser-produced plasmas, in: *Pulsed Laser Deposition of Thin Films*, D.H. Chrisey and G.K. Hubler, eds., John Wiley & Sons, Inc., New York.

Geohegan, D.B. and Puretzky, A.A., 1996, Laser ablation plume thermalization dynamics in background gases: Combined imaging, optical absorption and emission spectroscopy, and ion probe measurements, *Appl. Surf. Sci.*, 96–98:131.

Geohegan, D.B., Puretzky, A.A., and Rader, D.J., 1999, Gas-phase nanoparticles formation and transport during pulsed laser deposition of $Y_1Ba_2Cu_3O_{7-d}$, *Appl. Phys. Lett.*, 74:3788.

Gorbunov, A.A., Pompe, W., Sewing, A., Gaponov, S.V., Akhsakhalyan, A.D., Zabrodin, I.G., Kas'kov, I.A., Klyenkov, E.B., Morozov, A.P., Salaschenko, N.N., Dietsch, R., Mai, H., and Völlmar, S., 1996, Ultrathin film deposition by pulsed laser ablation using crossed beams, *Appl. Surf. Sci.*, 96–98:649.

Gottmann, J. and Kreutz, E.W., 1999, Pulsed laser deposition of alumina and zirconia thin films on polymers and glass as optical and protective coatings, *Surf. Coat. Technol.*, 116–119:1189.

Goyal, A., Norton, D.P., Kroeger, D.M., Christen, D.K., Paranthaman, M., Specht, E.D., Budai, J.D., He, Q., Saffian, B., List, F.A., Lee, D.F., Hatfield, E., Martin, P.M., Klabunde, C.E., Mathis, J., and Park, C., 1997, Conductor with controlled grain boundaries: An approach to the next generation, high-temperature superconducting wire, *J. Mater. Res.*, 12:2924.

Goyal, A., Ren, S.X., Specht, E.D., Kroeger, D.M., Feenstra, R., Norton, D., Paranthaman, M., Lee, D.F., and Christen, D.K., 1999, Texture formation and grain boundary networks in rolling assisted biaxially textured substrates and in epitaxial YBCO films on such substrates, *Micron*, 30:463.

Greer, J.A., 1994, Commercial scale-up of pulsed laser deposition, in: *Pulsed Laser Deposition of Thin Films*, D.H. Chrisey and G.K. Hubler, eds., John Wiley & Sons, Inc., New York.

Greer, J.A. and Tabat, M.D., 1995, Large-area pulsed laser deposition: techniques and applications, *J. Vac. Sci. Technol. A*, 13:1175.

Greer, J.A. and Tabat, M.D., 1995, On- and off-axis large-area pulsed laser deposition, in: *Film Synthesis and Growth Using Energetic Beams*, H.A. Atwater, J.T. Dickinson, D.H. Lowndes, and A. Polman, eds., Mater. Res. Soc., Pittsburgh.

Harshavardhan, K.S., Christen, H.M., Silliman, S.D., Talanov, V.V., Anlage, S.M., Rajeswari, R., and Claassen, J., 2001, Low-loss $YBa_2Cu_3O_7$ films on flexible, polycrystalline yttria-stabilized zirconia tapes for cryoelectronic applications, *Appl. Phys. Lett.*, 78:1888.

Hau, S.K., Wong, K.H., Chan, P.W., and Choy, C.L., 1995, Intrinsic resputtering in pulsed-laser deposition of lead-zirconate-titanate thin films, *Appl. Phys. Lett.*, 66:245.

Hayashi, N. and Yoshida, N., 1995, Method of fabricating oxide superconducting films by laser deposition, US Patent 5,447,910 (September 5).

Henyk, M., Vogel, N., Wolfframm, D., Tempel, A., and Reif, J., 1999, Femtosecond laser ablation from dielectric materials: Comparison to arc discharge erosion, *Appl. Phys. A*, 69(Suppl.):S355.

Hirayama, Y., Yabe, H., and Obara, M., 2001, Selective ablation of AlN ceramic using femtosecond, nanosecond, and microsecond pulsed laser, *J. Appl. Phys.*, 89:2943.

Horwitz, J.S. and Sprague, J.A., 1994, Film nucleation and film growth in pulse laser deposition of ceramics, in: *Pulsed Laser Deposition of Thin Films*, D.H. Chrisey and G.K. Hubler, eds., John Wiley & Sons, Inc., New York.

Huang, X.T., Wang, Y.Q., Wang, Q.L., and Chen, Q.M., 2000, In-plane aligned YBCO film on textured YSZ buffer layer deposited on NiCr alloy tape by laser ablation with only O^+ ion beam assistance, *J. Phys.: Condens. Matter*, 12:761.

Iijima, Y., Kakimoto, K., Takeda, K., and Saitoh, T., 2001, Long length IBAD process of fluorite type oxides for Y-123 coated conductors, presented at the *2001 International Workshop on Superconductivity, co-*

sponsored by ISTEC and MRS,"HTS Conductors, Processing and Applications," Honolulu, Hawaii, June 24–27.

Kabashin, A.V., Marine, W., Nikitin, P.I., and Sentis, M., 1996, Electromagnetic diagnostics during pulsed laser deposition, *Appl. Surf. Sci.*, 96–98:139.

Kelly, R. and Miotello, A., 1994, Mechanisms of pulsed laser sputtering, in: *Pulsed Laser Deposition of Thin Films*, D.H. Chrisey and G.K. Hubler, eds., John Wiley & Sons, Inc., New York.

Kerchner, H.R., Norton, D.P., Goyal, A., Budai, J.D., Christen, D.K., Kroeger, D.M., Specht, E.D., He, Q., Paranthaman, M., Lee, D.F., Sales, B.C., List, F.A., and Feenstra, R., 1997, Alternating current losses in biaxially textured $YB_2Cu_3O_{7-\delta}$ films on Ni tapes, *Appl. Phys. Lett.*, 71:2029.

Kim, J., Chrisey, D.B., Howitz, J.S., Miller, M.M., and Gilmore, C.M., 2000, Growth mechanism of $YBa_2Cu_3O_{7-\delta}$ thin films and precipitates on planar and vicinal $SrTiO_3$ substrates, *J. Mater. Res.*, 15:596.

Kinoshita, K., Ishibashi, H., and Kobayashi, T., 1994, Improved surface smoothness of $YBa_2Cu_3O_y$ films and related multilayers by ArF excimer laser deposition with shadow mask "Eclipse method", *Jpn. J. Appl. Phys. 2, Lett.*, 33:L417.

Koinuma, H., Kawasaki, M., Ohashi, S., Lippmaa, M., Nakagawa, N., Iwasaki, M., and Qiu, X.G., 1998, Nucleation and growth control in pulsed laser epitaxy of oxide thin films, *Proc. SPIE*, 3481:153.

Koren, G., Baseman, R.J., Gupta, A., Lutwyche, M.I., and Laibowitz, R.B., 1990, Particualtes reduction in laser-ablated $YBa_2Cu_3O_{7-\delta}$ thin films by laser-induced plume heating, *Appl. Phys. Lett.*, 56:2144.

Kusumori, T. and Muto, H., 2000, Effect of target composition on the crystallinity of $YB_2Cu_3O_y$ epitaxial films fabricated by Nd : YAG pulsed laser deposition, *Physica C*, 337:57.

Laube, S.J.P. and Stark, E.F., 1996, Hierarchical feedback control of pulsed laser deposition, US Patent 5,535,128 (July 9).

Laube, S.J.P. and Voevodin, A.A., 1998, Pulsed laser deposition, *in-situ* process modeling and feedback control for multi-layer deposition, *Surf. Coatings Technol.*, 99:14.

Lee, D.F., Paranthaman, M., Mathis, J.E., Goyal, A., Kroeger, D.M., Specht, E.D., Williams, R.K., List, F.A., Martin, P.M., Park, C., Norton, D.P., and Christen, D.K., 1999, Alternative buffer architectures for high critical current density YBCO superconducting deposits on rolling assisted biaxially-textured substrates, *Jpn. J. Appl. Phys.*, 38:L178.

Li, Q., Liu, S., Fenner, D.B., Luo, J., Hamblen, W.D., and Higis, J., 1995, *In-situ* optical diagnostics during pulsed laser deposition of high-T_c superconductor thin films, *IEEE Trans. Appl. Supercond.*, 5:1513.

List, F.A., Goyal, A., Paranthaman, M., Norton, D.P., Specht, E.D., Lee, D.F., and Kroeger, D.M., 1998, High J_c YBCO films on biaxially textured Ni with oxide buffer layers deposited using electron beam evaporation and sputtering, *Physica C*, 302:87.

Lorenz, M., Hochmuth, H., Natusch, D., Kusunoki, M., Svetchnikov, V.L., Riede, V., Stanca, I., Kästner, G., and Hesse, D., 2001, High-quality Y–Ba–Cu–O thin films by PLD-ready for market applications, *IEEE Trans. Appl. Supercond.*, 11:3209.

Lowndes, D.H., Geohegan, D.B., Puretzky, A.A., Norton, D.P., and Rouleau, C.M., 1996, Synthesis of novel thin-film materials by pulsed laser deposition, *Science*, 273:898.

Maguire, J.F., Busbee, J.D., Liptak, D.C., Lubbers, D.P., SeClair, S.R., and Biggers, R.R., 2000, Process control for pulsed laser deposition using Raman spectroscopy, US Patent 6,038,525 (March 14).

Millon, E., Albert, O., Loulergue, J.C., Etchepare, J., Hullin, D., Seiler, W., and Perriere, J., 2000, Growth of heteroepitaxial ZnO thin films by femtosecond pulsed-laser deposition, *J. Appl. Phys.*, 88:6937.

Miotello, A. and Kelly, R., 1999, Laser-induced phase exposion: new physical problems when a condensed phase approaches the thermodynamic critical temperature, *Appl. Phys. A*, 69(Suppl.):S67.

Muenchausen, R.E., Hubbard, K.M., Foltyn, S., Estler, R.C., Nogar, N.S., and Jenkins, C., 1990, Effects of beam parameters on excimer laser deposition of $YBa_2Cu_3O_{7-\delta}$, *Appl. Phys. Lett.*, 56:578.

Mukaida, M., Takano, Y., Chiba, K., Moriya, T., Kusunoki, M., and Ohshima, S., 1999, A new epitaxial $BaSnO_3$ buffer layer for $YBa_2Cu_3O_{7-\delta}$ thin films on MgO substrates, *Supercond. Sci. Technol.*, 12:890.

Mukherjee, P., Cuff, J.B., and Witanachchi, S., 1998, Plume expansion and stoichiometry in the growth of multi-component thin films using dual-laser ablation, *Appl. Surf. Sci.*, 127–129:620.

Nakagawa, M.N., Kawasaki, M., Ohashi, S., and Koinuma, H., 2000, Growth mode mapping of $SrTiO_3$ epitaxy, *Appl. Phys. Lett.*, 76:2439.

Noda, E., Suzuki, S., Morimiya, O., and Hayashi, K., 1992, Method for manufacturing an oxide superconductor thin film, US Patent 5,158,931 (October 27).

Norton, D.P., 1998, Science and technology of high-temperature superconducting films, *Annu. Rev. Mater. Sci.*, 28:299.

Ohashi, S., Lippmaa, M., Nakagawa, N., Nasagawa, H., Koinuma, H., and Kawasaki, M., 1999, Compact laser molecular beam epitaxy system using laser heating of substrate for oxide film growth, *Rev. Sci. Instrum.*, 70:178.

Okoshi, M., Higashikawa, K., and Hanabusa, M., 2000, Pulsed laser deposition of ZnO thin films using a femtosecond laser, *Appl. Surf. Sci.*, 154–155:424.

Park, C., Norton, D.P., Budai, J.D., Christen, D.K., Verebelyi, D., Feenstra, R., Lee, D.F., Goyal, A., Kroeger, D.M., and Paranthaman, M., 1998, Bend strain tolerance of critical currents for $YBa_2Cu_3O_7$ films deposited on rolled-textured (001)Ni, *Appl. Phys. Lett.*, 73:1904.

Park, C., Norton, D.P., Christen, D.K., Verebelyi, D.T., Feenstra, R., Budai, J.D., Goyal, A., Lee, D.F., Specht, E.D., Kroeger, D.M., and Paranthaman, M., 1999, Long length fabrication of YBCO on rolling assisted biaxially textured substrates (RABiTS) using pulsed laser deposition, *IEEE Trans. Appl. Supercond.*, 9:2276.

Park, S.M. and Moon, J.Y., 2001, Laser ablation of a $Pb(Zr_xTi_{1-x})O_3$ target in a pulsed oxygen jet, *Appl. Surf. Sci.*, 174:87.

Pique, A., Venkatesan, T., and Green, S., 1995, Pulsed laser passive filter deposition system, US Patent 5,458,686 (Oct. 17).

Puretzky, A.A., Geohegan, D.B., Jellison, G.E., Jr., and McGibbon, M.M., 1996, Comparative diagnostics of ArF- and KrF-laser generated carbon plumes used for amorphous diamond-like carbon film deposition, *Appl. Surf. Sci.*, 96–98:859.

Qian, F., Craciun, V., Singh, R.K., Dutta, S.D., and Pronko, P.P., 1999, High intensity femtosecond laser deposition of diamond-like carbon thin films, *J. Appl. Phys.*, 86:2281.

Reeves, J.L., Feldmann, D.M., Yang, C.-Y., and Larbalestier, D.C., 2001, Current barriers in Y–Ba–Cu–O coated conductors, *IEEE Trans. Appl. Supercond.*, 11:3863.

Riabkina-Fishman, M. and Zahavi, J., 1993, Effect of processing conditions on superconductive properties of laser-deposited YBCO thin films, *J. Mater. Sci.*, 28:3298.

Rijnders, G.J.H.M., Koster, G., Blank, D.H.A., and Rogalla, H., 1997, *In situ* monitoring during pulsed laser deposition of complex oxides using reflection high energy electron diffraction under high oxygen pressure, *Appl. Phys. Lett.*, 70:1888.

Roas, B., Endres, G., and Schultz, L., 1993, Method for manufacturing oxide superconducting films by laser evaporation, US Patent 5,212,148 (May 18).

Rode, A.V., Luther-Davies, B., and Gamaly, E.G., 1999, Ultrafast ablation with high-pulse-rate lasers. Part II: Experiments on laser deposition of amorphous carbon films, *J. Appl. Phys.*, 85:4222.

Samano, E.C., Machorro, R., Soto, G., and Cota-Araiza, L., 1998, *In situ* ellipsometric characterization of SiN_x films grown by laser ablation, *J. Appl. Phys.*, 84:5296.

Sato, Y., Matsuo, K., Takahashi, Y., Muranaka, K., Fujino, K., Hahakura, S., Ohmatsu, K., and Takei, H., 2001, Development of $YBa_2Cu_3O_y$ tape by using inclined substrate method, *IEEE Trans. Appl. Supercond.*, 11:3365.

Schenck, P.K., Vaudin, M.D., Bonnell, D.W., Hastie, J.W., and Paul, A.J., 1998, Particulate reduction in the pulsed laser deposition of barium titanate thin films, *Appl. Surf. Sci.*, 127–129:655.

Schey, B., Bollmeier, T., Kuhn, M., Biegel, W., and Stritzker, B., 1998, Large area deposition of $YBa_2Cu_3O_{7-x}$ films by pulsed laser ablation, *Rev. Sci. Instrum.*, 69:474.

Schwarz, H. and Tourtellotte, H.A., 1969, Vacuum deposition high-energy laser with emphasis on barium titanate films, *J. Vac. Sci. Technol.*, 6:373.

Shirk, M.D. and Molian, P.A., 2001, Ultra-short pulsed laser ablation of highly oriented pyrolytic graphite, *Carbon*, 39:1183.

Smith, H.M. and Turner, A.F., 1965, Vacuum deposited thin films using a Ruby laser, *Appl. Opt.*, 4:147.

Strikovski, M. and Miller, J.H., Jr., 1998, Pulsed laser deposition of oxides: Why the optimum rate is about 1 Å per pulse, *Appl. Phys. Lett.*, 73:1733.

Strikovski, M., Miller, J.H., Jr., and Wosik, J., 2000, Deposition rate as the key parameter in pulsed laser deposition of oxide films: A practical model and experiment, *Physica C*, 341–348:2349.

Szorenyi, T., Fogarassy, E., Fuchs, C., Hommet, J., and Le Normand, F., 1999, Chemical analysis of a-CN_x thin films synthesized by nanosecond and femtosecond pulsed laser deposition, *Appl. Phys. A*, 69(Suppl.):941.

Tanaka, H. and Kawai, T., 2000, Artificial construction of SrO/(La,Sr)MnO_3 layered perovskite superlattice by laser molecular-beam epitaxy, *Appl. Phys. Lett.*, 76:3618.

Tselev, A., Gorbunov, A., and Pompe, W., 2001, Cross-beam pulsed laser deposition: General characteristic, *Rev. Sci. Instrum.*, 72:2665.

Usoskin, A., Knoke, J., García-Moreno, F., Issaev, A., Dzick, J., Sievers, S., and Freyhardt, H.C., 2001a, Large-area HTS-coated stainless steel tapes with high critical currents, *IEEE Trans. Appl. Supercond.*, 11:3385.

Usoskin, A., Sturn, K., Knoke, J., Issaev, A., García-Moreno, F., Dzick, J., and Freyhardt, H.C., 2001b, Long-length YBCO coated stainless steel tapes produced by HR-PLD, presented at the *5th European Conf. on Appl. Supercond. (EUCAS, 2001)*, Aug. 26–30, Copenhagen, Denmark. To be published in *Physica C*.

Venkatesan, T. and Wu, X.D., 1991, Method and apparatus for pulsed energy induced vapor deposition of thin films, US Patent 5,015,492 (May 14).

Wagner, F.X., Scaggs, M., Koch, A., Endert, H., Christen, H.M., Knauss, L.A., Harshavardhan, K.S., and Green, S.M., 1998, Epitaxial HTS thin films grown by PLD with a beam homogenizer, *Appl. Surf. Sci.*, 127–129:477.

Wang, R.P., Zhou, Y.L., Pan, S.H., He, M., Lu, H.B., Chen, Z.H., Yang, G.Z., Liu, C.F., Wu, X., Wang, F.Y., Feng, Y., Zhang, P.X., Wu, X.Z., and Zhou, L., 2000, Deposition of high-temperature superconducting films on biaxially textured Ni(001) substrates, *Physica C*, 337:87.

Watson, T.A. and Rowan, C., 1996, Industrial excimer laser beam properties, *Appl. Surf. Sci.*, 96–98:532.

Weissmantel, S., Reisse, G., Keiper, B., and Schulze, S., 1999, Microstructure and mechanical properties of pulsed laser deposited boron nitride films, *Diam. Relat. Mater.*, 8:377.

Willmott, P.R. and Antoni, F., 1998, Growth of GaN(0001) thin films on Si(001) by pulsed reactive crossed-beam laser ablation using liquid Ga and N_2, *Appl. Phys. Lett.*, 73:1394.

Willmott, P.R. and Huber, J.R., 2000, Pulsed laser vaporization and deposition, *Rev. Mod. Phys.*, 72:315.

Witanachchi, S., Ahmed, K., Sakthivel, P., and Mukherjee, P., 1995, Dual-laser ablation for particulate-free film growth, *Appl. Phys. Lett.*, 66:1469.

Xiong, X. and Winkler, D., 2000, Rapid deposition of biaxially-textured CeO_2 buffer layers on polycrystalline nickel alloy for superconducting tapes by ion assisted pulse laser deposition, *Physica C*, 336:70.

Xu, S.Y., Ong, C.K., You, L.P., Li, J., and Wang, S.J., 2000, The growth mode and microstructure of Ag-doped $YB_2Cu_3O_{7-\delta}$ thin films prepared by dual beam pulsed-laser deposition, *Physica C*, 341–348:2345.

Yang, G.Z., Lu, H.B., Chen, F., Zhao, T., and Chen, Z.H., 2001, Laser molecular beam epitaxy and characterization of perovskite oxide thin films, *J. Cryst. Growth*, 227–228:929.

Zaitsev-Zotov, S.V., Martynyuk, R.A., and Protasov, E.A., 1983, Superconductivity of $BaPb_{1-x}Bi_xO_3$ films prepared by laser evaporation method, *Sov. Phys. Solid State*, 25:100.

Zander, W., Stritzker, B., and Frohlingsdorf, J., 1992, Apparatus for the ablation of material from a target and coating method and apparatus, US Patent 5,084,300 (January 28).

Chapter 6

METHODS OF HTS DEPOSITION: THERMAL EVAPORATION

Werner Prusseit

THEVA Dünnschichttechnik GmbH
Hauptstr. 1b
85386 Eching-Dietersheim
Germany

6.1 INTRODUCTION

Among physical vapor deposition techniques thermal evaporation (TE) is the one with the longest standing tradition. However, during the last 30 years of booming semiconductor industry which involves a great deal of thin film technology, deposition techniques like CVD (chemical vapor deposition) or sputtering which often offer unquestionable advantages have been developed to perfection and TE has largely been replaced in production lines. On the laboratory scale, due to their simplicity, techniques like PLD (pulsed laser deposition) or sputtering were much more promising to realize fast results.

Consequently, when the new superconductors emerged only a handful of research groups like Berberich (1989), Terashima (1988), Kwo (1989), Prakash (1990), and Chew (1990) performed deposition trials based on TE. However, as time went by it became clear that the performance of the various deposition methods strongly depends on material issues as well as economic aspects. With progressive commercialization, cost effective volume production and reproducibility became the driving forces and the intrinsic advantages of TE turned the scales. In this sense, the story of high temperature superconductor (HTS) film deposition can serve as an example how this technique, sometimes regarded as old-fashioned, still bears a high potential for innovation and surprising efficiency.

In the following chapter we will review the basic features of TE when applied for HTS deposition in detail. The specific advantages and problems will be discussed in view of metal tape coating to estimate whether this technique provides a long term perspective for the fabrication of coated conductors.

6.2 GENERAL FEATURES OF THERMAL EVAPORATION FOR HTS DEPOSITION

TE is the classical technique applied for metal–plating of glass or plastic surfaces, like, e.g., aluminium coatings widely used for capacitors, plastic wrappings, and as barrier against water diffusion. It is evident that the deposition of quaternary metal–oxide compounds imposes quite different requirements to the technique and will go far beyond the rudimentary concept of evaporating a single metal in a vacuum chamber.

The necessary features of a conventional HTS deposition system are depicted in Figure 6.1. The metal species the superconductor is composed of are evaporated in high vacuum ambient. Usually, the chamber is pumped to 10^{-6} mbar background pressure. During deposition, the distance between the sources and the substrate which is in the range of several ten centimeters sets the scale for the required mean free path. Ballistic propagation of the vapor requires a residual gas background below 10^{-4} mbar, even when an oxygen flow is introduced.

To control the film composition the evaporation rates have to be online monitored individually by some kind of sensor heads. Feedback loops to the sources serve to stabilize the evaporation rates.

The substrate is mounted in or on a heater element and kept at a temperature which promotes epitaxial film growth. Since HTS require elevated oxygen pressure for their formation the introduction of oxygen is an essential but tricky task. To avoid flooding of the chamber with oxygen the main chamber is permanently pumped at, while the reactive gas has to be introduced close to the substrate. In the early days, a housing as described by Baudenbacher (1990) served to confine the gas around the substrate. Its small opening towards the sources constitutes an enhanced gas flow resistance and allows differential pumping to a certain extent.

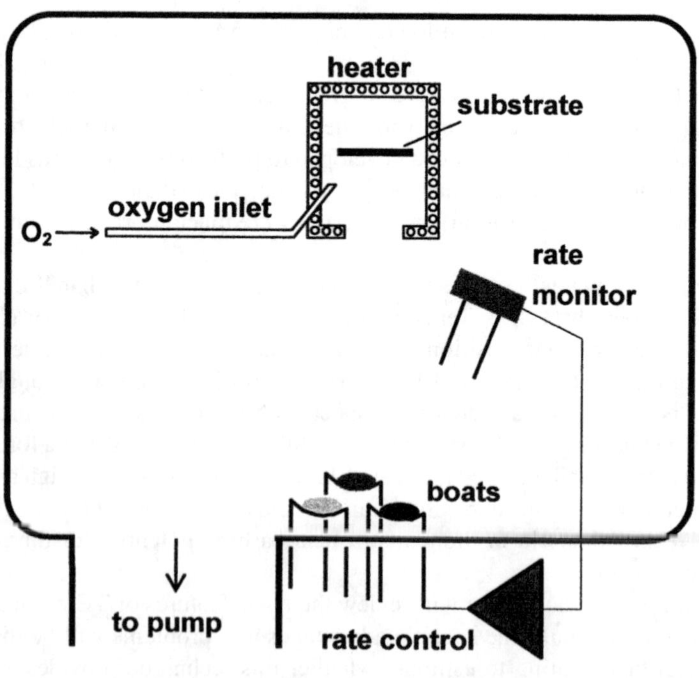

Figure 6.1. Main features of a conventional HTS evaporation system.

Since the height of the region with enhanced oxygen pressure is comparable to its lateral extension, i.e. the dimensions of the substrate, this concept can be applied only for small chips (typically in the order of 10–20 mm). Otherwise, with increasing thickness of this oxygen zone the metal vapors start scattering. Since the scattering cross section depends on the molecular weight, as a first consequence the composition gets extremely pressure dependent. If the gas pressure or the length of the path through the oxygen is further increased, the vapor does not penetrate this barrier and cannot reach the substrate any more.

6.3 PROS AND CONS OF THERMAL EVAPORATION

6.3.1 System Requirements

In comparison to most other techniques (PLD, CVD, sputtering) operating at forepump pressure level TE requires a high vacuum ambient and—if a reactive gas flow is introduced—also high pumping speed. These requirements can be met by oil diffusion—cryo—or turbo molecular pumps. Since performance and reliability has been largely improved in the last decade, turbo pumps are by far the most convenient solution especially if chambers have to be opened for substrate changing or if the process requires a backfilling with oxygen as in the case of HTS films. Since turbo pumps can be operated comfortably in a stop and go mode the higher investment costs pay because large gate valves necessary to disconnect continuously working pumps from the deposition chamber are not required.

On the other hand the vacuum specifications are mostly relaxed in comparison with semiconductor deposition processes which often take place under UHV conditions. Due to the low carrier densities in semiconductors even tiny impurity levels have significant if not detrimental effects on the electrical properties. In this respect, handling HTS deposition systems is a lot easier and there are only few precautions to avoid contamination when the system is opened or serviced.

6.3.2 Evaporation Sources

For industrial scale deposition Knudsen cells (K-cells) and electron guns (e-guns) are the most common evaporation sources. Both types can hold a rather large volume of deposition material which usually renders them a good choice for long term deposition. However, with respect to reactive co-evaporation of HTS films there are also severe limitations.

Due to the reactive gas flow into the chamber in the long run metal-oxide skins build up on the surface in K-cells. Since the deposition rate is mainly controlled by the temperature of the cell the oxide formation reduces the vapor flow rate—in the extreme case to zero if the skin cannot be penetrated any more. As a consequence, even K-cells will require an online rate monitoring system which has to be very sensitive, since typical deposition rates of K-cells are an order of magnitude lower than those of, e.g., boat sources and e-guns. To make things even worse the oxide growth of yttrium and other rare earths elements is not a self limiting process but the oxide thickens until the K-cell eventually fails. For this reason K-cells have been mainly used, e.g., by Bozovic (1990) and Eckstein (1989, 1992) in MBE-type deposition systems with extremely low background pressure in connection with activated oxygen confined at the substrate location. Beyond that, large capacity K-cells are bulky which limits the uniformity of the film composition across a given area.

Table 6.1. The advantages and disadvantages of boat sources

Pros	Cons
Compact	Limited capacity
Easy switching between many elements	Not all elements available for evaporation
Material cheap	
High rates available	
Fine tuning of composition	Rate control necessary

Some of these complications can be overcome by e-guns as used, e.g., by Naito (1991) or Chew (1990). However, due to their short time constants they also put higher requirements to the speed of the rate control. Multi-component film material cannot be evaporated from a single source since most compounds are cracked by the e-beam and will not evaporate stoichiometrically. As a consequence, co-evaporation requires several independent sources. However, large e-guns are very bulky and cannot be positioned as close to each other as would be desirable for realizing a quasi-point source. Only a large source to substrate distance can compensate for this disadvantage, which in turn reduces the material efficiency. In this respect, e-guns and K-cells encounter the same problem. There are quite compact multi-liner e-guns available, but for the sake of space the liner capacity is very limited and the original advantage of the large material volume gets lost again.

Among the possible evaporation sources metal "boats" are the simplest but have yet turned out as very effective. Positive and negative aspects are summarized in Table 6.1 and shall be discussed in more detail.

Boat sources are very compact and can be placed quite close to each other to approach the multiple species point source ideal which guarantees homogeneous film composition. The reverse of this medal is the limited capacity of such boats and in standard configurations the maximum YBCO film thickness achievable with one filling lies between 500 and 1000 nm. In view of long term operation, *in situ* refilling of such sources is a must. Fortunately, even quantitative refilling from reservoirs can be accomplished by such means as, e.g., vibrator conveyors. As the main advantage of this splitting into reservoir and evaporation source these large reservoirs can be placed independently at convenient locations and don't impede the boats.

Typically, evaporation employs pure metals which are available in various forms like grains, turnings or even powder. They are usually stable and quite cheap compared to sputter targets or their complex metal–organic counterparts used for CVD.

Boat sources provide a rather high degree of flexibility. Deposition rates can range from a few Å/s to quantitative flash evaporation. The film composition can be tuned over a wide range to optimize the desired properties. In turn, the multi-source deposition requires a very accurate and stable control of the deposition rates as described below.

As a disadvantage the maximum temperature of the boat sources is limited by the boat material itself, usually, tungsten, tantalum or molybdenum. Consequently, elements with very high melting points or low vapor pressure are not available for TE. Nevertheless, most homologous 123-compounds are easily accessible replacing yttrium by other rare earth elements and without much changing the deposition parameters.

6.3.3 Composition Control

No matter which sources are employed co-evaporation requires a close, accurate, and stable rate control of each channel. To be even more specific, this means an in situ online rate control with a time constant appropriate for a feedback loop to the evaporation sources.

An analysis of the growing film can be performed by RHEED or EDX and a combination thereof, like demonstrated by Kamei (1992). However, due to the limited signal to noise ratio and the strong thickness dependence of the signal the EDX in situ analysis can be regarded as exotic and is not applicable as standard manufacturing tool. RHEED and LEED are not sensitive to the elementary composition but were used by Eckstein (1990, 1992) and Naito (1995) to monitor composition-dependent surface properties of the growing film on a fixed and well oriented substrate. However, this technique is restricted to some fine tuning and does not replace other monitoring. The use of moving or polycrystalline substrates and for high deposition rates is practically impossible.

For this reason, only the analysis of the vapor composition in front of the substrate offers a chance for a reliable online rate control.

Note: The evaporation rate is a product of the vapor density times propagation velocity. Since the vapor originates from a melt the density is governed by an Arrhenius law, i.e. it depends exponentially on the temperature of the source, whereas the mean velocity (Maxwell distribution) scales with the square root of the temperature. Consequently, the velocity of the atoms in the vapor can be regarded as practically constant and rate changes are essentially caused by changes of the vapor density.

Due to the emission characteristics of the boats and temperature dependent sticking coefficients on the substrate surface these methods cannot provide absolute values but have to be calibrated empirically. However, they allow a quite accurate relative control to keep the rates and film composition constant over the deposition time.

Sensing the composition of metal vapor inside a vacuum chamber can be performed by a quadrupole mass spectrometer which is able to select and analyze the metal species due to their different atomic masses. This technique has been established and used by Chang (1973) for alloys and by Chew (1990) and Baudenbacher (1990) for YBCO. Usually, the quadrupole system has to be pumped differentially to ensure a long collision free path between the quadrupole rods and to save the filament of the electron source. The vapor is ionized by an electron beam in a cross beam configuration. This rate control has been successfully employed mainly for low rate deposition. To compensate drift it has to be re-calibrated against quartz crystal monitors.

EIES (electron impact emission sensors) employ a quite similar ionizing arrangement by an electron beam but detect the characteristic light emission from the metal species. Described by Lu (1977) it has been used for HTS deposition by Naito (1991, 1995). As in the case of the quadrupole, the emission filament in connection with the background oxygen constitutes a handicap in view of long term tape deposition.

The common, commercially available rate monitoring is based on frequency shifts of oscillating quartz plates due to the deposited material. These quartz crystal monitors are quite robust and easy to handle. Even for rates of a few Å/s the accuracy is in the range of a few percent and stability as well as reproducibility is excellent. Their only shortcoming becomes evident when used for long term processes like tape coating. When a certain material thickness has been deposited onto the quartz, the frequency shift is so large that the quartz fails oscillating. Multiple, exchangeable quartz heads may relax but not solve this problem.

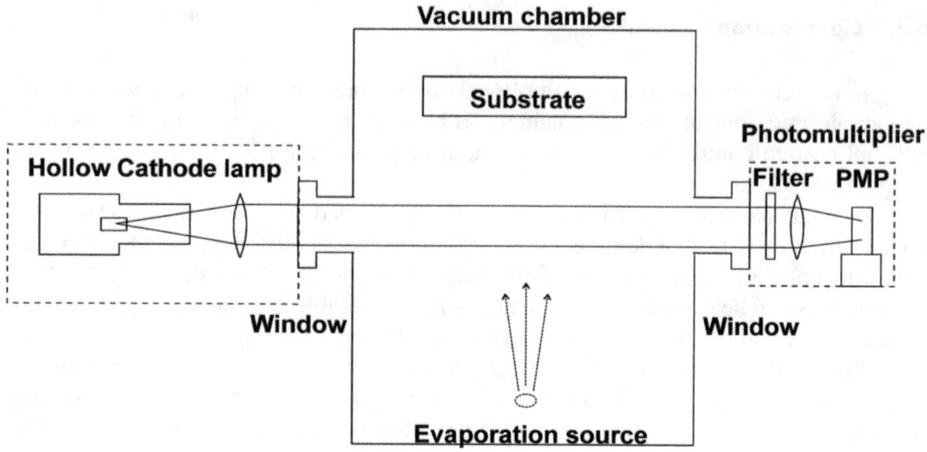

Figure 6.2. Setup of rate control by atomic absorption spectroscopy (AAS).

For this reason, in recent years atomic absorption spectroscopy (AAS) has become an increasingly interesting alternative to monitor the vapor composition. It has been established by Klausmeier (1992), Shinohara (1991), and Lu (1995) for HTS film deposition. The basic operating principle as depicted in Figure 6.2 is intriguingly simple. Light of a spectral line of the element to be analyzed is generated outside the chamber, e.g., by a hollow cathode lamp or a laser, passes through the chamber and is detected by a photomultiplier on the other side. The light beam can be coupled into the chamber by lenses and apertures or optical fibers.

The weakening of this light by resonant absorption in the metal vapor is a direct measure of the vapor density and hence the deposition rate. The technique offers the advantages that all components can be placed outside the vacuum chamber, have long lifetime and allow easy access.

In spite of the apparent simplicity, however, the problems arising with AAS are manifold and it took quite a lot of additional technical efforts to guarantee the required long term stability. Some setups described by Lu (1995), Klausmeier (1992), and Utz (1996) use special precautions and reference beams to compensate for long term drifts of the light sources and detectors. To name only a few, difficulties can also arise from the absorption probabilities and wavelength of the transition, which can range from IR to UV. Tunable semiconductor laser sources have to emit a single mode, intensity stabilized beam of light. To reach frequencies in the UV or blue part of the spectrum, their output frequency has to be doubled or even tripled accompanied by a loss of orders of magnitude of intensity.

On the other hand hollow cathode lamps emit spectral lines broadened by collisions in the gas filling and are susceptible to aging effects. Beyond that, at higher deposition rates saturation effects in the absorption of the spectral lines by the vapor can render AAS control difficult.

In summary, the composition control which is an essential ingredient of TE systems can be handled today, and what seemed to be a disadvantage in the early days of coevaporation has turned into more flexibility for the optimization of composition dependent film properties. The best choice of the monitoring method, however, depends to a great extent on process requirements like deposition time and—rates.

6.3.4 Distance between Sources and Substrate

TE is a non-directional deposition method. The source to substrate distance can be chosen very large and automatically results in a high uniformity of the deposited films. This inherent uniformity is certainly one of the essential and unquestionable advantages of TE over other methods. Due to the fact that a large area can be coated simultaneously a second important consequence is the high volume deposition rate.

The emission characteristics of the boats can be calculated directly and the larger the source to substrate distance the better is the resulting uniformity of the film thickness, composition, and the physical properties. This is in direct contrast to all short range plasma techniques (PLD, sputtering) where inhomogeneities of the plasma or target are directly mapped onto the substrate.

Obviously, there is a trade off between uniformity and material exploitation which inevitably drops quadratically with increasing distance. To achieve acceptable homogeneity for fixed substrates and a deposition area of about 400 cm^2 the material efficiency is in the order of about 5%. For a tape coating arrangement it seems feasible to increase this efficiency up to 15%.

The fact that a lot of material is deposited all over the chamber bears another problem of massive contamination and flaking, since during long term operation quantities in the order of kilograms are vaporized. All components directly exposed to the vapor will be coated and since the material is hygroscopic flakes start to peel off when the system is opened giving rise to dust problems. Capturing material which propagates in wrong directions and thus confining the vapor in a "deposition chimney" can be realized by metal screens.

6.3.5 Substrate Heater and Oxygen Supply

The most critical component of the deposition system is the substrate holder. Epitaxial growth of HTS films requires high substrate temperature as well as reactive gas ambient. Consequently, the substrate holder comprises both heating elements to obtain a homogeneous temperature distribution across the entire substrate area and a reactive gas supply unit.

Optimum temperature uniformity is realized by black body radiation from heated walls surrounding the sample. Backside heater plates which are often used for plasma techniques to allow access to the substrate surface suffer from changes in the IR transparency of the substrate during film growth and also depend strongly on the substrate material. The arrangement which comes closest to the ideal consists of a furnace with a small opening to allow access of the vapor as already depicted in Figure 6.1. Such a small opening can also serve as a flow resistance for the reactive gas introduced in the neighborhood of the substrate, cf. Baudenbacher (1990).

However, due to the small opening and the argumentation above this simple concept is only applicable for small substrates and the main advantage of TE gets lost. Two ways to resolve this dilemma have been pursued successfully:

(a) The use of highly reactive gas at such low pressure that the vapor propagation is not impeded. For HTS films ozone and atomic oxygen activated by downstream plasma sources or in a discharge, and even directional oxygen beams from RF sources have been employed by, e.g., Terashima (1988), Kwo (1989), Prakash (1989, 1990), Chew (1990), Eckstein (1989), Utz (1995), and Sato (1997). However, the activated gas supply bears a lot of technical problems, like the low degree of activation, recombination at walls before the substrate is reached, and inhomogeneous distribution across large areas. Using, e.g., an ozone distillery to enrich the gas may work on a laboratory scale but due to the high danger of explosions it is not a safe solution for industrial scale production. Besides, even when using activated oxygen for large area deposition the gas flow into the chamber is still very high and requires enormous pumping speed to maintain an acceptable background pressure.

For the deposition of bismuth-based superconductors, however, it has turned out that only activated gas as employed by Eckstein (1990) and Utz (1995) can supply the necessary oxidation potential and is the only practicable way to grow BSCCO- or BKBO-films by TE.

(b) Molecular oxygen at high pressure can be used when confined in a tricky way around the substrate. In this respect, the breakthrough was obtained by Berberich (1993, 1994). The discussion of the basic idea and the design of this heater shall be saved for the next section. It should just be noted here that it enables deposition on very large wafers of more than 8 inches in diameter and has become a real production tool for fabrication of the greater part of the large YBCO films worldwide.

This section should not be concluded without noting that deposition by TE employs the lowest substrate temperatures of all deposition techniques, typically around 670–690°C compared to 750–850°C for other techniques. It could be demonstrated by Baudenbacher (1997) that the YBCO lattice is formed even at temperatures as low as 490°C. The lower substrate temperature may turn out as a crucial point in view of tape coating since diffusion and unwanted substrate reactions are temperature activated processes which are slowed down if not halted at lower temperature. The probable reason for the lower film growth temperature lies in the higher surface mobility of metal atoms compared to metal oxide molecules which are already present in the gas phase for those deposition techniques involving hot plasma conditions.

6.4 LARGE AREA, LONG TERM DEPOSITION

6.4.1 Spatial Separation of Deposition and Oxidation

To a great extent large area YBCO-films are produced by vacuum evaporation employing molecular oxygen in an arrangement first developed at the Technical University of Munich in 1992 by Berberich (1993, 1994). The basic idea goes back to a RHEED-film growth study. In this survey, the intensity variations of the specular reflected e-beam were analyzed as a function of deposition time. Layer by layer growth gives rise to well known RHEED oscillations as a consequence of modulations of the surface roughness, first observed by Terashima (1990). Since the overall disorder inevitably increases with deposited film thickness these oscillations are damped. When closing a shutter in front of the substrate to stop deposition, however, the signal intensity recovers to a certain extent. It can be interpreted that the YBCO surface is still under construction even after metal atoms have stopped to impinge upon the substrate. From such measurements Baudenbacher (1997) could determine a time constant in the range of a second until the metal atoms are getting to rest at their final positions in the YBCO lattice. If metals are moving around for about a second until they are eaten up by chemical reactions (oxidation) they will not experience any difference if oxygen is supplied a few tenths of a second after they had reached the substrate surface. Consequently, the substrate can be shifted between a high vacuum metal deposition zone and a high pressure reaction zone as sketched in Figure 6.3.

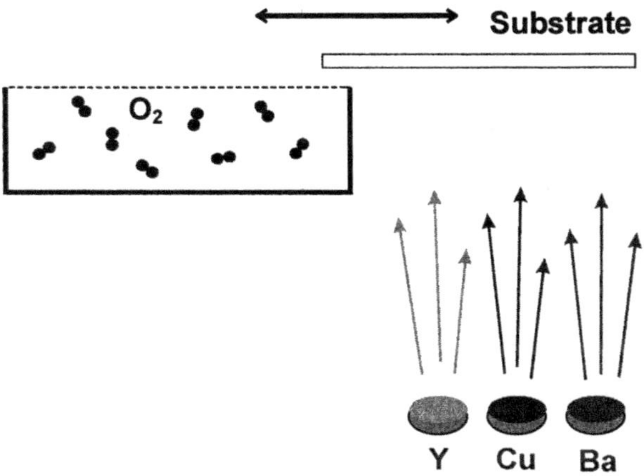

Figure 6.3. Basic principle of the oxidation pocket.

As a main advantage of this arrangement the oxidation conditions can be chosen more or less independently from the vapor deposition conditions and a much wider pressure range becomes accessible.

6.4.2 Turntable Substrate Holder for Wafer Deposition

Since it is hardly possible to move the substrate from one chamber to another on such a short time scale the substrate itself has to be a part of the chamber. Placing the substrate quasi as a lid with a narrow spacing on top of the oxygen pocket the slit around the edge constitutes a high gas flow resistance. The efficiency of such a "slit seal" improves quadratically with decreasing slit width.

The practical solution has been realized in the form of a turntable substrate holder as depicted in Figure 6.4. The substrate is mounted on a plate spinning with a frequency of 5–10 cycles per second. The oxygen pocket is placed closely underneath the substrate and covers a sector of the total area while metal vapors from the sources below have access to the substrate in the open area. By a proper mechanical arrangement the slit width can be maintained between 0.3 and 0.5 mm across a deposition area of 9 inches in diameter, resulting in a pressure drop of three orders of magnitude between the oxygen pocket and the vacuum chamber, cf. Utz (1997a). Consequently, the metals deposited in 10^{-5} mbar background pressure are oxidized in 10^{-2} mbar pure oxygen which suffices to build up the YBCO crystal lattice. Except for the deposition opening the substrate is surrounded by heater filaments and heat shields, which is a quite close approach of the ideal furnace geometry described above.

Present day deposition rates are typically in the range of 20–30 nm/min. However, no efforts have been made yet to speed up the process further.

6.4.3 The Oxygen Shuttle Concept

Although this turntable concept has been proven successful and became the standard technique for film production on single crystal wafers the movement restricts its use to substrates of a certain size. Especially, when thinking of very large rectangular plates or metal tape substrates it becomes immediately clear that they won't allow a spinning motion.

For this reason the principle has been inverted by Kinder (1996). Changing the frame of reference leaves the substrate fixed and moves the oxygen pocket instead.

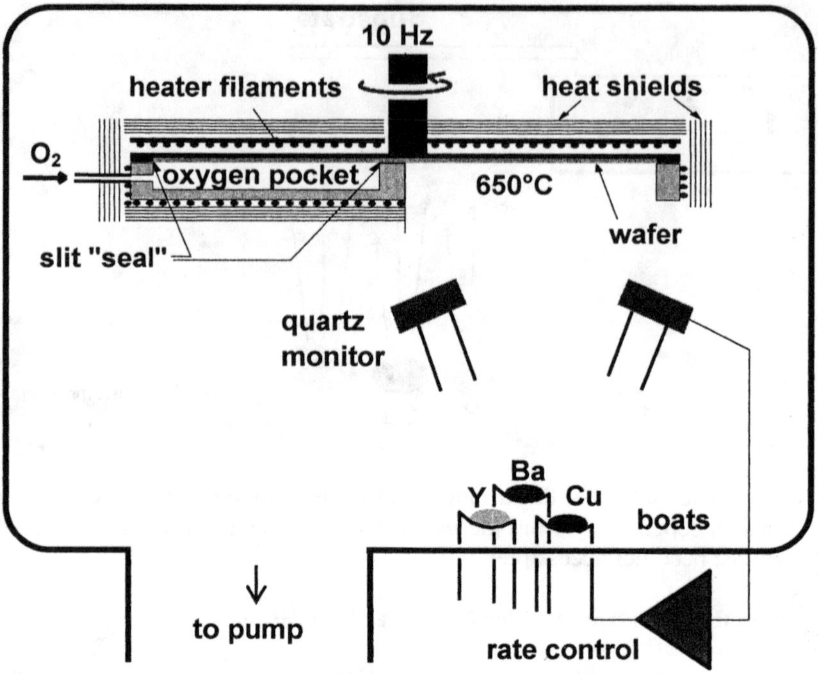

Figure 6.4. Operating principle of the turntable substrate holder.

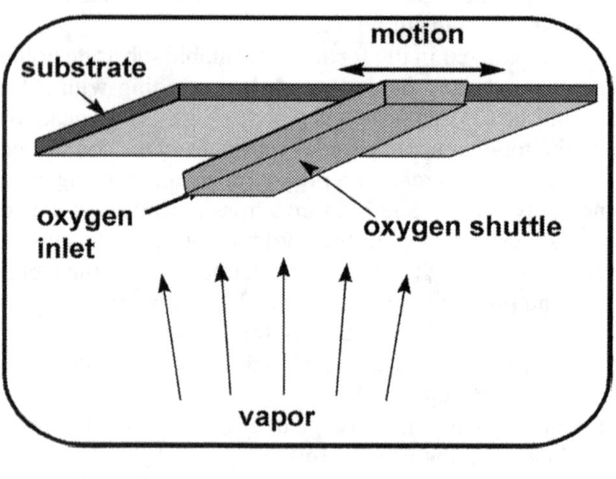

Figure 6.5. Oxygen shuttle arrangement.

Again, the slit seal around the oxygen pocket allows a homogeneous high pressure oxygen shower covering the entire deposition area several times each second. A schematic view is depicted in Figure 6.5 showing the oscillating oxygen shuttle beneath the substrate plate.

It should not be concealed that this kind of arrangement leads to distinct disadvantages compared to the turntable heater. One which is immediately evident is that it will result in a somewhat more open heater configuration deviating from the ideal closed geometry favored above. However, with respect to tape coating this problem is relaxed since the metal substrate is non-transparent and has a high thermal conductivity which improves the temperature uniformity.

In summary, the choice of the heater arrangement largely depends on the application and the involved substrate material. In most cases when wafers have to be coated the turntable heater is certainly the optimum solution. However, when aiming at long lengths tape coating there is no alternative to the oxygen shuttle. Fortunately, some of the disadvantages of this configuration are reduced when using metal substrates.

6.4.4 Long Term Continuous Operation

In recent years several improvements have been achieved to allow a long term operation of the TE deposition.

Since the capacity of the boat sources is limited an in situ refilling scheme is an absolute necessity. A variety of solutions seem feasible in this respect. One of the most elegant ways which does not involve much mechanics inside the vacuum chamber are vibrator conveyors driven by electromagnets as introduced by Utz (1997b). They can be placed at convenient positions and are very reliable standard tools used in industrial processes for hauling granular material. Several TE systems have been successfully equipped with conveyors already.

Another essential item is the rate control. Since the quartz crystal monitors get exhausted, the AAS-based rate monitoring is the only practical alternative to control long term processes like tape coating. All components are situated outside the vacuum chamber, offer long lifetime and easy access for servicing. However, practical long-term tests of even some hours are still lacking.

6.5 TAPE COATING

6.5.1 Substrates

Since large angle grain boundaries are detrimental for the current carrying capacity YBCO films have to grow highly oriented on a macroscopic scale. There are various techniques to achieve a high degree of alignment in the YBCO films by texturing the underlying substrate and buffer layers. They are described in detail in specific chapters of other authors and should only be mentioned here with respect to evaporation.

A nickel or nickel alloy based metal substrate can be textured by mechanical deformation (RABiTS). The subsequent buffer and superconductor layers will then adapt to the long range order of the substrate crystal grains similar to epitaxial film growth on single crystals. By ion beam assisted deposition (IBAD) and inclined substrate deposition (ISD) a buffer layer which is deposited onto a polycrystalline or even amorphous substrate is aligned due to an impinging ion beam or due to the growth kinetics under oblique material deposition.

The RABiTS approach works best with metals like nickel and Ni-alloys. Since the sensible material is easily damaged, handling and the tape transport during film deposition is crucial. Beyond that, most metals form stable oxide surface layers which can prevent the film growth with the desired orientation. Consequently, the chemistry and thermodynamics under deposition conditions play an important role.

The IBAD technique is based on a selective growth of crystal grains with the desired orientation by interaction with an ion beam which is incident on the substrate during deposition. This process is usually very slow. Nonetheless, IBAD is the most advanced alignment technique and still yields the best results for the critical current density comparable to films grown on single crystals. As another advantage IBAD does not depend on the substrate material and can be applied even to stainless steel. Although TE has been demonstrated to yield high quality YBCO films on IBAD buffers by Bauer (1998), the deposition of the IBAD buffer layers itself has been performed by PLD or sputtering.

The ISD technique depends on the growth kinetics at very high deposition rates. PLD as well as evaporation are capable to achieve the required rates. Its great potential originates from the fact that ISD produces textured buffers in a very short time. Although the results still fall behind those achieved by IBAD, it is certainly the most economic process to grow textured buffer layers.

6.5.2 Deposition on Short Samples

We have used evaporation in connection with RABiTS as well as ISD to grow buffer layers and YBCO films on short metal tape substrates up to a length of 12 cm. Results obtained with both techniques are comparable and promising.

Following the ORNL route and adapting it to evaporation, we have deposited a 100 nm thick CeO_2 buffer layer on Ni alloy tape in a two step process, cf. Egly (2000). To reduce the native and randomly aligned nickel oxide on top of the substrate surface the tape is heated in forming gas atmosphere to the deposition temperature of 650°C followed by deposition of a 10 nm thin CeO_2 layer in reducing ambient. The residual 90 nm of CeO_2 are deposited after pumping away the forming gas and switching to an oxygen gas flow. Due to the lower substrate temperature applied for thermal evaporation even this 100 nm thick CeO_2 buffer layer is sufficient to prevent interdiffusion. Consequently, this processing step is shorter and cheaper than the multiple buffer layer deposition usually used for PLD and other techniques requiring higher substrate temperatures. The alignment results for the Ni-substrate and YBCO film are depicted in Figure 6.6.

In the case of ISD buffer layers thermal evaporation employs MgO; cf. Bauer (1999); instead of YSZ which is used for PLD by Hasegawa (1998). The deposition of the textured buffer layer is performed at a rate of about 500 nm/min and takes only a

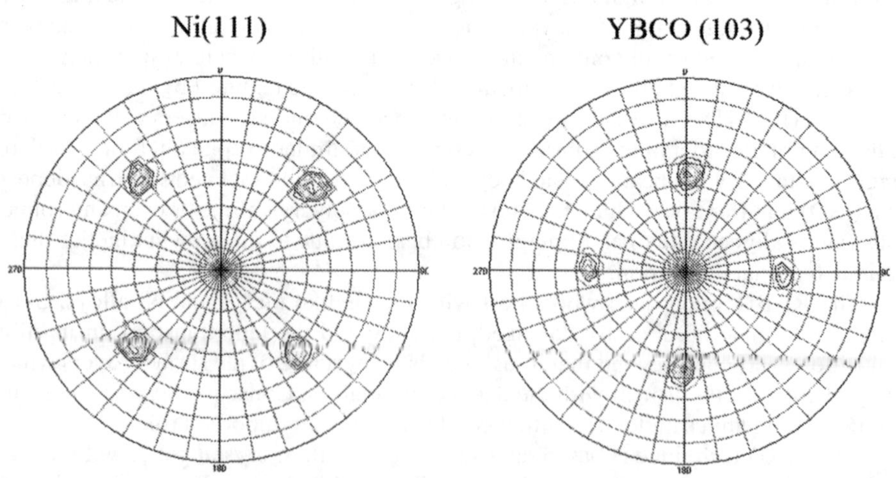

Figure 6.6. XRD pole figures of Ni (111) and YBCO (103).

few minutes. The details and peculiarities of this technique are described in a separate chapter of this book.

The subsequent YBCO film is deposited under standard conditions described above. In situ metallization with gold or silver can be applied to ensure a low resistivity contact to the superconductor.

Our deposition trials on ISD and RABiTS buffers resulted in an in-plane alignment of the YBCO films with a FWHM of the phi-scans or pole figures in the range of 7–10° (cf. Figure 6.6). 500 nm thick YBCO films on short samples (3–10 cm) exhibit transition temperatures between 86–87 K and critical current densities of 0.5–0.8 MA/cm^2 in zero field at 77 K.

A common problem of *ex situ* YBCO annealing techniques, like the BaF_2- or TFA-process, is the degradation of the superconducting properties of thicker films. By our *in-situ* process, we deposited YBCO-films up to 1.4 μm thickness by TE on 12 cm long tape samples. There was no indication of any degradation of the current carrying capacity and current values of 60–100 A in this 1 cm wide tape conductor translated into current densities of 0.5–0.7 MA/cm^2.

6.5.3 YBCO Deposition on Long Length Tape

The short sample results obtained so far were achieved with the turntable heater. Although the substrate holder arrangement cannot be scaled to a length larger than about 20 cm, the processing steps and deposition conditions were developed to be compatible with continuous long length tape deposition. Consequently, the main task to be solved consists of demonstrating a proper configuration for long length tape deposition and an adequate tape transport mechanism.

6.5.4 Tape Transport

It is rather straightforward to employ the oxygen shuttle arrangement for a tape coating apparatus. However, there are different ways to handle the tape transport through the deposition zone. Efficient tape coating should make use of the large area which can be covered simultaneously to achieve a high volume production rate. The first approach would imply the use of an about 20 cm wide metal foil covering the width of the area and a simple linear reel to reel transport. However, since the distance between oxygen shuttle and substrate surface is very narrow and critical, the requirements on the flatness of the substrate are very tough. Internal stresses in the metal can easily result in warping when the metal is heated to the deposition temperature. For this reason, splitting the broad metal sheet into parallel, narrow tape tracks which can be guided at the edges appears to be the better solution.

However, even such an arrangement has its disadvantages. To name only a few, e.g., in a given span of time only relatively short tape pieces can be produced and variations in the temperature profile and the composition across the deposition area will result in different quality of the tape tracks.

To obtain longer tape and to average over process conditions it is advantageous to wind a single tape back and forth to cover the whole deposition area in multiple turns. The principle is sketched in Figure 6.7. However, more work has to be done to make this scheme working.

In summary, thermal evaporation has already been demonstrated to handle the whole tape deposition process from aligned buffer layers to thick YBCO films and

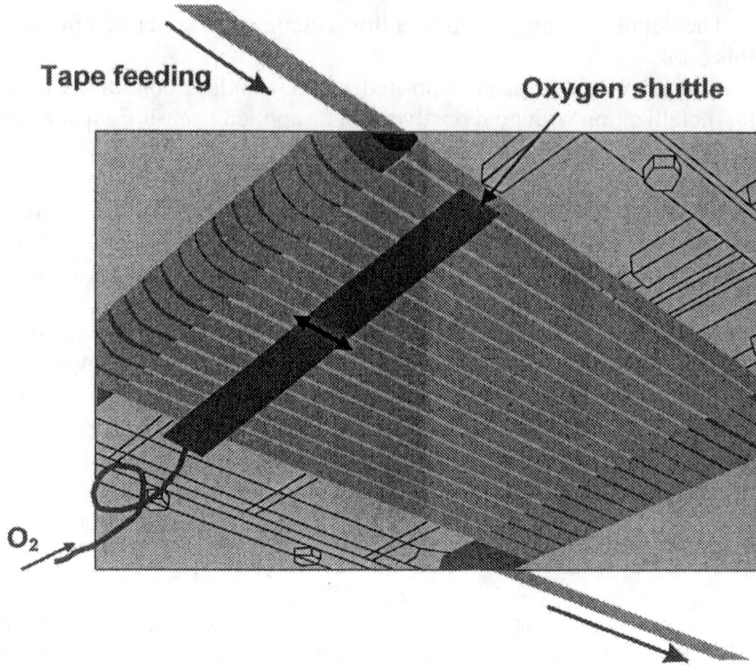

Figure 6.7. Multipass tape transportation in combination with oxygen shuttle.

in situ metal coatings on short samples. The extension to long length continuous deposition of several ten meters is currently under development. Cost calculations indicate that this technique may be a competitive solution. What remains to be done with respect to the mechanical arrangement is the demonstration of feasibility of longer lengths, in conjunction with reliability and long term stability. Yet, the most crucial remaining task is the enhancement of the critical current density to over 1 MA/cm^2 at 77 K for the most economic deposition routes.

REFERENCES

Baudenbacher, F., Karl, H., Berberich, P., and Kinder, H., 1990, RHEED studies of epitaxial growth of YBCO films prepared by thermal co-evaporation, *J. Less Common Met.*, 164&165:269.

Baudenbacher, F., 1997, Oberflächenmorphologie und Wachstumsmodus epitaktischer YBCO-Filme, Ph.D. thesis, Shaker Verlag, Aachen.

Bauer, M., Semerad, R., Kinder, H., Wiesmann, J., Dzick, J., and Freyhardt, H.C., 1998, Large area YBCO films on polycrystalline substrates with very high critical current densities, *IEEE Trans. Appl. Supercond.*, 9:2244.

Bauer, M., Semerad, R., and Kinder, H., 1999, YBCO films on metal substrates with biaxially aligned MgO buffer layers, *IEEE Trans. Appl. Supercond.*, 9:1502.

Berberich, P., Tate, J., Dietsche, W., and Kinder, H., 1989, Low-temperature preparation of superconducting YBCO films on Si, MgO, and $SrTiO_3$ by thermal coevaporation, *Appl. Phys. Lett.*, 53:925.

Berberich, P., Assmann, W., Prusseit, W., Utz, B., and Kinder, H., 1993, Large area deposition of YBCO films by thermal evaporation, *J. Alloys & Comp.*, 195:271.

Berberich, P., Utz, B., Prusseit, W., and Kinder, H., 1994, Homogeneous high quality YBCO films on 3″ and 4″ substrates, *Physica C*, 219:497.

Bozovic I., Eckstein, J.N., Schlom, D.G., and Harris, J.S., Jr., 1990, in: *Science and Technology of Thin Film Superconductors II*, R. McConnell and S. Wolf, eds., Plenum Press, New York, p. 267.

Chang, L.L., Esaki, L., Howard, W.E., and Ludeke, R., 1973, *J. Vac. Sci. Technol.*, 10:11.

Chew, N.G., Goodyear, S.W., Edwards, J.A., Satchell, J.S., Blankinsop, S.E., and Humphreys, R.G., 1990, Effect of small changes in composition on the electrical and structural properties of YBCO thin films, *Appl. Phys. Lett.*, 57:2016.

Eckstein, J.N., Schlom, D.G., Hellman, E.S., von Dessonneck, K.E., Chen, Z.J., Webb, C., Turner, F., and Harris, J.S., Jr., 1989, Epitaxial growth of high temperature superconducting thin films, *J. Vac. Sci. Technol. B*, 7:319.

Eckstein, J.N., Bozovic, I., von Dessonneck, K.E., Schlom, D.G., Harris, J.S., Jr., and Baumann, S.M., 1990, Atomically layered heteroepitaxial growth of single crystal films of superconducting BSCCO, *Appl. Phys. Lett.*, 57:931.

Eckstein, J.N., Bozovic, I., Klausmeier-Brown, M.E., Virshup, G.F., and Ralls, K.S., 1992, Atomically layered growth and properties of high temperature superconducting single-crystal films and superlattices, *Thin Solid Films*, 216:8.

Eckstein, J.N., Bozovic, I., Klausmeier-Brown, M.E., Virshup, G.F., and Ralls, K.S., 1992, Control of composition and microstructure in high temperature superconductors at the atomic level by molecular beam epitaxy, *MRS-Bulletin*, 17(8):27.

Egly, J., Nemetschek, R., Prusseit, W., Holzapfel, B., and DeBoer, B., 2000, YBCO-deposition on metal tape substrates, *Proceedings of the EUCAS'99*, IOP, London.

Hasegawa, K., Fujino, K., Mukai, H., Konishi, M., Hayashi, K., Sato, K., Honjo, S., Sato, Y., Ishii, H., and Iwata, Y., 1998, Biaxially aligned YBCO film tapes fabricated by all pulsed laser deposition, in: *Applied Superconductivity*, Vol. 4, Elsevier Science, Amsterdam, p. 487.

Kamei, M., Aoki, Y., Usui, T., and Morishita, T., 1992, *In situ* X-ray chemical analysis of YBCO films by RHEED-TRAXS, *Jpn. J. Appl. Phys.*, 31:1326.

Kinder, H., Semerad, R., Berberich, P., Utz, B., and Prusseit, W., 1996, Very large area YBCO film deposition, *Proc. SPIE*, 2697:154.

Klausmeier-Brown, M.E., Eckstein, J.N., Bozovic, I., and Virshup, G.F., 1992, Accurate measurement of the atomic beam flux by pseudo-double-beam atomic absorption spectroscopy for growth of thin-film oxide superconductors, *Appl. Phys. Lett.*, 60:657.

Kwo, J., Hong, M., Trevor, D.J., Fleming, R.M., White, A.E., Mannaerts, J.P., Farrow, R.C., Kortan, A.R., and Short, K.T., 1989, *In situ* growth of YBCO films by molecular beam epitaxy with an activated oxygen source, *Physica C*, 162–164:623.

Lu, C., Lightner, M.J., and Gogal, C.A., 1977, Rate controlling and composition analysis of alloy deposition processes by electron impact emission spectroscopy, *J. Vac. Sci. Technol.*, 14:103.

Lu, C. and Guan, Y., 1995, Improved method of nonintrusive deposition rate monitoring by atomic absorption spectroscopy for physical vapor deposition processes, *J. Vac. Sci. Technol. A*, 13:1797.

Naito, M., 1991, A study of the compositional dependence of the quality of *in situ* grown YBCO films in e-beam coevaporation, *Physica C*, 185–189:1977.

Naito, M. and Sato, H., 1995, Stoichiometry control of atomic fluxes by precipitated impurity phase detection in growth of $(Pr,Ce)_2CuO_4$ and $(La,Sr)_2CuO_4$, *Appl. Phys. Lett.*, 67:2557.

Prakash, S., Umarjee, D.M., Doerr, H.J., Deshpandey, C.V., and Bunshah, R.F., 1989, Superconducting films grown *in situ* by the activated reactive evaporation process, *Appl. Phys. Lett.*, 55:504.

Prakash, S., Chou, K., Potwin, G., Deshprandey, C.V., Doerr, H.J., and Bunshah, R.F., 1990, Superconducting films grown by activated reactive evaporation for high frequency device applications, *Supercond. Sci. Technol.*, 3:543.

Sato, H., Naito, M., and Yamamoto, H., 1997, Superconducting thin films of La_2CuO_4 by oxygen doping using ozone, *Physica C*, 280:178.

Shinohara, K., Matijasevic, V., Rosenthal, P.A., Marshall, A.F., Hammond, R.H., and Beasley, M.R., 1991, Anomalous compositional dependence in *in situ* growth of YBCO films at low oxygen pressure, *Appl. Phys. Lett.*, 58:756.

Terashima, T., Bando, Y., Iijima, K., Yamamoto, K., and Hirata, K., 1988, Epitaxial growth of YBCO thin films on (110) $SrTiO_3$ single crystals by activated reactive evaporation, *Appl. Phys. Lett.*, 53:2232.

Terashima, T., Iijima, K., Yamamoto, K., Bando, Y., and Mazaki, H., 1988, Single crystal YBCO thin films by activated reactive evaporation, *Jpn. J. Appl. Phys.*, 27:L91.

Terashima, T., Bando, Y., Iijima, K., Yamamoto, K., Hirata, K., Hayashi, K., Kamigaki, K., and Terauchi, H., 1990, RHEED oscillations during epitaxial growth of high temperature superconducting oxides, *Phys. Rev. Lett.*, 65:2684.

Utz, B., Prusseit, W., and Kinder, H., 1995, Epitaxial $Ba_{1-x}K_xBiO_3$ thin films on various substrates, in: *Applied Superconductivity 1995*, Conference series No. 148, D. Dew-Hughes, ed., IOP Publishing, Bristol, p. 823.

Utz, B., Semerad, R., Bauer, M., Prusseit, W., Berberich, P., and Kinder, H., 1997a, Deposition of YBCO and NdBCO films on areas of 9 inches in diameter, *IEEE Trans. Appl. Supercond.*, 7:1272.

Utz, B., Rieder-Zecha, S., and Kinder, H., 1997b, Continuous YBCO film deposition by optically controlled reactive thermal co-evaporation, *IEEE Trans. Appl. Supercond.*, 7:1181.

Chapter 7

SPUTTERING OF $Y_1Ba_2Cu_3O_{7-\delta}$

R. Krupke*, M. Azoulay, and G. Deutscher

School of Physics and Astronomy, Raymond and Beverly
Sackler Faculty of Exact Science, Tel Aviv University
69978 Tel Aviv
Israel

7.1 THE SPUTTERING-TECHNIQUE

7.1.1 DC Sputtering

Cathodic sputtering is a term which describes the erosion of a target due to its bombardment with positive ions under reduced pressure. Out of the many ways to produce positive ions only two are relevant for this review, the self-sustained and the non-self-sustained gas discharge.

The basic principle of the self-sustained gas discharge is to attach a DC power supply between the conducting target and the target shielding. The polarity is chosen such, that the target is the cathode and the shielding the anode. Above a specific voltage V_z, a plasma is generated. Positive ions, generated within the plasma are accelerated in the cathodic dark space towards the target. On impact, particles are sputtered off the target surface either by direct impact or by impact cascades, depending on the ion momentum and the target material (Behrisch, 1964).

The threshold energy for sputtering is several tens of eV. The particle flux sputtered off a metallic target consists of 95% neutral particles, the rest are positive and negative ions.

In addition secondary electrons are generated by direct momentum transfer or neutralization of the approaching ions (Auger-effect). Those electrons are accelerated away from the target and generate new positive ions by impact ionization.

To start and operate a self sustained gas discharge it is necessary to apply a significantly higher voltage than the threshold energy. The ignition voltage V_z is a function of gas pressure p, distance between the electrodes d, ionization energy E_i and secondary

* Since 04/2000 at the Institut für Nanotechnologie, Forschungszentrum Karlsruhe, Postfach 3640, 76021 Karlsruhe, Germany. E-mail: krupke@int.fzk.de

electron emission coefficient γ (Konuma, 1992). For most gases the minimum discharge voltage is several hundreds of volt at $p \cdot d \approx 1$ Torr·cm. V_Z increases at higher and lower pressure (Paschen curve) (Engel, 1965). At lower pressure, the mean free path of electrons increases and the probability of collision with gas molecules before reaching the anode drops. At higher pressure the mean free paths shortens and the electrons do not gain enough energy from the electrical field for the ionization of gas molecules.

7.1.2 RF Sputtering

DC sputtering is applicable only if the target is a conducting material. In the case of an insulating target, radio frequency (RF) sputtering has to be applied. The trick is to use a high-frequency potential at the target whereby the positive charge, which accumulates on the target, is periodically neutralised with plasma electrons during a portion of a cycle. The potential will self-bias negatively at an amount V_B, since the electron mobility is much higher than the ion mobility and no net current can flow through the dielectric target (Andersen et al., 1962).

When sputtering an oxide, 20–30% of the emitted particles are negatively charged. Those negatively charged particles, mainly oxides, are accelerated away from the target and can cause heavy bombardment of the growing film (Shintani et al., 1975).

While simple oxides are rather insensitive to such a bombardment, it is of crucial importance to avoid a bombardment of the growing $Y_1Ba_2Cu_3O_{7-\delta}$ layer, otherwise T_c and other properties will be compromised.

There exist several ways to reduce negative ion bombardment. One option is to increase the operating pressure P at a given distance target to substrate, d, to insure thermalisation of the fast particles (Westwood, 1988). Multiple gas collisions occur, if the mean free path l is significantly smaller than the distance target to substrate, with $l \sim P^{-1}$. One may as well increase the distance target to substrate at a given operating pressure. Both adjustments however do have an impact on the deposition rate r, with $r \approx P^{-1}$ and $r \sim d^{-2}$. An optimisation is therefore necessary to satisfy the specific needs.

An easier starting point to reduce bombardment is the use of magnetron sputtering.

7.1.3 Magnetron Sputtering

7.1.3.1 Planar Magnetron Sputtering Magnetron sputtering is a technique, where the plasma is confined by a strong magnetic field. As a result the plasma volume and surface are strongly reduced and so is the operating voltage. A low operating voltage reduces the kinetic energy of negative ions and therefore the bombardment. Magnetron sputter guns can be equally operated with DC and RF power supplies. The most simple magnetron sputter gun consists out of a planar target, hooked onto a ring and button magnet. The magnetic field lines confine the plasma in a torus due to Lorentz forces. Sputtering occurs therefore only on a rather small area of the target (Figure 7.1). Also the heat dissipation is large in this regions. If the heat conductivity of the target is low, this has to be taken into account by using a sufficient small operating power in order to avoid arcing. The inhomogeneous erosion of the target can be an obstacle.

7.1.3.2 Inverted Cylindrical Magnetron Sputtering The inverted cylindrical magnetron sputtering device (ICM) is shown schematically in (Figure 7.2). The sput-

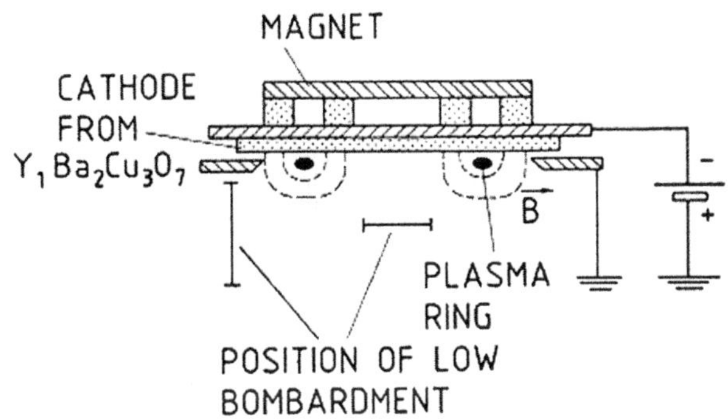

Figure 7.1. Planar Magnetron Sputtering (Geerk et al., 1989). The left substrate position is off-axis position.

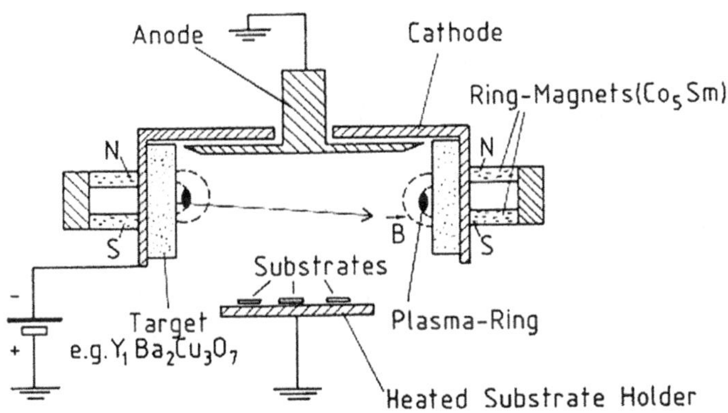

Figure 7.2. Inverted Cylindrical Magnetron (Geerk et al., 1989).

tering target is a hollow cylinder pressed into the cathode. The attached magnets confine the plasma in a torus similar to the planar magnetron device. The cylindrical configuration has the advantage that potentially existing high energetic negative ions do contribute to the sputtering of the opposed target area without damaging the growing film which has to be mounted on a so called on-axis position.

7.1.3.3 Electron Cyclotron Resonance Supported Sputtering Low operating voltages can also be achieved by electron cyclotron resonance supported sputtering. In that technique the plasma is generated by an electron cyclotron resonance independently from the target voltage (Figure 7.3). This allows the dc power supply at the targets to be operated over a wide range. Sputtering at a voltage just above the threshold value is possible. Low voltage sputtering reduces considerably the amount of negative ion bombardment. This method is however restricted to conducting targets only.

7.1.4 On-Axis Geometry

In general the substrate is positioned opposite to the target, since in the so called on-axis geometry the deposition rate is usually the largest. Unfortunately also the negative ion bombardment is at this position at a maximum, since the negative ions are accelerated away from the target, parallel to the macroscopic target surface normal.

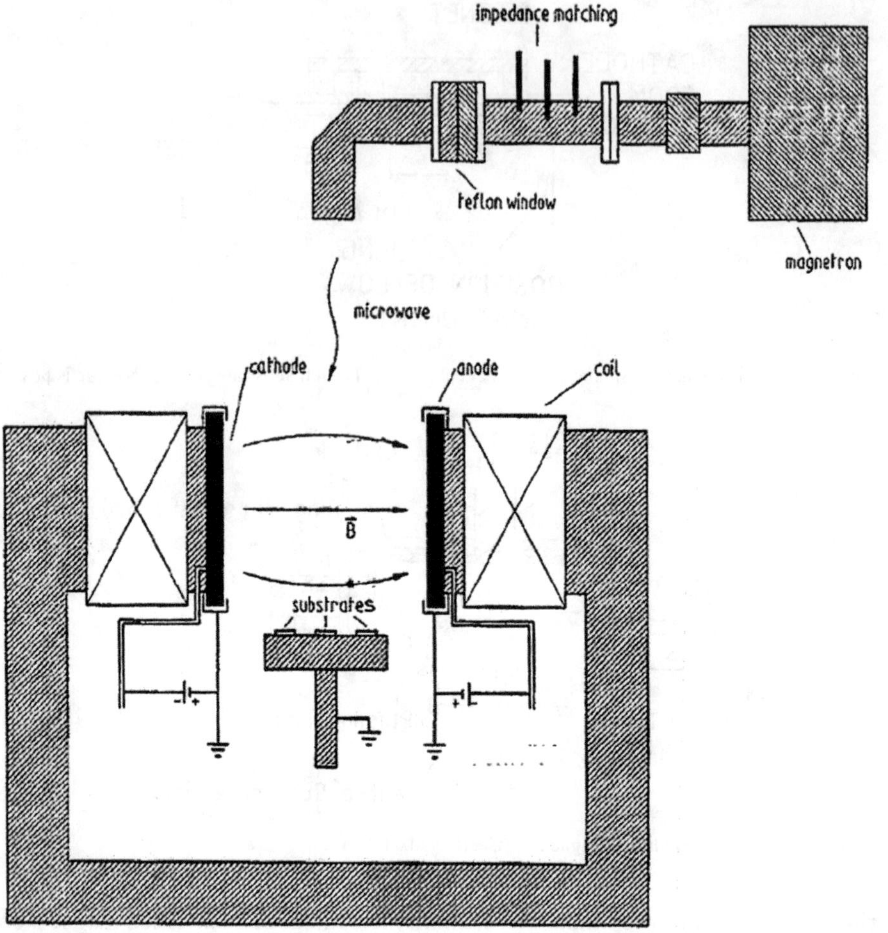

Figure 7.3. Schematic cross-sectional view of an ECR supported sputtering device with two facin planar targets and off-axis substrate placement showing also the magnet coils arrangement and the microwave supply (Krupke et al., 1997a).

However sputtering at higher pressure reduces the bombardment due to thermalisation by enhanced scattering (Poppe et al., 1988; Krüger et al., 1993).

7.1.5 Off-Axis Geometry

Moving the substrate in the off-axis geometry avoids this bombardment (Figure 7.1, left substrate position) (Eom et al., 1989). The inverted cylindrical magnetron sputtering is a device, where the substrates are intrinsically placed in off-axis geometry. Whether the rate is significantly reduced depends on the surface morphology of the target. In fact microscopic surface roughness can lead to an undercosine angular rate distribution, which maximises the rate at the off-axis position (Krupke et al., 1997b).

7.2 TARGET MATERIAL

$Y_1Ba_2Cu_3O_{7-\delta}$ can be grown by sputtering in many ways. For instance metallic targets and alloys can be used with oxygen added to the sputtering gas. Sputtering several oxide targets is possible as well. However a major problem with these concepts is that the material flux of each sputter gun and the gas composition have to be adjusted very carefully, otherwise the growing films are not stochiometric.

Rather expensive and drift sensitive feedback systems can be avoided if a single, stochiometric target is used. Although the sputter coefficients are different for each element, the material flux is stochiometric after some time of presputtering. This is due to a compensation of the larger sputter coefficient by a surface depletion.

7.3 SUBSTRATES

Standard substrates for the growth of *c*-axis oriented $Y_1Ba_2Cu_3O_{7-\delta}$ are perovskites such as $LaAlO_3$, $SrTiO_3$ or $(Y)ZrO_2$. For technical applications however Al_2O_3 and Si are of great importance. For instance Sapphire has very low tangent losses which makes it a good substrate for microwave devices, whereas Si is the base of the semiconducting industry and therefore unavoidable for the integration of HTS electronics. Of importance is also the availability of large wavers and their low cost, which is given for Si and Sapphire (Table 7.1). Unfortunately YBCO can not be grown directly onto these technical substrates.

A specific problem with Si as substrate is the easy diffusion of Si atoms into the YBCO thin film, which results in degradation of the superconducting film. In addition the Si lattice does not match well with that of YBCO, resulting in poor crystalline structure of the YBCO thin film. In this case, synthesis of a multilayer YBCO/M/Si offers a solution, where a buffer layer M is used to prevent the diffusion of Si from the substrate into YBCO and to offer a better-matched lattice for epitaxial growth of the YBCO thin film. YSZ is considered one of the most suitable buffers between YBCO thin film and Si substrate, with critical current density $J_c > 10^6$ Acm^{-2} (at 77 K) and transition temperature $T_c = 86$–88 K (Fork et al., 1990; Tian et al., 1999).

Similar problems are encountered with YBCO deposition on sapphire. The lattice mismatch is accompanied by chemical interaction between YBCO and Al_2O_3 taking place at elevated temperatures and yielding the formation of an uncontrollable $BaAl_2O_4$ interfacial layer. Also here the concept of a buffer layer is helpful. Among a large number of candidates MgO, YSZ (ZrO_2 stabilized with ~9 mol% Y_2O_3) and

Table 7.1. Properties of substrates suitable for the preparation of YBCO films. Thermal expansion α for YBCO: $\parallel c = 1 \times 10^{-5}$ K^{-1}, $\perp c = 3 \times 10^{-5}$ K^{-1}

Material	Thermal expansion α [K^{-1}]	Size [mm]	Price [€]
$SrTiO_3$	9×10^{-6}	Ø 50	845
$LaAlO_3$	1×10^{-5}	Ø 50	185
$Y{:}ZrO_2$(9.5% Y)	8×10^{-6}	Ø 50	160
MgO	8×10^{-6}	Ø 50	300
Al_2O_3	$\parallel c = 5.4 \times 10^{-6}$ $\perp c = 6.2 \times 10^{-6}$	Ø 76	90
Si	2.3×10^{-6}	Ø 150	45

CeO_2 are the most attractive candidates. To date, the highest structural quality of YBCO has been obtained on CeO_2 buffer layers with superconducting properties comparable with those of optimised films on structurally well-matched substrates (Wördenweber, 1999).

A serious problem is the crack formation in YBCO films on CeO buffered sapphire due to differences in the thermal expansion between film and substrate (Table 7.1). The issue has been addressed by Zaitsev and co-workers (1997) and the maximum crack free film thickness was raised from 300 nm to 700 nm. So far, the maximum waver size covered with YBCO by sputtering is 3″.

More recently also metallic Ni tapes have been used successfully as substrates (Goyal et al., 1998).

7.4 HEATER

Accurate and homogenous heating of substrates is difficult. The most simple concept is the use of a resistive heater plate onto which the substrate is glued by silver paint. Good thermal contact assures a substrate temperature being close to the heater plate temperature. Although the heater plate temperature can be stabilised by a temperature controller, the substrate temperature may change with time. For instance, if the heater plate is shiny metallic prior to growth, then the emissivity of the heater plate does change during growth, due to the deposition of black $Y_1Ba_2Cu_3O_{7-\delta}$ onto the plate. The enhanced radiation losses are compensated by the temperature controller, however the thermocouple is located within the heater plate, which has a finite heat conductance. Therefore the vertical temperature gradient increases and a drift of the substrate temperature can not be avoided completely. For small samples, glued onto a larger heater plate, the drift can be minimised, if the complete plate except for the substrate area is covered by a black material, either by not cleaning those areas or by covering them with a black material, such as $Y_1Ba_2Cu_3O_{7-\delta}$ covered Al_2O_3 ceramics.

For substrates larger than 1 cm^2, gluing is not practical, since it is difficult to remove the samples from the heater plate without breaking them.

A semi closed radiation furnace with a rotating inner plate is an alternative concept, which avoids gluing or clamping. Although the deposition process is continuous, local areas of the waver experience an interval deposition. Only during the time where such an area is not heated from above, the material deposition takes place. Careful optimisation of the heater temperature, the rotation speed, the plate material and target heater distance are necessary to achieve conditions yielding a high quality film.

For all heater types the combination of high temperature and reactive oxygen environment is problematic. Well-tried materials are Alumina for shieldings, steel for heater plates and gearwheels, and Cr–Ni heating elements sealed in quartz glass tubes.

An unconventional method is gluing the substrate to the heater plate with Indium foil, which becomes liquid above 150°C. Oxidation and evaporation of the indium is avoided by sealing the sample edged with silver paint.

7.5 DEPOSITION PARAMETERS

Deposition parameters of an experimental deposition system have to be optimised either because of the system's uniqueness or because of a desired specific sample property.

Table 7.2. Typical maximum deposition rates for different sputtering types

Type	Planar off-axis sputtering	ECR off-axis sputtering	High pressure on-axis sputtering	ICM on-axis sputtering
Rate in nm s^{-1}	0.03	0.1	0.1	0.2

Some parameters are quite well understood, while others need further clarifications. One example is the complex interaction between target power, oxygen pressure, hydrogenous impurities, which do influence the film morphology, the superconducting transition temperature and the target lifetime. The problem described below is not limited to sputtering only but rather to any deposition technique using ceramic targets $Y_1Ba_2Cu_3O_{7-\delta}$. The problem is triggered by the necessity to maximise the deposition rate such that growths, suitable for applications, can be terminated within a reasonable time (Table 7.2).

7.5.1 Target Lifetime, Film Morphology and Atomic Oxygen

In general, sputtering is favourable for large scale industrial applications because of its low cost components and no need of direct feedback control. After optimising the deposition parameters, films with high critical temperature and high critical current density are routinely grown by sputtering. Yet the reproducibility can be limited. It occasionally happens, that sputtering from a prolonged-time used target produces films with reduced T_c and holes (Figure 7.4(a)). In order to preserve a high film quality, a cost intensive replacement of the target is necessary, before the target lifetime is reached. Detailed investigations revealed, that the degraded film properties are correlated with an Yttrium rich film composition (Krupke et al., 1999), triggered by a reduced oxygen content of the target. Holes result from the nucleation of Y_2O_3, onto which YBCO does not grow (Figure 7.5).

The impact of a low atomic oxygen pressure is a reduced oxidation of the sputtered elements Barium and Copper (Lecœr et al., 1995), which leads to a reduced sticking coefficient of these elements and hence an excess of Yttrium.

A loss of target oxygen does not occur, if the target surface is in dynamic equilibrium with a sufficiently high atomic oxygen produced in the sputtering plasma.

In a gas discharge, the dissociation and recombination rates of molecular oxygen are extremely sensitive to nitrogenous or hydrogenous impurities (Kaufmann and Kelso, 1960; Costa et al., 1979). For instance each introduced hydrogen atom produces about 300 extra oxygen atoms. In a sputtering system, the amount of atomic oxygen is usually not under direct control, since the inserted sputter gas is a mixture of argon and molecular oxygen. However, a deliberate supply of hydrogenous impurities can raise the atomic oxygen pressure to such an extent, that the target surface is stabilised.

A simple way to add such impurities to the sputter gas is to evaporate H_2O molecules into the sputtering system (Gavaler et al., 1991). A few milliTorrs H_2O added to the Ar/O_2 mixture as a catalyst is sufficient to avoid hole formation and excess yttrium (Figure 7.4(b)).

The impact of the catalyst on the target surface can be observed directly by monitoring the DC-bias. As mentioned before, the DC-bias V_{DC} is a voltage, which builds itself up in a rf-sputtering system and is sensitive to many sputter parameters including the target surface. It is therefore useful to verify identical conditions. Any deviation

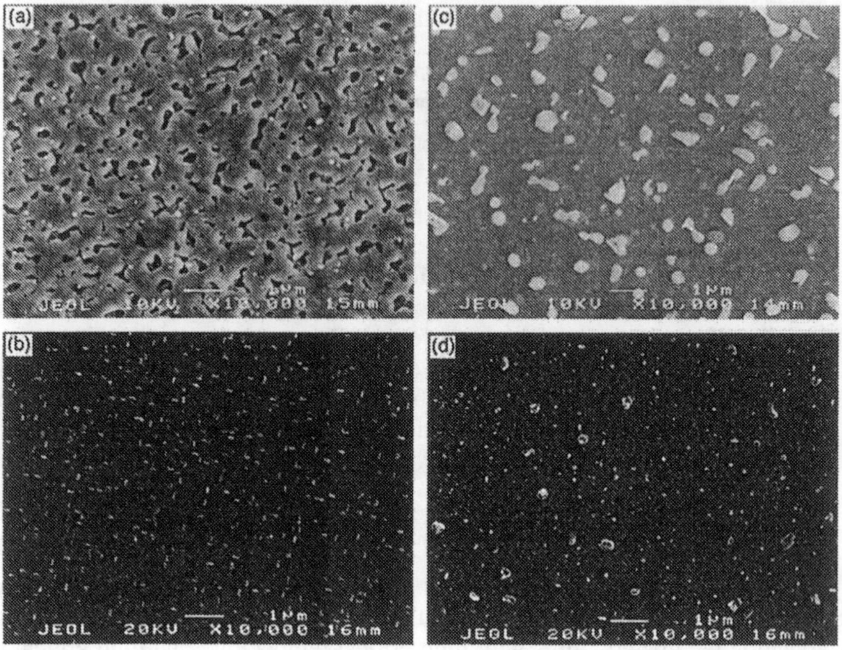

Figure 7.4. Scanning electron microscope pictures showing the surface morphology of YBCO films grown under different conditions: (a) films with holes and $T_c = 84$ K grown with a prolonged used target; (b) film without holes and $T_c = 85$ K after optimisation of rf-power and H_2O-pressure; (c) film with CuO-particles and $T_c = 90$ K grown at high oxygen pressure; (d) film without CuO-particles or holes and $T_c = 90$ K grown with the PST process.

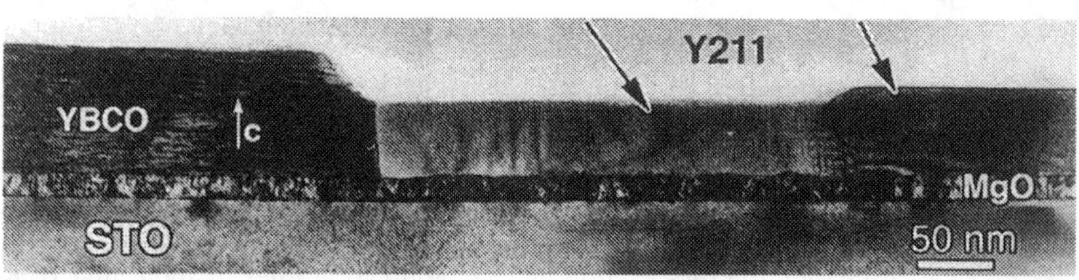

Figure 7.5. Y_2O_3 precipitates (indicated by the left arrow) embedded in the YBCO film (indicated by the right arrow). The YBCO film does not cover the precipitates (Uccio, 1999).

from its original value signals a drift of at least one parameter, including oxygen loss of the target. An oxygen depletion of the target surface results in a lower DC-bias. This can be understood since it is known that loss of oxygen does drive $Y_1Ba_2Cu_3O_{7-\delta}$ from the metallic state into the insulating state. Furthermore it is also known, that an insulator has in general a greater secondary electron emission coefficient than a metal, which gives a smaller DC-bias for insulating targets than for metallic targets.

Indeed when sputtering a prolonged-time used target in pure Ar/O_2 gas, a decreasing voltage is observed. On the contrary, the presence of the catalyst H_2O stabilises V_{DC} (Figure 7.6).

The specific amount of atomic oxygen (or catalyst), necessary to stabilise the $Y_1Ba_2Cu_3O_{7-\delta}$ target surface, depends on the sputtering power. The higher the power,

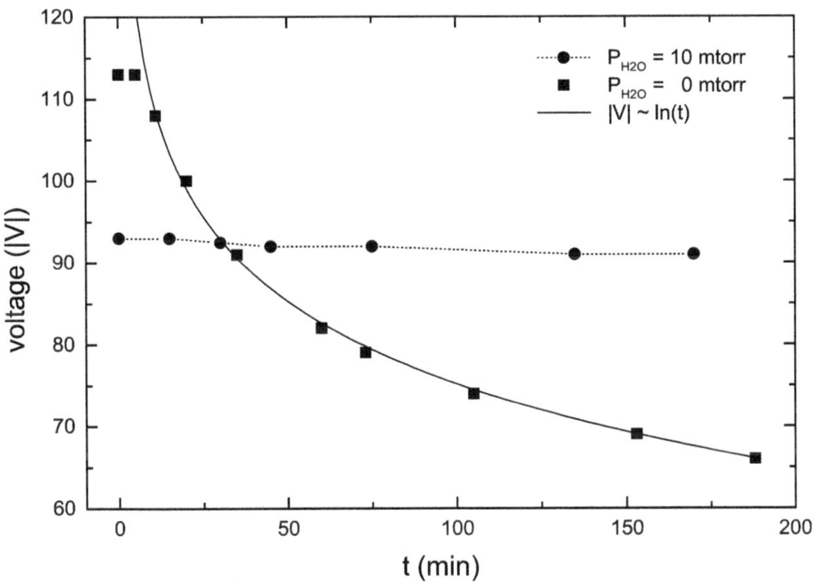

Figure 7.6. Bias voltage vs. time, with and without H_2O additive. RF-power was set to 175 W. The data obtained without H_2O additive is fitted to a logarithm (—).

the hotter the target surface and the higher the internal oxygen pressure and diffusivity. Using a magnetron cathode, V_{DC} can be represented as a first approximation by $V_{DC} = P_{rf}/(P_0 + P_{rf})$, reaching a constant value for $P_{rf} > P_0$, P_0 is a constant. This is not observed, when sputtering $Y_1Ba_2Cu_3O_{7-\delta}$. Instead V_{DC} decreases monotonously with higher power. This behaviour is very pronounced for sputtering without H_2O. With increasing H_2O pressure the curve is getting more flat, approaching the expected behaviour.

In summary, the target stability is important for the growth of high quality thin films hence the atomic oxygen pressure has to be matched with the sputtering power.

7.5.2 Avoiding Outgrowths—Pressure Template

Outgrowths in $Y_1Ba_2Cu_3O_{7-\delta}$ films, such as CuO particles, are obstacles for multilayer structures and lithography. Yet no negative effect on T_c or J_c has been reported. It was found, that the amount of the secondary phase CuO as well as T_c depend on the molecular oxygen pressure P_{O_2} used during growth (Figure 7.7). Films grown at low oxygen pressure close to the stability line of $Y_1Ba_2Cu_3O_{7-\delta}$ have fewer CuO outgrowths, but also a reduced T_c (Figure 7.8). On the other hand, close to the stability line of CuO, the density of CuO outgrowths increases as does T_c.

As a new strategy, the pressure template process was introduced, where the initial growth of $Y_1Ba_2Cu_3O_{7-\delta}$ starts at low pressure, in order to avoid CuO particle formation (Krupke et al., 1999). After the formation of a few monolayers of $Y_1Ba_2Cu_3O_{7-\delta}$, the pressure is raised into the high T_c regime. $Y_1Ba_2Cu_3O_{7-\delta}$ films free of CuO particles can as well be grown if quasi-equilibrium conditions are chosen, which is the case at ultra-low deposition rates (Tazoh and Miyazawa, 1993). The pressure template process is however less time consuming and therefore cheaper.

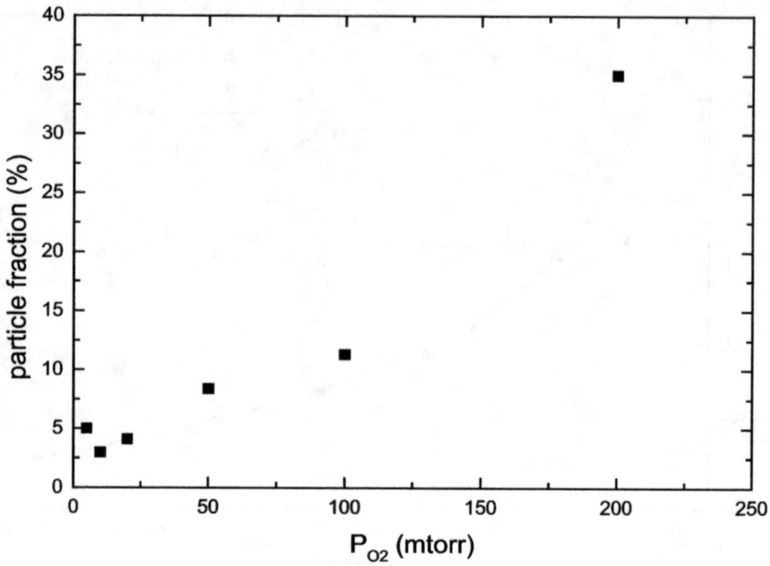

Figure 7.7. Particle fraction vs. Oxygen pressure P_{O_2}.

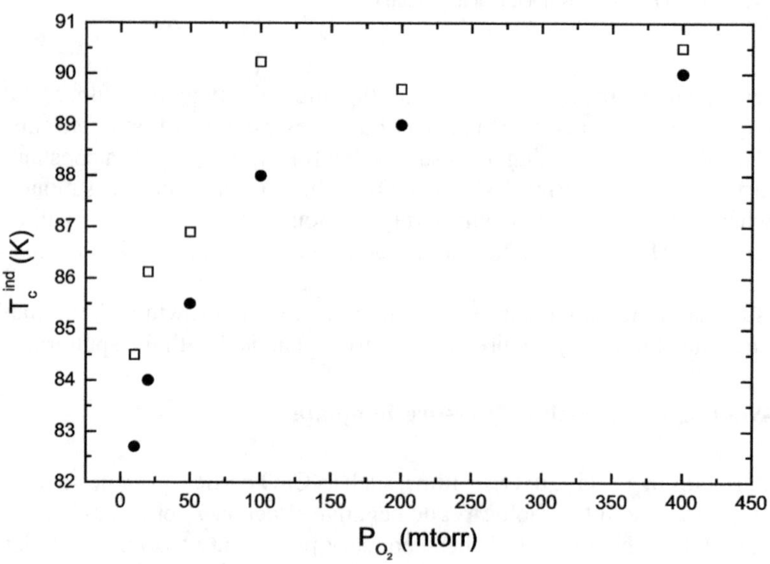

Figure 7.8. Onset (□) and zero-point (•) of the inductively measured superconducting transition, T_c^{ind} vs. the oxygen pressure $P_{O_2} \cdot T_{heater} = 750°C$.

7.5.3 Phase Diagram

The equilibrium phase diagram is used as a reference for growth conditions, although thin film growth at standard rates, is a non-equilibrium process. Still the phase diagram works as far as it correctly predicts that YBCO can not be grown below the decomposition line (Figure 7.9). Yet YBCO films grown close to the stability line, are inferior, and high quality samples are only grown at much higher pressure, close to the

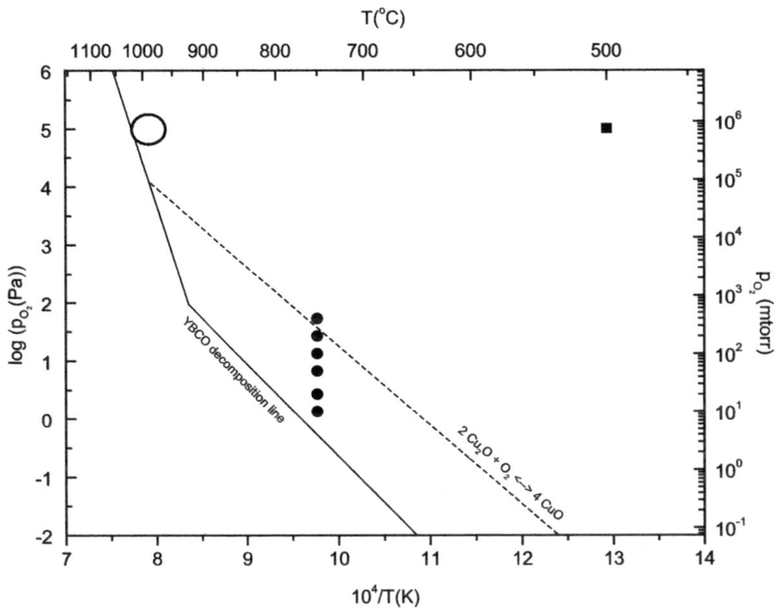

Figure 7.9. Growth conditions for YBCO single crystals (○) and films (•) plotted in a log P_{O_2} vs. $1/T$ diagram, included are the annealing condition (—), and the thermodynamic stability of the YBCO-123-phase (...) and CuO (Lindemer et al., 1991). The T_c of the films is shown in Figure 7.4.

Table 7.3. Temperature–pressure correlation for the growth of YBCO at the stability line of CuO

T_{Heater} [°C]	710	730	750	770
P [mtorr]	80	150	280	500

stability line of CuO single crystals. From Figure 7.8 and Table 7.3 it is obvious, that an accurate control of temperature and pressure is important. For instance, to obtain a superconducting transition width of ~0.1 K requires a uniformity in temperature of ~1 K with a pressure variation smaller than 10 mtorr.

The origin of the enhanced CuO formation on perowskite substrates is obviously a high density of CuO nucleation sites on the substrate surface.

7.5.4 Annealing or Quenching

It was widely accepted, that $Y_1Ba_2Cu_3O_{7-\delta}$ films grow at the limit of the tetragonal phase stability ($O_{6.0\text{–}6.1}$) during *in situ* formation, and that the complete (O_7) oxygenation takes place only during the cool down in an oxidizing atmosphere (Hammond and Borman, 1989). However, in 1997 Lopez and co-workers have shown that $Y_1Ba_2Cu_3O_{7-\delta}$ films do grow fully oxygenated. It turns out, that the high oxygen content can only be preserved, if films are cooled down much faster than the oxygen out or in-diffusion kinetics. Otherwise the films have to be reoxidised at low temperature in an oxidizing atmosphere. Although both procedures yield films with similar superconducting transition temperatures, the microscopic structure of the films is different. It seems, as if the surface resistance of quenched $Y_1Ba_2Cu_3O_{7-\delta}$ is somewhat increased, yet the cool down is fast and therefore competitive.

ACKNOWLEDGMENTS

This work was partially supported by the Heinrich Hertz-Minerva Center for High Temperature Superconductivity and by the Oren Family chair for Experimental Solid State Physics.

REFERENCES

Andersen, G.S., Mayer, Wm.N., and Wehner, G.K., 1962, *J. Appl. Phys.*, 33:2991.

Behrisch, R., 1964, *Festkoerperzerstaeubung durch Ionenbeschuss aus Ergebnisse der exakten Naturwissenschaften*, Band 35, Springer-Verlag.

Costa, M.D., Zuliani, P.A., and Deckers, J.M., 1979, *Can. J. Chem.*, 57:568.

Eom, C.B. et al., 1989, *Appl. Phys. Lett.*, 55:595.

Fork, D.K. et al., 1990, *Appl. Phys. Lett.*, 57:1161.

Gavaler, J.R. et al., 1991, *J. Appl. Phys.*, 70:8.

Geerk, J., Linker, G., and Meyer, O., 1989, *Mat. Sci. Reports*, 4:193.

Goyal, A. et al., 1998, *Physica C*, 302:87.

Hammond, R.H. and Borman, R., 1989, *Physica C*, 162–164:703.

Kaufmann, F. and Kelso, J.L., 1960, *Chem. Phys.*, 32:301.

Konuma, M., 1992, *Film Deposition by Plasma Techniques*, Springer-Verlag.

Krüger, U., Kutzner, R., and Wördenweber, R., 1993, *IEEE Trans. Appl. Supercond.*, 3:1687.

Krupke, R. et al., 1997a, *Physica C*, 279:153.

Krupke, R., Barkay, Z., and Deutscher, G., 1997b, *Physica C*, 289:146.

Krupke, R., Barkay, Z., and Deutscher, G., 1999, *Physica C*, 315:99.

Lecœr, Ph., Mercey, B., and Murray, H., 1995, *J. Appl. Phys.*, 78:1247.

Lindemer et al., 1991, *Physica C*, 178:93.

Lopez, J.G. et al., 1997, *Physica C*, 275:65.

Poppe, U. et al., 1988, *Solid State Commun.*, 66:661.

Scotti di Uccio, U., 1999, *Physica C*, 321:162.

Shintani, Y. et al., 1975, *J. Appl. Phys.*, 14:1875.

Tazoh, Y. and Miyazawa, S., 1993, *Appl. Phys. Lett.*, 62:408.

Tian, J.Y. et al., 1999, *Appl. Phys. Lett.*, 74:1302.

von Engel, A., 1965, *Ionized Gases*, 2nd edition, Clarendon, Oxford.

Westwood, N.D., 1988, *MRS Bull.*, 13:47.

Wördenweber, R., 1999, *Supercond. Sci. Technol.*, 12:R86.

Chapter 8

PULSED ELECTRON-BEAM DEPOSITION OF HIGH TEMPERATURE SUPERCONDUCTING FILMS FOR COATED CONDUCTOR APPLICATIONS

K.S. Harshavardhan and M. Strikovski

Neocera, Inc.
10000 Virginia Manor Road
Beltsville, MD 20705
USA

8.1 INTRODUCTION

In recent years, significant progress has been made towards the development of 'Coated Conductors' based on High-Temperature Superconducting (HTS) films (Coated Conductor Technology Development Roadmap, 2001; Foltyn et al., 2001; Iijima et al., 2001; Usoskin et al., 2001; Rupich et al., 2001; Solovyov et al., 2001; Paranthaman et al., 2001). A critical component in the coated conductor development is the deposition technique and methodology chosen for the HTS film depositions. Several vapor deposition methods are currently under use. These include physical vapor deposition techniques such as Pulsed Laser Deposition (Foltyn et al., 2001), Co-evaporation by e-beams (Kinder et al., 1997), Sputtering (De Winter et al., 2001) and Chemical Solution deposition techniques using trifluoroacetate precursor approach (Paranthaman et al., 2001; McIntyre et al., 1995; Smith et al., 1998). Of all the deposition methods used, Pulsed Laser Deposition (PLD) has been the front-runner in its ability to achieve a high HTS materials quality (Foltyn et al., 2001; Usoskin et al., 2001; De Winter et al., 2001). The 'pulsed' nature of the deposition technique, leading to a very high degree of stoichiometric fidelity in the deposited films, has been the most significant feature of this technique, lending to the excellent materials performance. Using PLD, a critical current density (J_c) of $\sim$2–3 $\times$ 10^6 A/cm^2 at 77 K with critical currents exceeding 200 A/cm have been achieved in thick (2–3 micron) YBCO films (Foltyn et al., 2001; Usoskin et al., 2001). An important milestone is the demonstrated potential in depositing these high quality films ($J_c > 10^6$ A/cm^2) with high throughputs (Foltyn et al., 1999).

Eventhough critical milestones have been achieved in the coated conductor materials research by Pulsed Laser Deposition (PLD), from the view point of a large scale commercial production of these HTS tapes, severe cost constraints impede the effectiveness of this technique. In order to obtain high throughputs (~100's of kilometer lengths/year or more), industrial pulsed excimer lasers with average powers in the range of 200–300 W (with a pulse repetition rate of ~300 Hz) are required. The deposition equipment may incorporate several of such lasers to meet the throughput requirements. The capital equipment costs and costs of ownership associated with the industrial lasers alone is sufficiently high, approaching $0.75–1.0 Million per each industrial laser capable of delivering about 200–300 W average power. To meet the needs of utility industry (requiring 1000's of kilometers of HTS cable) multiple $100 Millions in investments are therefore required by the private industry. With these high cost figures it may be very difficult to achieve an economically viable HTS cable which is required to meet a cost figures around $30–50/kA-m, in the immediate future (Coated Conductor Technology Development Roadmap, 2001).

An alternative method that could retain the benefits of the pulsed nature of the deposition process and at the same time is cost effective would have a significant impact on the development of coated conductors and is expected to be readily accepted by the industry. Pulsed Electron-beam Deposition has the potential to become such a technique.

This chapter has two main components. Pulsed Electron-beam Deposition (PED) technique is introduced and described in the first part to the extent that it provides the basic understanding of the working principle of the technique as well as some basic differences between PED and PLD. In the second part, data obtained on HTS films are presented establishing the effectiveness of this novel technique for coated conductor fabrication.

8.2 PULSED ENERGY DEPOSITION TECHNIQUES

As multi-component materials such as high temperature superconductors and a variety of complex metal–oxides continue to occupy a significant portion of the technologically important materials spectrum, the need for a suitable deposition technique enabling stoichiometric deposition of the films becomes obvious. Pulsed energy deposition techniques such as Pulsed laser deposition (PLD) accomplish this objective successfully.

Pulsed energy techniques facilitate unique materials processing methodologies. The short pulse length (~10 ns) facilitates a high power density at the target surface leading to a decoupling of the thermo-physical properties of the individual elements (Paine and Bravemen, 1990). Due to a short penetration depth of the beam into the target, rapid non-equilibrium heating results, leading to the formation of a highly forward directed, stoichiometric plasma plume. Deposition is possible in the presence of a reactive gas or in the presence of an inert gas.

8.2.1 Pulsed Electron-Beam Deposition

The main requirement of pulsed energy deposition is the ability to create a high power density at the target surface. Even though pulsed excimer lasers with nanosecond pulse widths can successfully meet this requirement, only recently it has become clear that a cost-effective method using pulsed electron beams can also achieve this objective (Hoebel et al., 1990). This novel technology is based on the pulsed electron beams generated in a transient discharge, with currents over 1000 A and electron energy over 10 kV, and thus are able to attain a beam power density of ~10^8 W/cm^2 at the target surface, an essential feature for rapid non-equilibrium heating at the target surface, leading to congruent evaporation.

Pulsed Electron-beam Deposition is a process in which a pulsed high power electron beam is incident on the surface of a target, penetrates approximately 1–2 μm resulting in a rapid evaporation of target material (Hoebel et al., 1990). The ablation process lasts about 100 ns leading to non-equilibrium heating, which in turn facilitates stochiometric preservation of the target composition in the deposited film. Most solid state materials can be deposited as thin films with PED. The energetic beam of pulsed electrons are created in a low pressure gas discharge known as "Pseudo-spark" or "Channel-spark" in the literature (Hoebel et al., 1990; Christiansen and Schultheiss, 1979; Muller et al., 1995; Dediu et al., 1995) and are briefly discussed below.

8.2.2 Pulsed Electron Beams from Pseudo-Spark and Channel-Spark Discharges

The fast low-pressure gas discharge called "pseudo-spark" was first reported in 1979 (Christiansen and Schultheiss, 1979) and the mechanism of the phenomenon has not been fully understood even now. The low-pressure gas discharge occurs between a planar anode and a hollow cathode with unique charged particle emission characteristics (Muller et al., 1995). The resulting electron beam is generated in a plasma inside a hollow cavity and travels through a background gas, forcing the beam to be magnetically pinched tightly as it propagates. Pseudo-Spark and Channel Spark discharges are quite similar in nature even though certain differences exist in the source design. The main difference however, is the conversion efficiency of stored electrical energy into beam energy at the target in these two cases which is about 4% in the case of pseudo-spark discharge and about 30% in the case of channel-spark discharge. This improvement in the electrical energy conversion efficiency in channel spark discharge is a direct consequence of modifications carried out in the accelerator part of the discharge source. The term 'Pulsed Electron Beams' used in the following text represents electron beams generated in a channel spark discharge. Schultheiss and Hoffman discovered that such pseudo-spark/channel spark discharges producing well-pinched electron beams with beam diameters as small as several mm^2 in size leading to high beam brightness $\sim 2 \times 10^{11}$ A/m^2-rad^2 (current densities up to 10^6 A/cm^2) are capable of producing power densities up to 10^9 W/cm^2 with a pulse duration of $\sim$100 ns (Schultheiss and Hoffman, 1990). This power density is comparable to the power densities obtained with pulsed lasers and can therefore be used for materials processing.

8.2.2.1 Pulsed Electron-Beam Source A special type of hollow cathode is used for generating high currents. The hollow cathode is a metal tube, positioned in front of a planar anode (Muller et al., 1995). The electrons are generated at the inner walls of the hallow cathode either by means of ion impact or by photo effect. Due to the reduced electric field inside the hollow cathode the electrons move slowly, increasing the probability for ion impact ionization. Special cathode geometries facilitate oscillations around the cathode axis, ionizing the gas in the cavity effectively, before escaping the cathode along the axis towards the anode. The current density of the axially focused electron beam are of the order of 1 A/cm^2. The hollow cathode operates in the DC.

For very high currents the hollow cathode is modified in such a way that it operates only in a 'pulsed' mode. This is called transient hollow cathode, consisting of a similar cathode but only with a narrow exit (Muller et al., 1995; Mittag et al., 1990). The transient hollow cathode is a low pressure gas discharge electron source, which develops a well focused electron flow with currents of even up to a few kilo-amperes with pulse widths ∼100 of nanoseconds. Current densities of 1 kA/cm^2 are attainable in this case. A pre-discharge trigger circuit, which affectively controls the discharge voltage, is used to ignite the plasma.

8.2.2.2 Acceleration

Pseudo-Spark Discharge In the pseudo-spark discharge, occurring in the high voltage low pressure region of the Paschen curve (Muller et al., 1995), electron acceleration is affected by a set of parallel electrodes separated by insulators. The special arrangement of hollow cathode and electrodes leads to an electric field gradient that focuses electrons along the central axis. At a given breakdown voltage (pressure dependent) the low pressure gas discharge escalates into (Hoebel et al., 1990) a very fast spark-like discharge characterized by an over exponential current rise. The discharge leads to the formation of intense pulsed electron beam along the central axis of the accelerator which can be extracted out of the cavity through an opening. The pulse width of the electron beam is ∼100 ns and the current density in the beam is about 5000 A/cm^2. The electrical energy of the pseudo-spark chamber is stored in a variable number of high voltage ceramic capacitors which can deliver up to about 3–5 J/pulse. For a more detailed description of Psuedo-spark discharges the reader is referred to excellent papers available in the literature (Hoebel et al., 1990; Jiang et al., 1993).

In the case of psuedo-spark discharge the transfer efficiency of electrically stored energy to beam energy at the target is rather low and is about 4%, as seen in the film deposition experiments of Hobel et al. (1990), and Jiang et al. (1993), on YBCO and is similar to the transfer efficiency (∼3%) of electrical to optical power by the excimer lasers (Basting, 1991).

Channel-Spark Discharge Even though the psuedo-spark discharge based pulsed electron beams have been able to produce good quality films, there have been some problems associated with the stack of metal disks used in the accelerator. The metal discs were subject to oxidation, changing properties of the discharge over a period of time, thereby calling for a frequent cleaning of the discs as a routine maintenance requirement (Jiang et al., 1994). To overcome these difficulties, Jiang et al. have developed a novel scheme in which the a dielectric 'channel' replaces acceleration section consisting of alternating metal and insulator discs.

The channel-spark discharge is quite similar to pseudo-spark discharge. The electron beams are generated from similar hallow cathodes and the beams are also magnetically self pinched and accelerated through specially designed electrodes providing the required electric field gradients. The major advantage of channel-spark is realized in generated electron beams which are found to be much more stable relative to those generated in the pseudo-sparks. More importantly, the efficiency of energy transfer from beam to target is about 30% which is about 7–8 times higher than in the case of pseudo-spark discharges (Jiang et al., 1994). Table 8.1 presents typical electron beam parameters.

The above parameters of the pulsed electron beam are very attractive for materials processing applications. The pulsed nature of the energy source (electron beam) facilitates the ability to deposit complex metal–oxides such as HTS films and presents an opportunity to develop a low cost alternative to Pulsed laser deposition technology for coated conductor manufacturing applications.

Table 8.1. Typical Pulsed electron-beam parameters

Maximum discharge voltage	20 kV
Electrical efficiency	30%
Stored energy	3 J
Gas pressure	4–30 mTorr
Discharge time	~100 ns
Repetition rate	1–100 Hz
Electron energy in the beam	⩽15 keV
Electron current	1.5 kA
Electron beam diameter at the target	~2–3 mm
Beam current density	⩽10^5 A/cm^2
Power in the beam	⩽15 MW
Power density in the beam	⩽500 MW/cm^2
Maximum range of electrons	0.4 μm
Specific energy	60 kJ/g

8.2.3 Pulsed Laser Deposition vs. Pulsed Electron-Beam Deposition

In several respects, the Pulsed Electron-beam Deposition, schematically shown in Figure 8.1 can be compared to Pulsed Laser Deposition. The main differences lie in the higher efficiency of the ablation process in the case of pulsed electron-beam and significantly different beam-target interaction mechanisms. While the range of (248 nm) excimer laser beam in YBCO target for example is typically of the order of 20 nm, that of pulsed electron beam can be controlled and varied between 10 nm and 2000 nm. In pulsed laser deposition, the amount of ablation is very sensitive to the optical absorption coefficient of the target material. In pulsed electron beam deposition, ablation is independent of the optical properties of the target materials. Nearly all solid state materials (including those such as SiO_2 that are transparent to 248 nm excimer lasers) can be deposited as thin films with pulsed electron beam deposition. Further, the deposition rate can be chosen to be much higher in the case of pulsed electron beams over a wide range. From an economic viewpoint, the pulsed electron-beam deposition method is expected to offer significant cost-performance advantages relative to pulsed laser deposition.

8.2.4 Electron Beam—Target Interaction and Plasma Plume Formation

When a high power density energy beam (electron beam) is incident on a solid surface, the energy is absorbed by the surface, leading to a rapid increase of surface temperature facilitating material evaporation and plasma plume formation. Continuation of this process depends critically on the amount of energy absorbed by the expanding plasma plume. If no absorption occurs, the evaporation front moves deeper into the target with velocity proportional to the beam intensity. If strongly absorbed by the plasma plume (by ineleastic free-electron scattering), the beam cannot reach the solid surface to facilitate evaporation. As the initial plasma expands away from the target, the free electron density decreases, the plasma becomes transparent and the energetic beam can interact with the target again.

This scenario is typical for PLD where the shielding effects of the plasma plume strongly affect the evaporation dynamics and the energy spectrum of the plasma plume. The shielding is wavelength (λ) dependent (Geertsen and Mauchien, 1995). Since the absorption coefficient of the plasma scales as λ^2, the shielding is less pronounced for

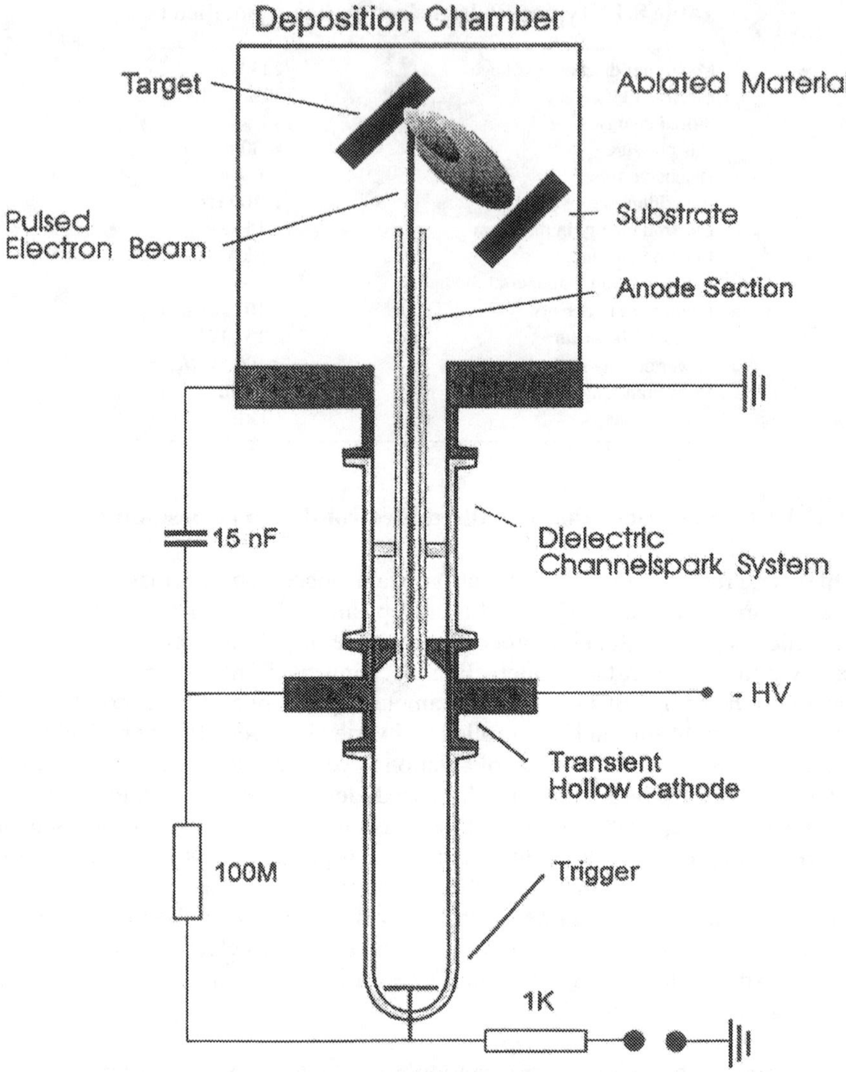

Figure 8.1. Schematic of pulsed electron-beam deposition system (from Muller et al., 1995).

UV radiation (193 or 248 nm) and results in a very efficient evaporation (ablation) process. In the case of longer wavelength radiation (1060 nm) however, the absorption of the radiation by the plasma generated is sufficiently large, thereby leading to a much smaller amount of material ablation from the target material.

In the case of PED, mechanisms of beam-solid and beam-plasma interactions are quite different and define the basic differences between PLD and PED. Similar to the case of pulsed laser deposition, in the case of pulsed electron-beam deposition, initial heating of the target surface to the evaporation temperature is controlled by a balance of the heat 'in-flow' via beam energy dissipation and heat 'out-flow' via thermal conductance. The absorption depth is therefore an important parameter in this regard. The pulsed-electron beam incident on the target loses energy through inelastic collisions with the electrons of the target atoms. After a short travel, energetic electrons experience strong Coulomb scattering, and tend to diffuse into the material, rather than proceed in a linear path. This effect helps to confine the entire 10–20 keV of dissipated energy to within about 1 μm of surface layer in a typical solid (of density, $\rho \sim 6.1$ g/cm^3 for YBCO). Thus the projected range R_p of electrons is an order of magnitude larger than the absorption length of the laser radiation. For the $\tau \sim 100$ ns long electron pulses, the heat diffusion spreads the energy to approximately about $(a\tau)^{1/2} \sim 1$ μm deep into the target (assuming a thermal conductance of $\sim 10^{-1}$ cm^2/s). As a result, beam intensity threshold Q_{th} for evaporation for PED is quite similar to PLD ($\sim 5 \times 10^7$ W/cm^2).

An important feature of electron beam ablation originates from the fact that energetic electron–atom interaction does not depend on whether it is bonded to a crystal lattice (in the solid) or whether it is in the vapor state. As a result, the effective cross section for electron scattering by the target atom does not change before and after ablation. In the case of PED therefore, all target materials within a certain layer (~ 1 μm) can be ablated if the beam has sufficient intensity. This is quite in contrast to PLD, where the amount of ablated material modulates with respect to the screening properties of the plasma plume.

Another feature of PED is that the mass of the ablated material is practically independent of the target material (for a given energy density of the beam and provided that thermal conductivity losses can be neglected (Muller et al., 1995)). This results from the fact that the electron range R_p is inversely proportional to the target density, and thus the mass (m) within the R_p layer ($m = \rho R_p$) is constant. Specifically for YBCO ($\rho \sim 6.1$ g/cm^3) and electrons of about 15 keV energy (with an $R_p \sim 0.8$ μm) one can expect to ablate $\sim$500 μg/pulse from each cm^2 of the ablated spot area. Values of 6–12 μg/pulse have already been obtained experimentally (M. Strikovski et al., private communication). This is at least 10 times greater than the typical mass ablated ($\sim$0.6 μg/pulse) by PLD using a 193 nm ArF excimer laser (Geertsen and Mauchien, 1995).

General condition for stoichiometric ablation requires the material to attain temperatures much higher than needed for equilibrium evaporation. Thus, electron beam intensity Q (W/cm^2) of several times the evaporation threshold Q_{th} is needed to provide optimal ablation conditions. Given that the number of evaporated atoms depends weakly on the intensity Q (at constant electron beam voltage) one might expect that the energy spectrum of the ablated species in the case of PED would significantly be higher than in the case of PLD.

8.2.5 Plume Propagation in a Background Gas

When a pulse of energetic electrons is incident on the target, a dense layer of overheated plasma is created that expands in the direction of the maximum pressure gradient. Although the physics of the ablation process in the case of PED is quite similar to PLD, there are some important differences originating from the fact that the operating pressure in the case of PED is quite different than PLD. For HTS and related film depositions for example, the oxygen partial pressure for PED is about 5–15 mTorr, more than an order of magnitude lower than used in the case of PLD ($\sim$200 mTorr). A simple model (Strikovski and Miller, 1998) considering plasma expansion, deceleration, and thermalization of the ablated material can be used to interpret the differences arising as a consequence of lower pressures. The model predicts that the plume range L_0 (close to the optimal substrate-to-target distance for film deposition) depends on the number of ablated atoms N_0 and the gas pressure p as

$$L_0 \sim \left(\frac{N_0}{p}\right)^{1/3}.$$

As N_0 may be ~10 times greater, and p is about 10 times lower in PED relative to PLD, the PED plume range is expected to be factor ~4.6 greater. If the common PLD finding—that an optimal substrate location for film growth is in the vicinity of plume-end is still valid for PED, this rise in L_0 indicates possible uniform deposition on large area substrates as the plume transversal dimensions scale with L_0 as well. Less than 10% decrease in deposited film thickness (relative its maximum value) is observed within 20-degree plume cone (Muller et al., 1995). A direct consequence of this feature is that deposition of films with a reasonable thickness uniformity over 2–3″ diameter seems possible without complicated substrate motion control schemes.

It is noteworthy that experiments carried out on the existing PED systems (including those in operation at Neocera, Inc.) have already indicated that higher deposition rates [g/pulse] relative to PLD are possible. A deposition rate of an order of magnitude higher seem feasible with PED. To facilitate further increase in the ablation rates however, pulsed electron beam sources delivering higher beam currents need to be developed.

8.2.6 Beam and Plasma Diagnostics

For nearly a decade, work in pulsed electron beams has been limited primarily to the research of the basic phenomenon, physics of the discharge, and the electron beam parameters. The discharge can exist in quite specific conditions of gas pressure, device geometry, and excitation parameters. Temporal and spatial development of the discharge plasma has been investigated revealing a very complex dynamics of the discharge plasma. The electron beam profile and energy distribution studies showed that the pulse consists of electrons of different energy with the leading edge consisting of the most energetic electrons (Muller et al., 1995; Stark et al., 1995). Spectroscopic studies revealed that the PED-produced target erosion plasma is characterized by a higher electron temperature (high degree of ionization of the ablated species) than the laser-produced plasmas. In contrast, average kinetic energy of the species in the PED produced plasma is estimated to be less than that found in PLD-produced plasmas. However, no detailed measurement of the energy distribution has been done. Also, no practical model for the electron beam-produced plasma dynamics exists as of now. For more details on the investigation of plasmas produced by PED, the reader is directed to excellent work of Witke et al. (1995), Dewald et al. (1997) and Gilgenbach et al. (1999).

8.2.7 HTS Films by Pulsed Electron-Beam Deposition

Deposition of superconducting YBCO thin films by pulsed electron beam ablation was first reported by Hobel et al. (1990). These early workers used electron beams generated in pseudo-spark discharges (Figure 8.2) and obtained encouraging results on YBCO deposited on single crystalline substrates such as yttria stabilized zirconia (YSZ) and $SrTiO_3$ substrates. C-axis textured YBCO films with superconducting transition temperatures of 85 K and a critical current density of 1.1×10^5 A/cm^2 were achieved in these experiments (Hoebel et al., 1990). Rutherford back-scattering (RBS)

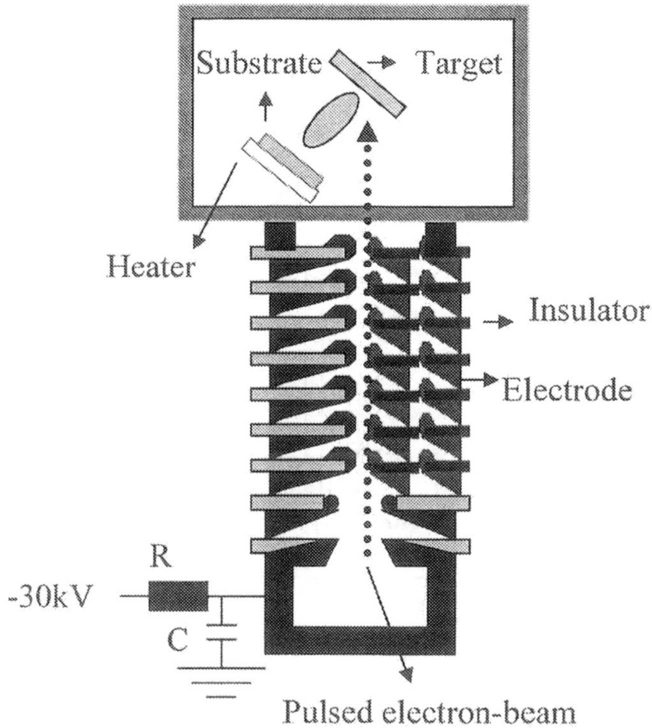

Figure 8.2. Pulsed electron beam deposition system using pseudospark-discharge (adopted from Hoebel et al., 1990).

measurements have revealed that the target composition is retained in the deposited thin film to a large extent and deviations in the film stoichiometry in most cases were within about 10% of the target composition. An important observation is that, film composition has been found to be less sensitive to the variations in the energy density at the target, than in the corresponding case of PLD deposited films, a feature of critical importance from a commercial view point. X-ray diffraction studies pointed out strong c-axis texture in the films. The x-ray data also indicated the presensce of undesirable impurity phases (trace amounts) in a predominantly 1-2-3 phase. The lower critical current densities of $\sim 10^5$ A/cm^2 at 77 K may be a direct consequence of the presence of these secondary phases which may decorate the grain boundaries forming weak-links. Though the YBCO quality is non-optimal, these preliminary results formed the basis for further exploration.

The experiments of Hobel et al., using pulsed electron beams generated in a psuedo-spark discharge were repeated by Dediu et al. (1995), using pulsed electron beams generated in channel spark discharge (shown in Figure 8.3(a). The details of the channel-spark are shown in Figure 8.3(b). These workers have carried out film depositions with YBCO, GdBCO and $Gd_xEu_{1-x}BCO$ targets. $NdGaO_3$ and $SrTiO_3$ are used as growth substrates in these cases and are maintained at a substrate temperature of 750°C and oxygen partial pressure of about 15 mTorr. By varying the geometry of the channel section, the voltage-pressure conditions for optimum deposition conditions of superconducting films were obtained by these workers. T_c's of 87–89 K for YBCO and 91–92.6 K in the case of GdBCO and Gd/EuBCO were achieved. Critical current densities measured on specially prepared bridges were quite impressive and are of the

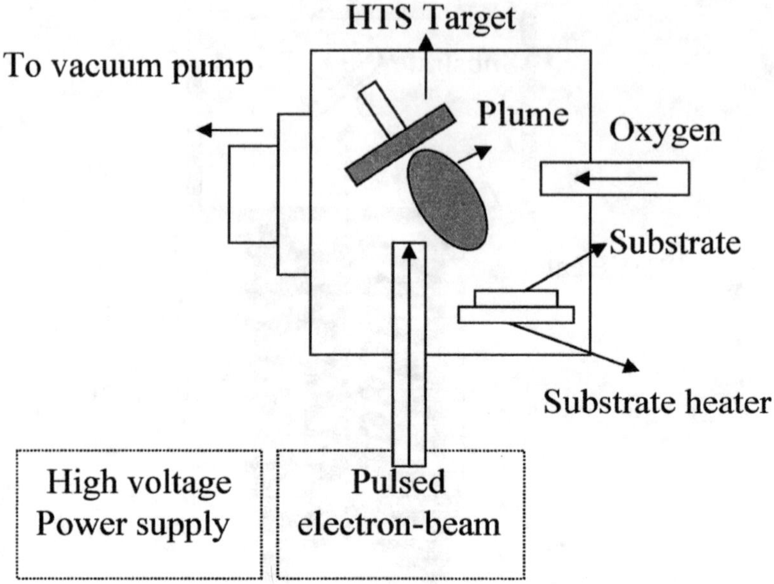

Figure 8.3a. Schematic of Pulsed Electron-beam Deposition system using Channel-spark discharge (used in Dediu et al., 1995).

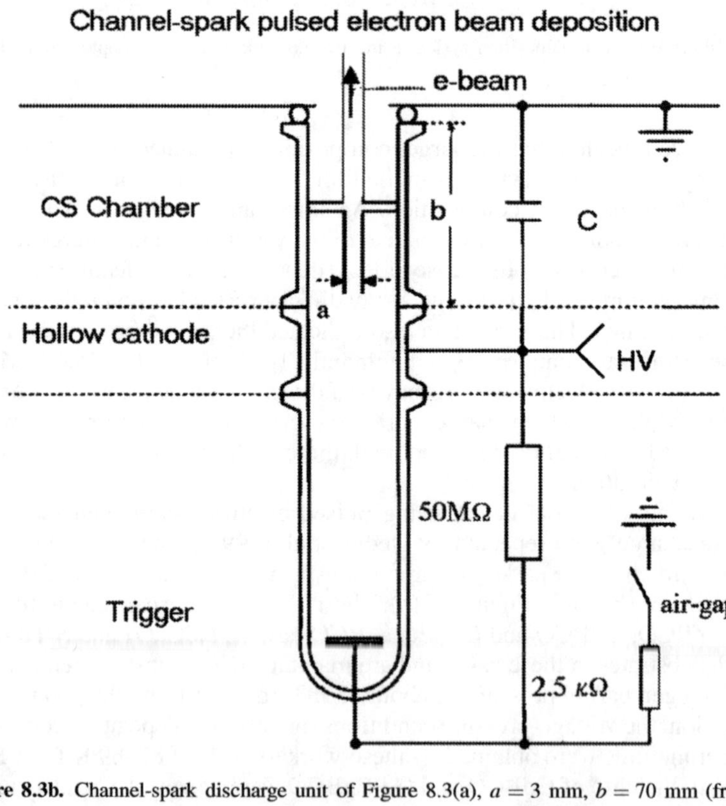

Figure 8.3b. Channel-spark discharge unit of Figure 8.3(a), $a = 3$ mm, $b = 70$ mm (from Dediu et al., 1995).

order of 3×10^6 A/cm^2 at 77 K at zero field. These experiments also pointed out that deviation in stoichiometry of YBCO films is minimal in the angular range of $\pm 30°$ from the target normal. Muller et al. (1995) have extended the applicability of PED for thin film deposition of materials other than HTS. These workers have shown that PED is a generic thin film deposition technique and can be applied to the deposition of complex materials such as HTS films, as well as sodium silicate and ionic conducting glasses. Films such as alumina, biocompatible hydroxyapatite and polymeric materials such as hydro- or fluoro-carbon polymer films (PTFE) films were also deposited in their study. It is shown that thin films of certain materials such as SiO_2 that are transparent to 248 excimer laser radiation of PLD, can be deposited by PED. For more details, the reader is referred to an excellent review by Muller et al. (1995).

8.2.8 Pulsed Electron-Beam Deposition of HTS Films for Coated Conductor Applications

The demonstrations of high critical current densities in HTS films on single crystalline substrates and the demonstrations of PED as a generic thin film deposition technique clearly points out the potential of PED. On the commercial front, significant cost advantages of PED relative to PLD appear to be possible particularly in high volume manufacturing processes. The cost effectiveness of the PED technique over the PLD is realized when expressed in terms of price of useful energy ($/Joule). For PED it is in the range of $0.01/Joule which is orders of magnitude lower than that of PLD. The low costs in the case of PED are a direct consequence of the higher beam generating efficiency of the technique which is about 30% (electrical to electrical), whereas in the case of PLD it is around 3% (electrical to optical). The pulsed electro-beam deposition technique therefore is expected to be very cost effective when scaled to a high volume manufacturing environment, a critically important requirement for coated conductor development program. Despite the enormous cost savings of PED over PLD (the PED source is expected to cost a fraction of excimer laser cost), this method has not gained popularity due to the non-availability of commercial pulsed electron beam sources until very recently. Neocera has licensed this technology from FZK and is now a commercial vendor of PED sources and PED systems.

Integration of pulsed electron-beam source with a film growth chamber is relatively simple. The pulsed electron-beam source is flange mounted and integrated with a standard thin film growth chamber that is conventionally used for PLD thin film growth. The angle between the incident electron beam and the target normal is typically around 45°. The self focusing (magnetically pinched) electron beam is guided to the target by an alumina beam guide. A standard Neocera designed PED source is shown in Figure 8.4. A complete PED deposition system with an integrated PED source is shown in Figure 8.5. It is to be noted that the power supply is in a 19″ rack-mount and therefore the whole PED system has much smaller foot-print than a typical PLD laboratory (including the laser). The deposition chamber incorporates a substrate heater capable of heating the substrate upto a maximum temperature of 950°C, a 6-target carousel carrying HTS and buffer layer targets and a gas (oxygen and other gases) manifold. High density YBCO (and GdBCO) targets are used for film depositions. Data obtained on the commercial version of the deposition equipment are presented in the following section. Results are presented for epitaxial HTS films on single crystalline $LaAlO_3$ substrates as well as films deposited on RABiTS substrates with CeO_2/YSZ/CeO_2 buffer layers. The RABiTS substrates were obtained from Oak Ridge National Laboratory.

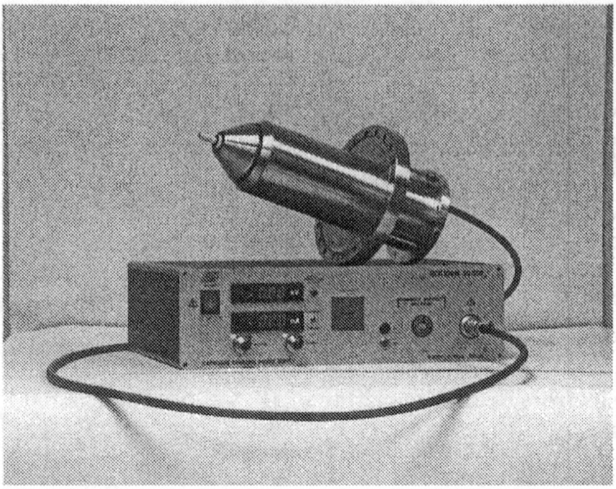

Figure 8.4. Neocera-designed pulsed electron-beam source with power supply.

YBCO and GdBCO were chosen as representative materials systems. Films of YBCO (and GdBCO) are deposited at a substrate temperature of 850°C. The oxygen partial pressure during HTS film deposition is about 10–15 mTorr. Film thicknesses are in the range of 3000–4000 Å. The typical pulse repetition rate used is around 6–8 Hz. The beam voltage is maintained at 17 keV. After deposition, the substrates are cooled down to room temperature in a background oxygen pressure of about 500 Torr.

The superconducting properties of HTS films are presented here. The HTS films were evaluated for their transition temperatures (T_c's) by AC susceptibility measurements. The structural properties were analyzed by 4-circle x-ray diffraction. Critical currents were measured by standard I–V measurements.

8.2.8.1 HTS Films Deposited on Single Crystalline $LaAlO_3$ Substrates This subsection summarizes data obtained on HTS films deposited on single crystalline $LaAlO_3$ substrates. Data obtained on HTS films deposited by pulsed electron beam deposition on buffered RABiTS substrates are presented in Subsection 8.2.8.2.

Figure 8.6 shows the AC susceptibility data of YBCO films on $LaAlO_3$ substrates. The films show transition temperatures (T_c's) around 89–90 K with transition widths of 1 K or less.

Figure 8.7 shows the AC susceptibility data of GdBCO films on $LaAlO_3$ substrates. The films show transition temperatures (T_c's) around 90–92 with transition widths of 1 K or less.

The critical current densities of the films were measured by standard I–V measurements. YBCO and GdBCO films on single crystalline substrates were patterned to 40 micron wide bridges by standard photolithographic patterning and wet chemical etching in 0.5% phosphoric acid. Figure 8.8 shows the critical current data for GdBCO films. From these measurements the critical current density for GdBCO films was estimated to be 1.1×10^6 A/cm^2 at 77 K, in self-field.

The critical current density obtained on YBCO deposited on $LaAlO_3$ substrates is also measured and is 1.1×10^6 A/cm^2 at 77 K and zero field. These values compare well with state-of-the-art films deposited by PLD and other deposition techniques.

Both YBCO and GdBCO films exhibit critical current density values of about 1×10^6 A/cm^2. In case of GdBCO however, the reproducibility of film properties

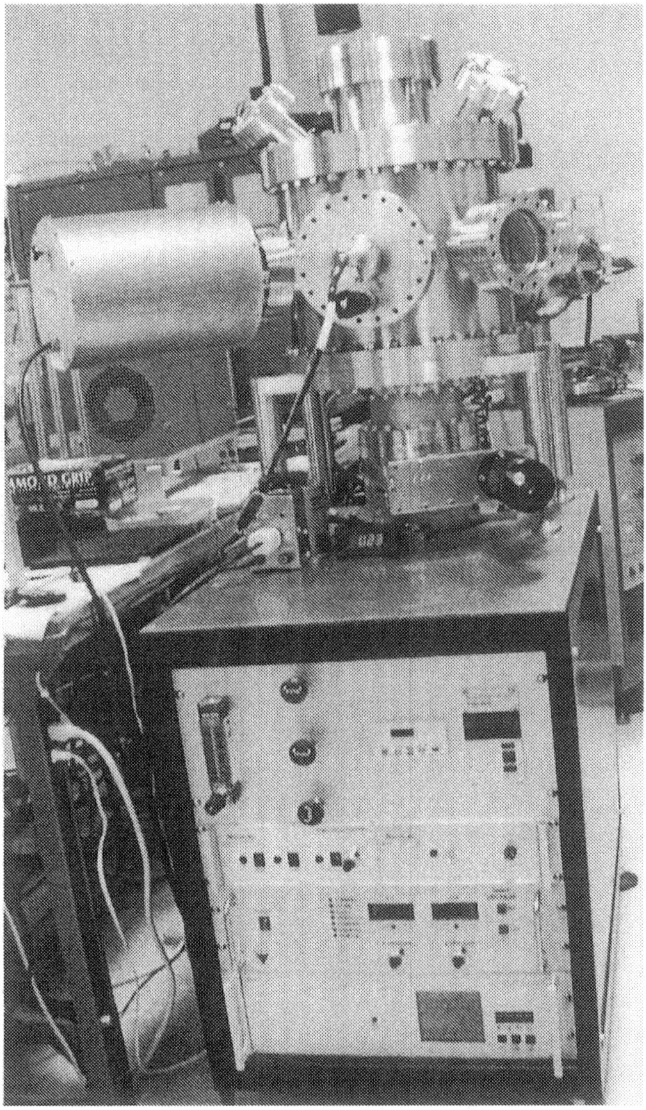

Figure 8.5. Prototype Pulsed Electron-beam Deposition system with an integrated electron-beam source (top left) with a 19-inch rack-mountable power supply (lower half).

appear to be higher relative to YBCO. In YBCO case, the optimum deposition parameter space which provides highest quality films (high T_c's and high J_c's) appears to be rather narrow and is relatively less tolerant to process condition variations. The deposition parameter space for GdBCO however appears to be rather wide, and this flexibility of GdBCO is not fully understood at the present time.

Figures 8.9–8.11 show the x-ray diffraction data for YBCO on $LaAlO_3$ substrates. The figures show θ–2θ (Figure 8.9), rocking angle (ω) scans (Figure 8.10) and Φ-scan (Figure 8.11) data for YBCO. The data indicate that the films are predominantly c-axis oriented with an ω-scan (rocking angle) FWHM of about 0.55° degrees. Good in-plane epitaxy is also evident by the Φ-scan data. The x-ray θ–2θ data however indicate the

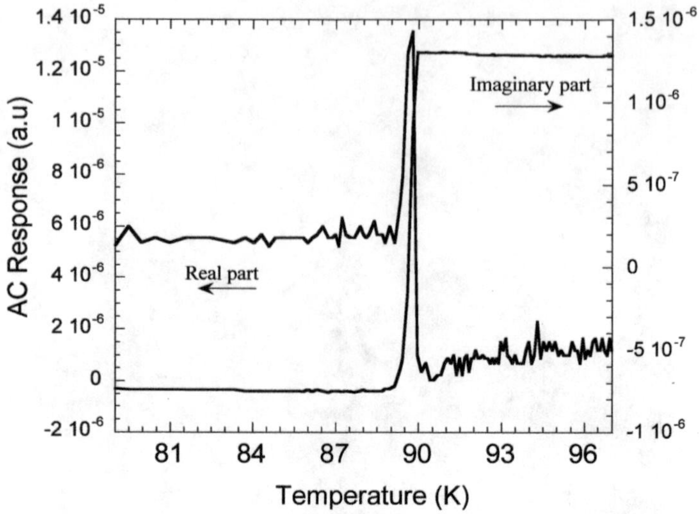

Figure 8.6. AC susceptibility data of YBCO on $LaAlO_3$ substrate.

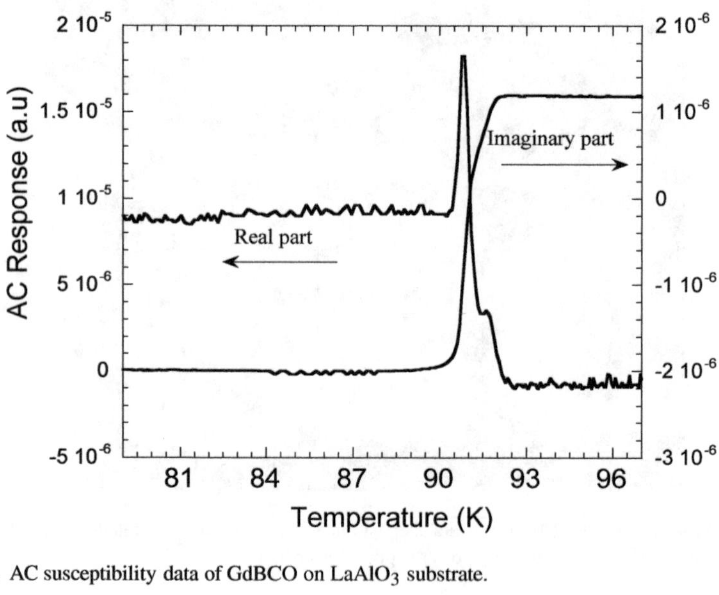

Figure 8.7. AC susceptibility data of GdBCO on $LaAlO_3$ substrate.

presence of some minor and yet unknown impurity phases. Such minor trace amounts were also reported by the earlier researchers (Hoebel et al., 1990). The origin of these phases is the subject of current investigations, and progress will be made to eliminate these second phases.

The x-ray data on GdBCO is shown in Figures 8.12–8.14. The θ–2θ data (Figure 8.12) indicates lower amounts of impurity phases relative to YBCO. The ω-scan data reveal that GdBCO films are structurally superior to YBCO as revealed by the rocking angle FWHM data (Figure 8.13). The in-plane epitaxy of the GdBCO films is revealed in the data presented in Figure 8.14.

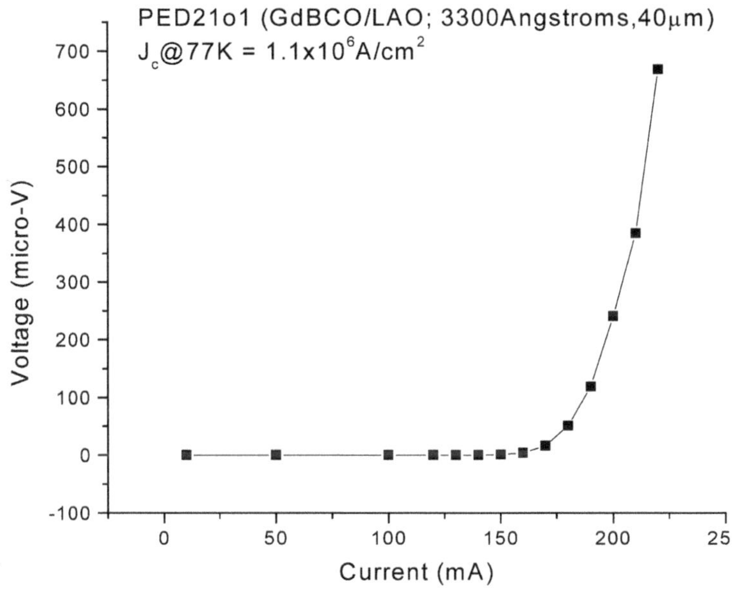

Figure 8.8. Critical current data of GdBCO on $LaAlO_3$ substrates.

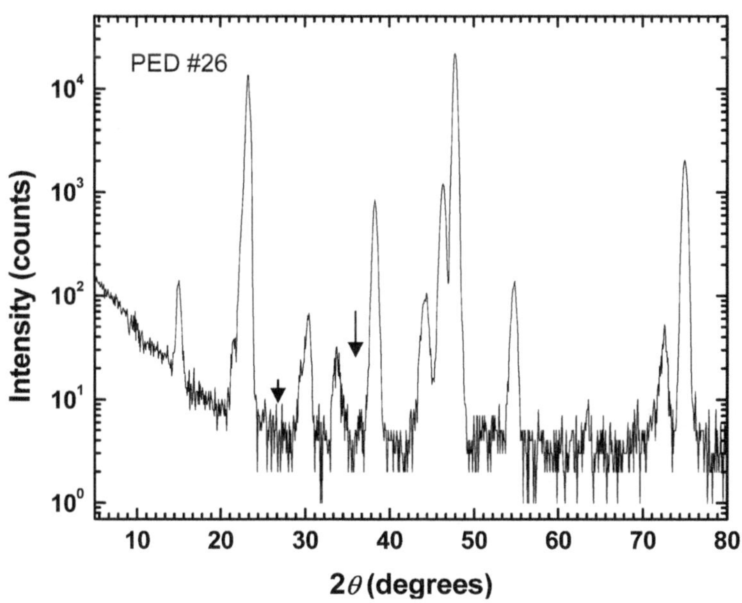

Figure 8.9. X-ray θ–2θ data for YBCO films grown by pulsed electron beam deposition on $LaAlO_3$ substrates (arrows indicate minor impurity phases).

8.2.8.2 HTS Films Deposited by Pulsed Electron-Beam Deposition on Buffered RABiTS Substrates The buffered (CeO_2/YSZ/CeO_2) RABiTS substrates used in this work were provided by Oak Ridge National Laboratory. The HTS (YBCO, GdBCO) films were evaluated for their transition temperatures T_c's, structure and transport (critical current densities). This section summarizes the data.

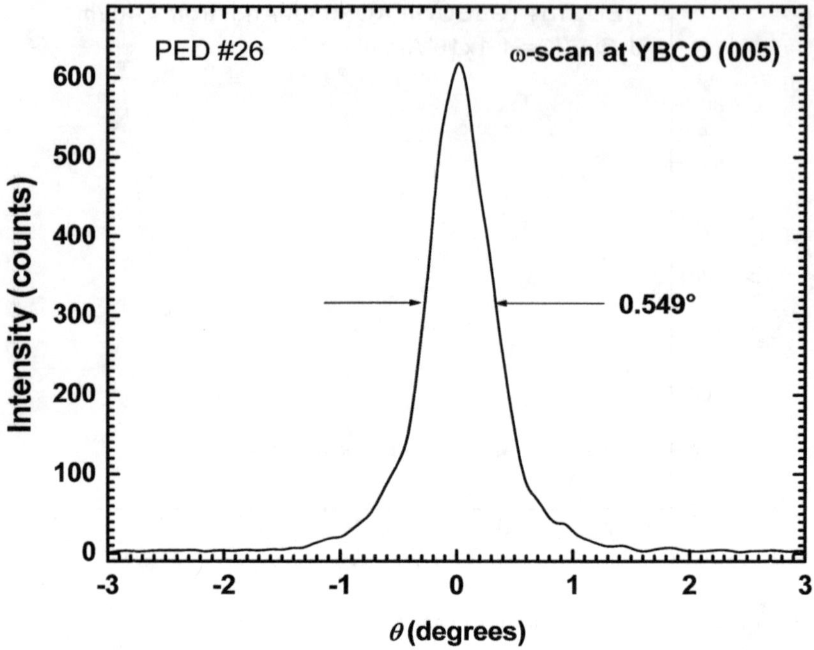

Figure 8.10. ω-scan data of YBCO on $LaAlO_3$ substrates.

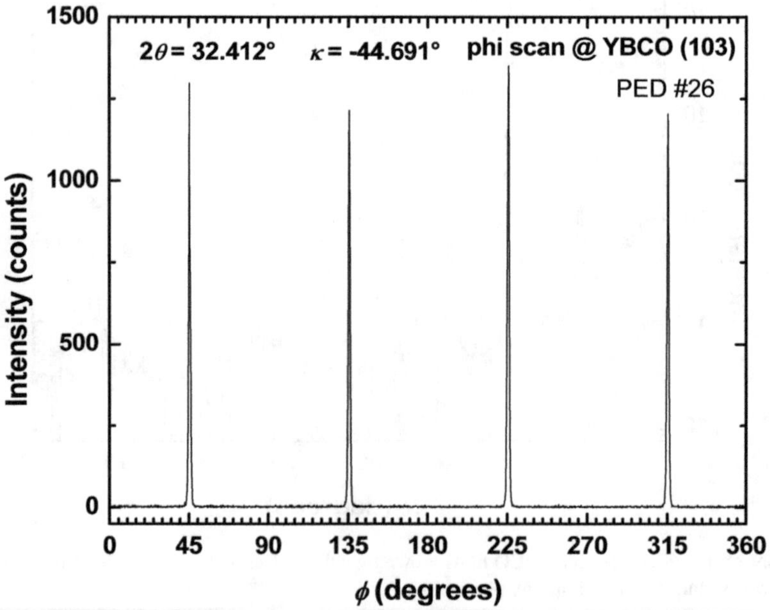

Figure 8.11. X-ray Φ-scan data of YBCO on $LaAlO_3$ by pulsed electron beam deposition.

Figure 8.15 shows data for YBCO on RABiTS substrates. The T_c's are in the range of 90 K with transition widths around 1 K. It is important to mention that this is the first experimental demonstration of HTS (YBCO) film deposition on RABiTS substrates

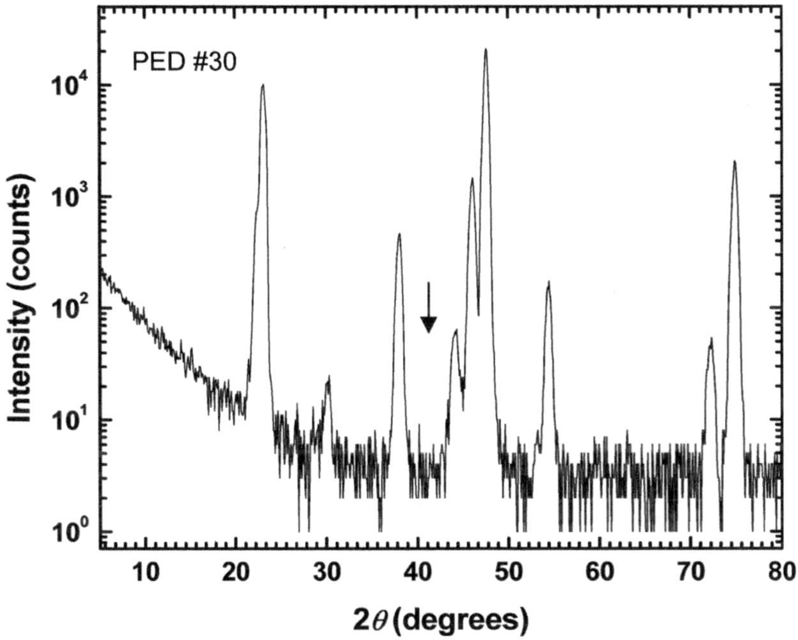

Figure 8.12. X-ray θ–2θ data for GdBCO films grown by pulsed electron beam deposition on $LaAlO_3$ substrates (arrow indicates traces of impurity phase).

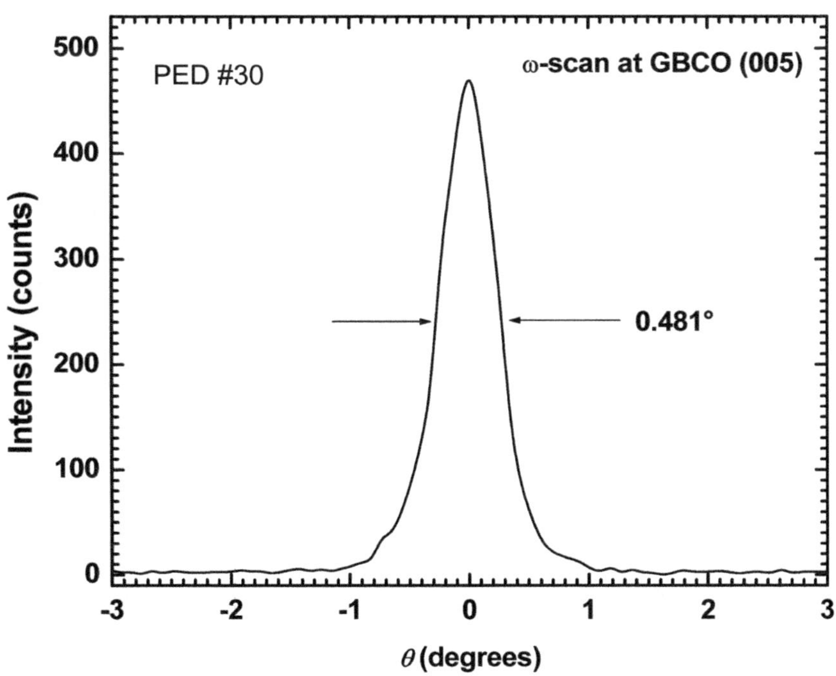

Figure 8.13. ω-scan data of GdBCO on $LaAlO_3$ substrates.

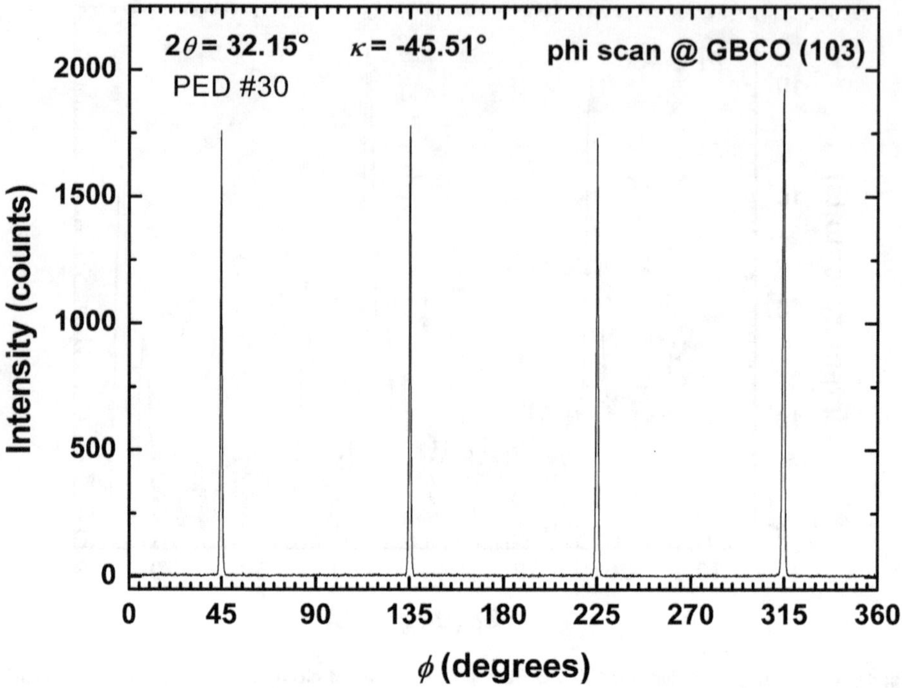

Figure 8.14. x-ray Φ-scan data of GdBCO on $LaAlO_3$ by pulsed electron beam deposition.

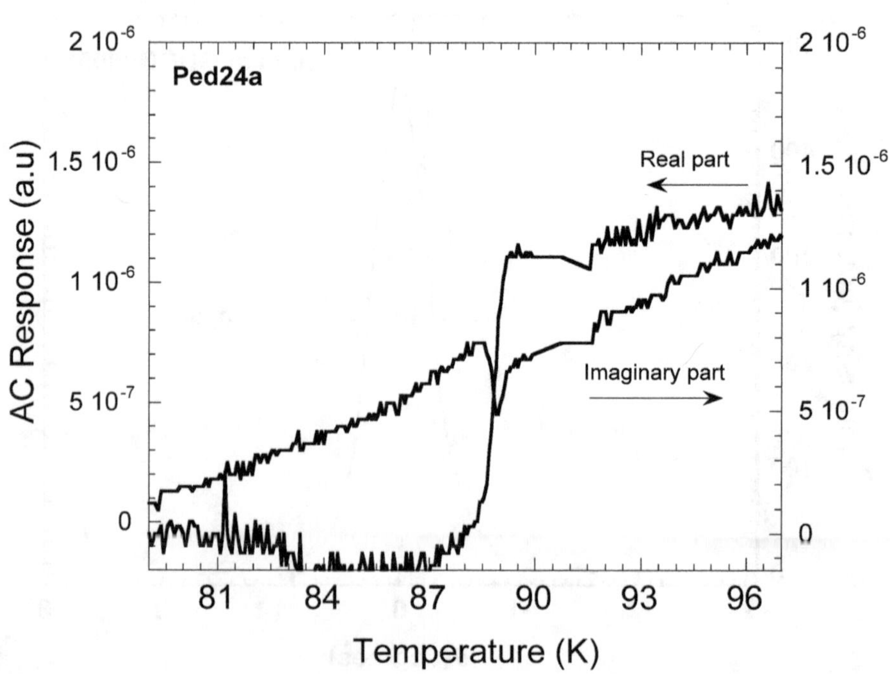

Figure 8.15. AC susceptibility data of YBCO on RABiTS substrate.

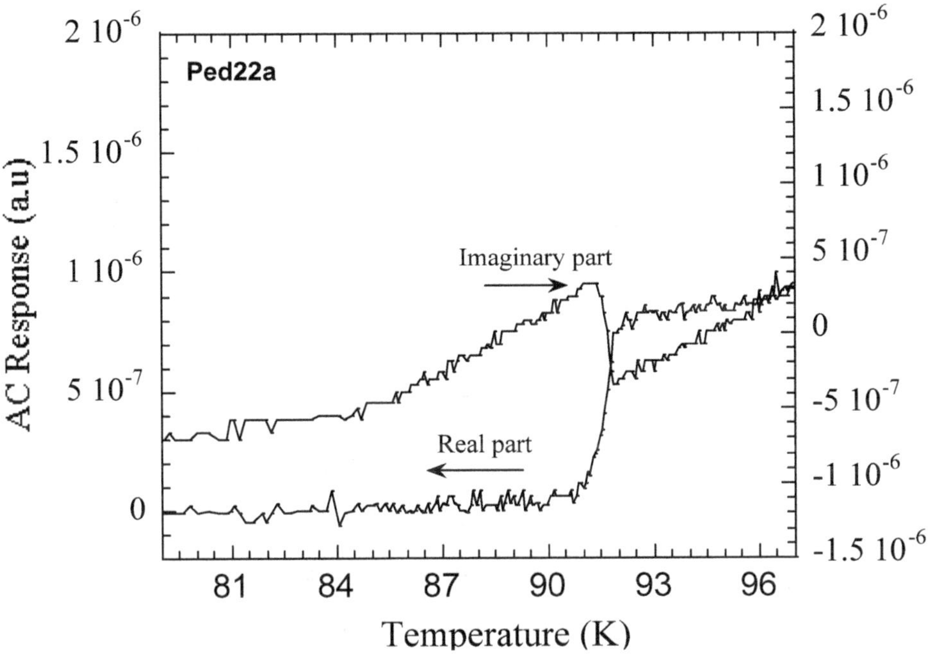

Figure 8.16. AC susceptibility data of GdBCO on RABiTS substrate.

by PED. It may be noted that the qualitative appearance of the AC suscsptibility data for RABITS substrates is different from the data obtained on $LaAlO_3$ substrates. This difference is a consequence of the shielding effects arising from the magnetic substrate.

Figures 8.16 shows data for GdBCO on RABiTS substrates. The T_c's are in the range of 91–92 K with transition widths around 1 K.

8.3 STRUCTURE AND TRANSPORT OF HTS FILMS ON RABiTS SUBSTRATES

Figures 8.17 and 8.18 show x-ray Φ-scan data of a typical RABiTS substrate and the HTS (GdBCO) film grown on this substrate respectively. The good in-plane epitaxy of GdBCO films is revealed in Figure 8.18 and is in good registry with the textured RABiTS substrate. The in-plane FWHM values are 11.1° and 9.6° for the RABiTS substrate (Figure 8.17) and the GdBCO film (Figure 8.18) respectively.

The critical current densities of GdBCO films on RABiTS substrates were measured by transport measurements. Silver was deposited for contact pads and the measurements were carried out using the entire 5 mm wide sample width. The thickness of GdBCO is 3200 Å. The tapes carry a critical current of 5.2 A at 77 K and from these values, the calculated critical current density (J_c) is 0.32×10^6 A/cm^2 in self-field.

The magnetic field dependence of the critical current density is presented in Figure 8.19. The field is applied perpendicular to the c-axis of the GdBCO film. The J_c-H behavior of HTS films deposited by pulsed electron-deposition is qualitatively similar to those observed on typical HTS film on RABiTS substrates.

It is important to mention here that at the time this manuscript is written, the pulsed electron beam technology for coated conductor applications is at a very early stages

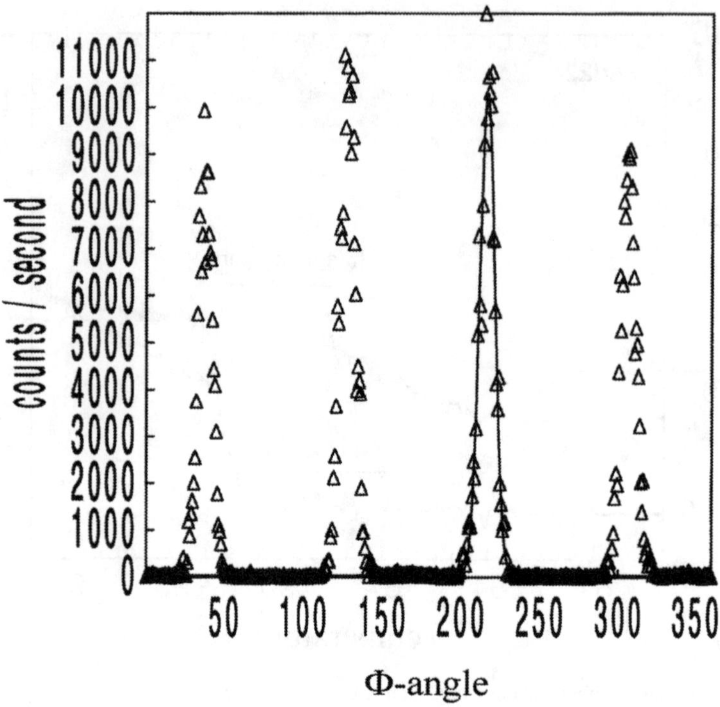

Figure 8.17. X-ray Φ-scan data for the RABiTS substrate with an in-plane FWHM of 11.1°.

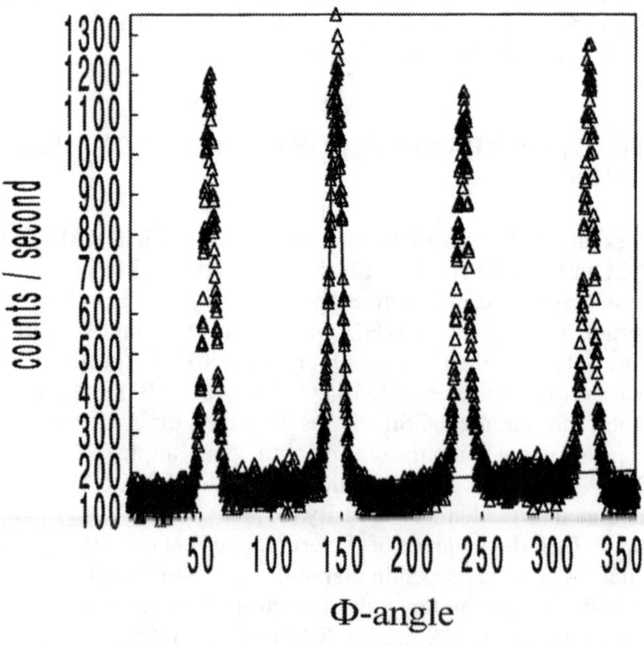

Figure 8.18. X-ray Φ-scan data for GdBCO films on RABiTS substrates of Figure 8.17. The in-plane FWHM values are 9.6° for the films.

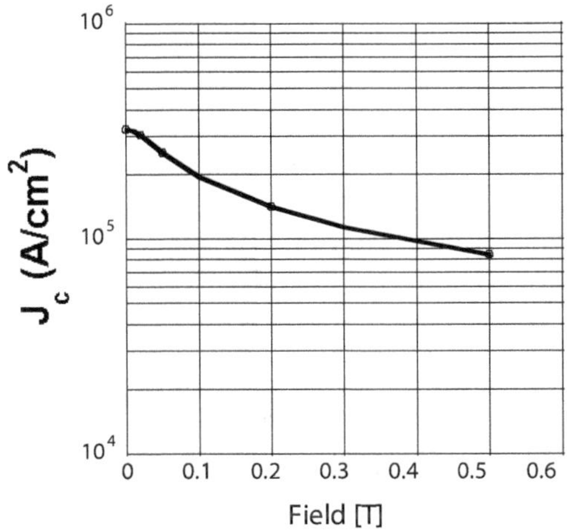

Figure 8.19. J_c-H behavior of HTS (GdBCO) films on RABiTS substrate. The data are taken at 77 K.

of development. The experiments and the results presented here are therefore preliminary and do not represent the optimum results. The data presented however points out the potential of this technology in becoming a cost effective alternative to pulsed laser deposition in high volume manufacturing of coated conductors. Developments in pulsed electron beam source and HTS process technologies are currently underway at Neocera, Inc.

8.3.1 High-Throughput Manufacturing of Coated Conductors by Pulsed Electron-Beam Deposition

Film deposition rate and thickness profile (uniformity) of the deposit on the substrate are two very important parameters that are of direct relevance to cost effective high volume manufacturing. These important issues are briefly discussed in the following section.

The rate of deposition—the amount of material deposited on the substrate per pulse in the case of pulsed electron beam depends on (i) electron beam source parameters such as discharge voltage determining the power density and range of electrons at the target, (ii) density of the target material and (iii) thermodynamic properties of the target material and (iv) the deposition geometry such as the distance between the substrate and the target. For a solid of density 6.1 gm/cm^3 (YBCO), Figure 8.20 shows the electron range. This range is typically between 1–2 microns for beam voltages in the range of 15–20 kV. An increase in the electron energy by a factor of two increases the range of electrons in the solid by a factor of four.

As in pulsed laser deposition, the ablated material escapes normal to the target surface in the case of pulsed electron beam deposition. Numerical simulation work of Muller et al. (1995) on materials such as SiO_2, Al, Cu and C and YBCO which are significantly different in their thermodynamic properties indicate that the amount of ablated mass in terms of μg/pulse for SiO_2, Al and YBCO all lie nearly on a common curve, which increases approximately quadratically with discharge voltage. Using a standard distance of 40 mm between the target and the substrate, the deposition rates for approximately all tested materials lie in a range between 0.5 to 10 Å per pulse.

Muller et al. have also measured the angular distribution of the evaporated species and found that the distribution can be approximated to $\cos^{2.5}\varphi$ where φ is the an-

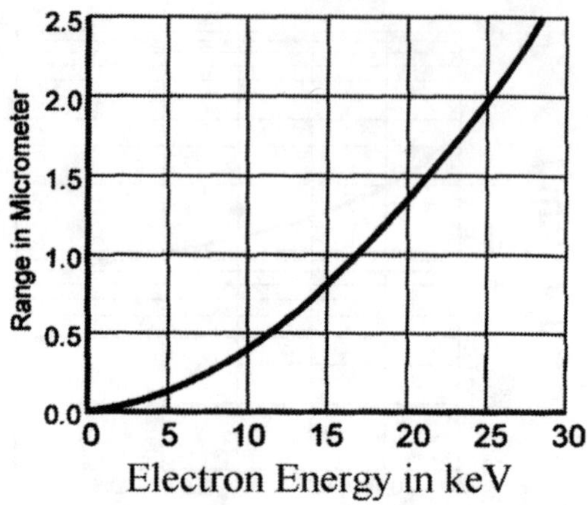

Figure 8.20. Electron range in a solid with density of 6.1 gm/cm^3 (from Muller et al., 1995).

gle between the target normal and the evaporating species (Muller et al., 1995). For substrates positioned parallel to the target surface, the evaporation profile results in a film thickness distribution proportional to $\cos^{5.5}\varphi$. The deposition rate and thickness homogeniety are therefore a function of the substrate position and size and for typical target to substrate distance of about 6 cm, thickness uniformity of about 4% are typically achieved in the case of small substrate sizes (about 1 cm $\times$ 1 cm).

In pulsed electron beam deposition, the effectiveness in converting electrical energy into beam energy could be as high as 30%, if the source is fully optimized. For evaluation of the source effectiveness for HTS film depositions, one can define a figure of merit (*FMP*) as

$$FMP = \frac{hS}{E},$$

where h [nm/pulse] is HTS film thickness deposited per pulse, S [cm^2] is effective area of the film deposition, and E [J] is the energy stored in the capacitor bank.

The deposition effectiveness (figure of merit) obtained with a prototype pulsed electron source is presented in Figure 8.21 as function of the source operating voltage (M. Strikovski et al., private communication). These results show that deposition rates with effectiveness parameter *FMP* as high as ~10 (nm/cm^2/J) are feasible if the source is optimized. It means that deposition of 1 μm thick films on 1 cm^2 area would require ~100 Joules of stored energy. As the typical stored energy is ~3 J, the productivity of a feasible apparatus, at repetition rate of 30 Hz, is about 1 cm^2 per second at a cost of about 100 W in electric power.

These experiments are very significant in the sense that they point out to a possibility of obtaining a film deposition rate of atleast an order of magnitude higher than in the case of PLD, a feature very attractive for high volume manufacturing of coated conductors.

Pulsed Electron Beam Arrays The observed film deposition profiles with an approximate $\cos^{5.5}\varphi$ dependence indicate that for coated conductor applications special geometries are necessary, not only in substrate manipulation for large area coverage, but also in the geometrical arrangement of the source itself. This is required to maximize the benefits of a high deposition rate possible with pulsed electron beams and at the same time to realize a cost economic high volume scale-up process.

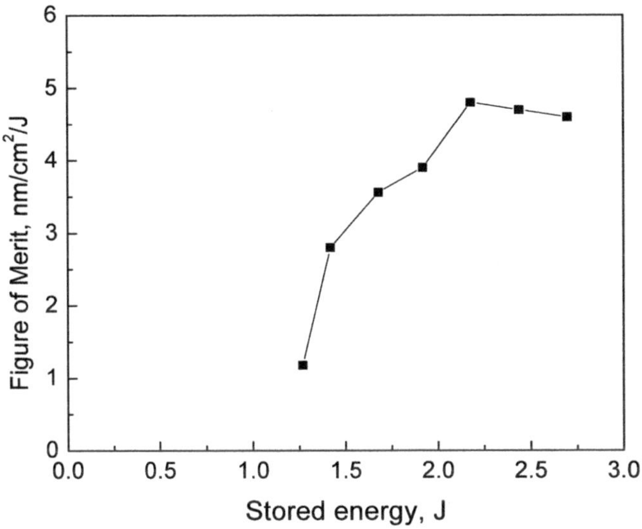

Figure 8.21. Figure of merit (nm/cm^2/J) as a function of stored energy (M. Strikovski et al., private communication).

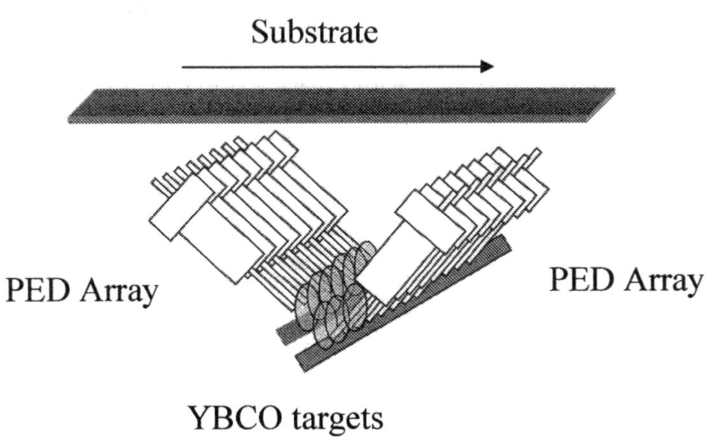

Figure 8.22. Schematic of pulsed electron beam deposition arrays for coated conductor applications.

Pulsed Electron Beam arrays incorporating a linear array of electron sources could offer advantages in high volume scale-up of coated conductors. A major advantage of this approach lies in the possibility of realizing the required high throughputs without incurring huge capital and running costs, that are intrinsic to high throughput scale-up of PLD, requiring multiple, high power excimer lasers. A schematic representation of such an array scheme is presented in Figure 8.22. It is possible to incorporate 10–15 electron beam sources that can be operated with a single power supply. Multiple arrays are also a feasibility as shown in the Figure 8.22. The economic benefits of such a scale-up process will be realized when cost savings associated with savings in manufacturing floor space, capital equipment, running and maintenance costs are aa taken into consideration. It is also noteworthy to mention that PED is an 'environmentally friendly' process and does not demand the usage of corrosive industrial gases—an added bonus.

8.4 CONCLUSIONS

Pulsed Electron-beam Deposition (PED) is a relatively obscure but potentially a powerful technique for cost effective manufacturing of high temperature superconducting (HTS) films for Coated Conductor applications. This chapter is intended to introduce the reader to this novel deposition technique with a hope of creating sufficient interest among the coated conductor researchers, accelerating progress in this field. The PED technique is compared and contrasted with a popular and well established HTS deposition method, the Pulsed Laser Deposition (PLD). The detailed mechanism of the phenomena and all the features that contribute to the success of the PED technique are not clear at this point of time, reminiscent of the early stages of the historic trail of the evolution of PLD during the last 14 years. Several improvements are anticipated in the immediate future, both in the pulsed electron source performance and PED-based HTS process technologies to close the gap between the current research phase and the commercial scale-up stage. Rapid progress is expected, since the 'technology road-map' for coated conductors has already been established, with PLD as the fore-runner, identifying critical milestones that need to be realized, both from a cost and performance points of view. HTS films deposited on single crystalline substrates as well as textured RABiTS substrates have already indicated a film quality close to what is needed for practical applications. With its intrinsic cost-effectiveness, PED could very well become the deposition method of choice for high volume manufacturing of coated conductors.

ACKNOWLEDGMENTS

The authors thank several of the early researchers whose work has largely contributed to the content of this chapter. Special thanks are due to C. Schultheiss and G. Muller of Forschungszentrum Karlsruhe, Juelich for stimulating discussions on the subject, FZK for technology licensing, F.C. Matacotta of ICTP for helpful discussions, R.A. Hawsey, A. Goyal and H.M. Christen of Oak Ridge National Laboratory for RABiTS substrates, film evaluations and providing financial support. We also thank several of our collegues at Neocera, T. Venkatesan, J. Kim, G. Doman, and J. Matthews for technical help, encouragement and constructive criticism.

REFERENCES

Basting, D., ed., 1991, *Industrial Excimer Lasers, Fundamentas, Technology and Maintenance*, 2nd edition, Lambda Physik GmBH, Gottingen.

Christiansen, J. and Schultheiss, C., 1979, *Z. Physik A*, 290:35.

Coated Conductor Technology Development Roadmap, 2001, Prepared by Energetics, Inc. for US Department of Energy, Superconductivity for Electric Systems Program, August.

De Winter, G., Denul, J., and De Gryse, R., 2001, Deposition of biaxially aligned yttria stabilized zirconia layers on metal tape by modified magnetron sputtering, *IEEE Trans. Appl. Supercond.*, 11(1):2893–2896.

Dewald, E., Frank, K., Hoffmann, D.H.H., Stark, R., Ganciu, M., Mandache, B.N., Nistor, M.G., Pointu, A.M., and Popescu, I.I., 1997, Pulsed intense electron beams generated in transient hallow cathode discharges: Fundamentals and applications, *IEEE Trans. Plasma Sci.*, 25(2):272–278.

Dediu, V.I., Jiang, Q.D., Matacotta, F.C., Scardi, P., Lazzarino, M., Nieva, G., and Civale, L., 1995, Deposition of $MBa_2Cu_3O_{7-x}$ thin films by channel-spark method, *Supercond. Sci. Technol.*, 8:160.

Foltyn, S.R., Jia, Q.X., Arendt, P.N., Kinder, L., Fan, Y., and Smith, J.F., 1999, Relationship between film thickness and the critical current of $YBa_2Cu_3O_{7-\delta}$ coated conductors, *Appl. Phys. Lett.*, 75:3692.

Foltyn, S.R., Jia, Q.X., Dowden, P.C., Arendt, P.N., Smith, J.F., Holesinger, T.G., Kung, H., Coulter, J.Y., Gibbons, B.J., Jan, D.B., DePaula, R.F., Stan, L., and Groves, J.R., 2001, *High Current Coated Conductors Cased on IBAD YSZ and Thick YBCO/Sm-123 Multilayers*, Superconductivity Program for Electric Systems, Annual Peer Review, August 1–3, Washington, DC.

Geertsen, C. and Mauchien, P., 1995, in: *Application of Particles and Laser Beams in Materials Technology*, P. Misaelidis, ed., NATO ASI Series. E: Applied Sciences, Vol. 283, Kluwer, p. 237.

Gilgenbach, R.M., Kovaleski, S.C., Lash, J.S., Ang, L.K., and Lau, Y.Y., 1999, Science and applications of energy beam ablation, *IEEE Trans. Plasma Sci.*, 27(1):150–158.

Hoebel, M., Geerk, J., Linker, G., and Schultheiss, C., 1990, Deposition of superconducting YBCO thin films by pseudospark ablation, *Appl. Phys. Lett.*, 56:973.

Iijima, Y., Kakimoto, K., Kimura, M., Takeda, K., and Saitoh, T., 2001, Reel-to-reel continuous formation of Y-123 coated conductors by IBAD and PLD method, *IEEE Trans. Appl. Supercond.*, 11(1):2816–2821.

Jiang, Q.D., Matacotta, F.C., Masciarelli, G., Fuso, F., Arimondo, E., Konijnenberg, M.C., Mueller, G., and Schultheiss, C., 1993, Characterization and *insitu* fluorescence diagnostic of the deposition of $YBa_2Cu_3O_{7-x}$ thin films by pseudo-spark electron beam ablation, *Supercond. Sci. Technol.*, 6:567.

Jiang, Q.D., Matacotta, F.C., Konijnenberg, M.C., Mueller, G., and Schultheiss, C., 1994, Deposition of $YBa_2Cu_3O_{7-x}$ thin films by channel-spark pulsed electron beam ablation, *Thin Solid Films*, 241:100.

Kinder, H., Berberich, P., Prusseit, W., Rieder-Zecha, S., Semerad, S., and Utz, B., 1997, YBCO film deposition on very large areas upto 20×20 cm^2, *Physica C*, 282(28):107–110.

McIntyre, P.C., Cima, M.J., and Roshko, A.J., 1995, Epitaxial nucleation and growth of chemically derived $Ba_2YCu_3O_{7-x}$ thin films on (001) $SrTiO_3$, *J. Appl. Phys.*, 77(10):5263–5272.

Mittag, K., Choi, P., and Kaufman, Y., 1990, *Nucl. Instr. and Methods, A*, 292:465.

Muller, G., Konijnenberg, M., Kraft, G., and Schultheiss, C., 1995, Thin film deposition by means of pulsed electron beam ablation, in: *Science and Technology of Thin Films*, F.C. Matacotta and G. Ottaviani, eds., World Scientific, p. 89.

Paine, D.C. and Bravemen, J.C., eds., 1990, *Laser Ablation for Material Synthesis*, Mater. Res. Soc. Proc. 191, Pittsburgh, PA.

Paranthaman, M., Chirayil, T.G., Satyamurthy, S., Beach, D.B., Goyal, A., List, F.A., Lee, D.F., Cui, X., Lu, S.W., Kang, B., Specht, E.D., Martin, P.M., Kroeger, D.M., Feenstra, R., Cantoni, C., and Christen, D.K., 2001, Fabrication of long lengths of YBCO coated conductors using a continuous reel-to-reel dip coating unit, *IEEE Trans. Appl. Supercond.*, 11(1):3146–3149.

Rupich, M.W., Li, Q., Annavarapu, S., Thieme, C., Zhang, W., Prunier, V., Paranthaman, M., Goyal, A., Lee, D.F., Specht, E.D., and List, F.A., 2001, Low cost Y–Ba–Cu–O-coated conductors, *IEEE Trans. Appl. Supercond.*, 11(1):2927–2930.

Schultheiss, C. and Hoffman, F., 1990, *Nuclear Instruments and Methods in Physics Research, B*, 51:187.

Smith, J.A., Cima, M.J., and Sonnenberg, N., 1998, High critical current density thick MOD derived YBCO films, *IEEE Trans. Appl. Supercond.*, 9(2):1531–1535.

Solovyov, V.F., Wiesmann, H.J., Li-jun Wu, Yimei Zhu, and Suenaga, M., 2001, *Ex-situ* post-deposition processing for large area $Y_1Ba_2Cu_3O_7$ films and coated tapes, *IEEE Trans. Appl. Supercond.*, 11(1):2939–2942.

Stark, R., Christiansen, J., Frank, K., Muecke, F., and Setter, M., 1995, *IEEE Trans. Plasma Sci.*, 23(3):258.

Strikovski, M. and Miller, J.H., 1998, *Appl. Phys. Lett.*, 73:1733.

Usoskin, A., Knoke, J., Garcia-Moreno, F., Issaev, A., Dzick, J., Sievers, S., and Freyhardt, H.C., 2001, Large-area HTS coated stainless steel tapes with high critical currents, *IEEE Trans. Appl. Supercond.*, 11(1):3385–3388.

Witke, Th., Lenk, A., Schultrich, B., and Schultheiss, C., 1995, Investigation of plasma produced by laser and electron pulse ablation, *Surface Coatings and Technology*, 74–75:580–585.

Chapter 9

BaF_2 POST-DEPOSITION REACTION PROCESS FOR THICK YBCO FILMS

M. Suenaga, V.F. Solovyov, L. Wu, H.J. Wiesmann, and Y. Zhu

Division of Materials and Chemical Sciences
Energy Sciences and Technology Department
Brookhaven National Laboratory
Upton, NY 11973-5000
USA

9.1 INTRODUCTION

The basic processes of the so-called BaF_2 process for the formation of $YBa_2Cu_3O_7$, YBCO, films as well as its advantages over the in situ formation processes are discussed in the previous chapter. The process and the properties of YBCO films by this process were also nicely described in earlier articles by Feenstra et al. (1991a, 1991b). Here, we will discuss two pertinent subjects related to fabrication of technologically viable YBCO conductors using this process. These are (1) the growth of thick ($\gg$1 μm) c-axis-oriented YBCO films and (2) their growth rates. Before the detail discussions of these subjects are given, we first briefly discuss what geometrical structure a YBCO-coated conductor should be. Then, we will provide examples of simple arguments for how thick the YBCO films and how fast their growth rates need to be. Then, the discussions in the following two sections are devoted to: (1) the present understanding of the nucleation and the growth process for YBCO, and why it is so difficult to grow thick c-axis-oriented films (>3 μm), and (2) our present understanding of the YBCO growth-limiting mechanism and methods to increase the growth rates. The values of critical-current densities J_c in these films are of primary importance for the applications, and the above two subjects are intimately related to the control of J_c of the films. In general, the lower the temperatures of the YBCO formation are the higher the values of J_c of the films (Feenstra et al., 1991a, 1991b; Solovyov et al., 1999). Thus, the present discussion is limited to those films which are reacted at ~735°C. This is the lowest temperature at which c-axis-oriented YBCO films (1–3 μm thick) are comfortably grown. It is also well known that the non-c-axis oriented YBCO platelets are extremely detrimental to the values of J_c such that their effects on J_c dwarf essentially all of other microstructural effects which control J_c.

Hence, the discussion given below is mainly focused on how to avoid the growth of these crystallites when the films are thick and/or the growth rates are high.

9.2 THE STRUCTURE OF THE YBCO CONDUCTORS

The YBCO conductors are only made in tape shapes since the bi-axially textured YBCO formation is required for attaining high values of J_c for the films. Hence, we expect the structure of the completed YBCO conductors to be a similar one to that of the early Nb_3Sn tapes which were used for a 100-m power transmission cable test (Forsyth, 1988; Bussiere et al., 1977). A sketch of the cross section of an expected YBCO tape is shown in Figure 9.1 following the geometry of the composite Nb_3Sn conductor. It consists of a central core of a metallic substrate with appropriate buffer layers (not shown in the sketch) and a YBCO layer on each side of the substrate, and then thin ($\sim$25–50 μm) Cu sheets are added for the protection of the superconductor from its accidental transformation to the normal state. The thickness of Cu will depend on the applications and has to be determined for each case. Also, it can be made different on each side so that YBCO layers can be placed in compression when the tape is bent. Additionally, a high strength alloy such as a thin stainless steel tape is added on one side of the tape for mechanical strengthening. This will not be required if a high-strength alloy is used for the substrate for the YBCO layers. All of these components will be soldered to the superconductor after a thin metallic layer such as Ag is deposited on the YBCO surfaces. The thickness of the unreacted Nb core in the Nb_3Sn was only 10–15 μm thick, yet it was possible to process the tape through the chemical etching of the surface layers and the soldering Cu and stainless tapes onto it without damaging the brittle Nb_3Sn (Forsyth, 1988; Bussiere et al., 1977). Thus, one expects that the thickness of the substrate can be as thin as or less than $\sim$25 μm as long as wide ($\gg$10 mm) tapes are processed for the YBCO deposition. After these processes are completed, the composite tapes are slit to desired widths.

9.3 THE REQUIRED THICKNESS OF YBCO

Now, what would be the required thickness of YBCO in a tape as shown in Figure 9.1? Of course, it will depend on the particular application, but we can obtain a general idea about an approximate required thickness by the following example for the requirement in a power transmission cable. Currently, model cables which have been made and tested for transmission of electrical power employ four layers of Bi(2223)/Ag tapes. If we want to replace them with YBCO tapes, we expect to use

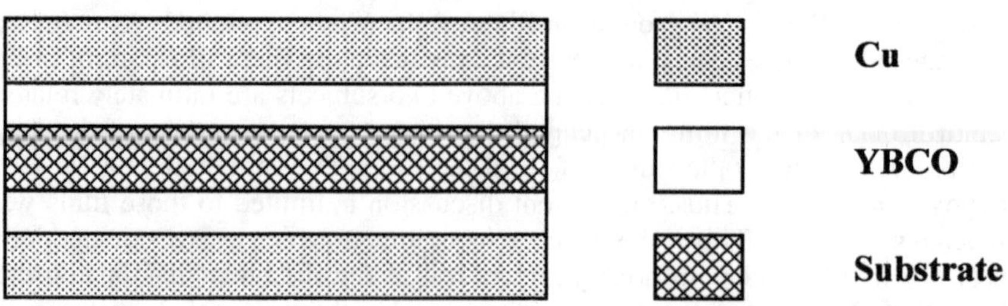

Figure 9.1. A schematic sketch of a YBCO conductor cross section.

only two layers of the conductors (as it was for the Nb_3Sn cable) instead of four. Then, in order to achieve the same critical current I_c in a YBCO cable as for a Bi(2223)/Ag cable, I_c per width of the YBCO tape needs to be ~70 A/mm at 77 K under zero magnetic field and since a recent Bi(2223)/Ag tape can carry ~100–120 A for ~3.5 mm width for its critical current. We also assume that the achievable critical-current density, J_c, of a YBCO layer is 10,000 A/mm^2 at 77 K and 0 Tesla. Then, the required total thickness is ~7 μm or ~3.5 μm per side. If one wants to be conservative or to have an added protection for fault currents, perhaps the thickness should be ~5 μm on each side of a tape. Some other applications require higher critical currents. For example, high-energy particle-accelerator applications demand >10 μm on both sides while retaining J_c of 10,000 A/mm^2 which is equivalent to ~10^5 A/mm^2 at ~15 K and 12 T. The above simple example clearly illustrates the necessity that the film thickness be made significantly greater than those currently fabricated (~1 μm or less).

9.4 THE YBCO GROWTH-RATE REQUIREMENT

We estimate the growth rate needed for processing of a YBCO tape with ~3.5 μm thick YBCO layers within a practical time by assuming that we need a 1-km tape and use a 10-m-long furnace. We also assume a continuous reel-to-reel process. Finally, we also assume that we need to heat treat this tape for less than a two-week period. This is an arbitrarily chosen period, but commercial Nb_3Sn wires currently require similar periods for their reaction heat treatment. Under these assumed constraints, the rate of the YBCO growth to meet this heat treatment duration is ~0.5 nm/s. For some applications, it is possible to heat treat the one sided conductors in a batch process, but this will be more difficult for processing tapes with YBCO on both sides. However, I believe that with some efforts it is possible to design a batch-process reactor for the tapes with YBCO on both sides if the reactor incorporates a subatmospheric pressure process which is described below. In this case, the growth rates of an order of 0.1 nm/s will be sufficient and at this growth rate a 5 μm thick film will be heat treated in less than a day. If such reactors are designed, the requirement for the increased growth rates will be considerably eased.

The above "required" thickness and growth rate (excluding a batch processing) are substantially greater than those which are being currently achieved, and meeting these requirements simultaneously is a difficult challenge for which we have to address our R and D efforts. Most of the reported thickness and the growth rates are of the order of 1 μm or less and ~0.1 nm/s, respectively. The thickest YBCO film, which was made in this process while retaining $J_c > 10{,}000$ A/mm^2, was 5-μm-thick films on $SrTiO_3$, STO, substrates (Solovyov et al., 1999). This was synthesized at atmospheric pressure using a very high partial pressure (150 Torr) of H_2O with a growth rate of ~0.2 nm/s at 725°C. However, we feel the use of such a high water partial pressure and an atmospheric pressure process is impractical as a commercial fabrication process. The main difficulty in achieving the thickness and growth-rate requirements are common, and it is the growth of the non-*c*-axis-oriented YBCO when the thickness of the YBCO is increased beyond 2–3 μm and/or the growth rate exceeds 0.2–0.3 nm/s in these films. An example of such undesirable growth of the non-*c*-axis-oriented YBCO, in this particular case the *a*-axis crystallites, is shown in Figure 9.2 from a film which was grown at a very high growth rate, ~1 nm/s. Here, large platelets of the *a*-axis oriented YBCO are grown out of the substrate while the *c*-axis-oriented YBCO exists at the substrate but much thinner than the former. When these non-*c* axis platelets of YBCO grow, the

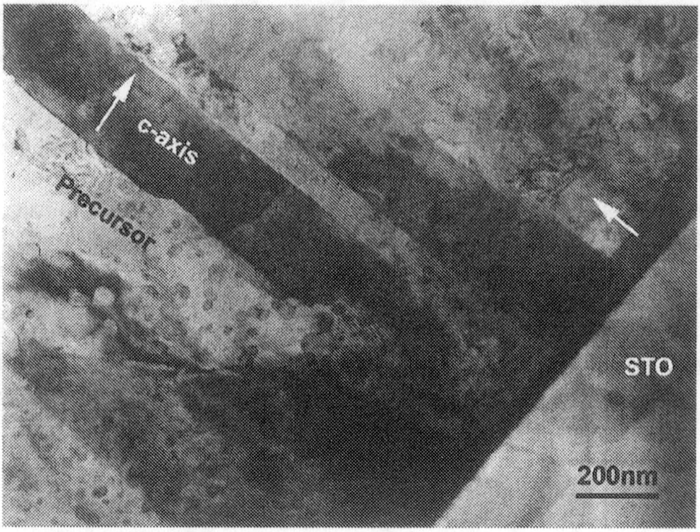

Figure 9.2. A cross sectional TEM image of a quenched YBCO film on STO illustrating the growth of the large *a*-axis oriented YBCO platelets.

critical currents of the films become disastrously smal due to the highly anisotropic nature of the superconducting current flow in YBCO.

In the following sections, we summarize the results of our studies to the present in the effort to gain understanding of the kinetics of the YBCO nucleation and growth in thick films as well as under the enhanced growth rates. This discussion will be limited to the growth of YBCO on STO from the precursor films which were prepared using electron-beam-evaporation techniques (Solovyov et al., 1998). The early stages of the growth of YBCO on CeO_2 have been previously described (Wu et al., 2001b) and will not be included here. Also, growing thick YBCO films is a major difficulty in the precursor films which are derived from sol gel processes such as the so-called trifluoroacetates (TFA) process. As will be discussed in another chapter, the nature of the difficulty about the growth of the non-*c*-axis YBCO is similar to what is discussed here. However, there are a number of important differences which exist in the growth processes for YBCO films form the TFA and the physical vapor deposition derived precursors.

9.5 THICKNESS: NUCLEATION KINETICS FOR THICK YBCO FILMS

In this section, we will describe our current understanding of the nucleation processes of YBCO from thick precursor films which were deposited onto STO by the electron-beam-evaporation technique (Solovyov et al., 1998). As described earlier (Wu et al., 2001b), an as-deposited precursor film consists of fine ($\sim$10 nm) grains of Y, Cu, and BaF_2. Upon heating, the precursor film quickly turns to a mixture of Cu_2O and a (Y, Ba) oxy-fluoride, and then from this mixture the *c*-axis oriented YBCO is grown on a substrate. The approximate composition of the oxy-fluoride, which plays a key role in the formation of YBCO as well as its nucleation, is $(Y_{0.3}Ba_{0.7})(O_{0.15}F_{0.85})_2$. It is important to note that this phase's crystal structure is identical with BaF_2 and its lattice parameter is very close to that of BaF_2. Thus, when only the standard θ–2θ x-ray diffraction, XRD, measurements are used to study the films, it is often mistakenly reported as BaF_2. Only the combined measurements of its chemical composition and its structural properties by transmission electron microscopy, TEM, techniques can properly identify this new phase. This new oxy-fluoride is clearly identifiable in TEM, particularly in the EDX compositional analysis. However, it is still puzzling how the replacement of Ba and F by Y and O in BaF_2 is accomplished without changing its lattice parameter and crystal structure when the atomic sizes of Y and Ba are significantly different. Thus, the clarification of this puzzle is one of the current subjects of our studies. The mechanism for the *c*-axis growth of YBCO layers is shown to be the epitaxial precipitation of YBCO onto the existing *c*-axis-oriented YBCO layer from a thin liquid layer containing Y, Ba, Cu, and O (Wu et al., 2001b; Solovyov et al., 2000). The Y–Ba oxy-fluoride decomposes through the reaction with H_2O at its interface with the liquid releasing F and H to the process gas. At the same time, Cu_2O or CuO, depending on the duration of heat treatment, decomposes into the liquid although the exact mechanism for this is not clear. These decomposition processes provide the necessary cations and oxygen to the liquid and then to the growth of YBCO. Note that this liquid layer only forms on an existing *c*-axis YBCO layer after a thin (a few tens of nm) layer of YBCO nucleates and covers the substrate, while the growth of the non-*c*-axis-oriented YBCO is not associated with this liquid layer. This liquid does not form on the non-*c*-axis-oriented grains. The details of the growth process are discussed elsewhere (Wu et al., 2001b). Here, in order to illustrate such a growth process taking place, a cross-sectional TEM image, which was taken from a quenched 3-μm-thick film, is shown in Figure 9.3. This shows a thin *c*-axis-oriented YBCO film on STO and a very thin ($\sim$7 nm) liquid layer separating the YBCO and the as-yet unreacted precursor. This type of growth is observed when there is no large lateral HF gradients over the film. When large HF gradients exist, the YBCO films appear to grow laterally. This can be seen by purposely designing such gradients over the films (Solovyov et al., 2001c) or at the edges of the precursor films (R. Feenstra, private communication). It is still unknown whether the liquid is present in these cases.

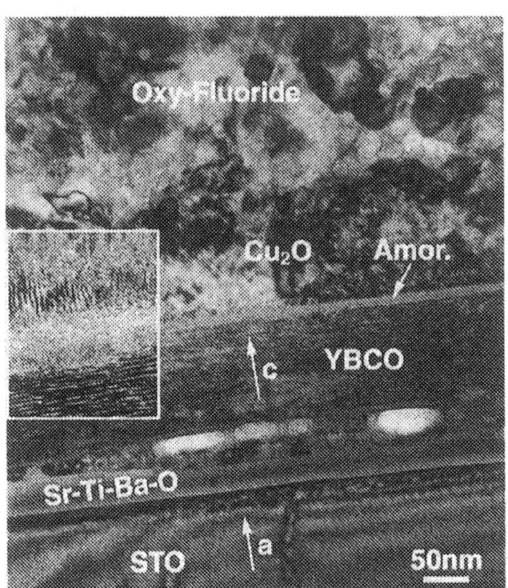

Figure 9.3. A cross sectional TEM image showing the liquid (amorphous) layer between the existing YBCO layer and the unreacted precursor from a 3-μm-thick quenched film. The inset image is the enlarge section of the region containing the liquid layer.

From the above description of the YBCO formation process, the basic chemical reaction for the formation of YBCO in the BaF_2 process can be expressed as follows (Wu et al., 2001b; Solovyov et al., 2000):

$$Y + Cu + BaF_2 + O_2 + H_2O \Rightarrow Cu_2O + (Y, Ba)\text{ oxy-fluoride} + O_2 + H_2O \quad (1a)$$

$$\Rightarrow YBa_2Cu_3O_{\sim 6.1} + HF\uparrow \quad (1b)$$

Also, as will be discussed below, the basic growth-rate-limiting process is determined by the removal rate of the reaction product HF in Equation (1b) from the surface of the film into the reaction atmosphere (Solovyov et al., 2001c). Furthermore, this reaction is nearly in equilibrium. This is demonstrated in the reversal of the reaction by controlling the partial pressure of HF in the atmosphere. (This was determined by the in situ measurements of the film's conductance during the reaction process (Wu et al., 2001a).) Hence, the general process for the YBCO growth in the BaF_2 is well understood. However, the precise mechanism of the nucleation of YBCO at the substrate surface is yet unknown. Thus, understanding this nucleation process and finding the means to minimize the non-*c*-axis-oriented YBCO are particularly important for the fabrication of technologically viable thick YBCO film conductors. This is because the growth of the non-*c*-axis YBCO becomes prevalent as the thickness of the films increase beyond ~2–3 μm, and is very detrimental to keeping high values of superconducting current densities. Thus, in this section, we will summarize the results of our recent study toward the clarification of the mechanism of YBCO nucleation in the BaF_2 process.

In order to understand the nucleation mechanism of YBCO in this process, we heat treated and quenched a set of the precursor films with thicknesses ranging from 1 to 5 μm (Solovyov et al., 2001b). All of the specimens were heated in a flowing processing-gas mixture of 100 mTorr of O_2, 25 Torr of H_2O, and N_2 at atmospheric pressure. After 10 min none of the films showed any indication of YBCO formation by XRD and TEM. At this time, only the Y–Ba oxy-fluoride and Cu_2O was observed and some of the oxy-fluoride grains' (111) planes were aligned with the STO's (001) plane on its surface as shown earlier (Wu et al., 2001a). The heating duration required for the initiation of the epitaxial nucleation of YBCO at the buried interface, a precursor film and STO, turned out to depend on the film thickness. In a 1-μm-thick film, a thin layer of *c*-axis oriented YBCO was observed by XRD after 20 min at 735°C. However, it took an additional 20 min of heat treatment for a 3- and 5-μm-thick film to develop a clearly detectable indication of YBCO being present by XRD. However, in the 5-μm film the strong diffraction lines were those belonging to the (h00) planes, or the *a*-axis-oriented grains, while the diffraction lines for a 3-μm film was mostly from the (00l) planes as shown earlier (Wu et al., 2001b).

A cross-sectional TEM study of the above 1-μm-thick film revealed that the thin (~60 nm) *c*-axis-aligned YBCO layer covered a good fraction of the surface of the substrate, and the remaining areas were occupied by the (111)-plane-aligned oxy-fluoride. In Figure 9.4, the alignment of the oxy-fluoride's (111) with STO's (001) plane is shown adjacent to a *c*-axis aligned YBCO nucleus. The dimensions of YBCO nuclei in the basal plane were far greater than along the *c*-direction. This is in accordance with the highly anisotropic growth rates of YBCO, i.e., once it nucleates, its growth in the basal plane was much faster than in the *c*-axis direction. Energy-dispersive spectroscopy analysis, EDX, shows that the YBCO layers at this stage were still deficient in Cu, and the amount varied significantly from one location to another. The nuclei's composition was in the range of $YBa_2Cu_{1.5\sim 2}O_x$.

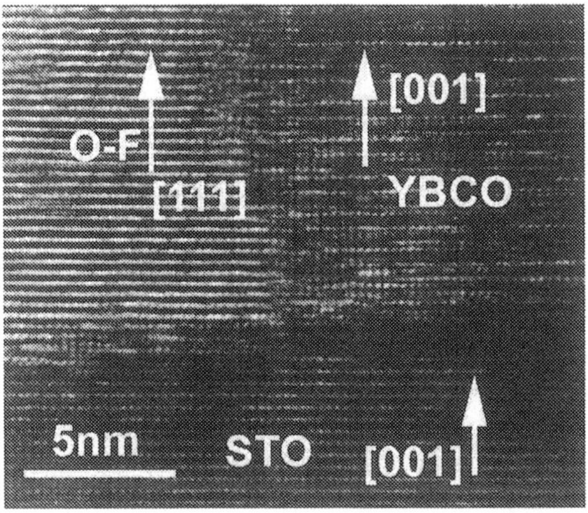

Figure 9.4. A cross sectional TEM images showing the YBCO nucleation at STO surface after 20 min at 735°C: the (001)- and the (111)-plane-aligned YBCO and the oxy-fluoride, respectively.

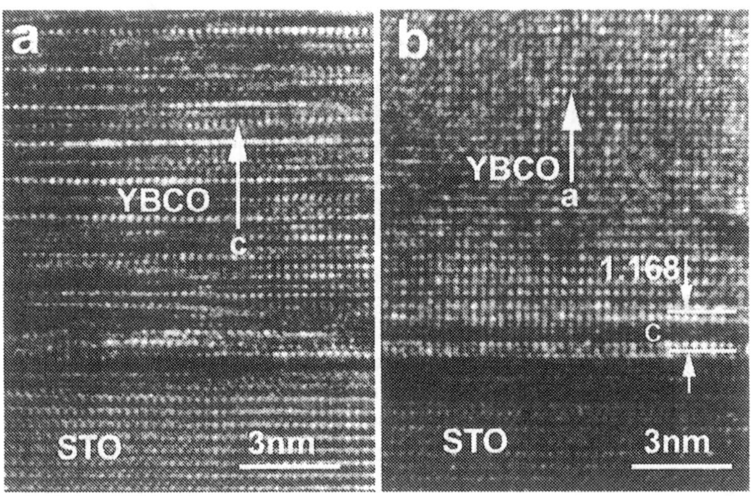

Figure 9.5. High resolution TEM images for: (a) c-axis and (b) a-axis oriented YBCO nuclei from a 5-μm-thick film heated for 30 min.

When the 5-μm-thick specimen was heated for 30 minutes, the unordered and the ordered oxy-fluorides were observed as before (Wu et al., 2001a, 2001b). The ordered oxy-fluoride has a lattice parameter which is three times that of the unordered one along its c-axis. In addition, what was new in this specimen was the observation of two types of YBCO nuclei which were oriented with their c-axis perpendicular and parallel to the STO surface, i.e., the c-axis and the a-axis grains, respectively, as shown in Figures 9.5(a) and (b). The amount of these nuclei was so small that XRD measurements did not detect the presence of YBCO. At this moment it is not clear how these two differently oriented nuclei form. After an additional 10 min of the heat treatment, the a-axis platelets were grown to exhibit strong intensities for the (h00) planes as observed in x-ray diffractograms. Thus, this indicated that once nucleated, the *a*-axis-oriented nuclei grew extremely rapidly along their *a*/*b*- as well as *c*-directions while the *c*-axis-oriented grains grew slowly from the substrate along the *c*-axis as shown in Figure 9.2.

The important question, which we need to answer, is what difference in the local conditions at the interface caused the nucleation of differently oriented YBCO nuclei. In attempts to answer this, it is often mentioned that in the thick films the diffusion of H_2O into and of F and H out of the interface region become more difficult as the thickness increases. It is possibly true that the slower diffusion of these species in thick films will delay the nucleation process, but it does not clarify why two types of the nuclei formed. Also, since it was found that the growth rates of the *c*-axis YBCO are independent of the precursor thickness (Solovyov et al., 1999), it is doubtful that the diffusion rates for H_2O and F and H are significantly different for thin and thick films. As mentioned above, the presence of the non-*c*-axis YBCO platelets in the films are so detrimental that finding the mechanism for and the means to prevent this type of the nucleation is extremely important for the successful fabrication of commercially viable YBCO conductors.

9.6 GROWTH KINETICS: ATMOSPHERIC AND SUBATMOSPHERIC PRESSURE PROCESSES

In order to develop a fabrication process for YBCO-coated conductors, it is very important to understand the basic kinetic mechanism(s) for the growth-limiting process in the BaF_2 process. This understanding will help to guide us in developing the required manufacturing process or the reactor designs. In order to accomplish this, we will first treat the growth of YBCO for the process gas flow at atmospheric pressure. Also, we choose a particular set of the conditions, i.e., a small specimen size and a slow gas flow for which an analytical solution is derivable (Solovyov et al., 2001b). Then, with similar simplifications, the analysis is extended for the reaction of a long conductor. Some of the implications from such analyses are discussed relative to processing long lengths of the tapes. We will also discuss a processing method where the heat treatment is carried out in partial vacuum (or subatmospheric pressures) (Solovyov et al., 2001a, 2001c). It was shown that not only the growth rates could be greatly enhanced in such a processing condition, but also the designs of the reactor operating at subatmospheric pressures could be significantly simplified in comparison with the atmospheric reactor. These findings are very important since in the atmospheric pressure reaction (1) the growth rate of YBCO will be very slow, ~0.1 nm/s, unless the gas flow is greatly increased to such high levels that it becomes impractical, and (2) the reactors can be very complicated for long tape fabrications due to the length of the tape, which can be uniformly reacted, is very limited, e.g., a few tens of mm. The advantages in the use of the subatmospheric pressure processing are described in the second part of this section.

9.6.1 Atmospheric-Pressure Process

As described previously, in the atmospheric pressure reactor a mixture of an inert gas, e.g., N_2, and oxygen is humidified and then passed into the reactor at atmospheric pressure. Such a simple design is sufficient to achieve good properties for YBCO films on short buffered tapes or single-crystalline substrates. In order to model the growth kinetics in such a system, we initially assumed that the diffusion of H_2O and/or HF through the unreacted precursor film was the growth-rate-limiting process (Solovyov et al., 2000). However, the analysis of the growth rate based on such an assumption did not yield the linear growth rate with heating time which was experimentally observed (Solovyov et al., 1999). This result and the observations by others (the attempts to process ~0.5 m long tapes in the atmospheric reactor resulted in a very inhomogeneous growth from one end of the tape to the other (R. Feenstra, unpublished)) lead us to seek other rate-limiting causes. One possibility for the occurrence of such an inhomogeniety is due to water vapor starvation of the growing tape. A simple estimate shows that due to very high water vapor pressure and low growth rates, typically found in an atmospheric reactor, the supply of water vapor by diffusion only is more than enough to support stable growth. It is more likely that removal of HF, which is produced as a result of the (Y, Ba) oxy-fluoride decomposition as a part of reaction in Equation (1b), causes problems with growth inhomogeneity. Thus, with the following simplifying assumptions, we derived an analytical expression for the YBCO growth rate which predicts correctly all of the experimental observations on the growth rate with the heating time, the H_2O partial pressure, and the size of the specimen as variables (Solovyov et al., 2001b).

The assumptions used in the model are:

(1) The reaction, Equation (1b), is in a state of dynamic equilibrium at the growth front. If we neglect the spatial variations of $p(H_2O)$ due to high $p(H_2O)$ values, the partial pressure of HF, $p_i(HF)$, at the YBCO-precursor interface is related to the water-vapor partial pressure, $p(H_2O)$, by the equilibrium condition:

$$\frac{p_i(HF)}{[p(H_2O)]^{1/2}} = K, \tag{2}$$

where K is the equilibrium constant of the reaction equation (1b). If the growth rate is limited by HF removal from the film surface, a simple argument shows that $G \propto p_i(HF)$ but from Equation (2) $G \propto p(H_2O)^{1/2}$ in agreement with the experimental observation (Forsyth, 1988; Bussiere et al., 1977). Our estimates show that the equilibrium constant K is quite small, for instance the equilibrium HF partial pressure is only about 10 mTorr at a typical value of $p(H_2O) = 20$ Torr.

(2) The rate of HF removal is limited by gaseous diffusion in the processing atmosphere. As a consequence HF pressure at the film surface, $p_s(HF)$, can replace the partial pressure at the growth interface, $p_i(HF)$, in Equation (2). We believe that this is true, at least for films thinner than 5 μm. The most compelling evidence is that the growth rate does not depend on the precursor film thickness, but depends on lateral width W of the film, that is $G \propto 1/W$. Our interpretation of this observation is that HF flow is determined by the pressure gradient in the gas phase which is approximately $p_s(HF)/W$. Then, the kinetics of growth in the BaF_2 process is reduced to a classic problem of mass transfer from a surface by diffusion and convection of the carrier gas. We treat the case where the gas velocity is quite low. Then, flow is laminar and diffusion is the predominant mechanism of the HF transport from the film surface.

Using the above assumptions, the average YBCO growth rate of a square precursor film with small width W can be expressed as follows (Solovyov et al., 2001b):

$$G \approx \frac{VDK[p(H_2O)]^{1/2}}{kTS}, \tag{3}$$

where $V = 10^{-22}$ cm^3 is the volume of the YBCO unit cell, k is Boltzmann's constant, T the processing temperature (= 735°C) and D is the diffusion constant of HF in the processing atmosphere (≈2.5 cm^2/s at $T = 735$°C and $p_t = 760$ Torr, *the total pressure*). S is a tabulated sample shape factor which also depends on the gas flow rate F. It can be shown that for the case of a small sample, $F \gg WD$, $S \approx W$. Note that Equation (3) appears to predict the growth rate being inversely proportional to temperature contrary to a general intuition about the growth rates. However, since the equilibrium constant K is exponential depends on temperature exponentially, Equation (3) predicts the temperature dependence correctly, i.e., the higher the temperature the faster the growth.

The case for a long tape can be treated in a similar manner with some additional simplifying assumptions. For example, we have assumed the conductor being a cylinder coated with the precursor film so that an analytical solution for the growth rate can be found. However, an essential parameter in the long-tape case is the HF background pressure, p_b(HF). Since the local growth rate is proportional to the pressure gradient, $G \sim [p_s(\mathrm{HF}) - p_b(\mathrm{HF})]$, the growth slows down considerably when $p_s(\mathrm{HF}) \approx p_b(\mathrm{HF})$ in down stream (Solovyov et al., 2001b).

Three factors contribute to the spatial distribution of p_b(HF) along the reactor: (a) HF production by the tape, (b) convection due to carrier gas flow, and (c) gaseous diffusion. In order to simplify the problem the HF concentration gradients in the radial direction were neglected. Combining the three factors above we obtained the following one-dimensional equation:

$$D\frac{d^2n_b}{dx^2} - v\frac{dn_b}{dx} + D\frac{n_s - n_b}{S_r} = 0, \tag{4}$$

where $n_b(\mathrm{HF}) = p_b(\mathrm{HF})/kT$ is the background HF concentration at a distance x from the "upstream" end of the tape, S_r is the reactor cross-section, $v = F/S_r$ is the carrier gas velocity, F is the gas flow rate, and $n_s(\mathrm{HF}) = p_s(\mathrm{HF})/kT$ is the equilibrium HF concentration at the tape surface, defined by Equation (2). An analytical solution of Equation (4) gives the following functional dependence of the growth rate on distance, x:

$$G(x) = G(x = 0)e^{-x/\lambda}, \tag{5}$$

where $G(x = 0)$ is the growth rate at the upstream end of the cylinder. The parameter λ in Equation (5) defines the approximate reactor capacity, or maximum tape length, which can be processed in the reactor. λ values depend on the reactor geometry and the gas flow. As assumed in this approximation, if the gas flow is slow, $F \ll DS_r^{1/2}$, then $\lambda = S_r^{1/2}$, i.e., λ is equal to the reactor radius. An estimate of λ demonstrates that a simple tubular atmospheric-pressure reactor is not suitable for large-scale processing of coated tapes. For example, for our reactor $F = 0.20$ l/min and $\lambda = 70$ mm. To react a long tape, much faster gas flow rates are needed, and an exact numerical solution of the corresponding mass transfer equation is required. However, this simple analysis clearly points out the difficulty of the use of a standard linear-gas-flow reactor for heat treating the long tapes. Such considerations lead to a design of the reactor where the closely spaced in- and out-lets for the gases are needed and these are placed perpendicular to the length of the tape. For a large-scale production facility, this type of the reactor design not only leads to a undesirable complication, but also has other flaws such as a shadowing effect, and requires an excessive amount of the gases needed for the process. Shadowing can be caused by slight local curvatures in the tape or the tape not lying truly parallel to the gas flow. Since the shadowing effect results in nonuniform flow of the gas, it causes nonuniform reactions in some parts of the tape due to nonuniform HF removal rates. If the gas flow is increased to gain high growth rates, an excessive amount of the gases may be needed for the processing long tapes. As described below, in addition to the advantage of having the increased growth rates, the subatmospheric-pressure process can eliminate some of these difficulties by the nature of its process.

9.6.2 Subatmospheric-Pressure Process

In a subatmospheric-pressure reactor a pump is used to force gas transport through the reactor tube. The balance between pumping speed and gas-flow rate defines the absolute pressure in the reactor. The HF diffusion constant in the processing atmosphere depends on the total pressure, p_t, as $D \propto 1/p_t$. It follows from Equation (3) that the growth rate, G, should depend on p_t and $p(H_2O)$ as (Solovyov et al., 2001c):

$$G = A\frac{[p(H_2O)]^{1/2}}{p_t}, \tag{6}$$

where A is a constant which can be easily inferred from Equation (3). We carried out a set of growth runs in a wide range of conditions: total pressure was varied from 0.2 to 760 Torr and water-vapor pressure from 10^{-3} to 300 Torr. The samples in each case were 1-μm thick precursor films on 3×10 mm^2 substrates, and the processing temperature was 735°C. The results are summarized in Figure 9.6 where the growth rate, G, is plotted versus the water-vapor pressure, $p(H_2O)$, for various total-pressure values. Data for the atmospheric reactor are also presented as the curve corresponding to $p_t = 760$ Torr. Approximations to Equation (6), where $A = 12$ and the pressures are in Torr, are shown as solid lines in the figure. It is clear that Equation (6) well describes the growth kinetics.

It is evident from Figure 9.6 that the partial-vacuum reactor allows the achievement of very high growth rates at low total pressures. However, there are important limitations on the utilization of the high growth rates achieved in this process as far as the superconducting current densities are concerned. For 1 μm thick films, we found that for $G > 0.5$ nm/s, YBCO nucleated as a mixture of random and c-axis oriented phases. Stable c-axis growth was observed at $G < 0.2$–0.3 nm/s with J_c (0 T) $\approx$ 10,000 A/mm^2 and J_c (1 T, $H \| c$) $\approx$ 1 to 2×10^3 A/mm^2 at 77 K. Higher values of J_c were measured at lower growth rates. An obvious way to slow down the growth rate is to decrease $p(H_2O)$ and increase p_t. It turns out, however, that $p(H_2O)$ values much less than 10 Torr are detrimental to c-oriented growth in the films much thicker than 1 μm. On the other hand, a partial-vacuum reactor operated at high total pressure is converging towards an atmospheric pressure reactor with all the associated problems outlined in the previous section. Thus, a set of compromise pressures is needed to be determined where a reasonable growth rate G and lengths λ for the uniform YBCO formation, e.g., $G \sim 0.3$–0.5 nm/s and $\lambda \sim 20$–40 m, respectively, are

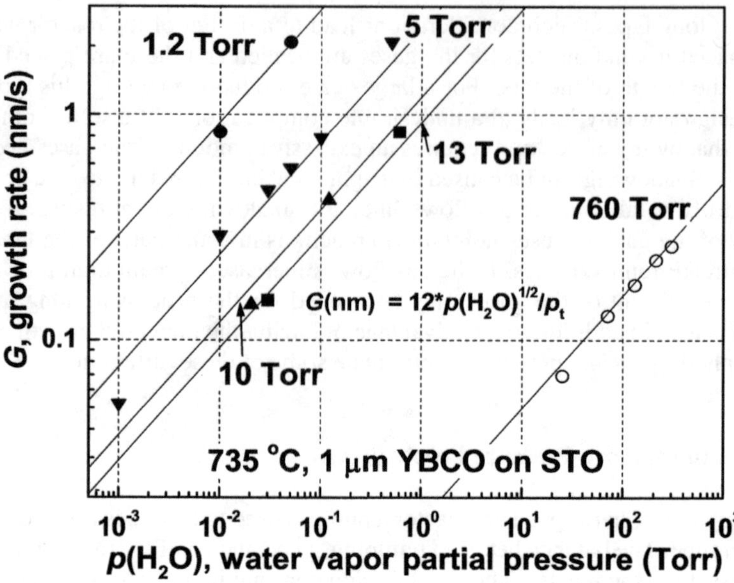

Figure 9.6. The growth rates as a function of the partial pressure of H_2O for various total reactor pressures are plotted for the sub-atmospheric BaF_2 process for 1 μm thick films.

attainable. One of the possible methods to accomplish this is to grade the total pressure along the reactor tube. For example, for a reel-to-reel system the growth rates at the front end of the tube is made slow by having high H_2O partial and total pressures, but growth rates will be made higher by decreasing the pressures toward the tail end of the reactor. In this way, the *c*-axis-oriented YBCO is nucleated while the growth rate is controlled to small values at the initial stage of the moving tape. Then, as it moves along the reactor, the pressures are gradually decreased such that the desired higher growth rates are achieved while retaining the *c*-axis orientation in the YBCO films. Such a reactor requires a careful design of the structure of the tube by a comprehensive study of the gas flow dynamics in the tube. This is one possible approach to solve this difficult problem of the avoidance of the non-*c*-axis crystallite growth. However, unfortunately, this may require undesirably complicated reactors, and may result in the increased cost for the fabrication of the YBCO conductors. An alternative approach to avoid this complication is to design a tape-holding fixture such that the tapes with YBCO on both sides can be reacted in a batch process. If a such fixture is made the required growth rate can be reduced significantly to more attainable values (e.g., ~0.1 nm/s or less). In both cases, some careful engineering reactor designs are needed.

9.7 SUMMARY

The BaF_2 process is a very attractive fabrication process for YBCO conductors since it allows the deposition and the formation steps of YBCO films to be separately performed, and it has been demonstrated that the films with high critical current densities can be fabricated on the buffered metallic substrates. However, we still face one major challenge for this process to be made as a commercially viable fabrication method. This is the avoidance of non-*c*-axis-oriented growth of YBCO when the thickness and the growth rates for the films are increased significantly beyond what are currently achieved. Meeting both the thickness and the growth-rate requirements simultaneously is extremely difficult. One possible way to make the rate requirement reduced is to find winding methods of the tapes in such ways that the both-sided tapes can be reacted in a batch process. Although the winding fixture will be more complicated than a simple dram, I believe that there are ways to construct such fixtures if the subatmospheric-pressure processing is incorporated in the design for reacting the wide precursor tapes. Then, we are only faced with the problem of growing the c-axis-oriented thick YBCO films. In order to overcome the latter difficulty, we need a detailed understanding of the YBCO nucleation kinetics in thick films under various atmoshperic conditions. Such an understanding is likely to lead to the development of new or modified precursor-deposition methods and/or reaction-heat treatment procedures to facilitate the growth of the thick YBCO layers without the non-c-axis-oriented YBCO growth.

ACKNOWLEDGMENT

The authors are grateful to D.O. Welch and R. Feenstra for stimulating discussions about the subject of this chapter during the coarse of the study. The work was performed under the auspices of Division of Materials Sciences, Office of Basic Energy Sciences, and Office of Hydrogen and Superconductivity, Office of Energy Efficiency and Renewable Energy, US Department of Energy under contract No. DE-AC02-98CH10886.

REFERENCES

Bussiere, J.F., Kovachev, V., Klamut, C., and Suenaga, M., 1977, *Adv. in Cryog. Engn.*, 24:449.

Feenstra, R., Christen, D.K., Budai, J.D., Pennycook, S.J., Norton, D.P., Lowndes, H.H., Klanbunde, C.D., and Galloway, M.D., 1991a, in: *Proc. of Sym. A-1 on High Temp. Supercond. Films at the Internat. Conf. on Adv. Mater.*, L. Correra, ed., Strasbury, France, North-Holland, Amsterdam, p. 331.

Feenstra, R., Lindemer, T.B., Bdai, J.D., and Gallorway, M.D., 1991b, *J. Appl. Phys.*, 69:6569.

Forsyth, E.B., 1988, *Science*, 242:391.

Solovyov, V.F., Wiesmann, H.J., Suenaga, M., and Feenstra, R., 1998, *Physica C*, 309:269.

Solovyov, V.F., Wiesmann, H.J., Wu, L., Suenaga, M., and Feenstra, R., 1999, *IEEE Trans. Appl. Supercond.*, 9:1467.

Solovyov, V.F., Wiesmann, H.J., Wu, L., Zhu, Y., and Suenaga, M., 2000, *Appl. Phys. Lett.*, 76:1911.

Solovyov, V.F., Wiesmann, H.J., and Suenaga, M., 2001a, in: *International Workshop on Superconductivity*, June 2001, Honolulu, Hawaii, an extended abstract.

Solovyov, V.F., Wiesmann, H.J., and Suenaga, M., 2001b, *Physica C*, 353:14.

Solovyov, V.F., Wiesmann, H.J., Wu, L., Zhu, Y., and Suenaga, M., 2001c, *IEEE Trans. Supercond.*, 11:2939.

Wu, L., Solovyov, V.F., Wiesmann, H.J., Zhu, Y., and Suenaga, M., 2001a, in: *International Workshop on Superconductivity*, June 2001, Honolulu, Hawaii, an extended abstract.

Wu, L., Solovyov, V.F., Wiesmann, H.J., Zhu, Y., and Suenaga, M., 2001b, *J. Mater. Res.*

Chapter 10

ISSUES AND PROGRESS RELATED TO THE CONTINUOUS *EX-SITU* BaF_2 PROCESSING OF LONG-LENGTH YBCO COATED CONDUCTORS

Dominic F. Lee, Keith J. Leonard, Song-Wei Lu, Donald M. Kroeger, and Fredrick A. List III

Metals and Ceramics Division
Oak Ridge National Laboratory
Oak Ridge, TN 37831-6116
USA

10.1 INTRODUCTION

For the past several years, an increasing amount of effort in the high temperature superconductor (HTS) community has been focused on the development of YBCO coated conductors. Very high values of critical current density (J_c) in excess of 10^6 A/cm^2 at 77 K and self-field have been reported on YBCO films grown on biaxially textured metallic substrates. These excellent results were typically obtained on a variety of short-length (less than 2cm) textured templates such as ion beam assisted deposition (IBAD) substrates (Iijima et al., 1992; Reade et al., 1992; Wu et al., 1995; Bauer et al., 1999b), rolling assisted biaxially textured substrates (RABiTS) (Goyal et al., 1996; Norton et al., 1996; Petrisor et al., 1999; Wang et al., 2000) and, to a lesser extend, inclined substrate deposition (ISD) substrates (Fukutomi et al., 1994; Bauer et al., 1999a). YBCO deposition methods employed to obtain these high J_c's include pulsed laser deposition (PLD) (Mathis et al., 1998), thermal evaporation (Bauer et al., 1999a), metalorganic chemical vapor deposition (MOCVD) (Ignatiev et al., 1996; Selvamanickam et al., 2000), BaF_2 precursor (Paranthaman et al., 2000; Feldmann et al., 2001) and TFA (Rupich et al., 2001).

While there are ample demonstrations of high J_c's on short-length YBCO coated conductors, there has been only limited investigation into the feasibility and issues related to the fabrication of meter-length YBCO coated conductors. This is mainly due to the limited availability of long-length substrates as well as the resources necessary to investigate the epitaxial YBCO formation process in a moving manner. Nevertheless, meter-length YBCO with high end-to-end J_c's have been produced on long-length substrates using the PLD technique (Foltyn et al., 1999; Iijima et al., 2000;

Sato et al., 2001). Although this *in-situ* technique is a well proven method, it requires simultaneous control of film composition and HTS epitaxial growth within a relatively narrow process window. Still, much knowledge has been gained in the continuous deposition of YBCO by PLD.

The *ex-situ* approach of YBCO formation, on the other hand, separates the precursor deposition step (i.e. compositional control) from the YBCO epitaxial growth process. By separating these steps, further simplification in processing procedures such as non-vacuum deposition of precursor film (Rupich et al., 2001) or batch conversion of YBCO is possible. Unfortunately, most of the *ex-situ* works have been performed on short-length samples to date. Consequently, issues related to the continuous *ex-situ* processing of YBCO are largely unknown and thus, the feasibility of this method in the manufacturing of long-length YBCO coated conductor cannot be evaluated. In this paper, we will present the progress and known issues that are important in the BaF_2 *ex-situ* processing of long-length YBCO coated conductor in a moving manner.

10.2 BaF_2 *EX-SITU* PROCESS

One of the better-known *ex-situ* YBCO formation techniques is the so-called BaF_2 process. In this process, stoichiometric amount of Y, Cu and BaF_2 is co-evaporated onto a substrate at room temperature. Then, the precursor film is converted into YBCO at an elevated temperature in a controlled atmosphere as shown schematically in Figure 10.1. Briefly, the precursor film is heated to a conversion temperature (e.g., 740°C) in a humidified (e.g., 70 Torr H_2O) reduced-oxygen environment (e.g., 130 mTorr O_2). It has been proposed that during this heat-up, the metal–oxifluoride precursor starts to decompose in the presence of H_2O and releases HF into the environment, and the decomposed precursor product is converted into YBCO according to the schematic reactions (Solovyov et al., 2000)

$$Y_2O_3 + CuO + BaF_2 \rightarrow CuO + (\text{Y–Ba–O–F}), \quad (1a)$$

$$CuO + (\text{Y–Ba–O–F}) + H_2O \rightarrow (\text{Y–Ba–Cu–O})_{\text{liquid}} + 2HF, \quad (1b)$$

$$(\text{Y–Ba–Cu–O})_{\text{liquid}} \rightarrow \text{YBCO}, \quad (1c)$$

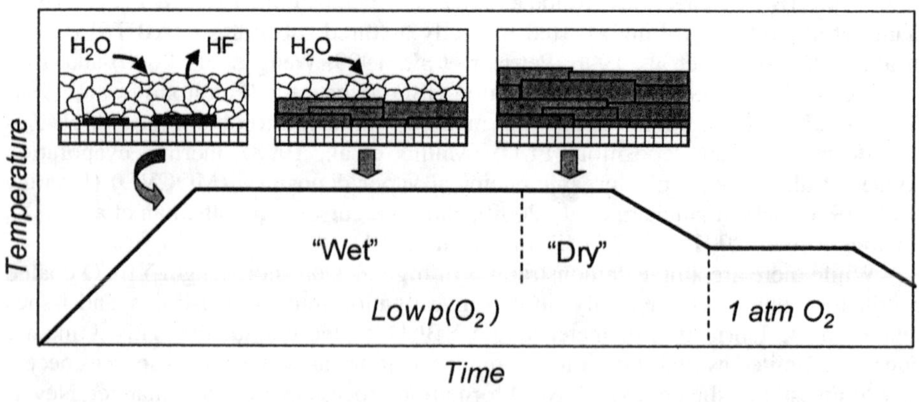

Figure 10.1. Schematic of typical oxifluoride decomposition/YBCO conversion schedule used in the *ex-situ* processing of YBCO conductors.

the critical step being that (Y–Ba–O–F) or BaF_2 is able to decompose totally and efficiently. That is, in order to achieve complete conversion of the precursor to epitaxial YBCO, it is necessary for the sample to be annealed in this humidified environment for a sufficient length of time such that the oxifluoride is fully decomposed. Once this has been accomplished, the environment is switched to "dry" to ensure that YBCO formation will proceed to completion without degradation due to moisture. Then, the sample is cooled to room temperature.

Using this general procedure and simple longitudinal-flow atmospheric reaction chambers (Figure 10.2), high quality YBCO films have been processed in short lengths at our laboratory and elsewhere (Paranthaman et al., 2000; Feldmann et al., 2001; Rupich et al., 2001; Solovyov et al., 2000; Smith et al., 1999). When we attempted to process samples of longer lengths, however, YBCO conversion was found to be non-uniform (Feenstra et al., 1999). The appearance of a 10 cm-long sample on RABiTS converted *ex-situ* in a simple longitudinal-flow geometry is shown schematically in Figure 10.2. It can be seen that while the upstream portion of the sample has been converted to YBCO, a substantial section at the downstream end remained unconverted. We believe that this non-uniformity in conversion is largely related to the local HF/H_2O concentration ratio near the sample surface. According to Equation (1b), HF is released into the environment as H_2O reacts with oxifluoride during its decomposition. Consequently, unless HF is efficiently removed from the sample surface, decomposition of the oxifluoride will slow or even terminate with time owing to a buildup of HF. When this occurs, a corresponding reduction and eventual stoppage of YBCO conversion will result. This is a possible scenario during conversion of lengths in a simple longitudinal-flow chamber where the gas velocity is typically low and the effective dimension of conversion is large (i.e. the sample length). During *ex-situ* annealing, the upstream portion of a sample will undergo the oxifluoride decomposition/YBCO conversion reaction first, while releasing HF into the environment. In the absence of efficient removal of HF, this will lead to a local increase in HF concentration downstream, effectively reducing the decomposition/conversion rate of the adjacent material. While this effect may be insignificant for short lengths, it is likely to be important as one increases the length of the sample leading to the non-uniformity in YBCO conversion that we have observed.

It is, of course, possible to increase the length of "wet" conversion time to obtain YBCO at the downstream portion of a long tape. However, doing so will result in the degradation of converted material at the upstream section of the tape due to excessive exposure to the humidified atmosphere. In addition, the time necessary to process a length of YBCO conductor in such a manner will be practically prohibitive. Therefore since 1998, we have engaged in the study of a two-prong approach that can minimize

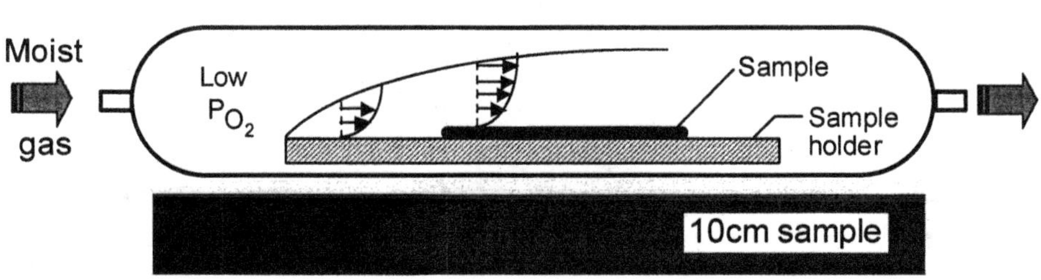

Figure 10.2. Schematic of a simple atmospheric longitudinal-flow quartz reaction chamber typically utilized in the *ex-situ* processing of short-length YBCO conductors. Also shown is the schematic of a typical 10 cm-long YBCO conductor on RABiTS converted in such a reactor.

the non-uniform conversion problem while maximizing the amount of material that can be converted at one time. This approach utilizes an extended transverse-flow geometry. By employing a transverse-flow geometry where the direction of gas flow is parallel to the width of the sample, the effective conversion distance is reduced significantly. Furthermore, the presence of an extended reaction zone along the sample length will enable the simultaneous conversion of a larger area of precursor, thereby reducing the total amount of processing time.

10.3 COMPUTATIONAL FLUID DYNAMICS SIMULATIONS OF TRANSVERSE-FLOW GEOMETRY

Prior to the design and fabrication of our extended transverse-flow reaction chamber, computational fluid dynamics simulations were performed to visualize the flow profile of a sample suspended in a transverse-flow geometry. In addition, simulations were performed to study the effects of flow rate on both gas velocity and HF concentration in the vicinity of the sample. Briefly, a commercially available program was used in the simulation. Square meshes were generated, which represent the transverse cross section of a 4.5 cm-wide by 1 cm-high rectangular reaction chamber (Figure 10.3). A 1 cm-wide tape is suspended at the center of this simulated chamber such that the longitudinal-axis of the tape, i.e. length of the tape, is perpendicular to the plane of the figure. Two 125 μm-high slots are located at either side of the chamber, with the slot on the left representing the gas inlet and the right one is the outlet thereby resulting in a transverse-flow geometry across the width of the sample.

Figure 10.3 shows the simulated flow profile within the transverse cross section of the reaction chamber. It can be seen that there is no unexpected stagnant region around the sample, and uniformity of gas flow velocity is rapidly achieved after a short distance from the gas inlet. Closer examination of gas velocity in the vicinity of the sample surface (Figures 10.3 and 10.4) reveals that gas flow is symmetric around the middle of the sample, with higher gas velocities at the leading and trailing edges of the sample as compared to that at the middle. As expected, Figure 10.4 shows that the boundary layer thickness increases with gas flow rate. More importantly, these simulations show that gas velocity is substantially increased at every location near the sample surface when a higher flow rate is employed. For example, at a location 100 μm above the sample surface and along the leading edge, gas velocity for the 5 l/min flow rate case is 4 times that of the 1 l/min case. This suggests that a high gas flow rate will

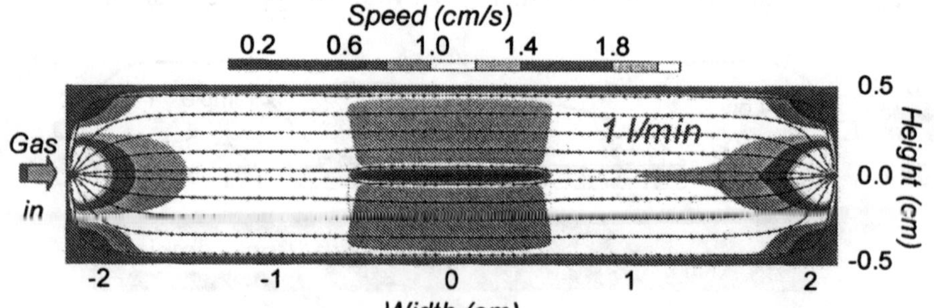

Figure 10.3. Simulated flow profile within the transverse cross section of a transverse-flow reaction chamber. Flow rate of the gas is 1 l/min and the 1 cm-wide sample is located at the center of the chamber.

be more effective in reducing the HF concentration near the sample surface by virtue of the higher gas velocity.

To examine the influence of gas flow rate on HF concentration near the surface of the sample, flow simulations were performed using the same mesh setup on a 1 cm-wide sample with precursor material on the top surface only. HF concentration maps were generated for the two flow rates (1 and 5 l/min) by assuming sufficient H_2O is present such that a 1 μm-thick precursor will be fully converted in 1 hour. It can be seen in Figure 10.5 that for a given gas flow rate, the highest HF concentration is located at the surface along the downstream portion of the sample. This is in agreement with our contention that as the oxifluoride decomposition/YBCO conversion reaction proceeds, HF concentration will increase downstream as HF is released at the leading edge. Then, it is up to the flowing gas to help sweep the HF away from the sample surface such that a sufficiently low HF/H_2O concentration ratio is present for the oxifluoride decomposition to proceed at a reasonable pace. The ability of a higher gas velocity to remove the HF more effectively can also be seen in the figure. Notice that not only is the HF concentration lower for the 5 l/min flow rate, but also that the region containing the highest HF concentration is much smaller. Figure 10.6 shows a more

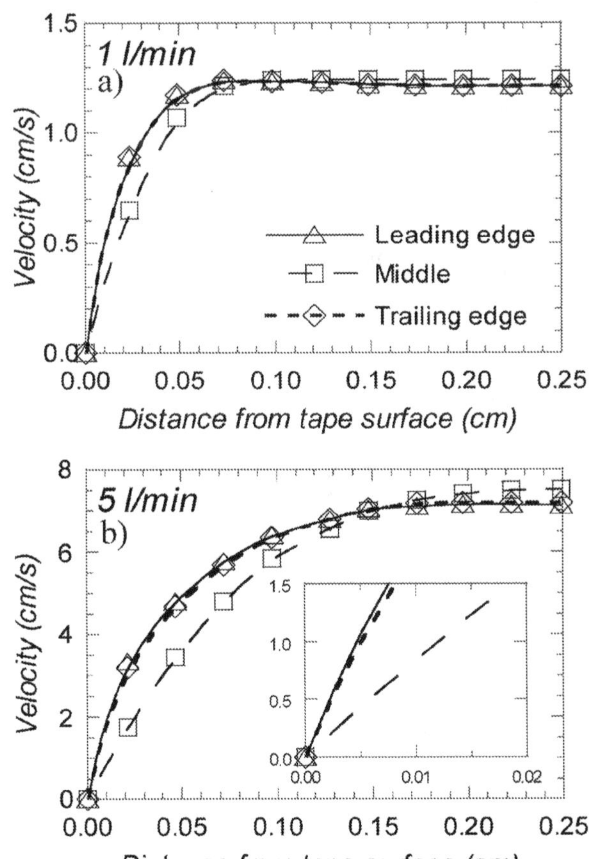

Figure 10.4. Variation in gas velocity with distance from tape surface at various locations along the sample width for flow rates of: (a) 1 l/min and (b) 5 l/min (inset shows enhanced detail near the tape surface in (b)).

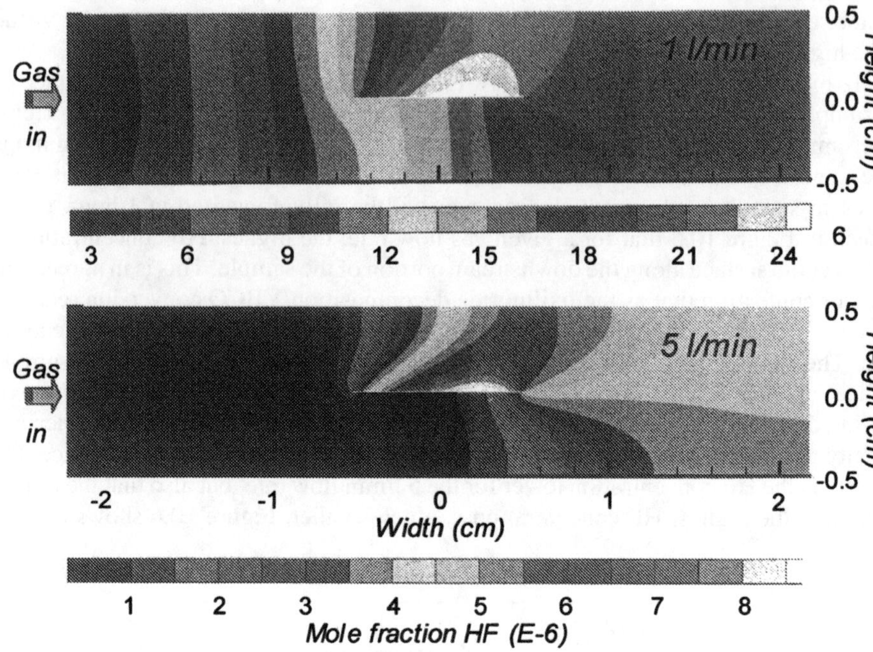

Figure 10.5. Simulated HF concentrate within the transverse cross section of a transverse-flow reaction chamber for gas flow rates of 1 l/min and 5 l/min.

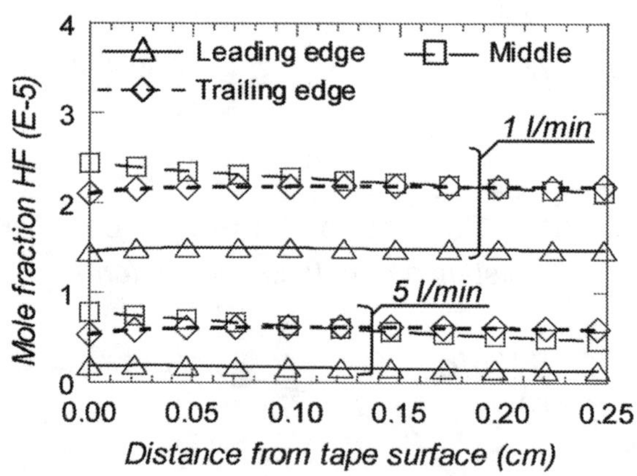

Figure 10.6. Variations in HF concentration with distance from tape surface at various locations along sample width for 1 and 5 l/min flow rates.

detailed analysis of changes in HF concentration near the surface of the sample. It can be seen from this figure that at any specific location along the width of the sample, the local HF concentration does not appear to vary greatly with distance from the tape surface. Instead, HF variation along the sample width appears to be more significant. More importantly, the HF concentration at all locations is much lower for the 5 l/min flow rate than the 1 l/min case, with the reduction as high as a factor of 6. Variation in HF concentrate at the surface of the sample is plotted as a function of distance from the leading edge in Figure 10.7. It can be seen more clearly from this figure that even

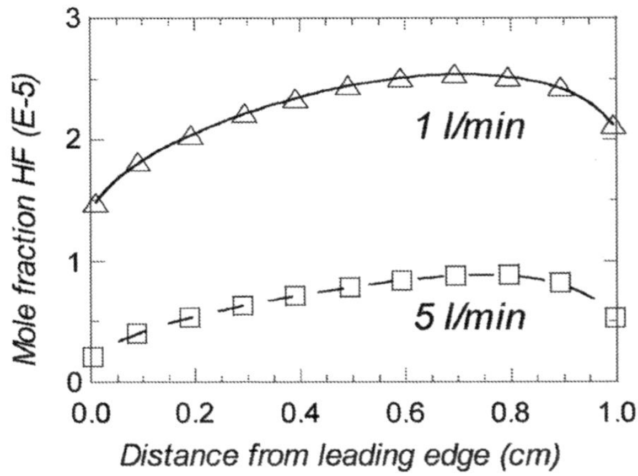

Figure 10.7. Changes in HF concentration at tape surface with distance from the sample leading edge for 1 and 5 l/min flow rates.

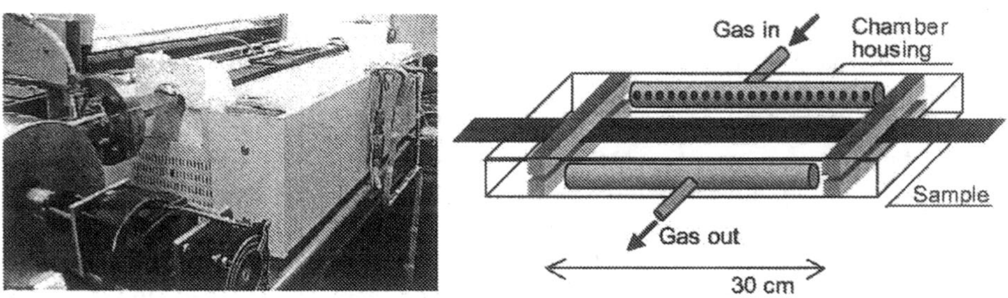

Figure 10.8. Photo of the single flow-module reaction chamber that sits in the cradle of a single-zone furnace. Also shown is a schematic of the 30 cm-long flow-module that is located in the center of the 1 meter-long reaction chamber housing.

though the tendency of HF concentrate to increase downsteam remains in effect, an increase in flow rate will reduce the HF buildup significantly. As such, an extended transverse-flow reaction chamber with a high flow rate capability should enable the *ex-situ* conversion of long-length YBCO coated conductors.

10.4 REEL-TO-REEL SINGLE-MODULE TRANSVERSE-FLOW REACTION CHAMBER: STATIONARY YBCO CONVERSION

10.4.1 Design of Single-Module Transverse-Flow Furnace

An extended single-module transverse-flow atmospheric reaction furnace was designed and built based on the flow simulation results and is shown in Figure 10.8. A reel-to-reel tape-handling capability was incorporated such that the sample can be freely suspended within the reaction chamber. This capability is desirable since the presence of a sample holder will alter the gas flow profile. Also, a tensioned reel-to-reel configuration will minimize any variation in orientation between sample and gas delivery system due to sample expansion on heat-up. In addition to the tape handling system, it was decided early on that the reaction chamber would be built with metal because of its superior mechanical integrity and dimensional stability. Selection of the metal was based primarily on it resistance to oxidation as well as HF corrosion. After literature survey and experimental testing, Inconel 601 was selected as the chamber material.

For the chamber design, a modular approach was selected such that the longitudinal dimension of the reaction chamber can be easily expanded if the experimental results are promising. Briefly, a single 30 cm-long flow-module was incorporated into a 5 cm-wide by 1 cm-high housing as seen schematically in Figure 10.8. Two 1 cm-diameter Inconel 601 tubes with 125 μm holes were used as gas inlet and outlet ports. These tubes were welded in place such that the holes are opposite their counterparts and inline with the sample suspended in between. Reels were connected to the ends of the 1 meter-long Inconel housing, which sits in the cradle of a single-zone furnace. Standard components such as gas supply, water bubbler and oxygen sensors were connected to the reaction chamber where necessary. In addition, pressure sensors were strategically located such that pressures at the gas inlet, gas outlet and within the chamber can be continuously monitored.

10.4.2 Sample Preparation

Samples used in the *ex-situ* conversion were prepared by co-evaporating Y, Cu and BaF_2 onto RABiTS fabricated either solely or jointly by Oak Ridge National Laboratory and 3M Company. Briefly, meter-length RABiTS were prepared by depositing ceramic buffer layers on textured Ni using a combination of electron-beam evaporation (seed-layer), sol-gel dip-coating (seed-layer), reactive (seed-, barrier- and cap-layers) or rf magnetron sputtering (barrier- and cap-layers) (W.B. Robbins, private communication; Cui et al., 1999; Lee et al., 2000; List, 2000; Dip-coating). A standard RABiTS architecture of CeO_2/YSZ/CeO_2 or Gd_2O_3/Ni was typically used in this study. Both Ni annealing and buffer deposition were performed in a moving configuration using reel-to-reel equipment Typical buffers and Ni thickness is 150 Å-CeO_2/2000 Å-YSZ/150 Å-CeO_2 or Gd_2O_3/50 μm-Ni, respectively. X-ray diffraction analyses performed on numerous standard architecture RABiTS confirmed that these substrates fabricated in an all-moving manner are highly textured with a single in-plane orientation. Occasionally, a thin Pd layer is deposited in-between Ni and the CeO_2 seed to facilitate seed-layer epitaxy by reactive sputtering (W.B. Robbins, private communication). In that case, we found a small fraction of second in-plane orientation originating from the CeO_2 seed that is rotated 45° from the major cube-on-cube orientation (Figure 10.9). The fraction of this secondary in-plane orientation, however, decreases with YSZ thickness and is typically not detectable in the YBCO film.

Once the textured RABiTS substrates were fabricated, they were loaded into a 3-gun reel-to-reel electron-beam evaporation system for precursor deposition. Detail of the system and deposition procedure is described elsewhere (Lu et al., 2001). Briefly, the system consists of a CVS-15 15 kW electron beam power supply with 3 electron-beam guns operating at a voltage of 8 kV. Source materials in the form of Y-metal, Cu-metal and BaF_2 crystal were situated in 3 separate pockets. Precursor deposition was carried out in a H_2O partial pressure of 1×10^{-5} Torr with a total pressure of less than 6×10^{-5} Torr, which were monitored by an SRS 100RGA mass spectrometer. It is critical to keep the water partial pressure constant since when varied, it affects the deposition rates of Y and Cu, and the oxygen content of the precursor film. At a tape traveling speed of 0.8 m/h, the precursor film was continuously deposited at a rate of 10 Å/s where the deposition rates of Y, BaF_2 and Cu were individually controlled by quartz crystal monitors. Precursor film deposited in such a fashion typically

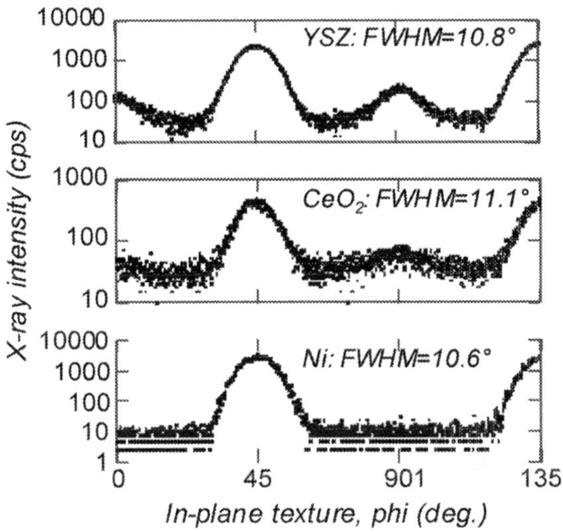

Figure 10.9. In-plane texture of CeO_2/YSZ/CeO_2/Pd/Ni RABiTS used this work. Phi-scan FWHM, of Ni, CeO_2 and YSZ layers are 10.6°, 11.1° and 10.8°, respectively.

has a mean thickness of approximately 3000 Å, with a standard deviation of ±2.5% over the entire meter length. RBS measurements at 10-cm increments showed that the "long-range" stoichiometry of the precursor is quite uniform, with a typical Y/Ba/Cu composition of roughly 1.05/2.1/3.0 and a standard deviation of less than ±2%. While the overall chemical composition of the precursor is excellent, occasional fluctuation and/or discharge in individual electron gun, however, have resulted in localized off-stoichiometry regions on the order of a few mm wide. The effect of these regions on YBCO tape performance will be discussed later. Meanwhile, it is sufficient to point out that at their worst, these regions led to very low or zero J_c whereas at their best, they contribute to the non-uniformity of the converted YBCO film.

10.4.3 YBCO Conversion in Stationary Mode

Since this is a single flow-module chamber in a single-zone furnace, continuous processing by moving the sample through the length of the furnace is not possible. Instead, conversion is performed under a schedule similar to that shown in Figure 10.1. During typical operation, a 1 cm-wide by 7 cm-long sample is spot-welded to Ni leaders, and loaded into the center portion of the flow module under a constant ~3 N tension force. The sample is set to oscillate around the central location at a speed of 30 cm/h and a distance of 7 cm to avoid any conversion inhomogeneity due to possible jetting of the gas. The chamber is then purged with N_2 before backfilling with N_2–O_2 mixed gas that is maintained at an oxygen partial pressure of 130 mTorr. Once the flow has stabilized, the furnace is heated to 740°C at a ramp-up rate of 25°C/min. When the furnace has reached 150°C, the mixed gas is diverted through a water bubbler to attain a 70 Torr H_2O wet environment. After the furnace has reached 740°C, the sample is kept at this wet environment for a designated length of time before the gas is diverted away from the bubbler and the environment is switched to dry. The sample is further annealed at this temperature for an additional one hour before being cooled to room temperature. Under this conversion schedule, even though the sample is in motion, the

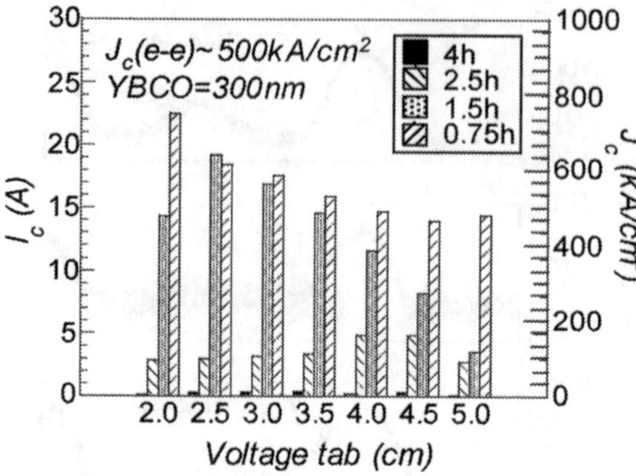

Figure 10.10. Sectional J_c's of *ex-situ* YBCO on RABiTS processed for various lengths of time in the single flow-module furnace.

mode of conversion is considered *stationary* because of the limited range of oscillating motion within a uniformly heated chamber.

Preliminary conversion runs were conducted in order to validate the design and operation of the single flow-module furnace. To ensure sample-to-sample uniformity, a set of four 1 cm-wide by 7 cm-long samples were cut from a 30 cm length of precursor on CeO_2 (rf sputtered)/YSZ (rf sputtered)/CeO_2 (e-beam)/Ni RABiTS where all depositions were performed in a moving manner. These samples were individually loaded into the reaction chamber and heated to processing temperature as described in the previous paragraph. The length of wet conversion time for various samples were set at 240, 150, 90 and 45 min whereas the dry annealing time was kept constant at 60 min. Other processing parameters were 740°C processing temperature, 130 mTorr oxygen partial pressure, 70 Torr H_2O content, gas flow rate of 4 l/min, and were kept constant for all four difference runs. Following conversion, XRD analyses of these samples did not reveal any BaF_2 diffraction peak thereby indicating that precursor conversions were completed in all the samples. This is in contrast to results of similar samples converted in an older low flow-rate perpendicular-flow quartz furnace where a wet conversion time of 120 to 180 min was necessary to convert precursors of identical thickness. For J_c characterization, Ag pads with 0.5 cm separation distance were deposited onto the samples such that sectional J_c values can be measured with the standard four-probe technique and 1 μV/cm criterion. Sectional J_c values of these samples are shown in Figure 10.10. It can be seen from this figure that while all the samples are superconducting at 77 K, there is a clear influence of "over-conversion" on the J_c's of these YBCO films. As the wet conversion time is reduced, the J_c increases reaching an end-to-end value of 500 kA/cm^2 for the 3000 Å-thick sample converted in 45 min.

Once the operation and effectiveness of the extended transverse-flow reaction chamber has been verified, we proceeded to examine the influences of various processing parameters on precursor conversion. Based on flow simulation and previous experimental observations, effects of flow rate and wet conversion time were investigated. A 1 meter-long precursor/CeO_2 (rf sputtered)/YSZ (rf sputtered)/CeO_2 (reactive sputtered)/Pd (reactive sputtered)/Ni sample was prepared in an all-moving manner. Individual 6.5 cm-long samples were cut from this stock to ensure sample-to-sample uniformity. A design-of-experiment approach based on a 2-factor Central Composite Design with continuous variables was selected for the experiment, and the parameters

Table 10.1. Conversion parameters and resulting sample characteristics of YBCO conductors processed in the design-of-experiment study

Run No.	Flow rate (l/min)	Wet conversion time (min)	$BaF_2(111)$ intensity (cps)	YBCO(006) intensity (cps)	Average J_c (S.D. < ±5%) (kA/cm^2)
3	0.5	143	12830	18654	0
7	1.0	45	80847	26569	0
11	1.0	180	0	56413	48.5
2	1.0	240	0	85703	102.1
8	2.5	45	6756	102880	567.1
12	2.5	60	0	124150	747.7
4	2.5	143	0	112515	438.6
6	2.5	200	0	108938	313.0
10	4.0	30	1633	85447	373.9
1	4.0	50	0	88852	387.3
5	4.0	90	0	106557	416.5
9	4.0	143	0	73371	269.5

are listed in Table 10.1. Individual sample was spot-welded onto Ni leaders, and was oscillated around the center of the chamber at a speed of 30 cm/h and a distance of 7 cm. Flow rate and wet conversion time for each experimental run was set according to the values listed in Table 10.1, while the remaining parameters were fixed as: ramp-up rate = 25°C/min, conversion temperature = 740°C, oxygen partial pressure = 130 mTorr, H_2O partial pressure = 70 Torr and dry conversion time = 60 min.

Following conversions, θ–2θ x-ray analyses were performed on each sample. All the YBCO diffraction patterns exhibited only the $(00l)$ peaks indicating that the films are well aligned in the c-axis. As to the extent of BaF_2 decomposition, integrated $BaF_2(111)$ intensities of the samples were analyzed and are listed in Table 10.1. As expected, BaF_2 signal in the samples disappeared as the conversional time for a given flow rate is increased. More importantly, the length of conversion time necessary for BaF_2 to decompose decreases with increasing gas flow rate, which is in agreement with the flow simulation results. In addition to BaF_2, integrated YBCO(006) intensities were also collected and are listed in the table. Variations in YBCO intensity with wet conversion time for the 4 different flow rates are shown in Figure 10.11. It can be seen from this figure that for a given flow rate, YBCO intensity initially increases with processing time. For higher flow rates (2.5 l/min and greater), the intensity reaches a maximum and then decreases on further processing. To better visualize the effect of changing processing parameters on YBCO intensity, response surface analysis was performed on the data. Surface and contour plots constructed from this analysis are shown in Figure 10.11. The experimental data as well as response surface analysis both indicate that for low flow rates (less than ~1.5 l/min), an extremely long wet conversion time is necessary to achieve high YBCO counts, which is likely to be unrealistic in practice. As the flow rate is increased to 2.5 l/min, integrated YBCO intensity increases rapidly until it reaches a maximum value at approximately 60 min, then decreases thereafter. This decrease is believed to reflect the degradation of YBCO phase in a moist environment, and therefore should be avoided. As the flow rate is further increased to 4 l/min, a maximum in YBCO intensity is again seen in at roughly the same conversion time as that of the 2.5 l/min flow condition. On the other hand, the maximum YBCO value found for the 4 l/min flow rate is lower than that of the 2.5 l/min condition. Since YBCO phase content, grain quality as well as degree of grain align-

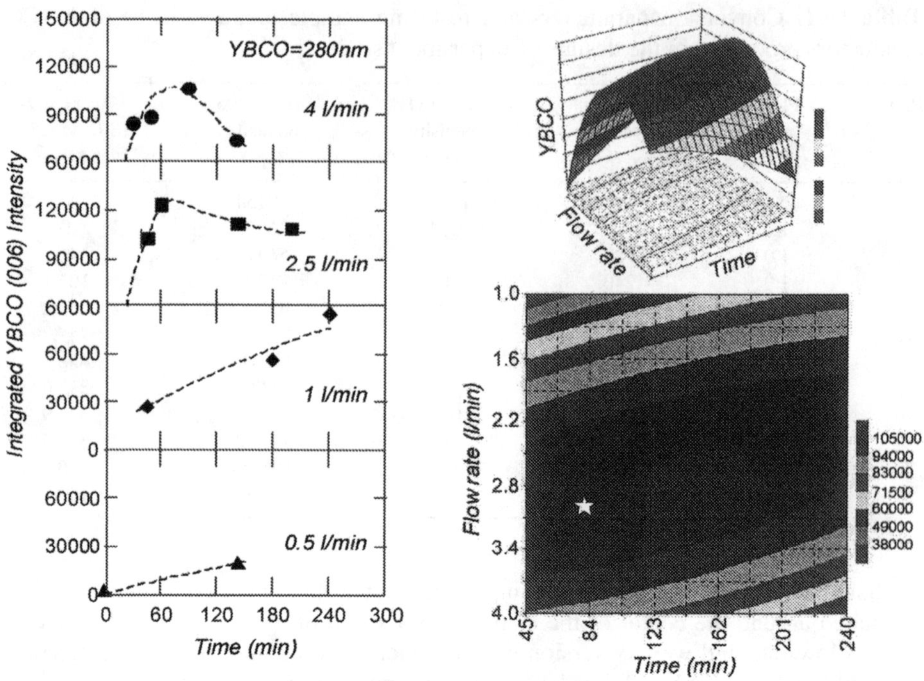

Figure 10.11. Variations in integrated YBCO(006) intensity with wet conversion time for the 4 flow rates examined in the design-of-experiment study. Also shown are surface and contour plots constructed from response surface analysis of the data.

ment all contribute to its intensity, this result may signify that an optimum growth rate exists above which degradation in grain quality or texture can occur. According to the response surface analysis, the optimum flow rate and processing time necessary to achieve maximum YBCO intensity (peak location marked by a star) are approximately 3 l/min and 60–80 min, respectively, for this reaction chamber, precursor thickness and other fixed processing parameters. If gas flow is further increased beyond 4 l/min, the oxifluoride decomposition/YBCO formation rate may become excessive such than a weaker YBCO intensity maximum is reached at a shorter time. Couple with this is the likelihood that YBCO degradation will occur rapidly resulting in degraded tape performance. More extensive investigations need to be performed to examine this and other predictions from the response surface analysis.

After X-ray diffraction analyses, multiple Ag contacts with a voltage-tap separation distance of 0.5 cm were sputtered onto individual sample. Using the standard four-probe technique and a 1 μV/cm criterion, seven sectional J_c measurements were performed on each sample covering a total distance of 3.5 cm. Average J_c values of this set of samples are listed in Table 10.1. Standard deviation of the J_c averages of all the samples is less than $\pm 5\%$. Variations in average J_c with wet conversion time for these 4 different flow rates are shown in Figure 10.12. It is immediately evident that for slow flow rates, J_c increases with processing time. At higher flow rates, the J_c reaches a maximum and then decreases on further processing. Thus, the characteristic of J_c as a function of wet conversion time is identical to the behavior of YBCO(00l) intensity, and strongly suggests that relative changes in YBCO intensity can be used as a feedback parameter to optimize the J_c during processing. Response surface analysis

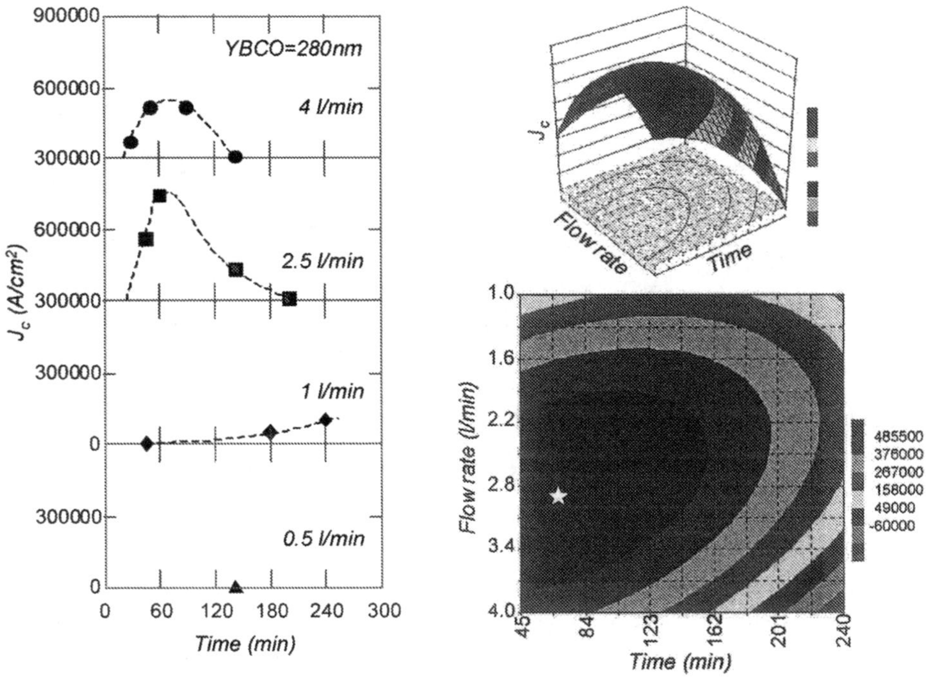

Figure 10.12. Variations in average J_c with wet conversion time for the 4 flow rates examined in the design-of-experiment study. Also shown are surface and contour plots constructed from response surface analysis of the data.

was again performed, and the surface and contour plots constructed from the analysis are shown in Figure 10.12. It can be seen from this figure that the response surface of J_c is similar to that of the YBCO intensity. In general, the J_c for a given flow rate increases with conversion time, reaches a maximum value and then decreases. In addition, the optimum conversion time decreases with flow rate. For the present reaction chamber, precursor thickness and fixed processing parameters, 3 l/min and 60–70 min of conversion (marked by a star) represent the optimum processing conditions for J_c. These predicted conditions are near identical to that of the YBCO intensity response surface, and further support the possibility of YBCO intensity as a feedback parameter. By comparing the two response surfaces, it can also be seen that J_c is more sensitive to changes in processing conditions than YBCO intensity; for processing parameters that are deviated greatly from the optimum conditions, J_c value is zero or very low even though the corresponding YBCO intensity may be sizable. Further design-of experiment studies are underway to elucidate the influence and relative importance of other processing parameters.

10.5 REEL-TO-REEL SEVEN-MODULE TRANSVERSE-FLOW REACTION CHAMBER: CONTINUOUS YBCO CONVERSION

10.5.1 Design of Seven-Module Transverse-Flow Furnace

Once the operation of the metallic single-module transverse-flow reaction chamber has been verified, a seven-module reaction chamber for continuous YBCO con-

Figure 10.13. Photo of reel-to-reel metallic seven-module transverse-flow reaction chamber for continuous conversion of BaF_2 precursor. The chamber is situated in a 2 meter-long 22-zone furnace.

version was designed and built. A picture of the continuous furnace is shown in Figure 10.13. Inconel 601 was again selected as the building material, and the chamber contains seven 30 cm-long flow modules identical to that of the single-module chamber shown in Figure 10.8. Each of the flow-module contains two 1 cm diameter Inconel 601 tubes with 125 μm holes, which serve as gas inlet and outlet ports. The tubes were welded in place such that the holes are opposite their counterparts and in line with the sample suspended in between. Blocks with 1.5 cm-wide slots were welded between neighboring flow modules to minimize cross talk while allowing the tape to pass through the slot. Standard components including gas supply, water bubblers and oxygen sensor were connected to the reaction chamber where necessary. In addition, multiple pressure sensors were strategically located such that pressures at gas inlets and gas outlets of different modules as well as that of the reaction chamber can be continuously monitored. For this reaction chamber, flow-modules 1 to 5 can be set to provide either a wet or dry gas environment, whereas modules 6 and 7 (downstream) remain dry at all times.

The 2.5 meter-long seven-module reaction chamber is placed within the cradle of a custom-built 2 meter-long 22-zone tube furnace. Temperature of the hot zones of this furnace can be individually controlled such that a wide variety of thermal profiles may be obtained. By continuously moving the sample at a chosen speed through a selected thermal profile, a prescribed conversion schedule such as that shown in Figure 10.1 for stationary conversion can be replicated. An example of a temperature profile that was employed in the 22-zone furnace is shown in Figure 10.14. When a tape traveling speed of 0.63 m/h is used, the equivalent stationary processing conditions can be approximated as 123 min wet conversion time and 36 min dry conversion time. In addition to conversion time, the temperature ramp-up rate, which can affect the texture and growth of YBCO, is also of interest. Since this custom built furnace is equipped with narrow hot zones at the sample entrance-end (hot zones 1 to 8), a sizable range of temperature gradients can be obtained by adjusting the settings of the hot zones. Figure 10.15 shows an example of temperatures gradients that can be achieved in this furnace. For a conversion temperature of 740°C, the ramp-up gradient can easily be varied from 14 to 70°C/cm. Thus, by adjusting the temperature profile, tape traveling

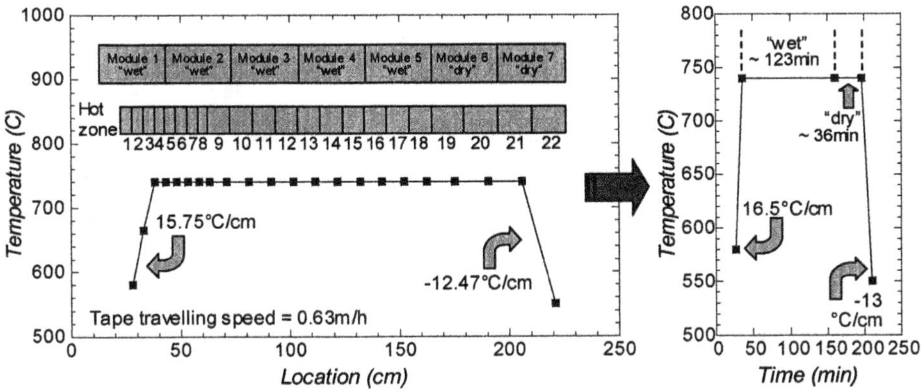

Figure 10.14. Example of temperature profile employed in the 22-zone furnace. Also shown is the equivalent stationary conversion schedule.

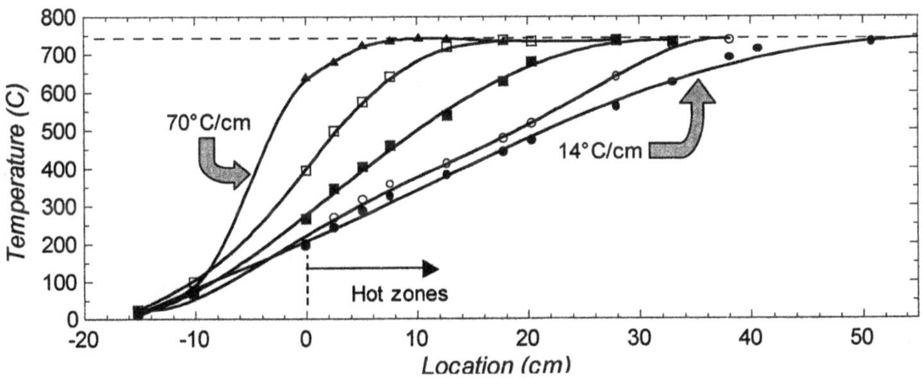

Figure 10.15. A sizable range of temperature ramp-up gradients can be obtained by adjusting the temperature settings of hot zones.

speed, gas flow rate as well as other processing parameters, a multitude of conversion conditions can be explored.

10.5.2 YBCO Conversion in Continuous Mode

Preliminary runs were first performed on short tapes to evaluate the conversion characteristics of moving BaF_2 precursor in the seven-module reaction chamber. A 1 meter-long precursor/CeO_2 (rf sputtered)/YSZ (rf sputtered)/CeO_2 (reactive sputtered)/Pd (reactive sputtered)/Ni was prepared in an all-moving manner. Several 1 cm-wide by 7 cm-long samples were cut from the long tape, and spot-welded to Ni leaders with the front-end of each sample situated outside of the furnace entrance prior to each run. Before commencement of conversion, the furnace was purged with N_2 gas and backfilled with N_2–O_2 mixed gas that was maintained at an oxygen partial pressure of 130 mTorr and a flow rate of 3 l/min. A temperature profile was then selected and the hot zones were heated to their respective set temperature. Using the same temperature profile, three conversion runs were conducted with targeted wet conversion times of 90, 60 and 45 min. In order to achieve these conversion times, three

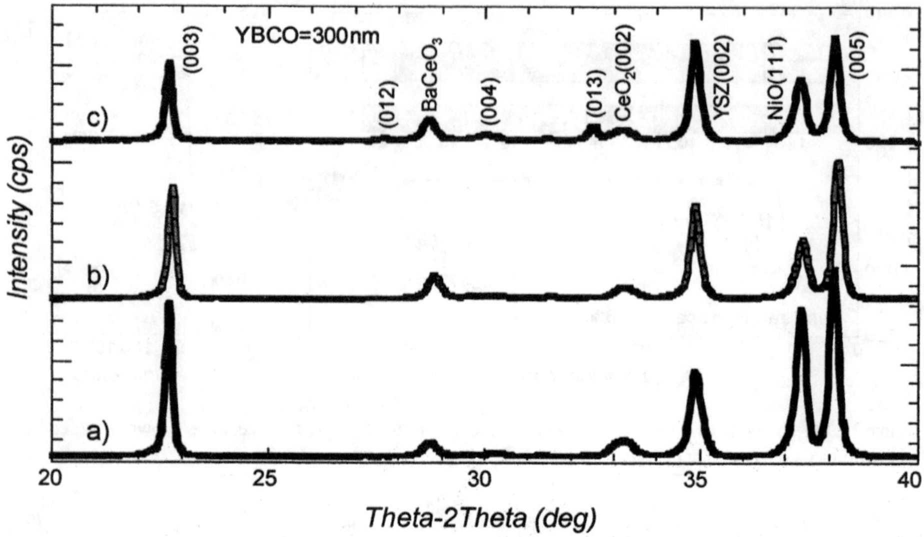

Figure 10.16. θ–2θ scans of samples converted continuously at: (a) tape speed = 0.86 m/h, ramp-up rate = 45°C/min, wet conversion time = 90 min; (b) tape speed = 1.3 m/h, ramp-up rate = 68°C/min, wet conversion time = 60 min; (c) tape speed = 1.73 m/h, ramp-up rate = 90°C/min, wet conversion time = 45 min.

tape traveling speeds of 0.86, 1.3 and 1.73 m/h with corresponding temperature ramp-up rates of 45, 68 and 90°C/min were necessary. Using these parameters, the tapes were continuously converted by moving each individual sample from the entrance to the exit end of the furnace. X-ray θ–2θ scans of these samples were performed and are shown in Figure 10.16. While YBCO(00*l*) peaks are found for all the samples, the θ–2θ scans revealed that when the ramp-up rate is increased beyond a certain limit, random YBCO grains are formed during conversion (see Figure 10.16(c) (103) peak). Sectional J_c measurements, with voltage-tap separation of 0.5 cm over a total distance of 3.5 cm, were performed. The average J_c's of the 90 min, 60 min and 45 min samples are 6.8 ± 0.5 A, 7.6 ± 0.4 A and 0.2 ± 0.2 A, respectively. These J_c results confirm that randomly oriented YBCO grains will degrade the tape performance considerably. Agreement between J_c and X-ray results show that the temperature ramp-up rate during conversion is a parameter that has to be considered if high performance coated conductors are to be fabricated by *ex-situ* techniques. Depending on the furnace design, the need to reduce the ramp-up rate and avoid random YBCO formation may place a limit on sample throughput unless the ramp-up temperature gradient can be lowered.

Following the initial runs, a 1 cm-wide by 30 cm-long tape cut from the same batch was continuously converted in the seven-module reel-to-reel chamber at a tape traveling speed of 0.7 m/h, conversion temperature of 740°C, P_{O_2} of 130 mTorr, P_{H_2O} of 70 Torr and gas flow rate of 5 l/min. A silver cap-layer was sputtered onto the sample for voltage and current contacts after conversion. End-to-end J_c measurement was performed at 77 K and self-field, which provided a disappointingly low J_c of 10 kA/cm^2. In order to determine the location(s) of the inferior YBCO film, sectional J_c measurements of 1 cm increment were conducted and the result is shown in Figure 10.17. It can be seen from this figure that the majority of J_c's in this tape range from ~100 to ~300 kA/cm^2, with a maximum sectional value of 400 kA/cm^2. While large portions

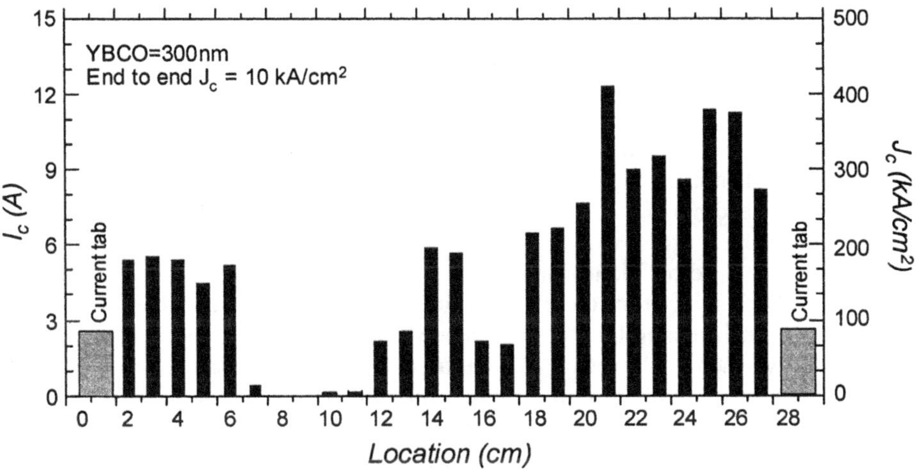

Figure 10.17. Sectional J_c's of the 30 cm-long YBCO (e-beam precursor)/CeO_2 (rf sputtered)/YSZ (rf sputtered)/CeO_2 (reactive sputtered)/Pd (reactive sputtered)/Ni sample at 77 K and self-field. Low J_c values are located between 7 and 11 cm sections in the YBCO film.

Figure 10.18. Optical image of the YBCO tape containing low J_c sections. Linear-features can be seen running across the entire width of the tape. These features correspond exactly to locations where discharges and accompanying system shutdown occurred during precursor deposition.

of this tape can sustain a good J_c, origin of the low end-to-end J_c can be traced to sections between 7 and 11 cm of the YBCO film. Following J_c measurements, the Ag cap-layer was chemically etched away such that the poor-quality sections in question can be examined. An optical picture of this low-J_c portion of tape is shown in Figure 10.18. It is immediately evident that linear-features that run across the entire width of the sample correspond exactly to the zero or low J_c sections. As mentioned previously, these off-stoichiometric line defects originated during precursor deposition, and can be attributed to two types of events that cause sudden changes in flux rates; these fluctuations can be caused either by arcing events in the high voltage supply of the guns or non-arcing events related to the thermal transfer through the crucible liner.

In the case of high voltage arcing, particle built-up in the deposition chamber can detach and fall within certain sensitive areas of the voltage supply to the guns. A critical area in the configuration of our system is the high-tension feed-through across the chamber wall. Loose particles that fell either during deposition or chamber cleaning can bridge the distance over the ceramic insulation, thereby shorting the high-tension lead to the chamber wall. When this happens, high voltage arcing will occur. Moreover, arcing of one gun will in general, interrupt the functioning of the other guns since all of them are connected to the same power supply in our system. This interruption will in turn, produces variations in the deposition rates, the severity of which depends on the material being deposited. For example, deposition of BaF_2 is accomplished by radiant heating of the target crystal by a tungsten cap over the crucible, which is heated by the e-gun. An instantaneous drop in voltage of the e-gun heating the tungsten cap has little effect due to its thermal mass. On the other hand, the Y-metal target of our system is only melted in the path of the e-beam. Therefore, a sudden change in the guns performance will create noticeable changes in the flux rate. In some cases, short-circuiting of a high-tension lead may be continuous thereby requiring system shut down to enable the removal of the short.

The second type of fluctuation originates from non-uniformities in the conductive heat transfer away from the crucible liner. Flaws such as scratches and pits on the surface of the crucible liner can produce variations in heat transfer from the target, especially when dealing with melted targets. For example, a sharp spike in the deposition rate of Cu, either as an increased flux rate to a value far above the set rate or a decreased flux to near zero, has been observed when using older crucible liners of poor surface quality. While a spike in deposition rate may recover within a couple of seconds on its own, reactions of the deposition controllers further exacerbate the problem. This occurs because on registering a large change in flux rate, the controllers immediately respond by either cutting or increasing the power to compensate for the initial rate change. This further compounds the problem of stabilizing the flux rate to the set value over time, thereby creating variations in precursor composition and thickness over as much as a few cm of the tape. In addition, variations in flux rate arising from the Cu can induce instabilities in the rates of both Y and BaF_2 as well by causing changes in the effective vapor pressures within the chamber. The adverse influence of precursor non-stoichiometry and non-uniformity on J_c points toward the need to improve the operational stability of our precursor deposition system. Work is underway to replace selected e-guns with more stable effusion cells, as well as general modification of system configuration to avoid shorting.

With the largely successful conversion of the 30 cm tape, a 1 cm-wide by 1 meter-long precursor/CeO_2 (rf sputtered)/YSZ (reactive sputtered)/CeO_2 (reactive sputtered)/Pd (reactive sputtered)/Ni sample was prepared, and continuously converted at a tape traveling speed of 0.65 m/h, conversion temperature of 740°C, P_{O_2} of 130 mTorr, P_{H_2O} of 70 Torr and gas flow rate of 5 l/min. The appearance of the tape was carefully examined following conversion. Since no significant arcing had occurred during precursor deposition of this tape, no linear-feature similar to those shown in Figure 10.18 was visually observed. There were, however, several instances of minor fluctuations in deposition rates, which led to local precursor non-uniformity and lower sectional J_c. In addition, a grayish region located at the gas outlet edge of the tape, a typical example of which is shown in Figure 10.19, has been observed. This feature is found along the entire 1 meter length of the tape although its width varied from location to location. X-ray diffraction analyses revealed that the grayish region is richer in BaF_2 when compared to the gas inlet edge and the center of the tape. This variation in BaF_2 content indicates the occurrence of non-uniform oxifluoride decomposition. A possible contributing factor to the decomposition inhomogeneity even across this reduced length-scale may arise from the upward curvature of the sample. Since a typical one-sided coated conductor is an anti-symmetric composite, sample curvature across the tape width will result owing to the different thermal expansions and stress states of YBCO, buffer(s) and metallic substrate. The sample curvature can be quite pronounced especially when a soft substrate such as annealed Ni is used.

The influence of sample curvature on gas flow profile was investigated by computational fluid dynamics simulations. Dimensions and mesh setup of the simulated

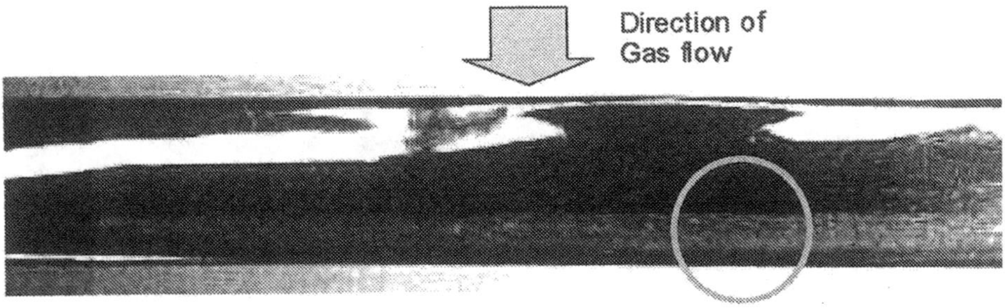

Figure 10.19. A grayish region is found at the gas outlet edge of the 1 meter-long sample. The width of this BaF_2-rich region varied from location to location, but can be seen along the entire length of the tape.

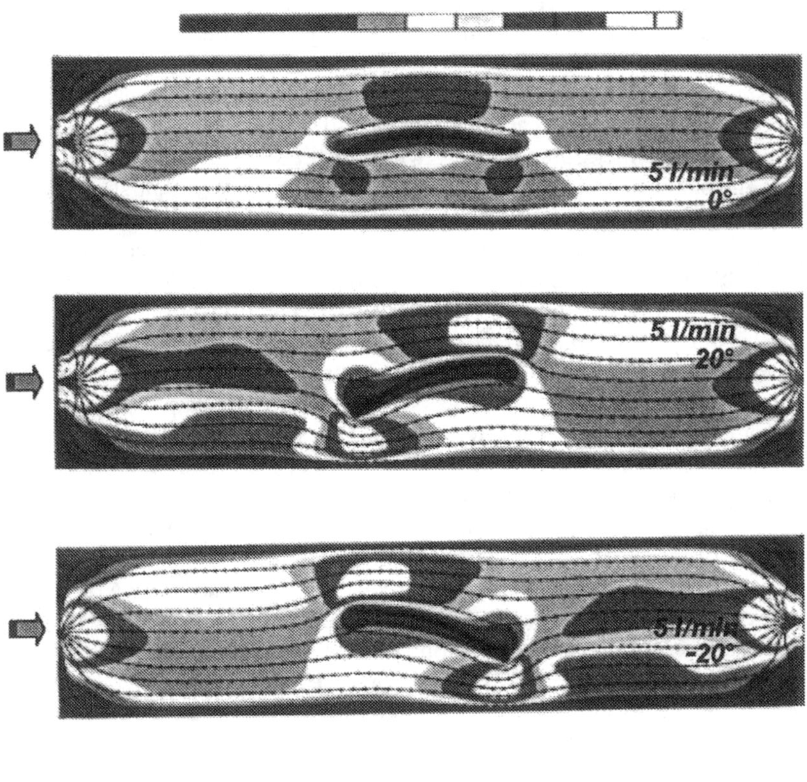

Figure 10.20. Simulated flow profile of a curved sample within the transverse cross section of a transverse-flow reaction chamber. Flow rate of the gas is 5 l/min for a sample that is: (a) leveled along the horisontal plane of the chamber; (b) tilted 20° toward gas inlet, and (c) tilted 20° away from the gas inlet.

reaction chamber were identical to those of the previous study, except a curvature was introduced across the width of the sample as shown in Figure 10.20. It can be seen in Figure 10.20(a) that gas flow at the vicinity of the curved surface is symmetric around the middle of the sample, just as in the case of a flat tape. However, along the precursor-side of the curved sample (top surface), gas velocities at the leading and trailing edges

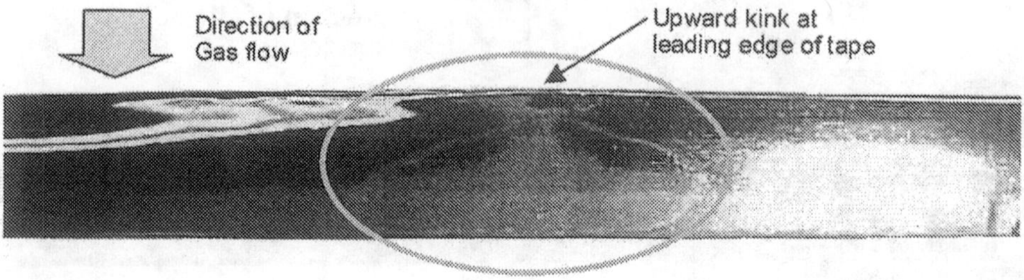

Figure 10.21. Section of the meter-long tape that contains an upward kink at the leading edge of the sample. Flowing gas was channeled around such a flaw leading to incompletely converted material that fans out behind the obstacle.

of the tape are now lower than that of the middle portion owing to the increased local clearance between sample surface and chamber wall. The HF/H_2O concentration ratio at the trailing edge of such a curved sample will likely be higher thereby resulting in slower oxifluoride decomposition/YBCO conversion such as that seen in Figure 10.19. In addition, since the ~3 N tension force on the tape is provided by reels that are situated more than 2.5 meters apart, and the tape is only supported at 30 cm increments within the reaction chamber, sample tilting with respect to the gas injection ports can occur. When the sample is tilted away from the gas inlet such as that shown in Figure 10.20(c), gas velocity at the trailing edge of the tape will decrease further leading to greater tendency toward incomplete decomposition/conversion. This can account for the varying width of the incompletely converted region along the length of the tape. Effort is underway to redesign the flow-modules such that gas injection is set at a slight angle with respect to the sample surface as well as alternating gas injection directions from module to module. This modification should ensure that part of the tape would always be in correct orientation with the gas flow regardless of sample tilting.

Apart from the incompletely reacted region seen at the downstream edge of the entire tape, there are other more striking examples of sample/gas orientational effects. These regions are again comprised of incompletely converted precursor, and are typically found at locations with dimensional flaws, i.e. kinks at edge of tape. Figure 10.21 shows a section of the meter-long tape containing an upward kink that is originated from tape to reel contact. Since the original reels that were used in RABiTS fabrication have very tight width tolerance, any misalignment of the reels or variation in sample width can result in reel-wall to sample contact during wind-up. This will lead to friction built-up between the reel-wall and sample, resulting in localized kinks in the soft Ni-based substrate during subsequent pay-out. This type of flaws not only tilts the sample away from the gas inlet but also acts as an obstacle where the gas is channeled to either side, resulting in incompletely reacted material that fans out from the flaw. Incompletely converted regions such as that shown in Figure 10.21 have been found to be significant contributors to J_c non-uniformity in continuously converted YBCO tapes.

In addition to these general features, one other blemish was found at a location 59 cm away from the front of the tape and is shown in Figure 10.22. This imperfection was first noticed before precursor deposition, and does not conform to the shape nor the appearance of known flaws generally associated with precursor deposition and conversion. In addition, local x-ray analyses conducted *a prior* confirmed that the spot is not associated with degradation in either the Ni or buffer texture. Rather, the appearance of the spot led us to believe that it is the consequence of contamination by

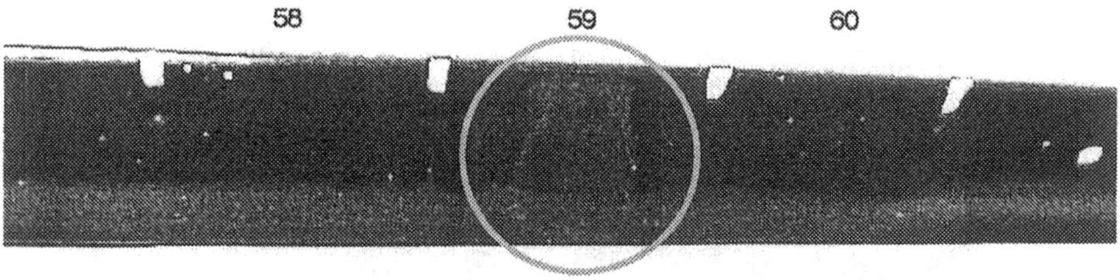

Figure 10.22. Section of the meter-long tape that contains an imperfection in YBCO located at 59 cm from the front of the tape. This blemish is not believed to be the result of anomalies during deposition or conversion.

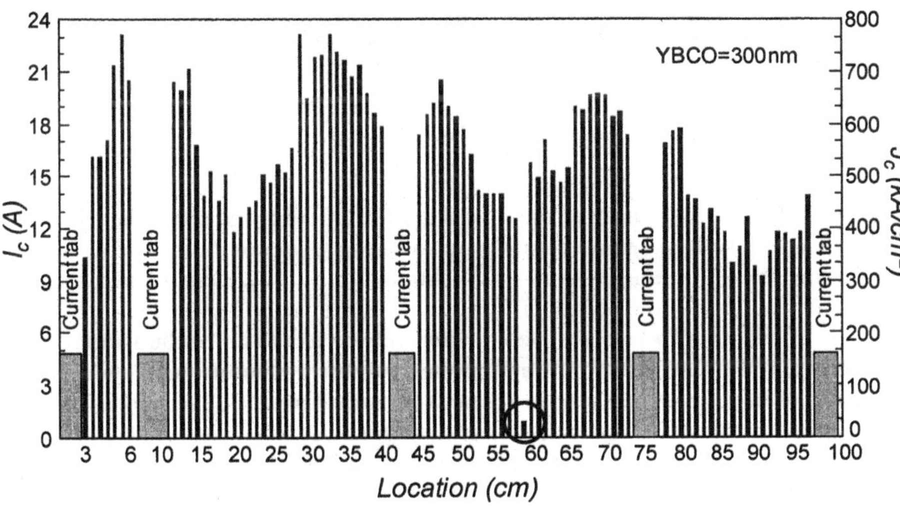

Figure 10.23. Sectional J_c's of the meter-long YBCO/CeO_2 (rf sputtered)/YSZ (reactive sputtered)/CeO_2 (reactive sputtered)/Pd (reactive sputtered)/Ni sample at 77 K and self-field. The sample was cut into shorter sections after conversion due to the dimensional constraints of our batch oxygen annealing and measurement equipment. Low J_c value is located at 59 cm from the front of the sample and is highlighted by a circle.

cleansing solvent after buffer deposition, and should be avoided by improving care in tape handling.

In order to evaluate the current carrying capability of the long tape, a silver cap-layer was first sputtered onto the YBCO surface. The meter-long tape was then cut into 2 sections of 93 cm and 7 cm each because of the limited length of our batch oxygen-annealing furnace. End-to-end J_c measurement was performed at 77 K and self-field on the 93 cm-long tape with a voltage separation distance of 89 cm. The end-to-end J_c of this tape is found to be 100 kA/cm^2 using the 1 μV/cm criterion, even though voltage first appeared at a current density of 33 kA/cm^2. Sectional J_c measurements were performed next to study the J_c uniformity of the sample. Since our sectional J_c apparatus is limited to samples of 30 cm long, the 93 cm tape was cut into 3 separate sections for the measurements. Sectional J_c values of these plus the 7 cm section are reconstructed and shown in Figure 10.23. It can be seen from this figure that the section of YBCO film located at 59 cm from the front of the tape possesses a low J_c value of 33 kA/cm^2. This corresponds exactly to the imperfection seen in Figure 10.22 as well as the current density value where voltage was first detected in the end-to-end

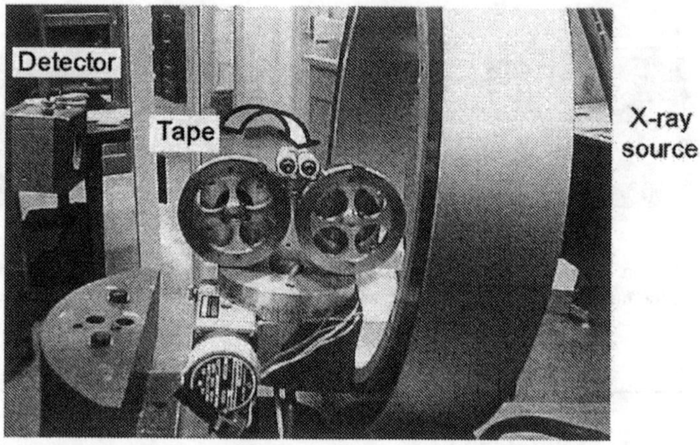

Figure 10.24. First generation reel-to-reel x-ray diffraction system at ORNL. Photo shows the large 4-circle diffractometer together with the modular reel-to-reel insert and location of the tape section being examined.

measurement, and is the cause of the low overall J_c of the long tape. Other than this low J_c spot, Figure 10.23 reveals that the sectional J_c values of the entire meter-long tape are quite good; J_c ranges from ~300 kA/cm^2 to ~770 kA/cm^2. Unfortunately, standard deviation of these sectional J_c's is at a significant value of 24%. The variation in sectional J_c can be correlated qualitatively to the width of the incompletely converted region found near kinks and at the gas outlet edge of the sample, which is believed to be a major contributor to the J_c non-uniformity. In addition, variation in precursor stoichiometry due to deposition fluctuation is likely to worsen this non-uniformity. Nevertheless, the maximum attained J_c of 770 kA/cm^2 is similar to that of a short 1.5 cm-long sample cut from the same stock and previously converted in a standard longitudinal-flow furnace, and meets the best expected performance based on the texture of the RABiTS. These results, in totality, indicate that there is much room for improvement through process optimization, tape handling and possible modification in reaction chamber design.

To further explore the origins of the J_c inhomogeneity as well as the low J_c section in the long sample, we made use of the correlation between x-ray characteristics and J_c found previously in the design-of-experiment results. For this task, we utilized our unique reel-to-reel x-ray diffraction system designed especially for the coated-conductor work (E.D. Specht, Reel-to-reel X-ray). The system, shown in Figure 10.24, consists of a large 4-circle diffractometer, a modular reel-to-reel insert for tape positioning, x-ray source and detector and standard computerized controls. At present, the nominal tape traveling speed is 200 m/h when the system is operating in a moving θ–2θ scanning mode at a fixed diffraction angle. This unique facility provides an opportunity to study relative changes in x-ray characteristics at various locations of a long tape; an investigation such as this is extremely difficult for short samples due to batch-to-batch variability. This reel-to-reel system has proven to be invaluable in the development of RABiTS. To prepare the long sample for reel-to-reel x-ray characterization, silver cap-layers were chemically etched off the 3 tape sections originally cut from the 93 cm sample. These sections were then spot-welded together with Ni spacers in between to account for sample lose during cutting and welding. The reconstructed tape was loaded into the x-ray reel-to-reel insert and two moving scans were performed at diffraction angles corresponding to YBCO(005) and YBCO(103), respectively. Fig-

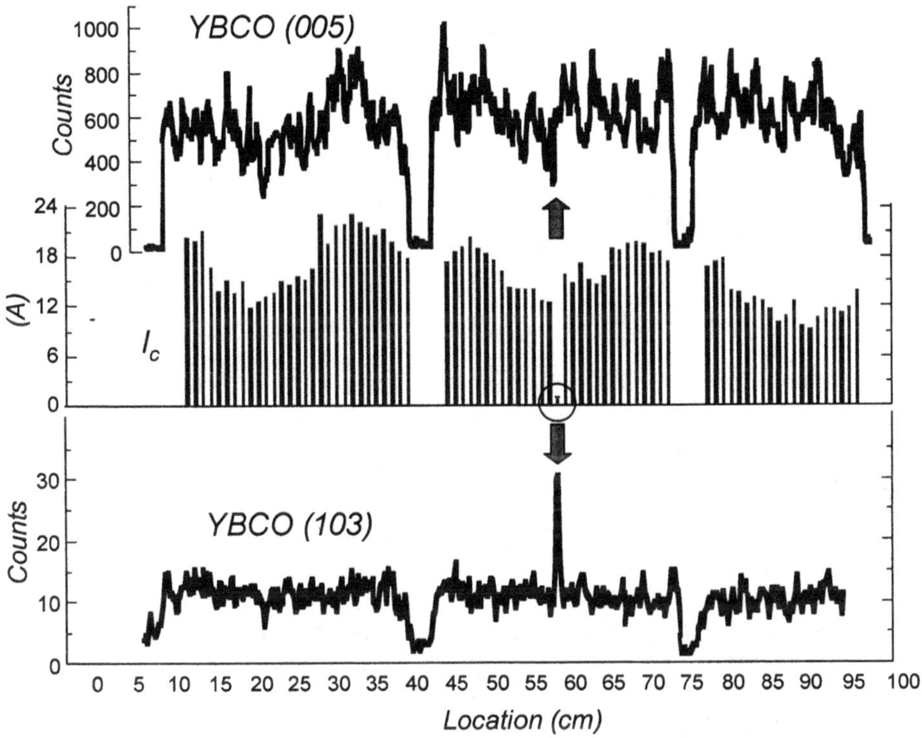

Figure 10.25. Variations in YBCO(005) and YBCO(103) x-ray intensities with tape location. Also shown is the corresponding I_c of the reconstructed tape. Arrows mark the locations of YBCO(005) minimum and YBCO(103) maximum. These anomalies occur exactly at the spot where the I_c is low and the identified imperfection is present.

ure 10.25 shows the changes in YBCO(005) and YBCO(103) diffraction intensities together with sectional I_c. Zero x-ray intensity values seen near 5, 40, 75 and 95 cm in the plot correspond to the Ni spacers where no YBCO was present. It can be seen from the figure that like the results of the design-of-experiment study, there is a good correlation between YBCO(005) intensity variation and the relative changes in measured sectional I_c values. While there is a local intensity minimum at 59 cm (the location where the imperfection was identified), another minimum of similar intensity can be seen at 21 cm which exhibited an I_c of ~12 A. Thus, it appears that YBCO(00l) intensity is a good but insufficient gauge of ultimate tape performance. Also shown in this figure is the variation in YBCO(103) intensity along the length of the reconstructed tape. Here, one can see that a single maximum, which indicates the presence of a sizable amount of randomly oriented YBCO grains, corresponds exactly to the identified imperfection of low I_c. Thus, while YBCO(00l) intensity indicates the quality and formation of epitaxial YBCO, YBCO(103) intensity can show the locations where breakdowns in YBCO epitaxy occur. Using these measurements in conjunction may provide a powerful online feedback parameter to optimize and control the quality of *ex-situ* processed YBCO coated conductors.

In the course of continuous reel-to-reel conversion of YBCO on RABiTS, we have been able to identify some factors that either limit the overall J_c value or contribute to the non-uniformity of J_c. Among these are:

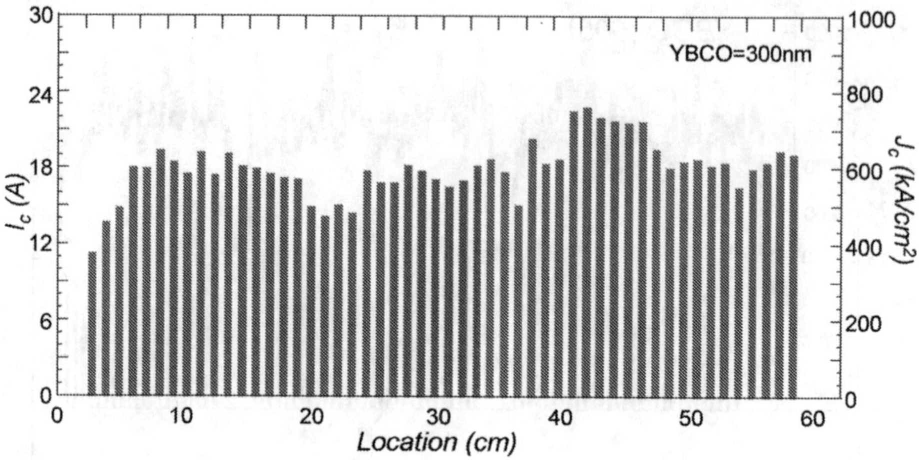

Figure 10.26. Sectional J_c's of the 60 cm-long YBCO/CeO_2 (rf sputtered)/YSZ (rf sputtered)/CeO_2 (e-beam)/Ni sample at 77 K and self-field.

(1) RABiTS texture, which mostly limits the overall J_c value,
(2) precursor non-stoichiometry, which affects both the end-to-end J_c as well as J_c uniformity,
(3) sample-gas orientation, which contributes to J_c non-uniformity,
(4) dimensional flaws, i.e. kinks, which also contributes to J_c non-uniformity.

As mentioned previously, we are currently implementing improvements and modifications to circumvent these issues. In particular, the origin of kinks has been traced to the tight tolerance of our reels being used in the fabrication of RABiTS. We have found that if the reel alignment is not perfect, the Ni substrate is not straight or there is variation in the width of the substrate, the substrate will be in contact with the reel-wall during sample wind-up. During pay-out, the sample will remain in contact with the reel-wall at those points until sufficient force is applied to overcome the frictional force, and the sample breaks free from the reel. It is this action that results in kinks in the soft Ni substrate. By utilizing wider reels and modifying the tape transfer procedure, it is now possible to routinely obtain RABiTS without dimensional flaws. While the presence of these flaws may seem trivial at first glance, their effect on J_c uniformity has proven to be dramatic. Before the removal of kinks, the standard deviation in sectional J_c was about 23% to 25%. As soon as the kinks are eliminated in the RABiTS, the standard deviation has dropped to a much lower value of about 10% to 13%!

An example of such an improvement is shown in Figure 10.26, where a 60 cm-long precursor/CeO_2 (rf sputtered)/YSZ (rf sputtered)/CeO_2 (e-beam)/Ni sample was continuously converted in the seven-module reel-to-reel chamber at a tape traveling speed of 0.65 m/h, conversion temperature of 740°C, P_{O_2} of 130 mTorr, P_{H_2O} of 70 Torr and gas flow rate of 5.5 l/min. Due to the length limitation of our batch oxygen-annealing furnace, the sample was instead annealed in the reel-to-reel furnace following Ag-cap layer deposition. With no significant discharge or fluctuation during precursor deposition, the end-to-end J_c of this 60 cm-long tape was found to be a respectable 400 kA/cm^2. In addition, sectional J_c measurements were performed using a modified setup such that it is no longer necessary to physically divide the sample into shorter portions for the measurements. It can be seen from Figure 10.26 that the sectional J_c uniformity has improved due to the elimination of kinks, leading to

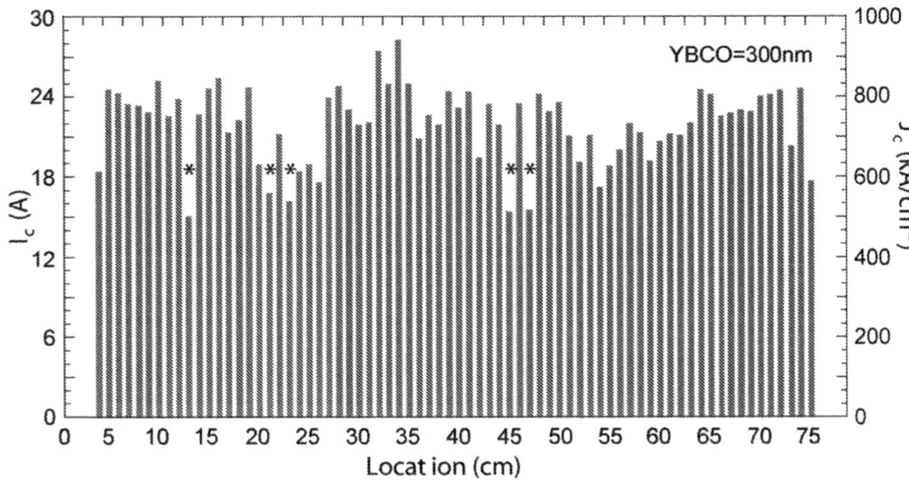

Figure 10.27. Sectional J_c's of the 80 cm-long YBCO/CeO_2 (rf sputtered)/YSZ (rf sputtered)/Gd_2O_3 (dip-coated)/Ni sample at 77 K and self-field. * denote locations where moderate discharges and faint line-features are visible.

an averaged J_c of 600 kA/cm^2 and a reduced standard deviation of 12.5%. Further improvement in overall sample performance is also being investigated by utilizing a dip-coated Gd_2O_3 seed layer instead of the e-beamed CeO_2 material (Dip-coating). By annealing the Ni substrate and recrystalizing the sol-gel Gd_2O_3 seed-layer at higher temperatures, both the out-of-plane and in-plane textures of a 1 meter-long RABiTS have been improved to FWHM of 11.2° and 9.4°, respectively. Unfortunate, moderate to severe discharges occurred during precursor deposition, forcing the precursor deposition run to be terminated after 80 cm. This 80 cm-long precursor/CeO_2 (rf sputtered)/YSZ (rf sputtered)/Gd_2O_3 (dip-coated)/Ni sample was continuously converted in the reel-to-reel chamber at a tape traveling speed of 0.65 m/h, conversion temperature of 740°C, P_{O_2} of 130 mTorr, P_{H_2O} of 70 Torr and gas flow rate of 5.5 l/min. Even with areas of non-stoichiometry associated with the discharges, the end-to-end J_c of this sample is found to be 625 kA/cm^2. Sectional J_c measurements were also performed on the tape and shown in Figure 10.27, where the asterisks represent locations of precursor discharges with associated visible line-features. It can be seen from this figure that despite the non-stoichiometric sections, J_c value as high as 950 kA/cm^2 has been obtained on this sample owing to the improved RABiTS texture. Also, the tape possesses a high averaged J_c of 750 kA/cm^2 with a low standard deviation of 10.5%.

10.6 SUMMARY

With the development of flexible biaxially textured substrates, fabrication of long-length high-J_c coated conductors may be possible. *Ex-situ* processing of epitaxial YBCO, which separates YBCO compositional control from the epitaxial growth step, is an attractive alternative to traditional vapor deposition techniques. Early on in our investigation of long-length *ex-situ* conversion of BaF_2 precursor on RABiTS, we found that YBCO conversion is inhomogeneous when a simple atmospheric longitudinal-flow reaction chamber is used. Since BaF_2 or oxifluoride has to decompose totally and efficiently for YBCO conversion to proceed to completion, it is our belief that the

HF/H_2O concentration ratio near the sample surface must remain low. That is, HF has to be effectively removed from the sample following oxifluoride decomposition. As such, an increase in gas flow rate should aid in the removal of HF.

In order to verify these assumptions, we performed numerical fluid dynamics simulations to visualize the gas flow profile of a sample suspended in a transverse-flow chamber. In addition, simulations were performed to study the effects of flow rate on both gas velocity and HF concentration in the vicinity of the sample. Simulated results revealed that the gas flow is symmetric around the middle of the sample, with higher gas velocities at the leading and trailing edges. More importantly, the results showed that gas velocity is significantly increased at all locations near the sample surface when a higher flow rate is utilized. Therefore, a higher flow rate should aid in the removal of HF. This is verified in the HF concentration simulations which revealed that while the highest HF concentration is always located at the surface along the downstream portion of a sample, both the volume and the concentration of HF can be substantially reduced when a higher flow rate is employed.

Based on these results, we selected a two-prong approach that can minimize the non-uniform conversion issue while maximizing the amount of precursor that can be converted at one time. The approach utilizes a transverse-flow geometry such that the effective conversion distance is the sample width, and an extended conversion zone to maximize the volume of material undergoing the oxifluoride decomposition/YBCO conversion reaction. In order to validate this approach, an extended single-module transverse-flow reaction furnace capable of high flow rates was designed and built. To ensure the mechanical integrity and dimensional stability of the reaction chamber, Inconel 601 was chosen as the building material. Since this chamber is located in a single-zone furnace, continuous conversion of YBCO is not possible, and the furnace essentially operates in a stationary mode. Design-of-experiment study was performed on short samples to investigate the effects of flow rate and conversion time on the characteristics of YBCO processed in this furnace. The results showed that YBCO(00*l*) x-ray intensity increases gradually with conversion time at slow flow rates. At higher flow rates, the x-ray intensity initially increases rapidly, reaches a maximum value and then decreases on further exposure to the moist conversion environment. The variation in J_c with flow rate and conversion time for these samples is remarkably similar to that of YBCO(00*l*) x-ray intensity. These results suggest that an efficient atmospheric reaction chamber should be able to sustain significant gas flow since the time that is necessary to completely convert a length of YBCO at low flow rates may be impractical. In addition, the results also showed that for a given furnace design, an optimum growth rate exists above which degradation in grain quality or texture can occur. Moreover, the similarity between YBCO(00*l*) x-ray intensity and J_c suggests that relative changes in YBCO x-ray intensity may be used as a feedback parameter to optimize the J_c during *ex-situ* processing.

Once the operation of the metallic single-module reaction chamber was validated, we proceeded to build a seven-module version of the reaction chamber. Inconel 601 was again chosen as the building material since after 1 year of operation, the monitored flow parameters as well as the ability of the single-module chamber to convert YBCO have remained constant, thus validating Inconel 601 as an excellent building material. The 2.5 meter-long chamber sits in the cradle of a 2 meter-long 22-zone furnace, and is connected to two reels thus enabling the continuous conversion of long-length YBCO. Preliminary studies of precursor conversion of short samples in a moving manner showed that randomly orientated YBCO can result if a large temperature ramp-up rate is employed. When a sizable amount of random YBCO is present, the YBCO film possesses very low J_c. Therefore, this influence of ramp-up rate on YBCO nucleation may place a limit on sample throughput unless the thermal gradient of a chosen furnace can be adjusted. Once the basic conversion parameter space of the seven-module furnace is known, a 30 cm-long sample was converted in a continuous manner. The YBCO film is found to contain linear-features at certain locations that run across the width of the sample. These locations support little or no J_c, and correspond exactly to sites where significant discharges and fluctuations occurred during precursor deposition, thereby resulting in local non-stoichiometric areas. While the end-to-end J_c of the processed sample is low, majority of the sectional J_c's is satisfactory which gave us the impetus to process longer tapes.

A 1 meter-long BaF_2 precursor on RABiTS was thus converted next. Careful examination of the YBCO appearance following conversion revealed that a BaF_2-rich region is present at the gas outlet edge along the entire length of the meter-long tape. The width of this BaF_2-rich region varies from location-to-location, and is believed to be the result of incomplete oxifluoride decomposition/YBCO conversion stemming from sample curvature and flaws due to tape mishandling. In addition, a blemish can be seen visually which is not believed to be the result of anomalies during buffer/precursor deposition or film conversion. Rather, it is believed that contamination of the tape occurred during sample transfer and thus should be possible to avoid in the future. End-to-end J_c of the meter-long tape sample is found to be 100 kA/cm^2, the less than stellar value due to the low J_c at the blemish spot. Other than this ~0.6 cm-wide low J_c spot, sectional measurements at 1 cm increment revealed J_c's that range from 300 to 770 kA/cm^2, but with a sizable standard deviation of 24%. Reel-to-reel x-ray scans of the sample showed that variation in J_c along the length of the tape corresponds qualitatively to the relative change in YBCO(00*l*) intensity. In addition, a sudden increase in YBCO(103) intensity is seen at the blemish spot. These results fortify the suggestion that x-ray measurements at selected diffraction angles may be used during sample conversion to optimize the tape performance.

During the course of continuous reel-to-reel YBCO conversions, various factors that can affect the overall J_c value and/or J_c uniformity have been identified. These include items such as RABiTS texture and uniformity, precursor stoichiometry and uniformity, sample to gas orientation, as well as dimensional flaws in RABiTS that can affect the gas flow pattern. Solutions to circumvent these issues have either been devised or have since been implemented. In particular, the consequence of eliminating dimensional flaws has been proven to be dramatic; while the standard deviation of sectional J_c's in samples with kinks were between 23% to 25%, samples without kinks typically exhibit a J_c standard deviation of 10% to 13%. For example, a 60 cm-long kink-free sample converted in an identical manner as that of the meter-long tape was found to possess an end-to-end J_c of 400 kA/cm^2, an averaged J_c of 600 kA/cm^2 and a reduced standard deviation of 12.5%. Efforts are also underway to improve the overall J_c value by enhancing the RABiTS texture. One of the approaches is to employ dip-coated sol-gel Gd_2O_3 seed-layer, where better RABiTS texture was achieved by annealing the Ni substrate and recrystalizing the seed layer at higher temperatures. Despite the occurrences of discharge during precursor deposition, a 80 cm-long kink-free sample converted on such a substrate was found to possess an end-to-end J_c of 625 kA/cm^2, highest sectional J_c of 950 kA/cm^2, an averaged J_c of 750 kA/cm^2 and a low standard deviation of 10.5%. Results obtained in this study have provided valuable insights into the *ex-situ* conversion process. Efforts are now underway to investigate the conversion of micron-thick YBCO precursors, and to improve the conversion process through parameter optimization and equipment modification.

ACKNOWLEDGMENT

The authors would like to thank W.B. Robbins from 3M Company for providing buffered RABiTS tapes. They would also like to thank K.W. Childs for performing the gas flow simulations, and P.M. Martin and E.D. Specht for their assistance in sample preparation and characterization. Research sponsored by the DOE Office of Energy Efficiency and Renewable Energy, Office of Power Technologies, under contract DE-AC05-00OR22725 with UT-Battelle, LLC, managing contractor for Oak Ridge National Laboratory.

REFERENCES

Bauer, M., Semerad, R., and Kinder, H., 1999a, YBCO films on metal substrates with biaxially aligned MgO buffer layers, *IEEE Trans. Appl. Supercond.*, 9(2):1502.

Bauer, M., Semerad, R., Kinder, H., Wiesmann, J., Dzick, J., and Freyhardt, H.C., 1999b, Large area YBCO films on polycrystalline substrates with very high critical current densities, *IEEE Trans. Appl. Supercond.*, 9(2):2244.

Cui, X., List, F.A., Kroeger, D.M., Goyal, A., Lee, D.F., Mathis, J.E., Specht, E.D., Martin, P.M., Feenstra, R., Verebelyi, D.T., Christen, D.K., and Paranthaman, M., 1999, Continuous growth of epitaxial CeO_2 buffer layers on rolled Ni tapes by electron beam evaporation, *Physica C*, 316:27.

Feenstra, R., List, F.A., O'Neill, D., and Hawsey, R.A., 1999, An *ex situ* processed YBCO coated conductor, in: *Proc. 1999 DOE Superconductivity Program for Electric Systems Annual Peer Review*, Vol. 1, p. 361.

Feldmann, D.M., Reeves, J.L., Polyanskii, A.A., Goyal, A., Feenstra, R., Lee, D.F., Paranthaman, M., Kroeger, D.M., Christen, D.K., Babcock, S.E., and Larbalestier, D.C., 2001, Magneto-optical imaging of transport currents in $YBa_2Cu_3O_{7-x}$ on RABiTS^{TM}, *IEEE Trans. Appl. Supercond.*, 11(1):3772.

Foltyn, S.R., Arendt, P.N., Dowden, P.C., DePaula, R.F., Groves, J.R., Coulter, J.Y., Jia, A., Maley, M.P., and Peterson, D.E., 1999, High-T_c coated conductors—Performance of meter-long YBCO/IBAD flexible tapes, *IEEE Trans. Appl. Supercond.*, 9(2):1519.

Fukutomi, M., Aoki, S., Komori, K., Chatterjee, R., and Maeda, H., 1994, Laser deposition of $YBa_2Cu_3O_y$ thin films on a metallic substrate with biaxially textured YSZ buffer layers prepared by modified bias sputtering, *Physica C*, 219:333.

Goyal, A., Norton, D.P., Budai, J.D., Paranthaman, M., Specht, E.D., Kroeger, D.M., Christen, D.K., He, Q., Saffian, B., List, F.A., Lee, D.F., Martin, P.M., Klabunde, C.E., Hatfield, E., and Sikka, V.K., 1996, Fabrication of long range, biaxially textured, high temperature superconducting tapes, *Appl. Phys. Lett.*, 69:1795.

Ignatiev, A., Chou, P.C., Zhong, Q., Zhang, X., and Chen, Y.M., 1996, Photo-assisted MOCVD growth of YBCO thick films for wire applications, *Appl. Supercond.*, 4:455.

Iijima, Y., Tanabe, N., Kohno, O., and Ikeno, Y., 1992, In-plane aligned $YBa_2Cu_3O_{7-x}$ thin films deposited on polycrystalline metallic substrates, *Appl. Phys. Lett.*, 60:769.

Iijima, Y., Kimura, M., Saitoh, T., and Takeda, K., 2000, Development of Y-123-coated conductors by IBAD process, *Physica C*, 335:15.

Lee, D.F., List, F.A., Cui, X., Martin, P.M., Specht, E.D., Goyal, A., Kroeger, D.M., Paranthaman, M., and Robbins, W.B., 2000, Progress in scaling up YBCO-coated conductor on RABiTS^{TM} using the BaF_2 precursor approach, in: *ORNL Superconducting Technology Program for Electric Power Systems: Annual Report for FY 1999*, Report No. ORNL/HTSPC-11:1–33.

List, F.A., 2000, Progress toward continuous processing of YBCO/RABiTS^{TM} tape, in: *ORNL Superconducting Technology Program for Electric power Systems: Annual Report for FY 1999*, Report No. ORNL/HTSPC-11:1–36.

Lu, S.W., List, F.A., Lee, D.F., Cui, X., Paranthaman, M., Kang, B.W., Kroeger, D.M., Goyal, A., Martin, P.M., and Ericson, R.E., 2001, Electron beam co-evaporation of Y–BaF_2–Cu precursor films for $YBa_2Cu_3O_7$ coated conductors, *Supercond. Sci. Technol.*, 14:218.

Mathis, J.E., Goyal, A., Lee, D.F., List, F.A., Paranthaman, M., Christen, D.K., Specht, E.D., Kroeger, D.M., and Martin, P.M., 1998, Biaxially textured $YBa_2Cu_3O_{7-\delta}$ conductors on rolling assisted biaxially textured substrates with critical current densities of 2–3 MA/cm^2, *Jpn. J. Appl. Phys. Lett. Part II*, 37:L1379.

Norton, D.P., Goyal, A., Budai, J.D., Christen, D.K., Kroeger, D.M., Specht, E.D., He, Q., Saffian, B., Paranthaman, M., Klabunde, C.E., Lee, D.F., Sales, B.C., and List, F.A., 1996, Epitaxial $YBa_2Cu_3O_7$ on biaxially-textured (001) Ni: An approach to high critical current density superconducting tapes, *Science*, 274:755.

Paranthaman, M., Park, C., Cui, X., Goyal, A., Lee, D.F., Martin, P.M., Chirayil, T.G., Verebelyi, D.T., Norton, D.P., Christen, D.K., and Kroeger, D.M., 2000, $YBa_2Cu_3O_{7-y}$-coated conductors with high engineering current density, *J. Mater. Res.*, 15:2647.

Petrisor, T., Boffa, V., Celentano, G., Ciontea, L., Fabbri, F., Gambardella, U., Ceresara, S., and Scardi, P., 1999, Development of biaxially aligned buffer layers on Ni and Ni-based alloy substrates for YBCO Tapes fabrication, *IEEE Trans. Appl. Supercond.*, 9(2):2256.

Reade, R.P., Berdahl, P., Russo, R.E., and Garrison, S.M., 1992, Laser deposition of biaxially textured yttria-stabilized zirconia buffer layers on polycrystalline metallic alloys for high critical current Y–Ba–Cu–O thin films, *Appl. Phys. Lett.*, 61:2231.

Rupich, M.W., Li, Q., Annavarapu, S., Thieme, C., Zhang, W., Prunier, V., Paranthaman, M., Goyal, A., Lee, D.F., Specht, E.D., and List, F.A., 2001, Low cost Y–Ba–Cu–O coated conductors, *IEEE Trans. Appl. Supercond.*, 11(1):2927.

Sato, Y., Matsuo, K., Takahashi, Y., Muranaka, K., Fujino, K., Hahakura, S., Ohmatsu, K., and Takei, H., 2001, Development of $YBa_2Cu_3O_y$ tape by using inclined substrate method, *IEEE Trans. Appl. Supercond.*, 11(1):3365.

Selvamanickam, V., Galinski, G.B., Carota, G., DeFrank, J., Trautwein, C., Haldar, P., Balachandran, U., Chudzik, M., Coulter, J.Y., Arendt, P.N., Groves, J.R., DePaula, R.F., Newnam, B.E., and Peterson, D.E., 2000, High-current Y–Ba–Cu–O superconducting films by metal organic chemical vapor deposition on flexible metal substrates, *Physica C*, 333:155.

Smith, J.A., Cima, M.J., and Sonnenberg, N., 1999, High critical current density thick MOD-derived YBCO films, *IEEE Trans. Appl. Supercond.*, 9(2):1531.

Solovyov, V.F., Wiesmann, H.J., Wu, L.J., Zhu, Y., and Suenaga, M., 2000, Kinetics of $YBa_2Cu_3O_7$ film growth by postdeposition processing, *Appl. Phys. Lett.*, 76:1911.

Wang, R.P., Zhou, Y.L., Pan, S.H., He, M., Lu, H.B., Chen, Z.H., Yang, G.Z., Liu, C.F., Wu, X., Wang, F.Y., Feng, Y., Zhang, P.X., Wu, X.Z., and Zhou, L., 2000, Deposition of high-temperature superconducting films on biaxially textured Ni(001) substrates, *Physica C*, 337:87.

Wu, W.D., Foltyn, S.R., Arendt, P.N., Blumenthal, W.R., Campbell, L.H., Cotton, J.D., Coulter, J.Y., Hults, W.L., Maley, M.P., Safar, H.F., and Smith, J.L., 1995, Properties of $YBa_2Cu_3O_{7-\delta}$ thick films on flexible buffered metallic substrates, *Appl. Phys. Lett.*, 67:2397.

Chapter 11

SOLUTION DEPOSITION OF $YBa_2Cu_3O_{7-\delta}$ COATED CONDUCTORS

Paul G. Clem

Materials Chemistry Department 01846
Sandia National Laboratories
Albuquerque, NM 87185-1411
USA

11.1 INTRODUCTION

Solution-based deposition routes for complex oxides have been developed over the past two decades due to their ease of incorporating multiple elements, good control of local stoichiometry, and feasibility for large area deposition. Beyond the widespread silica-based antireflective coating methods developed for auto and window glass, a variety of solution deposition routes have been reported for processing complex perovskite-based materials such as ferroelectric oxides and conductive electrode oxides. In this chapter, several solution deposition routes toward $YBa_2Cu_3O_{7-\delta}$ coated conductors are reviewed, and recent results are presented detailing an all solution deposition approach to >1 MA/cm^2 $YBa_2Cu_3O_{7-\delta}$ coated conductors on rolling-assisted, biaxially textured, (200) oriented Ni tapes. Process technologies and cost factors contributing to a viable coated conductor architecture such as that shown in Figure 11.1 are also discussed. The current development of solution-deposition methods appears to have promise to compete with vapor phase methods for superconductor electrical properties, with potential advantages for large area deposition and low cost per kiloamp-meter of wire.

11.2 CHEMICAL SOLUTION DEPOSITION

Chemical solution deposition (CSD) (Schwartz, 1997) can be divided into three categories: aqueous-based deposition, metal–organic decomposition, and sol-gel chemistry approaches. Aqueous deposition of materials may be conducted by hydrothermal synthesis (Lange, 1996) or evaporation of the solvent from dissolved salts,

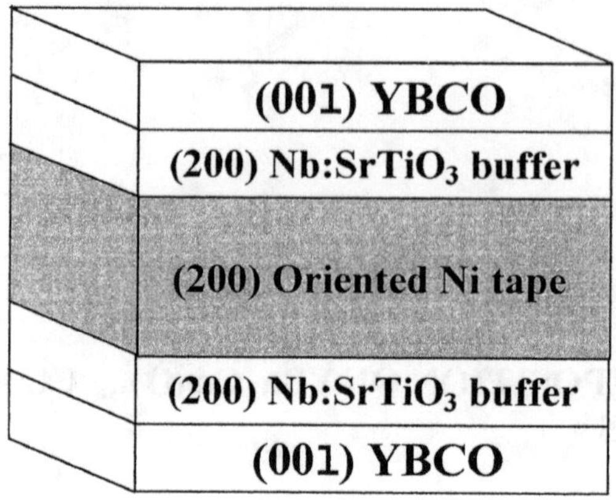

Figure 11.1. Potential architecture of double side dip-coated YBCO coated conductor. Sol-gel YBCO on sol-gel Nb-doped $SrTiO_3$ buffer layers, coated on (200) oriented, biaxially textured Ni tape.

followed by appropriate processing of the residual dissolved material. While the latter may be especially effective for depositing salts, it has not been widely used for processing of superconductors, since aqueous solubilities of the elements of interest are low, and YBCO stoichiometry control within a few percent is critical (Carlson et al., 1990). Metal-organic decomposition (MOD) is generally defined as a route using dissolved metal soaps in a non-polar solvent, such as ethylhexanoates in toluene or benzene (Chen et al., 1989). Following evaporation of the solvent to form a metalorganic film, the metalorganic may be decomposed in air to form a variety of complex oxides. This approach is often straightforward and robust, as the chemistry in solution may be very stable. On decomposition of the metalorganic deposit, a large volume change occurs due to burnout, or pyrolysis, of the organic materials into CO_2 and H_2O, so this step may need to be closely controlled to avoid extremely exothermic process conditions, or formation of large biaxial tensile stresses in the film that could lead to cracking (Lange, 1996).

Sol-gel chemistry is similar to metal–organic decomposition, except that the chemistry is generally designed to enable metal–oxide–metal chain formation through use of polycondensation reactions, as will be discussed below. Precursors for chemical sol-gel approaches generally are metal salts (acetates, nitrates, etc.) or metal alkoxides (i.e. titanium isopropoxide) dissolved in an alcohol-based solvent system. Specifics of sol-gel chemistry are described in detail elsewhere (Brinker and Scherer, 1990), but the general principles are formation of a stable chemical dispersion, or sol, and subsequent reaction of the sol to form a continuous gel network through hydrolysis and polycondensation reactions, as described in formulae (1) and (2). In these formulae, R represents an organic group such as an alkyl (i.e. methyl CH_3) or perhaps a salt (i.e. acetyl CH_3COO) group, and ROH could then represent a free alcohol or other liberated organic group. Through progression of the polycondensation reactions, a further more cross-linked M–O–M metal oxide gel network is formed, which possesses a high degree of local chemical homogeneity, and will display a decreased change in volume on solvent removal compared to MOD and aqueous CSD processes. This change in volume is particularly important for minimizing film tensile stresses. In pursuing thicker CSD films, low film tensile stress is critical, and methods of decreasing this will be presented later in this chapter.

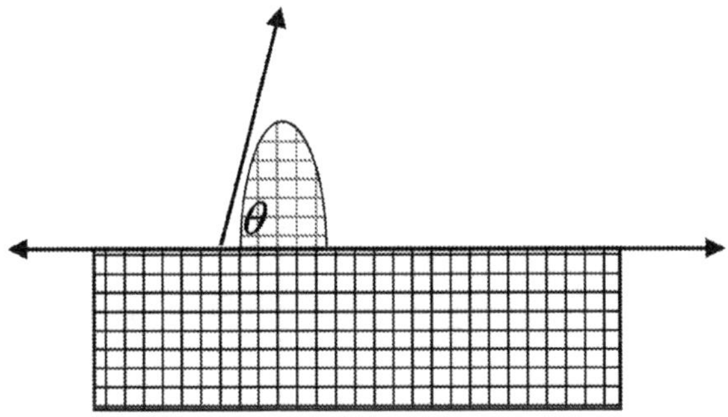

Figure 11.2. Heterogeneous nucleation on a surface, indicating the balance of surface energies and the origin of wetting angle θ.

(1) hydrolysis reaction of a metal organic $M(OR)_n$

$$M(OR)_n + H_2O \rightarrow M(OH)(OR)_{n-1} + ROH$$

(2) polycondensation reaction of a metal organic $M(OR)_n$

$$M(OR)_n + M(OH)(OR)_{n-1} \rightarrow (OR)_{n-1}M\text{–}O\text{–}M(OR)_{n-1} + ROH$$

The goal for complex oxides is then to convert the amorphous oxide gel formed as a result of polycondensation reactions into the desired phase and orientation of the final material of interest. In the case of complex oxides, particularly containing alkali, alkaline earth, and transition metals, the transformation pathway from metalorganic to metal oxide species is of critical importance. In particular, these metals are susceptible to formation of metal carbonates, which may be very stable species, and delay or prevent formation of the desired oxide species. The formation of barium carbonates is known to complicate the crystallization of both $BaTiO_3$ and $YBa_2Cu_3O_{7-\delta}$, making epitaxial or oriented crystallization of these materials very difficult (Frey and Payne, 1995, 1996; Malecki et al., 1995; Pak et al., 1992). Control of this chemical pathway enables or prevents the formation of epitaxial films. Generally, best results are obtained when a desired oxide crystalline phase is nucleated within a matrix of an identical composition, allowing an isochemical phase transformation, without need for evolving chemical species. In the simplest concept of oriented nucleation from an isochemical matrix, a nucleus would form at the film/substrate interface, nucleate a phase of lower free energy, and transform exothermically. This release of energy would theoretically provide the energy to continue sustained growth of the nucleus into the matrix, resulting in columnar grain growth. Standard nucleation and growth theory may be applied to develop the following relationships for heterogeneous (surface, hemispherical) nucleation and homogeneous (bulk, spherical) nucleation (Schwartz, 1997; Ohring, 1992), where r^* is the minimum stable nucleus size, γ is surface free energy, ΔG_v is the volume free energy of the nucleated phase, and θ is the "wetting angle" derived from the balance of matrix, nucleated phase, and substrate surface free energies, analogous to solid-liquid-air wetting angles (Figure 11.2).

(3) free energy of homogeneous nucleation

$$\Delta G_{\text{hom}}(r^*) = \frac{16\pi\gamma^3}{3(\Delta G_v)^2},$$

(4) free energy of heterogeneous nucleation for wetting angle θ

$$\Delta G_{\text{het}}(r^*) = \frac{16\pi\gamma^3}{3(\Delta G_v)^2}\left[\frac{1}{2} - \frac{3}{4}\cos\theta + \frac{1}{4}\cos^3\theta\right], \quad \text{or}$$

$$\Delta G_{\text{het}}(r^*) = \Delta G_{\text{hom}}\left[\frac{1}{2} - \frac{3}{4}\cos\theta + \frac{1}{4}\cos^3\theta\right]$$

$$\therefore \Delta G_{\text{het}}(r^*) < \Delta G_{\text{hom}}(r^*) \quad \text{for all } \theta \neq 90°.$$

Thus, heterogeneous nucleation would be thermodynamically favorable in all surface energy conditions. It may be shown that nuclei of similar structure to the interface would possess a lower "wetting angle" and, thus, nucleus orientations with an epitaxial match to a substrate would be most energetically favorable ($\theta \to 0$, $\Delta G \to 0$, enabling columnar epitaxial growth of a film on a near-lattice matched substrate.

In an alternative case in which nucleation is endothermic (as is proposed for decomposition of many metal carbonates) nucleation would require consumption of energy, resulting in multiple nucleation sites without continued growth, and a resulting polycrystalline or granular film structure. In the case of alkaline earth species, it has been shown that use of a volatile intermediate liquid phase, such as a fluoride, is also effective in producing epitaxial films from solution (Gupta et al., 1988; Clem et al., 2001). In this case, the thermodynamic stability of the alkaline earth fluoride prevents formation of carbonate phases, which are very stable and generally decompose endothermically. The decomposition of this fluoride species, however, is complicated, and has been studied by a number of groups for production of YBCO (Solovyov et al., 2000; McIntyre and Cima, 1994). The following sections present a review of the use of sol-gel processes for developing highly oriented, high J_c YBCO films, and discuss methods toward commercial scale-up.

11.3 SOL-GEL YBCO APPROACHES

Sol-gel deposition of $YBa_2Cu_3O_{7-\delta}$(YBCO) holds great potential as a fast and efficient method of producing large-scale, biaxially-textured superconducting films at a lower cost than physical and chemical vapor deposition techniques (Sheth et al., 1998). Two likely substrate candidates would likely be an oxide-buffered RABiTS (rolling-assisted, biaxially textured Ni-based) substrate (Goyal et al., 2000) or an IBAD (ion-beam assisted deposition) oxide buffered hastelloy substrate (Foltyn et al., 1999). Two main routes have been pursued toward high quality YBCO thin films: fluorinated and non-fluorinated. As discussed above, non-fluorinated films in many cases form intermediate barium carbonate phases due to decomposition of metal–organics, which liberates CO_2 and water as byproducts (Hirano et al., 1990). Fluorinated routes have been shown to avoid formation of such carbonates by forming a more thermodynamically stable BaF_2 phase instead of $BaCO_3$, and later decomposing this BaF_2 phase through reaction with water vapor. The crystallization behavior of this fluorinated sol-gel derived film appears to be similar to that of evaporated Y–BaF_2–Cu films.

Many sol-gel YBCO syntheses have used non-fluorinated sol-gel approaches (Gross et al., 1988; Rice et al., 1987), and in many recent cases achieved high T_c values and >100 KA/cm^2 J_c (77 K) values (Chu et al., 1993; Matsubara et al., 1999). The presence of barium carbonate during the processing of metal–organic derived YBCO has been reported by many authors (Hirano et al., 1990; Nonaka et al., 1988) as a

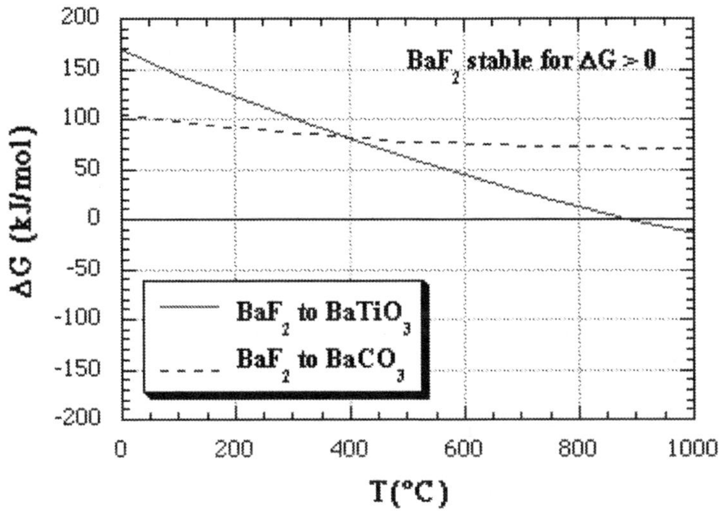

Figure 11.3. Thermocalc™ calculated thermodynamics of BaF_2 to $BaCO_3$ and BaF_2 to $BaTiO_3$ reaction vs. temperature, indicating the stability of the fluoride phase, and a means of avoiding $BaCO_3$ formation through its use.

potential hindrance to high quality films, as long anneal times and temperatures are often required to decompose this phase. Thermodynamic calculations of $BaCO_3$ stability in the YBCO system suggest stability of the carbonate phase (positive ΔG) until very high temperatures, and $BaCO_3$ has been observed experimentally to decompose gradually up to 1000°C (Malecki et al., 1995).

The decomposition reactions have been examined in detail by Manabe et al. (1995, 1997), who have demonstrated J_c (77 K) >1 MA/cm^2 by closely controlling the annealing atmosphere during YBCO crystallization. In this work, the stability of the carbonate phase was used constructively; a high $p(CO_2)$ atmosphere was used at low temperatures to suppress 123 nucleation and a-axis grain growth, but switched to a low $p(O_2)$, near zero $p(CO_2)$, at growth temperature to grow dominant c-axis YBCO on $SrTiO_3$ with excellent properties.

In the growth of YBCO films by vapor phase processes, similar carbonate concerns had been observed, and a solution was found in sputtering films containing BaF_2 instead of pure Ba, which may getter CO_2 and OH from the atmosphere (Mankiewich et al., 1987; Siegal et al., 1990b; Feenstra et al., 1991). The earliest research using fluorinated sol-gel precursors was reported by Gupta et al. (1988), who showed for the first time the implementation of trifluoroacetic acid as both a fluorine source, and as a solution stabilizer. In the synthesis of YBCO precursor solutions, three issues are often difficult: solubility of copper and barium precursors, solution stability, and avoiding carbonate degradation of oriented films. Trifluoroacetic acid (CF_3COOH) is a strong chelating or stabilizing agent, which aids in the solubility and stability of dissolved metal salts for these precursor solutions. Additionally, on decomposition of the sol-gel deposited film, the trifluoroacetic acid forms an yttrium–barium fluoride intermediary compound, which greatly aids in producing highly oriented films through formation of a transient liquid phase. Thermodynamic stability of a BaF_2 phase vs. a carbonate phase is shown in Figure 11.3 for the $BaCO_3/BaF_2/BaTiO_3$ perovskite system, which is more refractory than YBCO. BaF_2 is predicted to be more stable (positive ΔG) than $BaCO_3$ for all temperatures, but may be decomposed thermally or by reaction with water to form the desired perovskite phase. Gupta et al. developed a heat treatment cycle of slow heating in air to 400°C, followed by heating in water vapor-saturated helium gas to 850°C, a short dry He anneal at 900–920°C, and cooling in oxygen (Gupta et al., 1988). Resulting films displayed strong *c*-axis YBCO texture and sharp resistive transitions. TFA-based processes continue to be the basis of most current sol-gel YBCO research.

Understanding and optimization of this trifluoroacetic acid, or TFA, approach was developed thoroughly by McIntyre, Cima, and colleagues (McIntyre et al., 1990, 1992, 1995; McIntyre, 1993), who achieved J_c values as high as 5 MA/cm^2 for 80 nm thick films and demonstrated conclusively that growth occurs by transient formation of a liquid phase (McIntyre and Cima, 1994). Other notable advancements also include an understanding of the metalorganic pyrolysis regime, the growth rate vs. $p(O_2)$, suppressing *a*-axis grain density, film growth mechanisms, and microstructural origins of high film flux pinning densities (McIntyre et al., 1995; McIntyre, 1993). Recent work by colleagues (Smith et al., 1999) have demonstrated growth of TFA YBCO films to as thick as 0.5–1 μm with 1 MA/cm^2 J_c values through use of a tailored heat treatment schedule. Other efforts, in collaboration with American Superconductor, have demonstrated $J_c > 1$ MA/cm^2 values on (200) oriented RABiT Ni tape substrates buffered with e-beam deposited CeO_2/YSZ/CeO_2 buffer layers (Annavaparou et al., 2000). Research continues in this collaboration with ties to Oak Ridge National Laboratories (Malozemoff et al., 2000), with goals of long length processing and thick YBCO processing in pursuit of a \$10/kA-m price-performance target.

Several other groups are currently active in developing high quality YBCO films from TFA precursors. Araki et al. at ISTEC (Japan) have deposited ~0.25 μm YBCO films with J_c (77 K, 0 T) of 1.3–2.5 MA/cm^2 on IBAD CeO_2/YSZ/Hastelloy substrates. In their work, high viscosity solutions are used to deposit thicker per-layer coatings, where 2.31 and 2.78 mol/liter YB_2C_3 concentrations result in film thicknesses of 0.3 μm and 0.45 μm, for instance. Additionally, close attention is paid to processing atmosphere to obtain the best superconducting properties. Salama et al. (1998, 2000, 2001) are also pursuing TFA-based methods to YBCO films, with J_c values near 1 MA/cm^2 on single crystal substrates. Use of high molarity (2–4 M) solutions and control of heat treatment conditions to obtain dense and *a*-axis free films are notable in this work.

11.4 APPROACHES TO DECREASED YBCO PROCESS TIME

To obtain low cost coated conductor films, the metric of production cost per kiloamp-meter has been defined as the cost of plant operation divided by film quality times conductor volume per year, or:

$$\text{Cost} = \frac{\text{annual capitalization} + \text{labor}}{(\text{linear process speed})(\text{tape width})(\text{film thickness})(J_c)}.$$

One clear motivation for moving to solution deposition techniques is the capability for high linear rate of deposition, which is currently 2 cm/s in our lab setting and likely 10 cm/s in an industrial process. A complication to this, however is the historically slow rate of YBCO processing, which normally takes more than 24 hours of thermal treatments for sol-gel precursor solvent evaporation, metalorganic pyrolysis, YBCO crystallization, and YBCO oxygenation. Additionally, YBCO precursor preparation is

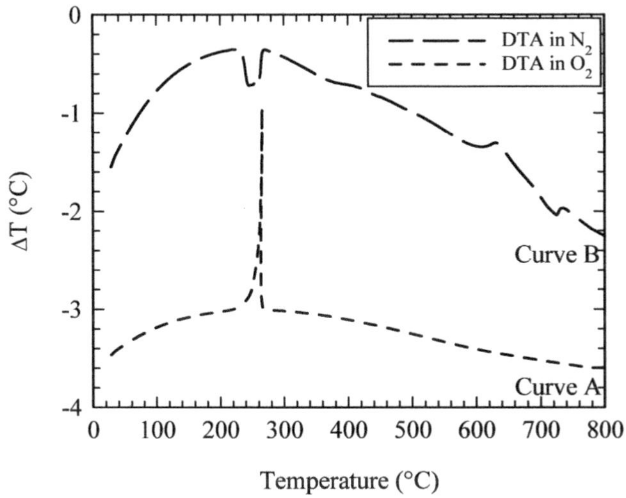

Figure 11.4. Differential thermal analysis (DTA) of copper acetate powders fired in O_2 (curve A) and N_2 (curve B), illustrating exothermic and endothermic precursor decomposition.

often slow, as many processes involve dissolving metal salts in water and TFA, evaporating the solvent completely, and redissolving the precursors in a solvent such as methanol before coating samples. Efforts at Sandia National Laboratories have focused on minimizing time of precursor synthesis, metalorganic pyrolysis, and YBCO crystallization. In addition, new precursor chemistries and processing cycles have been developed to enable thicker per-coating films, and to enable multilayering of YBCO to achieve multimicron thicknesses.

Speed of solution synthesis may be appreciably accelerated by use of alternate precursor chemicals. One approach is dissolution of metal acetates directly in the final solvent (Dawley et al., 2001a), rather than a carbonate or acetate dissolution/evaporation/redissolution cycle. As an example (Dawley et al., 2001a, 2001b), Ba-acetate may be dissolved in TFA at 60–70°C, followed by additions of Y-acetate tetrahydrate and anhydrous Cu-acetate to form a stock 0.6 M (mol YBCO/liter) solution with 1 : 2 : 3 (Y : Ba : Cu) molar ratios. The stock solution in this process is then diluted to either (1) 0.2 M with methanol (MeOH) or (2) 0.3 M with isopropanol and 1,3 propanediol (1 : 1 molar ratio). Such a simplified, 30 minute solution synthesis may be one approach to speeding processing time on the front end of coated conductor processing.

A second complication in processing time of YBCO coated conductors has historically been the need for a very slow metal–organic pyrolysis, or organic burnout step. This is due to the highly exothermic nature of copper acetate and trifluoroacetate precursor decomposition, shown as a plot of differential thermal analysis (DTA) in Figure 11.4, curve A. Rapid ramp rates through this exothermic, ~230°C transition often result in severe film roughening, blistering, and porosity, attributed to rapid evolution of gases within the film, and an accelerating exothermic burnout rate. Approaches to minimizing the processing time of pyrolysis have included a slow approach to the organic pyrolysis temperature, followed by slow ramp rates over the 200–240°C range, which can decrease the pyrolysis time from 12 to 3–6 hours. An alternate approach to avoiding this exothermic pyrolysis is suggested by curve B in Figure 11.4, DTA of a film pyrolyzed in an N_2 (nonoxidizing) atmosphere. In this case, film decomposition

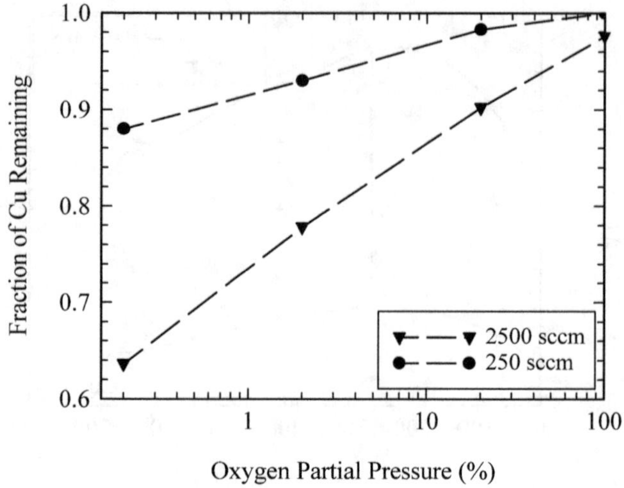

Figure 11.5. YBCO film copper retention vs. gas flow rate and oxygen partial pressure, illustrating predictable process metrics to control and compensate for copper volatility under low $p(O_2)$ pyrolysis conditions.

is endothermic, possibly by a process such as chain scission, in which organic groups are removed from the sample without oxidation. In such a condition, thermal runaway and gas evolution would not be expected, and films so processed remain smooth, rather than rough and porous.

This approach of a low $p(O_2)$ pyrolysis atmosphere has been used (Dawley et al., 2001a, 2001b) to reduce overall pyrolysis time to as little as 90 minutes, using conditions such as a 3°C/min ramp to 400°C in a 250 sccm flow of 0.2% O_2 in N_2. In practice, the use of low $p(O_2)$ conditions is complicated somewhat by the volatility of copper trifluoracetate in low $p(O_2)$ conditions (Dawley et al., 2001a, 2001b; Krupoder et al., 1995), illustrated in Figure 11.5. At the 0.2% O_2, 250 sccm condition, a copper loss of 10–11% was reproducibly observed from originally stoichiometric 1 : 2 : 3 films. This amount of Cu loss could easily be compensated for by simply adding an appropriate excess of Cu-acetate to the precursor solution. By tailoring ramp profiles, it is expected the use of low $p(O_2)$ conditions would ultimately allow pyrolysis times of 20–45 minutes, a significant improvement from 6–24 hour process times.

To carry sufficient current for power transmission applications, coated conductors will likely require at least 1–2 μm thick YBCO coatings. Low viscosity solutions, such as those containing high volume fractions of alcohols or ethers, typically form coatings that are only 0.1 μm thick (Schwartz, 1997). As a result, 10 to 20 coatings would be needed in order to achieve a 1–2 μm overall thickness on each side of the coated conductor. While tailoring solution concentration and processing conditions, such as coating rate and time, can increase film thickness, cracking caused by drying or decomposition stresses has typically prevented substantial increases in thickness. However, in recent years, several researchers showed that by adding high viscosity compounds, such as diols, triols, and high-molecular weight polymers such as polyvinylpyrrolidone, to sol-gel solutions, it is possible to increase the single coating thickness of lead zirconate titanate (PZT) and $BaTiO_3$ to 0.25–2 μm (Kozuka and Kajimura, 1999; Schwartz et al., 1997; Arscott et al., 1999; Liu and Mevissen, 1997). The high viscosity and high boiling point of these compounds allows rearrangement of species within the film during processing, which reduces drying stresses and helps inhibit cracking.

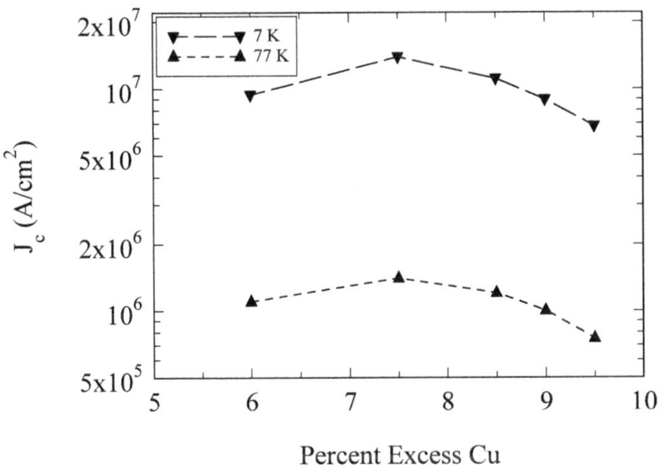

Figure 11.6. J_c vs. excess Cu content for 0.25 μm thick YBCO films fabricated via a diol route. The films were pyrolized using a rapid, low $p(O_2)$ process.

Applying such schemes to YBCO may provide an efficient means of producing thicker films, thereby reducing the required number of recoating steps. Fewer deposition cycles should also reduce the large-scale defects, such as pin-holes, voids, and microcracks, within the completed film stack and improve electrical properties.

In recent work, we developed a simplified trifluoroacetic acid (TFA)/metal acetate/methanol based solution route and used a rapid, low $p(O_2)$ pyrolysis process to produce high quality 0.1 μm YBCO films ($J_c = 3$ MA/cm^2 at 77 K) on $LaAlO_3$ (Dawley et al., 2001a). Following the high viscosity solution approach, we produced high quality ($J_c = 2$ MA/cm^2 at 77 K), 0.25 μm thick YBCO films using a diol (1,3 propanediol; one of a class of dihydroxyalcohols referred to as diols) solution route with the rapid, low $p(O_2)$ pyrolysis process (Dawley et al., 2001b). The properties of these films, as well as thick films (up to 1.5 μm) made by multiple coatings of methanol and diol solutions, were studied to gain insight into the effects of overall film thickness on phase development, microstructure, and J_c. It was observed that the J_c decreased by as much as 75–90% from single coating values when multiple coatings were deposited on previously crystallized YBCO coatings. Two effects were isolated that appear to control J_c for multilayered sol-gel samples: control of copper stoichiometry, and furnace atmosphere during multilayer crystallization.

A series of solutions with different amounts of excess Cu were synthesized to determine the optimum Cu excess for the diol films. Since precise, nondestructive composition measurement is non-trivial for 0.25 μm thick films, the response variables used to optimize the Cu excess were the J_c at 7 K and 77 K. The film with the highest J_c will correlate to the overall composition closest to 1 : 2 : 3 (Carlson et al., 1990). Figure 11.6 shows the effects of excess Cu composition on J_c. The plot shows that J_c is sensitive to slight changes in Cu content. For these thicker diol films, highest J_c values were found in films with less Cu excess than thinner MeOH films that underwent the same low $p(O_2)$ pyrolysis. Films fabricated from solutions with 7.5% excess Cu exhibit the highest J_c values (14 MA/cm^2 and ~2 MA/cm^2 at 7 K and 77 K, respectively). XRD patterns of those films indicate that they were phase pure 123. Films with different amounts of excess Cu had small fractions of secondary phases, such as 211 and CuO.

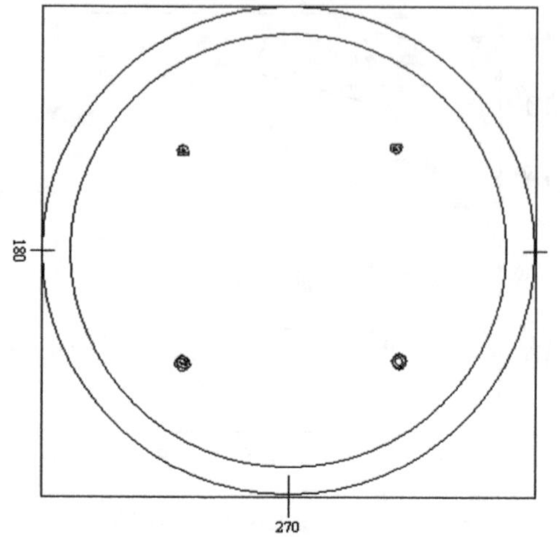

Figure 11.7. Pole figures for YBCO films fabricated from diol solutions verifying the biaxial texture and high quality of the films. Phi scans show no 45° variants around the YBCO (113).

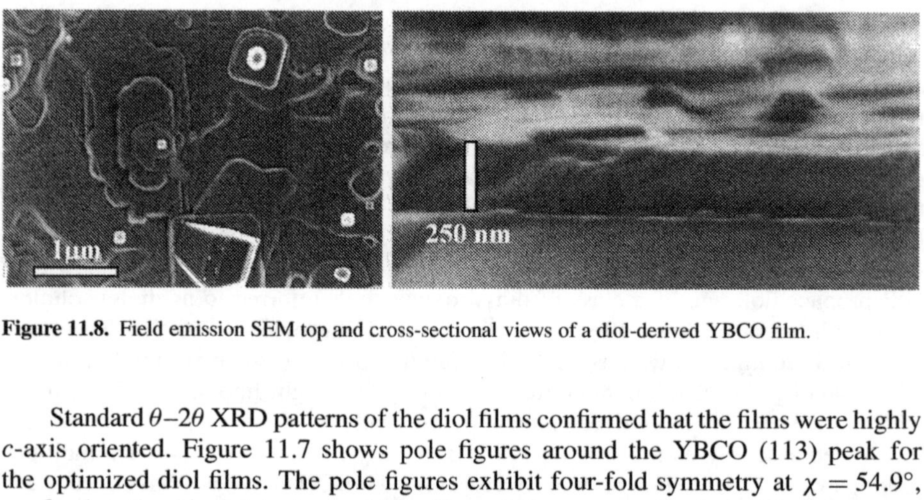

Figure 11.8. Field emission SEM top and cross-sectional views of a diol-derived YBCO film.

Standard θ–2θ XRD patterns of the diol films confirmed that the films were highly c-axis oriented. Figure 11.7 shows pole figures around the YBCO (113) peak for the optimized diol films. The pole figures exhibit four-fold symmetry at $\chi = 54.9°$, confirming the biaxial texture of the films. Phi scans verified the biaxial texture and that there were no 45° variants. The FWHM of the phi scans of the diol film was 1.2°, which indicates a high degree of grain-to-grain alignment. It should be noted that a 1° step size was used for the phi scan, therefore the 1.2° FWHM is likely an upper limit. Top view and cross-sectional FESEM images of a diol film are shown in Figure 11.8. The images reveal that the diol films are dense, with average YBCO grain size between 0.5 to 1 μm. The terraced morphology appears to be consistent with three-dimensional island growth, which has been observed by other researchers for *ex situ*, BaF_2 process, YBCO deposited using coevaporation or solution deposition, and *in situ* pulsed laser deposited films (Solovyov et al., 1999; Roshko et al., 1997). The J_c vs. temperature behavior of the diol films is shown in Figure 11.9(a). Compared to the MeOH films, there is a slight decrease in J_c at a given temperature. The drop in J_c with thickness is likely due to a difference in crystal quality, since the crystallization anneal was optimized for 0.1 μm thick films (Siegal et al., 1990a). However, the J_c value still remains above 1 MA/cm^2 until 80 K.

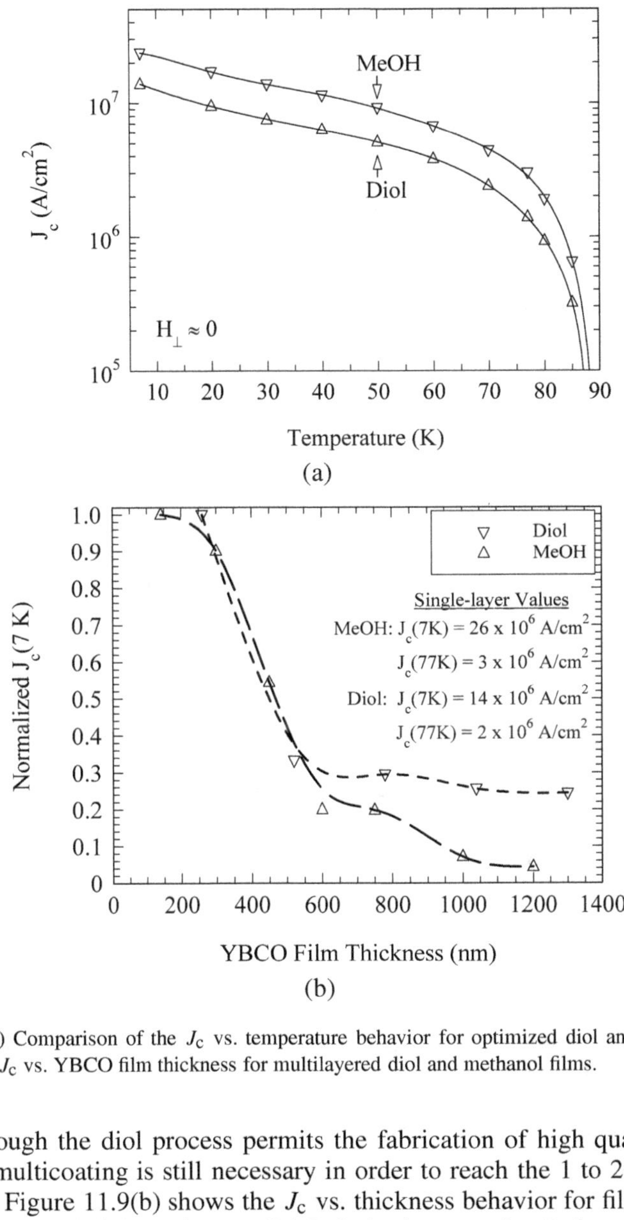

Figure 11.9. (a) Comparison of the J_c vs. temperature behavior for optimized diol and methanol films. (b) Normalized J_c vs. YBCO film thickness for multilayered diol and methanol films.

Even though the diol process permits the fabrication of high quality, 0.25 μm thick films, multicoating is still necessary in order to reach the 1 to 2 μm thickness requirement. Figure 11.9(b) shows the J_c vs. thickness behavior for films made from multiple MeOH and diol coatings on LAO. In both cases, the J_c drops sharply when the YBCO film thickness increases above 0.25 μm. Using the diol process, the highest $J_c \times$ thickness product we have obtained to date is 0.7 MA/cm^2 in a 1.3 μm sample, which equates to 91 A/cm width. XRD patterns of the films indicate that the quantity of a-axis and randomly oriented grains increases with thickness, as observed by others (Hsieh et al., 1990).

An increase of a-axis oriented grains has been linked to slow growth conditions during processing. Recent work by Suenaga and coworkers (Solovyov et al., 1999, 2000) examining the crystallization kinetics of thick YBCO films fabricated with the BaF_2 process, showed that for films with thicknesses >0.25 μm, the diffusion of water vapor into the film to aid in fluoride decomposition, and the removal of HF from film are the rate limiting steps for growth. Therefore, as the film thickness increases, the local concentration of HF in the film increases, slowing the growth rate and favoring a-axis oriented grain growth. The a-axis grains are well-known to transport electrical current much less efficiently than the c-axis oriented grains. Consequently, as the volume fraction of a-axis oriented grains increases, the connectivity of the film decreases, decreasing the J_c. Low absolute pressure (vacuum) crystallization processes were found to enhance the diffusion rate of HF out of the film, increasing the growth rate, and leading to high quality multi-micrometer thick, BaF_2 process, YBCO films.

In both vapor BaF_2 and sol-gel processes, it has been observed that a maximum per-layer thickness without significant a-axis formation is often around 0.3 μm. While research is underway at many locations to improve process methods and parameters to improve this figure, it appears that multilayering of these high quality ($J_c \sim$ 3–4 MA/cm^2) 0.3 μm layers is currently a critical area of need. To date, multilayering of high J_c films atop previously crystallized films has not been reported. A clear reason for this, in the case of evaporated or sol-gel BaF_2-type processes, is the corrosion of YBCO and other high T_c superconductors by the water vapor used in the crystallization process (Siegal et al., 1999; Zhou et al., 1997). This corrosion is readily observed as reduced YBCO XRD peak intensities, and the development of second phases. The corrosion that occurs in these humid processes, however, appears to be limited to cases where water vapor is introduced to the furnace atmosphere at low temperatures, where it possesses a high $p(H_2O)$. Figure 11.10(a) shows that the J_c for sol-gel YBCO films is not degraded when water vapor is introduced late in the process at temperatures up to 700°C. Using this approach, a single YBCO layer was crystallized, a second sol-gel layer deposited atop the first, and water vapor introduced during the second layer crystallization at 400°C. The magnetic J_c for single 120 nm layers was found to be 1.3 MA/cm^2, and the J_c for the composite 240 nm bilayer was 1.1 MA/cm^2, with doubled c-axis XRD intensities shown in Figure 11.10(b). This suggests a new route to high quality multilayers of YBCO: deposition of subsequent 0.3 μm high J_c films, using this delayed water vapor introduction to achieve multilayering (Dawley et al., 2001c). Current research suggests the further possibility of reducing crystallization times to as little as 3 minutes per layer of YBCO, which would dramatically decrease overall processing time. Estimated process time for YBCO layers may be below 1 hour in the future (20 minute pyrolysis, 10 minute ramp up, 3 minute crystallization, 10 minute ramp down, with a final 20 minute back-end oxygenation after all layers are crystallized) for a 2–10 cm/s solution deposition process. Key areas of current research include scale-up of current batch processes into continuous processing of tapes through static furnace hot zones, removal of reaction vapors from tapes during processing, and increasing the maximum YBCO thickness that may be processed in one step without a-axis formation.

A working all-solution deposition route to >1 MA/cm^2 coated conductors was first reported in August 2001 by our group using a 0.1 μm sol-gel YBCO film on a 0.3 μm sol-gel Nb : $SrTiO_3$ buffer layer on an Oxford Superconducting Technologies-produced RABiTS Ni(200) tape (Clem et al., 2001; Dawley et al., 2001d). Pole figures for the YBCO, $SrTiO_3$, and Ni(200) substrate tape are shown in Figure 11.11, indicating maintenance of the substrate biaxial orientation through the STO and YBCO layers. A transport J_c (77 K, 1 μV/cm) of 1.3 MA/cm^2 was measured for this composite. Critical steps to achieving this composite were (a) developing a dense, oxygen-blocking buffer layer (4% Nb-doped $SrTiO_3$, in this example), (b) suppressing non-(200) $SrTiO_3$ buffer layer orientations atop the (200) Ni tape, and (c) developing a smooth YBCO coating. $SrTiO_3$ was chosen as a buffer layer due to its compatibility

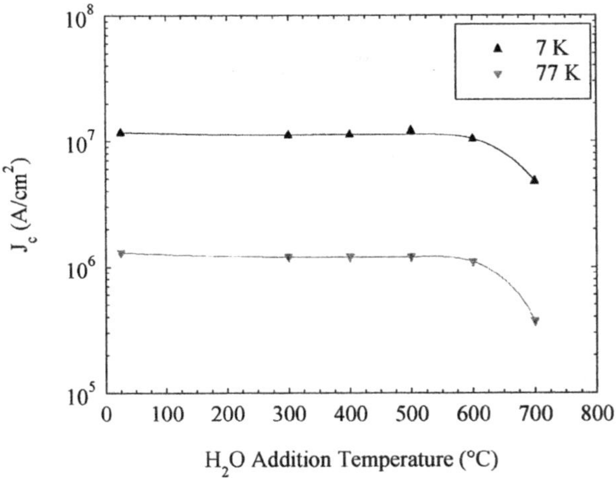

(a)

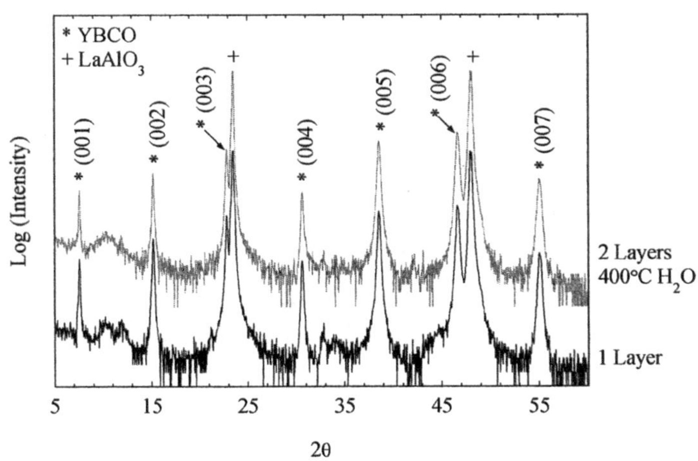

(b)

Figure 11.10. (a) Effect of water vapor introduction temperature on YBCO critical current density. (b) XRD intensities of single and bilayer YBCO using a 600°C water vapor introduction temperature to avoid corrosion of the first crystallized layer.

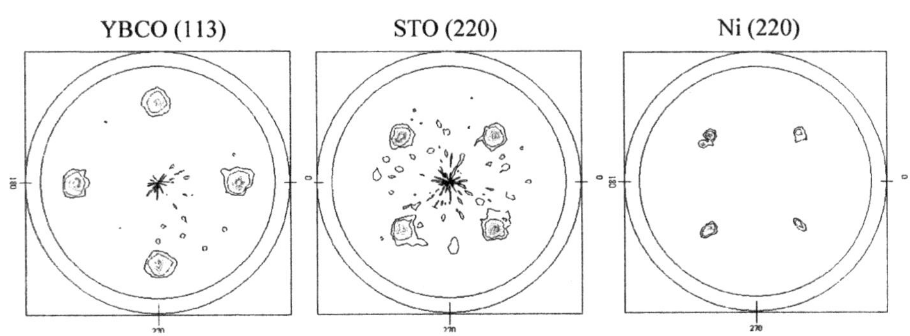

Figure 11.11. Pole figures of a YBCO/Nb : STO/Ni(200) composite which displayed a J_c of 1.3 MA/cm^2. A 2 μm silver top coating was deposited atop the sample; additionally the Nb : STO buffer layer is a semiconducting oxide.

with YBCO and its low oxygen diffusion rate at 800°C, which was confirmed by ^{18}O SIMS in-diffusion experiments. Additionally, a number of process modifications have been previously reported to enable growth of dense, highly oriented perovskite materials such as $SrTiO_3$ (Schwartz et al., 1999). For the YBCO overcoat, a dense buffer layer that serves as an oxygen and nickel diffusion barrier was found to be necessary. Current efforts are aimed at increasing the thickness of the YBCO grown on these buffered metal tapes, scaling up deposition to lengths beyond 25 cm, and ultimately to continuous processing. It is hoped a number of technologies developed for scale-up of evaporated BaF_2 process YBCO films may be also applied to sol-gel approaches.

11.5 CONCLUSIONS

Solution deposition of coated conductors shows potential to be the lowest cost process toward workable coated conductors, but is confronted with several issues. Among these are the need for demonstration of high J_c in multimicron thicknesses, and the need for scale-up of the crystallization anneal. While a value of 91 A/cm width has been obtained from a 6 layer, 0.7 MA/cm^2, 1.3 μm diol-derived YBCO film, a method for fabricating yet thicker films of higher J_c is certainly desired. It is hoped the delayed water introduction method, and improvements in crystallization kinetics will play a role in this development. Of key concern is development of a process to enable YBCO crystallization of layers greater than 0.3 μm without significant a-axis content; this remains an issue for both solution deposited and evaporated BaF_2 process films. An additional area of concern is the scale-up of the crystallization anneal, and ability of the anneal to remove reaction products uniformly from the film. It is hoped that the parallel studies of evaporated BaF_2 process film crystallization and scale-up will prove applicable to solution deposition due to the similarity of crystallization mechanisms in these *ex situ* processes.

REFERENCES

Annavaparou, S. et al., 2000, Progress towards a low-cost coated conductor technology, *Physica C*, 341–348:2319.

Arscott, S., Miles, R.E., Kennedy, J.D., and Milne, S.J., 1999, Rapid thermal processing of lead zirconate titanate thin films on Pt–GaAs substrates based on a novel 1,1,1-tris(hydroxymethyl)ethane sol-gel route, *J. Mater. Res.*, 14(2):494.

Brinker, C.J. and Scherer, G.W., 1990, *Sol-Gel Science, the Physics and Chemistry of Sol-Gel Processing*, Academic Press, San Diego.

Carlson, D.J., Siegal, M.P., Phillips, J.M., Tiefel, T.H., and Marshall, J.H., 1990, Stoichiometric effects in epitaxial YBCO thin films on $LaAlO_3$ (100), *J. Mater. Res.*, 5(12):2797.

Chen, Y.L., Mantese, J.V., Hamdi, A.H., and Micheli, A.L., 1989, Microstructure and superconducting properties of YBCO and YbBCO thin films formed by metalorganic decomposition, *J. Mater. Res.*, 4(5):1065.

Chu, P.-Y., Campion, I., and Buchanan, R.C., 1993, Processing effects on high T_c properties of YBCO films from carboxylate solution precursors, *J. Mater. Res.*, 8:261.

Clem, P.G. et al., 2001, Process for forming epitaxial perovskite thin film layers using halide precursors, US Patent 6,231,666.

Dawley, J.T., Clem, P.G., Siegal, M.P., and Overmyer, D.L., 2001a, High J_c $YBa_2Cu_3O_{7-d}$ films via rapid, low $p(O_2)$ pyrolysis, *J. Mater. Res.*, 16(1):13.

Dawley, J.T., Clem, P.G., Siegal, M.P., Overmyer, D.L., and Rodriguez, M.A., 2001b, Thick sol-gel derived YBCO films, *IEEE Trans. Appl. Supercon.*, 11(1):2873.

Dawley, J.T., Clem, P.G., Siegal, M.P., and Overmyer, D.L., 2001c, in press.

Dawley, J.T., Clem, P.G., Siegal, M.P., Overmyer, D.L., and Ong, R.J., 2001d, in press.

Feenstra, R., Lindemar, T.B., Budai, J.D., and Galloway, M.D., 1991, Effect of oxygen pressure on the synthesis of YBCO thin films by postdeposition annealing, *J. Appl. Phys.*, 69:6569.

Foltyn, S.R. et al., 1999, High-T_c coated conductors: Performance of meter-long YBCO/IBAD flexible tapes, *IEEE Trans. Appl. Supercond.*, 9(2/Pt.2):1519.

Frey, M.H. and Payne, D.A., 1995, Synthesis and processing of barium titanate ceramics from alkoxide solutions and monolithic gels, *Chem. Mater.*, 7(1):170.

Frey, M.H. and Payne, D.A., 1996, Grain-size effect on structure and phase transformations for barium titanate, *Phys. Rev. B*, 54(5):3158.

Goyal, A. et al., 2000, $YBa_2Cu_3O_{7-y}$-coated conductors with high engineering current density, *J. Mater. Res.*, 15(12):2647.

Gross, M.E., Hong, M., Liou, S.J., Gallagher, P.K., and Kwo, J., 1988, Versatile new metalorganic process for preparing superconducting thin films, *Appl. Phys. Lett.*, 52:160.

Gupta, A., Jagannathan, R., Cooper, E.I., Giess, E.A., Landman, J.I., and Hussey, B.W., 1988, Superconducting oxide-films with high transition-temperature prepared from metal trifluoroacetate precursors, *Appl. Phys. Lett.*, 52(24):2077.

Hirano, S., Hayashi, T., and Miura, M., 1990, Preparation of BYCO thin films with preferred orientation through an organometallic route, *J. Am. Ceram. Soc.*, 73(4):885.

Hsieh, Y.F., Siegal, M.P., Hull, R., and Phillips, J.M., 1990, Microstructure of epitaxial $YBa_2Cu_3O_{7-x}$ thin-films grown on $LaAlO_3$(001), *Appl. Phys. Lett.*, 57(21):2268.

Kozuka, H. and Kajimura, M., 1999, Achievement of crack-free $BaTiO_3$ films over 1 um in thickness via non-repetitive dip-coating, *Chem. Lett.*, 10:1029.

Krupoder, S.A., Danilovich, V.S., Miller, A.O., and Furin, G.G., 1995, Polyfluorocarboxylates 1. Copper (II) trifluoroacetate and its analogs, *J. Fluorine Chem.*, 73:13.

Lange, F.F., 1996, Chemical solution routes to single-crystal thin films, *Science*, 273:903.

Liu, D. and Mevissen, J.P., 1997, Thick layer deposition of lead perovskites using diol-based chemical solution approach, *Int. Ferro.*, 18(1–4):263.

Malecki, A., Oblankowski, J., and Labus, S., 1995, The role of $BaCO_3$ in high-temperature synthesis of electronic materials, *Mater. Res. Bull.*, 30(6):731.

Malozemoff, A.P. et al., 2000, Low-cost YBCO coated conductor technology, *Supercond. Sci. Technol.*, 13:473.

Manabe, T., Yamaguchi, I., Nakamura, S., Kondo, W., Kumagai, T., and Mizuta, S., 1995, Crystallization and in plane alignment behavior of YBCO films on MgO (001) prepared by the dipping-pyrolysis process, *J. Mater. Res.*, 10(7):1635.

Manabe, T., Yamaguchi, I., Nakamura, S., Kondo, W., Mizuta, S., and Kumagai, T., 1997, Carbon dioxide controlled annealing method for preparation of $YBa_2Cu_3O_{7-y}$ films by dipping-pyrolysis process, *Physica C*, 276:160.

Mankiewich, P.M., Schofield, J.H., Skocpol, W.J., Howard, R.E., Dayem, A.H., and Good, E., 1987, Reproducible technique for fabrication of thin films of high transition temperature superconductors, *Appl. Phys. Lett.*, 51:1753.

Matsubara, I., Paranthaman, M., Chirayil, T.G., Sun, E.Y., Martin, P.M., Kroeger, D.M., Verebelyi, D.T., and Christen, D.K., 1999, Preparation of epitaxial $YbBa_2Cu_3O_{7-\delta}$ on $SrTiO_3$ single crystal substrates using a solution process, *Jpn. J. Appl. Phys.*, 38:L727.

McIntyre, P.C., 1993, Sc.D. thesis, MIT.

McIntyre, P.C., Cima, M.J., and Ng, M.F., 1990, Metalorganic deposition of high J_c YBCO films from trifluoroacetate precursors onto (100) $SrTiO_3$, *J. Appl. Phys.*, 68(8):4183.

McIntyre, P.C., Cima, M.J., Smith, J.A., Jr., Hallock, R.B., Siegal, M.P., and Phillips, J.M., 1992, Effect of growth conditions on the properties and morphology of chemically derived epitaxial thin films of $Ba_2YCu_3O_{7-x}$ on (001) $LaAlO_3$, *J. Appl. Phys.*, 71(4):1868.

McIntyre, P.C. and Cima, M.J., 1994, Heteroepitaxial growth of chemically derived *ex-situ* BYCO thin films, *J. Mater. Res.*, 9(9):2219.

McIntyre, P.C., Cima, M.J., and Roshko, A., 1995, Epitaxial nucleation and growth of chemically derived BYCO thin films on (100) $SrTiO_3$, *J. Appl. Phys.*, 77(10):5263.

Nonaka, T. et al., 1988, BYCO thin films fabricated by dip coating using concentrated mixed alkoxide solution, *Jpn. J. Appl. Phys. Pt.* 2, 27:L867.

Ohring, M., 1992, *The Materials Science of Thin Films*, Academic Press, San Diego, p. 199.

Pak, S.S. et al., 1992, Solution-condensed YBCO superconductor thin films from thermosetting meta organic precursors, *J. Am. Ceram. Soc.*, 75:2268.

Rice, C.E., van Dover, R.B., and Fisanick, G.J., 1987, Preparation of superconducting thin films of BYCO by a novel spin-on pyrolysis technique, *Appl. Phys. Lett.*, 51:1842.

Roshko, A., Stork, F.J.B., Rudman, D.A., Aldrich, D.J., and Hotsenpiller, P.A.M., 1997, Comparison of heteroepitaxial $YBa_2Cu_3O_{7-\delta}$ and TiO_2 thin film growth, *J. Crystal Growth*, 174(1–4):398.

Sathyamurthy, S. and Salama, K., 1998, Processing of YBCO films by solution techniques using metal–organic decomposition, *J. Supercond.*, 11(5):545.

Sathyamurthy, S. and Salama, K., 2000, Fabrication of Y123 coated conductors using metal organic decomposition process, *Physica C*, 341–348:2479.

Sathyamurthy, S. and Salama, K., 2001, Application of solution deposition to fabricate YBCO coated conductor, *IEEE Trans. Appl. Supercond.*, 11(1):2935.

Schwartz, R., 1997, Chemical solution deposition of perovskite thin films, *Chem. Mater.*, 9(11):2325.

Schwartz, R.W., Reichert, T.L., Clem, P.G., Dimos, D., and Liu, D., 1997, A comparison of diol and methanol-based chemical solution deposition routes for PZT thin film fabrication, *Int. Ferro.*, 18(1–4):275.

Schwartz, R.W., Clem, P.G., Voigt, J.A., Byhoff, E.R., Van Stry, M., Headley, T.J., and Missert, N.A., 1999, Control of microstructure and orientation in solution-deposited $BaTiO_3$ and $SrTiO_3$ thin films, *J. Am. Ceram. Soc.*, 82(9):2359.

Sheth, A., Schmidt, H., and Lasrado, V., 1998, Review and evaluation of methods for application of epitaxial buffer and superconductor layers, *Appl. Supercond.*, 6(10–12):855.

Siegal, M.P., Phillips, J.M., van Dover, R.B., Tiefel, T.H., and Marshall, J.H., 1990a, Optimization of annealing parameters for the growth of epitaxial $Ba_2YCu_3O_{7-x}$ films on $LaAlO_3(100)$, *J. Appl. Phys.*, 68(12):6353.

Siegal, M.P., Phillips, J.M., van Dover, R.B., Tiefel, T.H., and Marshall, J.H., 1990b, Optimization of annealing parameters for the growth of epitaxial BYCO films on $LaAlO_3$, *J. Appl. Phys.*, 68(12):6353.

Siegal, M.P. et al., 1999, Remarkable properties of Tl–Ba–Ca–Cu–O thin films following post-growth high-temperature annealing, *IEEE Trans. Appl. Supercond.*, 9(2:pt.2):1555.

Smith, J.A., Cima, M.J., and Sonnenberg, N., 1999, High critical current density thick MOD-derived YBCO films, *IEEE Trans. Appl. Supercond.*, 9(2):1531.

Solovyov, V.F., Wiesmann, H.J., Wu, L.J., Suenaga, M., and Feenstra, R., 1999, High rate deposition of 5 um thick $YBa_2Cu_3O_7$ films using the BaF_2 *ex-situ* post annealing process, *IEEE Trans. Appl. Supercond.*, 9(2/pt.2):1467.

Solovyov, V., Wiesmann, H.J., Wu, L., Zhu, Y., and Suenaga, M., 2000, Kinetics of YBCO film growth by postdeposition processing, *Appl. Phys. Lett.*, 76(14):1911.

Yamagaiwa, K. et al., 2001, Epitaxial growth of $REBa_2Cu_3O_{7-y}$ films on various substrates by chemical solution deposition, *J. Crystal Growth* 229:353.

Zhou, J.P. et al., 1997, Environmental degradation properties of $YBa_2Cu_3O_{7-\delta}$ and $Y_{0.6}Ca_{0.4}Ba_{1.6}La_{0.4}$-$Cu_3O_{7-\delta}$ thin film structures, *Physica C*, 273:223.

Chapter 12

NON-FLUORINE BASED BULK SOLUTION TECHNIQUES TO GROW SUPERCONDUCTING $YBa_2Cu_3O_{7-\delta}$ FILMS

M. Parans Paranthaman

Chemical Sciences Division
Oak Ridge National Laboratory
Oak Ridge, TN 37831-6100
USA

12.1 INTRODUCTION

Major advances have been made in the last 14 years in the area of high-temperature superconductor (HTS) research, resulting in increasing use of HTS materials in commercial and pre-commercial applications. HTS conductors are expected to be useful for numerous electric power applications, including underground transmission cables, oil-free transformers, high efficiency motors, compact generators, and superconducting magnetic energy storage systems for smoothing voltage fluctuations in the power grid. Research on the $YBa_2Cu_3O_{7-\delta}$ (YBCO) based second generation conductors is now intensively carried out in the world. Recently, the US Department of Energy conducted a Coated Conductor Technology Development Roadmap Workshop in St. Petersburg, Florida. This roadmap identified specific near-term activities that are needed to advance techniques for continuous processing of high quality, low-cost coated conductors that will lead to industrial-scale commercial manufacturing (Energetics, Inc., Columbia, MD, 2001). The activities specified in this roadmap are focused on achieving the following vision. "*Low-cost, high-performance YBCO Coated Conductors will be available in* 2005 *in kilometer lengths. For applications in liquid nitrogen, the wire cost will be less than* \$50/kA-m, *while for applications requiring cooling to temperatures of* 20–60 K *the cost will be less than* \$30/kA-m. *By* 2010 *the cost-performance ratio will have improved by at least a factor of four*." One of the important critical needs that came out of this workshop was to develop alternative non-vacuum processes for fast, reliable and economic deposition of YBCO. The traditional *in-situ* process, in which pulsed laser deposition (PLD) of oxide or co-evaporation of Y, Ba, and Cu metals under appropriate oxygen atmospheres, could be used to fabricate YBCO films. However, it may be difficult to scale up these processes to produce low-cost conductors. This is mainly due to the initial investment of a high cost laser,

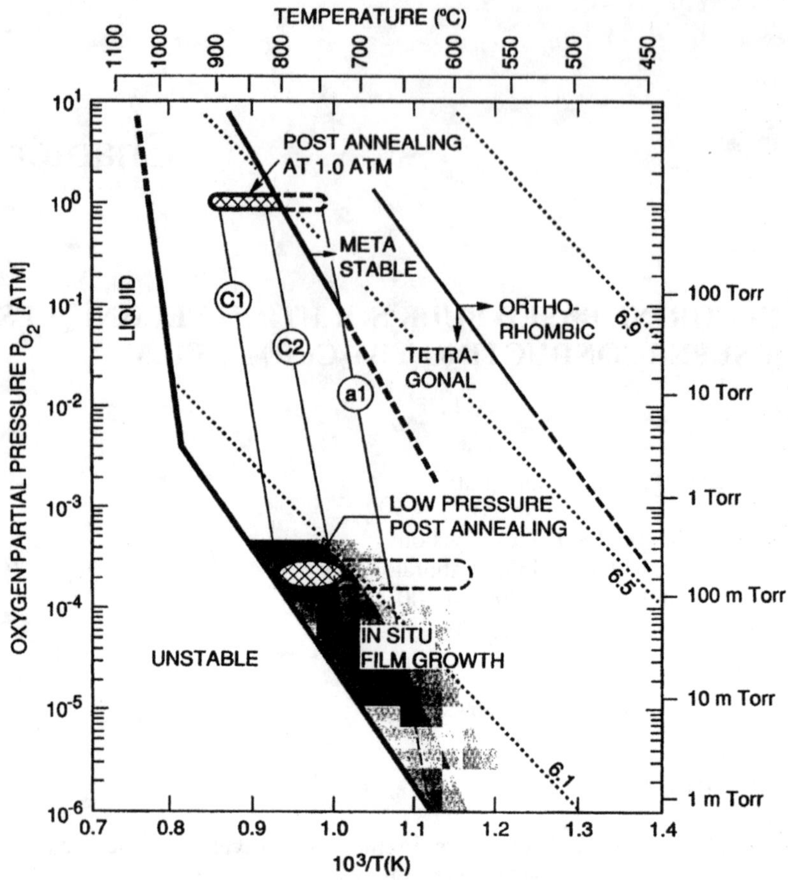

Figure 12.1. Oxygen partial pressure vs. temperature diagram showing liquid-phase boundaries, stability of YBCO, tetragonal-to-orthorhombic transition line, and lines of constant oxygen stoichiometry of $YBa_2Cu_3O_{7-x}$. *c*-axis aligned YBCO films are obtained between the c1 and c2 boundary lines; *c*- and *a*-axis aligned YBCO films between c2 and a1 boundary lines. Experimental and extrapolated ranges are indicated by solid and dashed lines, respectively (adapted from Feenstra et al., 1991).

the high cost of large vacuum chambers with pumping system, etc. The control of substrate temperature during deposition is also difficult. To circumvent this problem, *ex-situ* precursor processes can be used. Chemical solution epitaxy has emerged as a viable, low-cost, non-vacuum process for fabricating long lengths of YBCO coated conductors. In these processes, YBCO precursors can be deposited at room temperature and later post-annealed in a controlled atmosphere furnace. The advantages of *ex-situ* processes are the separation of the deposition and post-annealing steps, and a wider processing window by combining temperature, and oxygen partial pressures. The dependence of oxygen partial pressure and the YBCO process temperatures is shown in Figure 12.1. Also, the precursor stoichiometry, and dopant concentration can be easily controlled and the post-annealing step can be a batch process. The growth rate of YBCO generally varies from 1–3 Å/sec. This could be a rate-limiting step in these processes. However, it is possible to overcome these limitations by suitably modifying the furnace designs to process large quantities of wires (hence, large area) in a single step. The most commonly used bulk solution techniques to fabricate YBCO coated conductors are:

- (i) Sol-gel processing
  - (a) Sol-gel alkoxide approach
  - (b) Metal–organic decomposition (MOD)
- (ii) Electrodeposition
- (iii) Electrophoresis
- (iv) Spray (Aerosol) pyrolysis techniques
- (v) Chemical vapor deposition
  - (a) Combustion chemical vapor deposition (CCVD)
  - (b) Metal–organic chemical vapor deposition (MOCVD)
- (vi) Powder suspension techniques
- (vii) Liquid phase epitaxy (LPE)

Each film deposition process will have some maximum rate, beyond which defects or other problems such as supplying source material or removing by-products may become limiting. The properties of YBCO films will depend critically upon the microstructures that develop during the nucleation and growth of the films. These microstructures depend on the substrate properties, the particular deposition technique, the processing conditions and the film thickness. The Trifluoroacetate (TFA), MOCVD, CCVD, and LPE techniques are reviewed elsewhere in this book. In this review, we will report only the recent achievements in growing YBCO films using non-fluorine containing solution precursors. In addition, the recent developments in the buffer layer work at Oak Ridge National Laboratory are also highlighted.

12.2 SOL-GEL PROCESSING

The most commonly used chemical solution deposition may be grouped into three categories: (i) sol-gel processes that use 2-methoxyethanol as a reactant and solvent; (ii) hybrid processes that use chelating agents such as acetylacetonate or diethanolamine to reduce alkoxide reactivity, and (iii) metal–organic decomposition techniques that use high-molecular-weight precursors and water-insensitive carboxylates, 2-ethyl-hexanoates, etc. The sol-gel precursor route has been used to grow both oxide buffer layers and superconductors because of the ease of formation of epitaxial oxides at relatively lower temperatures, control over the polymeric viscous gel formation, and the relatively easy scale-up of the thickness of the films.

The sol-gel processing is a wet chemical route to synthesis of a colloidal suspension of solid particles or clusters in a liquid (sol) and subsequently to formation of a dual phase material of a solid skeleton filled with a solvent (wet gel) through sol-gel transition (gelation). When the solvent is removed, the wet gel converts to a Xerogel through ambient pressure drying or an aerogel through supercritical drying. The sol-gel process involves synthesis of a polymerizable solution (often referred to as sol) by mixing or reacting metal alkoxides and metal–organic salts in a common solvent. Alkoxides are referred to as $M(OR)_n$ where M is a metal, n is the valency of the metal, and R is an alkyl group. The most common solvent used in this process is 2-methoxyethanol. The complete hydrolysis of the sol will form a rigid gel that can be heat-treated to powders. Partial hydrolysis of the sol will produce a polymeric, viscous gel that can be deposited on substrates and heat-treated to crystallize. The polymeric network can be important for microstructure and phase development. The reactivity of alkoxide ligands with water is the driving force for the sol-gel process. This reaction must be controlled in order to promote the desired gelation, and the starting metal alkoxides must undergo complete ligand exchange to form methoxyethoxide ligands. The ligand exchange can be described easily by the equation:

$$M(OR)_n + nR'OH = M(OR')_n + nROH. \tag{1}$$

Due to the bidentate nature of the methoxyethoxide ligand, which ties up vacant coordination sites, it slows the rate of hydrolysis and thus more readily allows the formation of a gel rather than precipitate. In addition, the bidentate nature of the methoxyethoxide ligand allows the more facile formation of mixed-metal alkoxide complexes. The complete gelation process can be summarized by the following equations:

Hydrolysis of alkoxides:

$$M(OR)_n + H_2O = M(OR)_{n-1}(OH) + ROH. \tag{2}$$

Condensation:

$$M(OR)_{n-1}(OH) + M(OR)_n = M_2O(OR)_{2n-2} + ROH, \tag{3}$$

$$M(OR)_{n-1}(OH) + M(OR)_{n-1}(OH) = M_2O(OR)_{2n-2}. \tag{4}$$

The primary advantage of sol-gel processing over conventional ceramic processing is that the polymeric network formation of the metal–organic complexes leads to intimate mixing of the amorphous preceramic oxides, thus allowing a dramatic reduction in reaction temperatures and time.

Thin ($\sim$100 nm), uniform and crack-free films can be readily formed on various materials by spin, dip or spray coating; thick films can be obtained by multiple coatings. Spin coating involves the acceleration of a liquid puddle on a rotating substrate. The coating material is deposited in the center of the substrate either manually or by a robotic arm. The physics behind spin coating involves a balance between centrifugal forces controlled by spin speed and viscous forces, which are determined by solvent viscosity. The spin coating technique consists of three basic stages: (i) the polymer is dispensed onto the substrate, (ii) the polymer is spread across the substrate (by spinning at approximately 500 rpm), (iii) the substrate is then spun at a higher speed (2000–4000 rpm). Some of the variable process parameters involved in spin coating are: solution viscosity, solid content, angular speed, and spin time. The film-forming process is primarily driven by two independent parameters—viscosity and spin speed. A range of film thickness can be easily obtained by spin coating. For thicker films, high material viscosity, low spin speed, and a short spin time are needed. However, these parameters can affect the uniformity of the coat. In order to scale-up these techniques, a dip-coating process has to be developed. The dip-coated tapes could be processed in a batch or in a continuous mode. The advantage of the dip coating process is its ability to coat large areas, complex shapes, and double-sided tapes. Furthermore, material utilization is almost 100%. Solution derived films crack with increasing thickness due to the high volume shrinkage as organics are removed during the heat treatment process. Restricting the film thickness to a critical film thickness prevents cracking, and thicker films may be achieved by multiple coating and heat treatment procedures. In the dip-coating process, the substrate is usually withdrawn vertically from the coating bath at a constant speed. The film thickness is directly proportional to the withdrawal velocity. The inner layer of the coating solution moves upward with the substrate, while the outer layer is returned to the bath. The viscosity and surface tension of the coating liquid control the film thickness to a lesser extent. A schematic diagram of the reel-to-reel continuous dip-coating unit is shown in Figure 12.2. The sol-gel alkoxide and

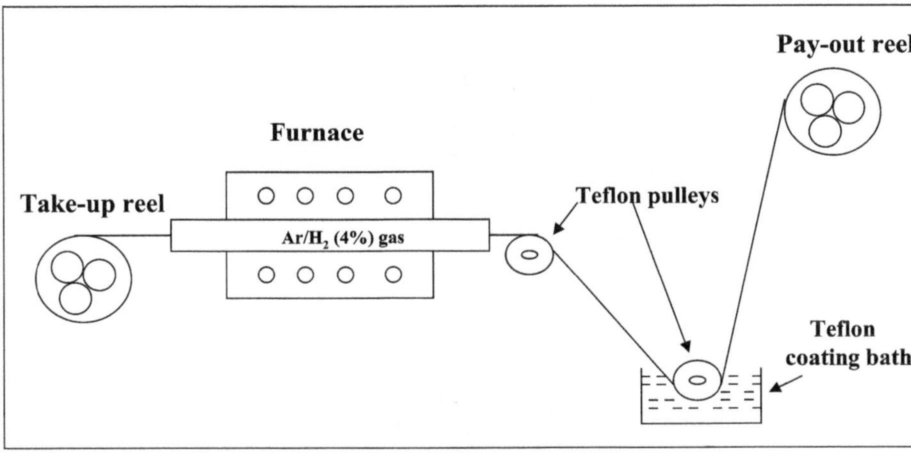

Figure 12.2. Schematic diagram of reel-to-reel continuous dip-coating unit.

MOD approaches to grow YBCO films are discussed in Sections 12.2.1 and 12.2.2, respectively. In the sol-gel alkoxide approach, the recent developments in the solution buffer layer work at Oak Ridge are described in Section 12.2.1.1.

12.2.1 Sol Gel Alkoxide Approach

12.2.1.1 Buffer Layers Chemical solution epitaxy has emerged as a viable, low-cost, nonvacuum process for fabricating long lengths of YBCO coated conductors (Brinker and Scherer, 1990; Brinker et al., 1992; Rupich et al., 1992; McIntyre et al., 1992; Paranthaman and Beach, 1995; Paranthaman et al., 1997b; Lange, 1996; Schwartz, 1997; Shoup et al., 1997). Rolling-Assisted Biaxially Textured Substrates (RABiTS) are ideal starting templates for this solution process (Goyal et al., 1996; Norton et al., 1996; Paranthaman et al., 1997a). For a film to function as an effective buffer, it is also essential to grow dense and crack-free films. To develop the solution process, buffer layers such as $REAlO_3$ (rare-earth aluminate; RE = La, Nd) and $BaZrO_3$, were initially grown epitaxially on $SrTiO_3$ (100) single-crystal substrates using sol-gel alkoxide precursors (Paranthaman et al., 1997b; Shoup et al., 1997). The $REAlO_3$ buffers grown on biaxially textured Ni (100)$\langle 001 \rangle$ substrates had a good out-of-plane texture but had two in-plane textures (Shoup et al., 1998; Beach et al., 1998; Paranthaman et al., 1999). Following this work, single cube-on-cube epitaxy of various RE_2O_3 (RE = Gd, Yb, and Eu) and $RE_2Zr_2O_7$ (rare-earth zirconates; RE = La, Nd) buffers were grown directly on textured Ni substrates using spin coating (Chirayil et al., 1999, 2000; Morrell et al., 2000; Paranthaman et al., 2000). To scale up this technique, a dip-coating process was developed. As seen from the Figure 12.3, both ends of the annealed nickel tape were electrically spot-welded to nickel leaders mounted on two reels. The take-up reel was driven continuously by a stepper motor, and the pay-out reel was tensioned by a variable torque motor. The travel speed of the tape could be varied up to 100 m/h. The reel-to-reel system has the capability of processing up to several meters of buffered tape.

Using the dip-coating process, epitaxial buffers of Eu_2O_3, Gd_2O_3, and $La_2Zr_2O_7$ (LZO) were grown on both Ni and Ni–W (3 at%) (strengthened and substrate with reduced magnetism) substrates. A 2-methoxyethanol solution of europium

Figure 12.3. The typical microstructure of 20 nm thick Gd_2O_3 seed layer on Ni–W substrate (Tolga et al., 2001).

methoxyethoxide/acetate, gadolinium methoxyethoxide/acetate or lanthanum zirconium methoxyethoxide was used for the dip-coating process. The details of the solution preparations were reported earlier (Chirayil et al., 1999, 2000; Morrell et al., 2000; Paranthaman et al., 2000). The concentration of the coating solution was typically 0.25–0.5 M. The Ni and Ni–W tapes from the pay-out reel were withdrawn from the coating bath at a constant speed of 1–10 m/h. The tapes were coated on both sides. The dip-coated tapes were then annealed in the furnace, which had been preheated to 1000–1100°C. The flow rate of Ar/H_2 (4%) gas purging the furnace was 2–4 l/min. The heat-treatment times typically varied from 10 min to 1 h in the hot zone. After heat treatment, the tapes were spooled on the take-up reel. The details of the optimized coating speed and annealing speed are reported elsewhere (Paranthaman et al., 2001a; Sathyamurthy et al., 2001; Tolga et al., 2001). Detailed X-ray studies indicated the presence of single cube texture of Eu_2O_3, Gd_2O_3, and LZO buffers were produced. The typical thicknesses of these buffers are 20–60 nm. The dip-coated seed layers are carbon-free, smooth, continuous, and crack-free. The typical microstructure of 20 nm thick Gd_2O_3 on Ni–W substrate is shown in Figure 12.3.

As shown in Figure 12.4, the AFM scan indicated the surface roughness of the Gd_2O_3 seed layer on Ni–W tape to be 3.2 nm. This proves that smooth buffers can be produced using the solution process. In addition, the substrate grain boundary is also completely covered by the Gd_2O_3 layer. One to two meter lengths of Eu_2O_3, Gd_2O_3, and LZO seed layers were produced. Attempts to grow YBCO films directly on the dip-coated buffer layers using pulsed laser deposition resulted in YBCO films with poor properties. Therefore, sol-gel chemistry for CeO_2 cap layers was also developed. On all solution buffers ((CeO_2 cap/Eu_2O_3 (5-coats)/Ni) or (LZO (4-coats)/Ni–W)), the highest J_c obtained to-date is 2×10^5 A/cm^2 at 77 K and self-field (Paranthaman

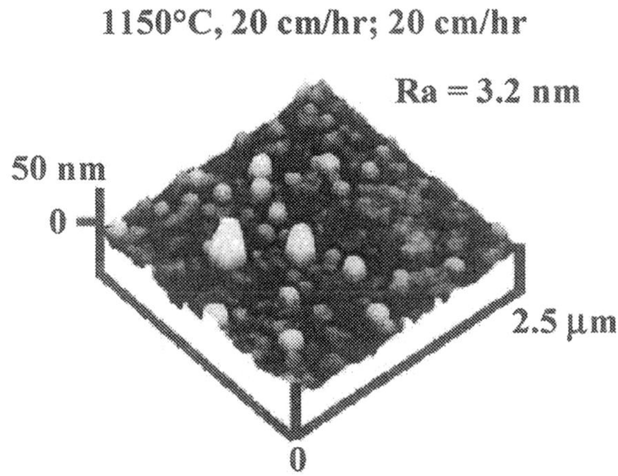

Figure 12.4. AFM scan of the 20 nm thick Gd_2O_3 seed layer on Ni–W substrate.

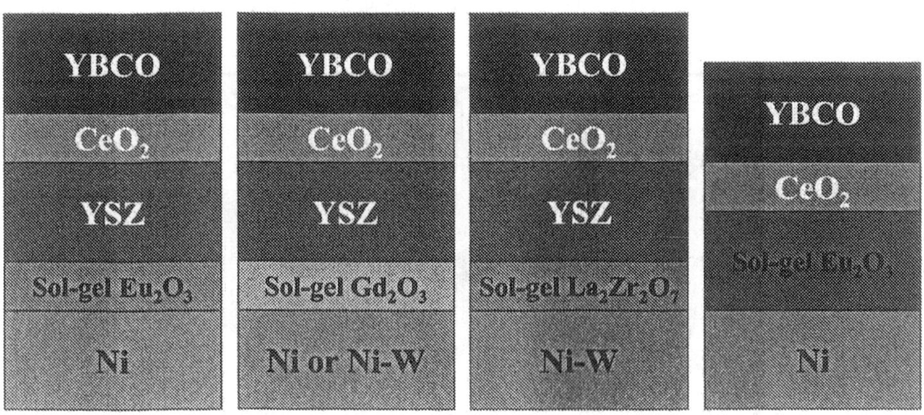

Figure 12.5. The various architectures developed using solution seed layers.

et al., 2001b). On either short or long tapes, both YSZ barrier layers and CeO_2 cap layers were deposited by rf magnetron sputtering at 780°C in 10 mTorr of Ar/H_2 (4%) gas. The plasma power was 75 W at 13.56 MHz. *Ex situ* YBCO precursors were deposited on CeO_2-buffered YSZ/dip-coated seed/Ni or Ni–W tapes using electron beam coevaporation of yttrium, copper, and BaF_2 in reel-to-reel configuration. The tapes were post-annealed in wet oxygen atmospheres. The details are reported elsewhere in this book by Lee et al. (2004). On fully buffered short tapes, YBCO films were also grown at 780°C and a $p(O_2)$ of 120 mTorr using pulsed laser deposition. The transport property measurements of the YBCO films grown on these dip-coated seed layers are shown in Figure 12.5. The field-dependence of J_c for YBCO films deposited on Gd_2O_3 and LZO seed layers with sputtered YSZ and CeO_2 cap layers is shown in Figure 12.6. YBCO films deposited on e-beam Y_2O_3 seed layers are also compared. A high J_c of

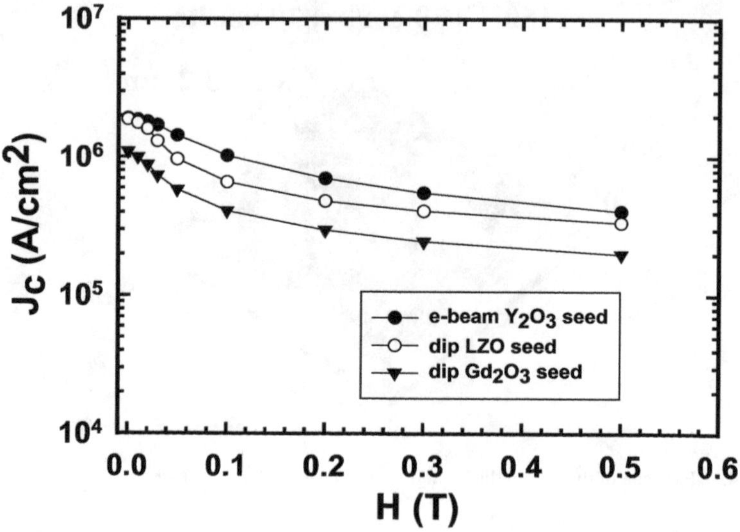

Figure 12.6. The field dependence of critical current density, J_c, for YBCO films grown on various seed layers. The architecture is YBCO/CeO_2/YSZ/dip-coated (or e-beam) seed layer/Ni–W.

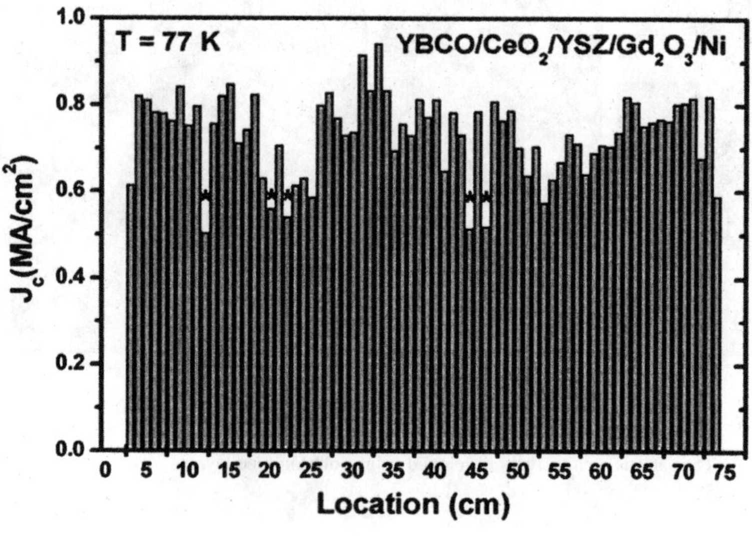

Figure 12.7. J_c data on the 0.8-meter long solution seed ORNL RABiTS (Tolga et al., 2001).

1.9 MA/cm^2 at 77 K and self-field was obtained on both Y_2O_3 and LZO seed layers. The performance of the solution seed layers approached that of the vacuum seed layers.

Very recently, on 0.8-meter long Gd_2O_3 seeded ORNL RABiTS, YBCO films with end-to-end J_c of 625 KA/cm^2 were produced using the reel-to-reel BaF_2 precursor approach (Lee et al., 2004). The average J_c is 750 KA/cm^2 with the standard deviation of only 10.5%. The J_c data on the 0.8-meter long solution seed is shown in Figure 12.7. This demonstrates that the high J_c YBCO films can be grown on the solution seed layers in lengths. This also promises a route for producing long lengths of

YBCO coated conductors using both vacuum and non-vacuum buffer layer technologies.

12.2.1.2 YBCO Films Highly oriented superconducting YBCO films were prepared by Rupich et al. (1992) on (100) $LaAlO_3$ (LAO) and (100) Yttria-stabilized Zirconium oxide (YSZ) single crystal substrates by sequential heat treatment in flowing Ar (<2 ppm oxygen) and oxidation of Y–Ba–Cu alkoxide precursor films at 730°C. The precursor solution was prepared from a pyridine solution of Y-methoxyethoxide, Ba-methoxyethoxide and $(CuO)_n$ in 2-methoxyethanol at room temperature. The YBCO films processed at 730°C showed metallic behavior in the normal state and a sharp resistive superconducting transition with T_c (zero) at 89.5 K. Critical current densities of 2×10^5 A/cm^2 at 77 K and self-field were obtained with YBCO films on YSZ substrates. The Cu-alkoxide precursors are generally insoluble in alcohol. However, Paranthaman and Beach (1995) have dissolved Cu-methoxide precursors in solvents like triethanolamine or diethanolamine in their alkoxide approach and grown highly aligned superconducting $TlBa_2Ca_2Cu_3O_{9-y}$ films on Ag substrates. Similarly Monde et al. (1988) have prepared the YBCO precursor solution by dissolving Yttrium butoxide, Barium methoxide and Copper methoxide in triethanolamine-methanol solution. However they obtained only a T_c of 56 K for YBCO films grown on polycrystalline YSZ substrates.

Masuda et al. (1991, 1992) have grown superconducting YBCO films with a T_c of 85 K on polycrystalline YSZ substrates using all alkoxide precursors at 920°C in pure O_2. The YBCO precursor solutions were prepared by dissolving copper acetate in di-methyl formamide (DMF) solvent and mixing the copper solution with the alcohol solution in which Yttrium-isopropoxide and Ba-ethoxides were dissolved. The films with a thickness of about 10 μm were produced by dip-coating the sols kept at temperatures 60 to 80°C and a viscosity of 5×10^{-3} to 1×10^{-2} Pascal.seconds (Pa.s). Highly oriented YBCO and $YBa_2Cu_4O_8$(Y-124) films were fabricated by Katayama et al. (1990, 1992) on Ag substrates. A Cu-methoxide solution was obtained by dissolving Cu-methoxide in 2-methoxyethanol and ethylenediamine to form a coordination compound of $Cu(OCH_3)_2(en)_2$. Y-isopropoxide solution was prepared by dissolving Y-isopropoxide in 2-methoxyethanol and ethyl acetoacetate to form a chelate compound of $Y(O\text{–}iC_3H_7)_2(EAA)$. The modified Cu alkoxide was partially hydrolyzed with equimolar water and mixed with Y and Ba alkoxide solutions to prepare a heterometallic alkoxide solution. Superconducting YBCO films were produced by dip-coating the YBCO precursor solution onto polycrystalline Nb-doped $SrTiO_3$ (STO) substrates and heat-treating up to 870°C in oxygen (Benavidez et al., 2000). The precursor solution was based on Y-acetate, Ba-alcoholate in 2-methoxyethanol, and Cu-butyrate in an alcoholic solvent. Niobium used as dopant in STO, inhibited the formation of Cu–Ti and Ba–Ti containing oxides, yielding better superconducting properties than for films deposited on undoped STO substrates. However, the YBCO films grown on these substrates had much lower T_cs. The summary of the growth of YBCO films using sol-gel alkoxide approach is shown in Table 12.1.

12.2.2 Metal–Organic Decomposition (MOD) Approach

In the MOD approach, various metal–organic precursors such as citrates, oxalates, neodecanoates, trifluoroacetates, halides, acetates, acetylacetonates, and naphthenates have been described in the literature. During the $YBa_2Cu_3O_7$ growth, it is possible

Table 12.1. Summary of the growth of YBCO films using sol-gel alkoxide approach

YBCO precursors	Substrates	Process conditions	Physical properties	Reference
Y-methoxyethoxide; Ba-methoxyethoxide; $(CuO)_n$ in 2 MeEtOH	YSZ xtal	730°C; 2 ppm O_2	$T_c = 89.5$ K; $J_c = 2 \times 10^5$ MA/cm^2 (77 K)	(Rupich et al., 1992)
Y-butoxide; Ba-methoxide; Cu-methoxide in triethanolamine and methanol	YSZ poly	800°C; air	$T_c = 56$ K	(Monde et al., 1988)
Y-isopropoxide; Ba-ethoxide; Cu-acetate in DMF	YSZ poly	920°C; O_2	$T_c = 85$ K	(Masuda et al., 1991, 1992)
Y-isopropoxide; Ba-alkoxide; Cu-methoxide in 2 MeEtOH/ ethylenediamine	Ag	800°C	Y-124 phase	(Katayama, 1990; Katayama et al., 1992)
Y-acetate/ Ba-alcoholate in 2 MeEtOH; Cu-butyrate	Nb-STO poly	870°C	low T_c Nb-STO inhibits Cu–Ti; Ba–Ti formation	(Benavidez et al., 2000)

that certain metalorganic precursors can lead to barium carbonate as the intermediate phases due to the formation of carbon dioxide during processing of the carbon-containing precursor ligands attached to the barium ions. Usually, the ligand is removed by hydrolysis with excess water followed by precipitation of the resulting barium hydroxide. However, fatty acid esters ligands can not be removed by simple oxidation or hydrolysis. It is necessary to process YBCO films above 900°C to decompose the intermediate barium carbonate, which can interfere kinetically with the formation of YBCO phase. For these reasons, non-carbon containing precursors are preferred. Trifluoroacetate (TFA) approach is the most popular one where high current density YBCO films have reproducibly grown on RABiTS. It is also possible to process non-fluorine containing precursors under reduced pressures to avoid the $BaCO_3$ formation as the intermediate phase.

The trifluoroacetate approach was originally developed by Gupta et al. (1988) to grow high quality YBCO films. McIntyre et al. (1992) later perfected the TFA process. Superconducting YBCO films with a J_c of 1–5 $\times 10^6$ A/cm^2 were grown on single crystal STO or LAO substrates by various groups (McIntyre et al., 1992; Paranthaman et al., 2001b; Sathyamurthy and Salama, 1998; Yamagiwa et al., 2001; Dawley et al., 2001a, 2001b; Clem et al., 2001; Li et al., 2001; Honjo et al., 2001). $NdBa_2Cu_3O_7$ films with a T_c of 89 K have also been prepared on single crystal STO substrates at 800°C and a $p(O_2)$ of 3×10^{-4} atm using the TFA approach (Honjo et al., 2001). The disadvantage of this approach is: the formation of HF during the YBCO film growth; the rate of HF removal determines the growth rate of YBCO; the YBCO growth rate is typically 1–3 Å/sec; thickness/coat; processing of thicker films take longer processing times; and slow burn-out step. Li et al. (2001) have processed YBCO films of 1.2 μm with a J_c of 2.5 MA/cm^2 on 20 nm CeO_2-buffered YSZ single crystal substrates us-

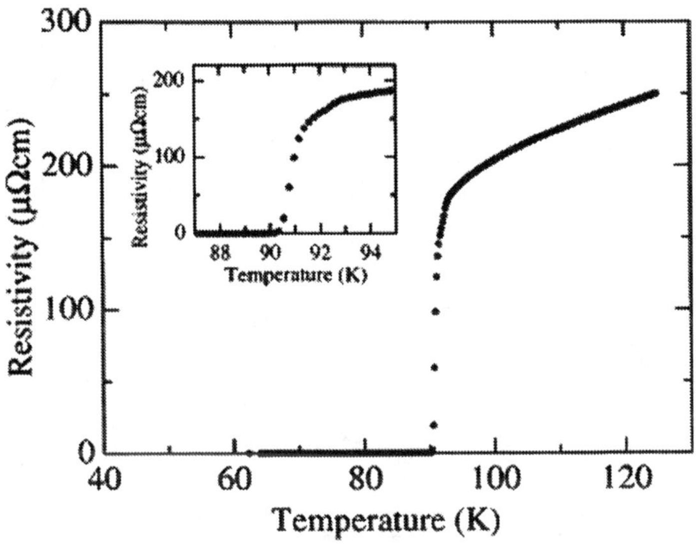

Figure 12.8. The temperature dependence of resistivity for YBCO film on STO (100) single crystal substrates grown using the iodide precursors (Matsubara et al., 1999b).

ing multiple coating (3 coats). In addition, they have also demonstrated the growth of YBCO films with a J_c of 1.7 MA/cm^2 on CeO_2 (sputtered)/YSZ (sputtered)/solution Gd_2O_3/Ni substrates. In this demonstration, both seed layers and the superconductors were grown by a solution process. It is also possible to modify the furnace design to post-anneal long lengths of thick and wider tape in a batch mode. Very recently Clem et al. (2001) have demonstrated the growth of YBCO films with a J_c of 1.3 MA/cm^2 at 77 K on all solution Nb-doped STO buffered Ni substrates in short lengths. This demonstration is the first step towards the fabrication of low-cost YBCO coated conductors. The details of the TFA process are explained in detail elsewhere in this book.

Similar to Fluoride precursors, solution based metal iodide precursor routes have been developed (Baney et al., 1992; Matsubara et al., 1999b). In this approach, the stoichiometric mixtures of YI_3, BaI_2, and CuI were dissolved in either acetonitrile/ethanol solutions or ammonium iodid/dimethylformamide/2-methoxyethanol solutions. The fact that BaI_2 can be decomposed without the use of water during YBCO film growth makes this approach very interesting. This process will eliminate the formation of highly reactive HI vapors. YBCO films with a T_c of 90 K and a J_c of 1.3×10^5 A/cm^2 have been grown on STO single crystal substrates. The films were processed at 800–830°C in 300 ppm oxygen. The temperature dependence of the resistivity for YBCO film on STO substrate is shown in Figure 12.8. The surface morphology of the YBCO film using the iodide process is shown in Figure 12.9. Enhanced plate-like grains with size of 1–3 μm with less porous microstructure are obtained. It may be possible to use the chloride precursors in the form of dichloroacetic acid or trichloroacetic acid to grow YBCO films.

Recently, Shi et al. (2001) have developed a non-fluorine solution route to grow YBCO films on LAO single crystals. In this method, stoichiometric yttrium trimethylacetate, barium hydroxide, and copper trimethylacetate powders were dissolved in a mixture of propionic acid/amine solvent. YBCO films with a T_c of 90 K and a transport J_c of 1×10^5 A/cm^2 at 77 K. This precursor approach was originally developed by Chu and Buchanan (1993). Matsubara et al. (1999a) have prepared epitaxial

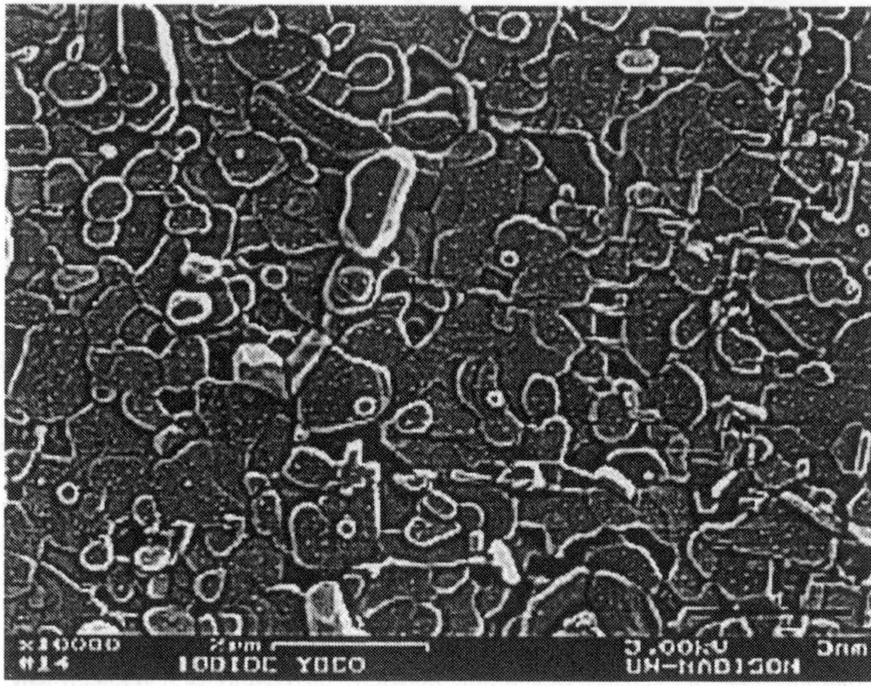

Figure 12.9. Scanning electron micrograph of YBCO film on STO (100) single crystals substrates grown using the iodide precursors (Matsubara et al., 1999a).

$YbBa_2Cu_3O_{7-y}$ films on STO single crystal substrates using ytterbium acetylacetonate, barium neodecanoate, and copper (II) 2-ethylhexanoate in a mixture of solvents containing toluene/pyridine/propionic acid. In this method, the YbBCO films were processed at 730–770°C and 100 ppm oxygen. The highest T_c obtained was 87.2 K. The measured transport J_c was 6.4×10^5 A/cm^2 at 77 K and self-field. The magnetic field dependence J_c of $YbBa_2Cu_3O_{7-y}$ is shown in Figure 12.10. Even though the J_c is little bit low, but it is extremely promising to use this precursor route for growing superconducting films on substrates developed using RABiTS, Ion-Beam Assisted Deposition (IBAD), and Inclined Substrate Deposition (ISD) processes. In the literature, various acetylacetonate (acac) precursor approaches have been described in detail (Rice et al., 1987, 1999; Manabe et al., 1991, 1997; Breeze and Wang, 1999; Nagano and Greenblatt, 1988; Hussain and Sayer, 1992). The highest J_c obtained on YBCO films grown using the acac approach is in the order of 10^5 A/cm^2 at 77 K.

Highly *c*-axis aligned superconducting $REBa_2Cu_3O_{7-\delta}$ (RE = Rare Earth) films were prepared on STO and LAO single crystal substrates using the corresponding metal naphthenates in toluene solution (Ma et al., 1999; Yamagiwa and Hirabayashi, 1998, 2000; Kumagai et al., 1993; Hiei et al., 2001; Shibata et al., 1999; Kanaya et al., 1997; Yokota and Abell, 1994; Shiohara and Hobara, 2000; Iijima and Matsumoto, 2000). In this approach, YBCO films with a J_c of 1.2×10^6 A/cm^2 were produced. The summary of the growth of YBCO films using metalorganic decomposition routes is shown in Table 12.2. Ottosson et al. (1989) have deposited superconducting YBCO films by chemical vapor deposition of the halide precursors YCl_3, BaI_2, and CuCl as metal sources in the presence of O_2/H_2O gas mixtures. The as-grown films had a T_c of 40 K and the T_c was increased to 70 K upon post-annealing in O_2 at 475°C.

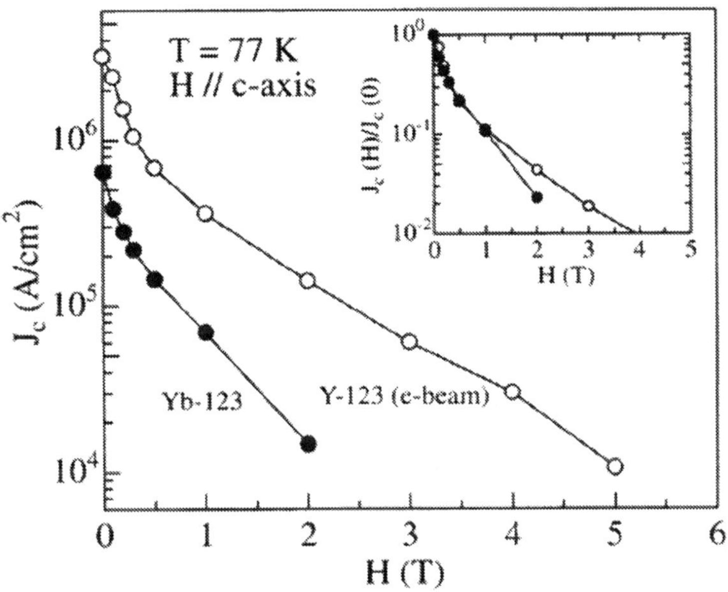

Figure 12.10. Magnetic field dependence of the transport J_c of $YbBa_2Cu_3O_{7-y}$ film deposited by solution process (solid circle) compared with YBCO film deposited by *ex situ* BaF_2 process (open circle) on STO (100) single crystal substrates. Inset is the magnetic field dependence of normalized J_c, $J_c(H)/J_c(0)$ (Matsubara et al., 1999a).

12.3 ELECTROCHEMICAL DEPOSITION

Electrochemical deposition involves reduction of metallic ions from aqueous, inorganic and fused salt electrolytes. The reduction process M^{n+} (solution) + $ne^- \rightarrow M$ (lattice) can be accomplished by the electrodeposition process in which n electrons are supplied by an external power supply or by electroless (autocatalytic) deposition process in which a reducing agent in the solution is the electron source (there is no external power supply). Thin films of hydroxide precursors of YBCO have been electrodeposited from aqueous solution at a potentiostatic voltage of −1.4 V (vs. Standard Calomel Electrode (SCE)) onto a Ag electrode by Monk et al. (1998). The compositions of the films were tailored by adjusting the deposition voltage and also by varying the relative amounts of precursor solutes in solution. The aqueous solution contained 0.0125 mol dm^{-3} of $Y(NO_3)_3$, 0.150 mol dm^{-3} of $Ba(NO_3)_2$, and 0.025 mol dm^{-3} of $Cu(NO_3)_2$. No further annealing of these hydroxide precursors was reported. Bhattacharya et al. (1991, 1992) electrodeposited YBCO precursors from an electrolyte solution consisting of nitrate salts of Y(III), Ba(II), and Cu(II) dissolved in an organic solvent dimethyl sulfoxide. Nominal deposition potentials ranged from −2.5 to −4 V (vs. Ag/$AgNO_3$ reference electrode). The post-annealed electrodeposited films show reproducible T_c (zero) of 74 K on Ni, 78 K on MgO, and 91 K on (100) YSZ substrate. The J_c obtained for YBCO films on YSZ substrate is 4000 A/cm^2 at 4 K and self-field and 360 A/cm^2 at 77 K and self-field. YBCO films with a T_c of 88 K were deposited on Ag substrates using electrodeposition of metal perchlorate salts, tetrabutylammonium perchlorate (TBAP) and acetonitrile (Rosamilia and Miller, 1988, 1989; Rubin et al., 1989). Abolmaalli and Talbot (1993) have deposited YBCO hydroxide precursors using metal nitrate salts in isopropanol and post-annealed in oxygen. The YBCO films grown from a single simultaneous deposition had a T_c of 79 K and 88 K for films

Table 12.2. Summary of the growth of YBCO films using MOD approach

YBCO precursors	Substrates	Process conditions	Physical properties	Reference
Y-acetate; Ba-acetate; and Cu-acetate in trifluoroacetic acid/methanol	LAO or STO xtals	725–800°C; 100–300 ppm O_2	$T_c = 91$ K; $J_c = 1\text{–}4 \times 10^6$ A/cm^2 thickness 0.4 μm (77 K)	(McIntyre et al., 1992; Paranthaman et al., 2001b; Sathyamurthy and Salama, 1998; Yamagiwa et al., 2001; Dawley et al., 2001a, 2001b; Clem et al., 2001; Li et al., 2001)
YBCO TFA	Nb-STO/Ni	3-coats	$J_c = 1.3 \times 10^6$ A/cm^2	(Clem et al., 2001)
$NdBa_2Cu_3O_7$ TFA precursors	STO xtal	800°C; $p(O_2)$ 3×10^{-4} atm	Nd123 phase $T_c = 89$ K	(Honjo et al., 2001)
YI_3, BaI_2, and CuI in acetonitrile/ Ethanol (or NH_4I/ DMF/2-MeEtOH	STO xtal	800–830°C 300 ppm O_2	$T_c = 90$ K; $J_c = 1.3 \times 10^5$ A/cm^2	(Baney et al., 1992; Matsubara et al., 1999b)
Y-trimethylacetate $Ba(OH)_2$; Cu-tri methylacetate in propionic acid/ amine solvent	STO xtal	800°C	$T_c = 90$ K $J_c = 1 \times 10^5$ A/cm^2	(Shi et al., 2001)
Yb-acac; Ba-neo decanoate, and Cu 2-ethylhexanoate in toluene/pyridine/ propionic acid	STO xtal	730–770°C 100 ppm O_2	Yb-123 $T_c = 87$ K $J_c = 6.4 \times 10^5$ A/cm^2	(Matsubara et al., 1999a)
YBCO acac precursors	STO xtal		$J_c = 1\text{–}5 \times 10^5$ A/cm^2	(Risse et al., 1999; Manabe et al., 1991, 1997)
YBCO naphthenate Precursors	LAO & STO	low $p(O_2)$	$J_c = 1.2 \times 10^6$ A/cm^2	(Ma et al., 1999; Yamagiwa and Hirabayashi, 2000)

grown form several successive depositions. In the literature, no information was found related to suitable buffer coating over a metal substrate, however, it is possible that this technique may work on any suitably textured substrate followed by post-annealing in low $p(O_2)$. The summary of the growth of YBCO films using electrodeposition is shown in Table 12.3.

12.4 SPRAY (AEROSOL) PYROLYSIS TECHNIQUES

In this technique, aqueous solutions of $Y(NO_3)_3.5H_2O$, $Ba(NO_3)_2$, and $Cu(NO_3)_2.3H_2O$ in stoichiometric amounts are prepared from high purity nitrates and de-ionized water (Chu et al., 1988; Derraa and Sayer, 1990; Jergel et al., 1992; Kodas et al., 1989; Jergel, 1995; Chung et al., 1996; Lovchinov et al., 1995; Kawai et al., 1987; Kullberg et al., 1991). The aqueous solution was sprayed onto a heated substrate maintained at a temperature in the range of 200–400°C. Typical thicknesses of the as-deposited samples vary from 2–5 μm. The films were then heat-

Table 12.3. Summary of the growth of YBCO films using electrodeposition

YBCO precursors	Substrates	Process conditions	Physical properties	Reference
Y, Ba, and Cu hydroxides were deposited using nitrate solutions	Ag	−1.4 V (vs. SCE)		(Monk et al., 1998)
Y, Ba, and Cu nitrates in dimethyl sulfoxide	Ni MgO YSZ xtal	post-annealed	$T_c = 74$ K (on Ni) $T_c = 78$ K (on MgO) $T_c = 91$ K (on YSZ); $J_c = 360$ A/cm^2 (77 K)	(Bhattacharya et al., 1991, 1992)
Y, Ba, and Cu perchlorates in TBAP and acetonitrile	Ag		$T_c = 88$ K	(Rosamilia and Miller, 1988, 1989; Rubin et al., 1989)
Y,Ba, and Cu hydroxides were deposited using nitrate solutions		post-annealed in O_2	$T_c = 88$ K	(Abolmaalli and Talbot, 1993)

Table 12.4. Summary of the growth of YBCO films using spray pyrolysis technique

YBCO precursors	Substrates	Process conditions	Physical properties	Reference
Y, Ba, and Cu nitrate solutions	various substrates	substrate at 200–400°C and heat-treated at 600–900°C	low J_c; film 2–5 μm	(Chu et al., 1988; Derraa and Sayer, 1990; Jergel et al., 1992; Kodas et al., 1989; Jergel, 1995; Chung et al., 1996; Lovchinov et al., 1995; Kawai et al., 1987; Kullberg et al., 1991)
Y, Ba, and Cu nitrate solutions	MgO xtal Al_2O_3 poly	post-annealed at 750°C	$T_c > 80$ K; film 1–10 μm; $J_c = 10^3$–10^4 A/cm^2 (77 K)	(Jergel et al., 1992)
Y, Ba, and Cu nitrate solutions		post-annealed in a fluorine atmosphere		(Suenaga, 2001)

treated in various partial pressures of O_2 and temperatures range from 600–900°C. Most of the groups obtained very low J_cs. Jergel et al. (1992) obtained YBCO films of 1–10 μm thickness with $T_c > 80$ K and J_c of 10^3–10^4 A/cm^2 at 77 K and self-field. To use this technique, it is necessary to develop suitable buffer layers and also process YBCO films at temperatures around 750°C. It is also essential to avoid the formation of $BaCO_3$ as the intermediate phase. To circumvent this problem, Suenaga (2001) have post-annealed spray pyrolysed films in a fluorine atmosphere to form BaF_2 as the intermediate phase. The summary of the growth of YBCO films using spray pyrolysis technique is shown in Table 12.4.

12.5 CONCLUSIONS

A detailed review of the summary of the various bulk solution techniques to grow superconducting YBCO films has led to the following conclusions.

1. It is essential to develop an alternative low-cost solution approach to grow both buffers and superconductors to fabricate low-cost YBCO coated conductors.
2. Solution seed layers such as Gd_2O_3, $La_2Zr_2O_7$ provide a good template to grow high current density YBCO films. The performance of the solution seed layers approached that of the vacuum seed layers.
3. 0.8-meter long YBCO tapes carrying an end-to-end J_c of 625 KA/cm^2 were produced by combining both non-vacuum and vacuum buffer layer technologies.
4. YBCO films with a J_c of 2×10^5 A/cm^2 were produced using sol-gel alkoxide precursors.
5. Very high current density ($J_c > 4 \times 10^6$ A/cm^2) YBCO films have been fabricated using the Trifluoroacetate (TFA) precursor approach. The TFA process has become the most popular non-vacuum route to YBCO so far. HF evolution during the film growth could limit the YBCO growth rate.
6. Iodide precursors are promising. This is mainly because this method does not need any water during processing. YBCO films with a T_c of 90 K and a J_c of 1.3×10^5 A/cm^2 have been obtained using the iodide precursors.
7. Non-fluorine solution precursor routes have been developed to deposit YBCO and YbBCO films with a J_c of up to 6.4×10^5 A/cm^2. This is another promising route.
8. Naphthenate precursors in toluene have been used to grow YBCO films with a J_c of 1.2×10^6 A/cm^2. Naphthenate precursors are only available in Japan. This could be a limiting factor in widespread of this method.
9. Electrodeposition and spray pyrolysis have resulted in low J_cs so far.

ACKOWLEDGMENTS

Thanks are due to T. Aytug, S. Sathyamurthy, and D.F. Lee (ORNL) for providing some of the unpublished data. This work was supported by US Department of Energy, Division of Materials Sciences, Office of Science, and the Office of Power Technologies-Superconductivity Program, Office of Energy Efficiency and Renewable Energy. This research was performed at the Oak Ridge National Laboratory, managed by UT-Battelle, LLC, for the USDOE under contract DE-AC05-00OR22725.

REFERENCES

Abolmaalli, S.B. and Talbot, J.B., 1993, Synthesis of superconductive thin films of YBCO by a nonaqueous electrodeposition process, *J. Electrochem. Soc.*, 140:443.

Baney, R.H., Bergstrom, D.F., and Justice, B.H., 1992, Metal Iodides: Novel solution precursors to cuprate superconductors, *Chem. Mater.*, 4:984.

Beach, D.B., Vallet, C.E., Paranthaman, M., Specht, E.D., Morrell, J.S., and Xue, Z.B., 1998, Sol-gel synthesis of rare earth aluminate films as buffer layers for high T_c superconducting films, *Mater. Res. Soc. Symp. Proc.*, 495:263.

Benavidez, E., Gonzalez Oliver, C.J.R., Caruso, R., and De Sanctis, O., 2000, Chemical method to prepare $YBa_2Cu_3O_{7-x}$ films by dipping onto $SrTi(Nb)O_3$ ceramics, *Mater. Chem. Phys.*, 62:9.

Bhattacharya, R.N., Noufi, R., Roybal, L.L., and Ahrenkiel, R.K., 1991, YBaCuO superconductor thin films via an electrodeposition process, *J. Electrochem. Soc.*, 138:1643.

Bhattacharya, R.N., Parilla, P.A., Noufi, R., Arendt, P., and Elliott, N., 1992, YBaCuO and TlBaCaCuO superconductor thin films via an electrodeposition process, *J. Electrochem. Soc.*, 139:67.

Breeze, S.R. and Wang, S.N., 1999, The preparation of YBCO epitaxial superconducting films by a chemical solution deposition process, *J. Mater. Sci.*, 34:1099.

Brinker, C.J. and Scherer, G.W., 1990, *Sol-Gel Science: The Physics and Chemistry of Sol-Gel Processing*, Academic Press, San Diego, CA.

Brinker, C.J., Hurd, A.J., Schunk, P.R., Frye, G.C., and Ashley, C.S., 1992, Review of sol-gel thin film formation, *J. Non-Cryst. Solids*, 147–148:424.

Chirayil, T.G., Paranthaman, M., Beach, D.B., Morrell, J.S., Sun, E.Y., Goyal, A., Williams, R.K., Lee, D.F., Martin, P.M., Kroeger, D.M., Feenstra, R., Verebelyi, D.T., and Christen, D.K., 1999, Epitaxial growth of Yb_2O_3 buffer layers on biaxially textured-Ni (100) substrates by sol-gel process, *Mater. Res. Soc. Symp. Proc.*, 574:51.

Chirayil, T.G., Paranthaman, M., Beach, D.B., Lee, D.F., Goyal, A., Williams, R.K., Cui, X., Kroeger, D.M., Feenstra, R., Verebelyi, D.T., and Christen, D.K., 2000, Epitaxial growth of $La_2Zr_2O_7$ thin films on rolled-Ni substrates by sol-gel process for high T_c superconducting tapes, *Physica C*, 336:63.

Chu, J.J., Liu, R.S., Kung, J.H., Wu, P.T., and Chen, L.J., 1988, Epitaxial growth of high T_c superconducting Y–Ba–Cu–O thin films on (001) MgO by a chemical spray pyrolysis method, *J. Appl. Phys.*, 64:2523.

Chu, P.-Y. and Buchanan, R.C., 1993, Reactive liquid phase sintering of solution-derived $YBa_2Cu_3O_{7-\delta}$ superconducting thin films: Part I. Ambient and precursor effects on BaO–CuO liquid phase formation, *J. Mater. Res.*, 8:2134.

Chung, Y.S., Auh, K.H., and Norman Hill, D., 1996, Effects of solution parameters on the deposition of YBCO phase prepared by aerosol feed method in a cold plasma reactor, *Mater. Lett.*, 27:201.

Clem, P., Siegel, M., Dawley, J., Ong, R., Overmyer, D., and Voigt, J., 2001, *Solution Deposition of YBCO Coated Conductors, DOE Superconductivity Annual Peer Review Meeting Presentations*, Washington, DC, August 1–3.

Dawley, J.T., Clem, P.G., Siegal, M.P., and Overmyer, D.L., 2001a, High J_c $YBa_2Cu_3O_{7-\delta}$ films via rapid, low $p(O_2)$ pyrolysis, *J. Mater. Res.*, 16:13.

Dawley, J.T., Clem, P.G., Siegal, M.P., Overmyer, D.L., and Rodriguez, M.A., 2001b, Thick sol-gel derived $YBa_2Cu_3O_{7-\delta}$ films, *IEEE Trans. Appl. Supercond.*, 11:2873.

Derraa, A. and Sayer, M., 1990, Superconducting Y–Ba–Cu–O thin films by spray pyrolysis, *J. Appl. Phys.*, 68:1401.

Energetics, Inc., Columbia, MD, 2001, Coated conductor technology development roadmap: Priority research & development activities leading to economical commercial manufacturing, Sponsored by U.S. Department of Energy, Superconductivity for Electric Systems Program, August.

Feenstra, R., Lindemer, T.B., Budai, J.D., and Galloway, M.D., 1991, Effect of oxygen pressure on the synthesis of $YBa_2Cu_3O_{7-x}$ thin films by post-deposition annealing, *J. Appl. Phys.*, 69:6569.

Goyal, A., Norton, D.P., Budai, J.D., Paranthaman, M., Specht, E.D., Kroeger, D.M., Christen, D.K., He, Q., Saffian, B., List, F.A., Lee, D.F., Martin, P.M., Klabunde, C.E., Hatfield, E., and Sikka, V.K., 1996, High critical current density superconducting tapes by epitaxial deposition of $YBa_2Cu_3O_{7-x}$ thick films on biaxially textured metals, *Appl. Phys. Lett.*, 69:1795.

Gupta, A., Cooper, E.I., Jagannathan, R., and Giess, E.A., 1988, Preparation of superconducting oxide films from metal trifluoroacetate solution precursors, in: *Chemistry of High-Temperature Superconductors II*, D.L. Nelson and T.F. George, eds., ACS Symp. Series, Vol. 377, American Chemical Society, Washington, DC, p. 265.

Hiei, H., Yamagiwa, K., Takahashi, Y., Kim, S.B., Yamada, Y., Shibata, J., Hirayama, T., Ikuta, H., Hirabayashi, I., and Mizutani, U., 2001, YBCO thin films on multilayers prepared by all-chemical solution deposition processing, *Physica C*, 357–360:942.

Honjo, T., Fuji, H., Huang, D., Nakamura, Y., Izumi, T., and Shiohara, Y., 2001, Preparations of $REBa_2Cu_3O_{7-y}$ films grown by metal trifluoroacetate precursors, *Physica C*, 357–360:999.

Hussain, A.A. and Sayer, M., 1992, Chemical fabrication of superconducting Y–Ba–Cu oxide films, *J. Supercond.*, 5:11.

Iijima, Y. and Matsumoto, K., 2000, High-temperature-superconductor coated conductors: Technical progress in Japan, *Supercond. Sci. Technol.*, 13:68.

Jergel, M., 1995, Synthesis of high-T_c superconducting films by deposition from aerosol, *Supercond. Sci. Technol.*, 8:67.

Jergel, M., Chromik, S., Strbik, V., Smatko, V., Hanic, F., Plesch, G., Buchta, S., and Valtyniova, S., 1992, Thin YBCO films prepared by low-temperature spray pyrolysis, *Supercond. Sci. Technol.*, 5:225.

Kanaya, H., Kaneyuki, T., Senoh, H., Cho, Y., and Awai, I., 1997, Microwave characteristics of YBaCuO coplanar waveguide resonators fabricated by the sol-gel process on polycrystalline MgO, *Jpn. J. Appl. Phys.*, 36:6311.

Katayama, S. and Sekine, M., 1990, *Better Ceramics Through Chemistry IV*, B.J.J. Zelinski, C.J. Brinker, D.E. Clark, and D.R. Ulrich, eds., Material Research Society, Pittsburgh, PA, p. 897.

Katayama, S., Sekine, M., Fudouzi, H., and Kuwabara, M., 1992, Highly oriented $YBa_2Cu_4O_8$ films fabricated at atmospheric pressure by the sol-gel method using metal alkoxide, *Appl. Phys. Lett.*, 60:118.

Kawai, M., Kawai, T., Mashuhira, H., and Takahasi, M., 1987, Formation of Y–Ba–Cu–O superconducting film by a spray pyrolysis method, *Jpn. J. Appl. Phys.*, 26:L1740.

Kodas, T.T., Engler, E.M., and Lee, V.Y., 1989, Generation of thick $YBa_2Cu_3O_7$ films by aerosol deposition, *Appl. Phys. Lett.*, 54:1923.

Kullberg, M.L., Lanagan, M.T., Wu, W., and Poeppel, R.B., 1991, A sol-gel method for preparing oriented $YBa_2Cu_3O_{7-\delta}$ films on silver substrates, *Supercond. Sci. Technol.*, 4:337.

Kumagai, T., Yamasaki, H., Endo, K., Manabe, T., Niino, H., Tsunoda, T., Kondo, W., and Mizuta, S., 1993, Critical current densities at 77 K in $YBa_2Cu_3O_{7-\delta}$/Ag films prepared by dipping-pyrolysis process, *Jpn. J. Appl. Phys.*, 32:L1602.

Lange, F.F., 1996, Chemical solution routes to single-crystal thin films, *Science*, 273:903.

Lee, D.F. et al., 2004, Reel-to-reel processing of YBCO coated conductors using a BaF_2 process, chapter in this book.

Li, Q., Zhang, W., Schoop, U., Rupich, M.W., Annavarapu, S., Verebelyi, D.T., Thieme, C.L.H., Prunier, V., Cui, X., Teplitsky, M.D., Fritzemeier, L.G., Riley, G.N., Jr., Paranthaman, M., Goyal, A., Lee, D.F., and Holesinger, T.G., 2001, Progress in solution-based YBCO coated conductor, *Physica C*, 357–360:987.

Lovchinov, V., Mihova, G., Nedkov, I., Razkazov, I., and Todorovska, R., 1995, Preparation of YBCO superconducting films by spray deposition and MTG processing, *Mater. Lett.*, 24:267.

Ma, X.L., Shibata, J., Hirayama, T., Yamagiwa, K., Hirabayashi, I., and Ikuhara, Y., 1999, Characterization of $YbBa_2Cu_3O_{7-\delta}$ superconducting thin films prepared by chemical solution deposition of STO (001) and LAO (001) substrates, *Phys. Stat. Sol. (a)*, 173:441.

Manabe, T., Kondo, W., Mizuta, S., and Kumagai, T., 1991, Preparation of high-J_c $YBa_2Cu_3O_{7-\delta}$ films on STO (100) substrates by the dipping-pyrolysis process at 750°C, *Jpn. J. Appl. Phys.*, 30:L1641.

Manabe, T., Yamaguchi, I., Nakamura, S., Kondo, W., Mizuta, S., and Kumagai, T., 1997, Carbon dioxide controlled annealing method for preparation of $YBa_2Cu_3O_{7-\delta}$ films by dipping-pyrolysis process, *Physica C*, 276:160.

Masuda, Y., Tateishi, T., Matsubara, K., Ogawa, R., and Kawate, Y., 1991, Preparation of $YBa_2Cu_3O_{7-x}$ superconductive film using sol-gel method, *Jpn. J. Appl. Phys.*, 30:1390.

Masuda, Y., Ogawa, R., Kawate, Y., Matsubara, K., Tateishi, T., and Sakka, S., 1992, Preparation of $YBa_2Cu_3O_{7-x}$ superconducting films through the sol-gel method using metal alkoxides as starting materials, *J. Mater. Res.*, 7:819.

Matsubara, I., Paranthaman, M., Chirayil, T.G., Sun, E.Y., Martin, P.M., Kroeger, D.M., Verebelyi, D.T., and Christen, D.K., 1999a, Preparation of epitaxial $YbBa_2Cu_3O_{7-\delta}$ on $SrTiO_3$ single crystal substrates using a solution process, *Jpn. J. Appl. Phys.*, 38:L727.

Matsubara, I., Paranthaman, M., Singhal, A., Vallet, C., Lee, D.F., Martin, P.M., Hunt, R.D., Feenstra, R., Yang, C.-Y., and Babcock, S.E., 1999b, Preparation of textured YBCO films using all-iodide precursors, *Physica C*, 319:127.

McIntyre, P.C., Cima, M.J., Smith, J.A., Hallock, R.B., Siegal, M.P., and Phillips, J.M., 1992, Effect of growth-conditions on the properties and morphology of chemical derived epitaxial thin films of $YBa_2Cu_3O_{7-x}$ on (001) $LaAlO_3$, *J. Appl. Phys.*, 71:1868.

Monde, T., Kozuka, H., and Sakka, S., 1988, Superconducting oxide thin films prepared by sol-gel technique using metal alkoxides, *Chem. Lett.*, 287.

Monk, P.M.S., Janes, R., and Partridge, R.D., 1998, Electrochemical deposition of the hydroxide precursors to $YBa_2Cu_3O_{7-\delta}$ and related phases, *J. Mater. Chem.*, 8:1779.

Morrell, J.S., Xue, Z.B., Specht, E.D., Goyal, A., Martin, P.M., Lee, D.F., Feenstra, R., Verebelyi, D.T., Christen, D.K., Chirayil, T.G., Paranthaman, M., Vallet, C.E., and Beach, D.B., 2000, Epitaxial growth of gadolinium oxide on roll-textured nickel using a solution growth technique, *J. Mater. Res.*, 15:621.

Nagano, M. and Greenblatt, M., 1988, High temperature superconducting films by sol-gel preparation, *Solid State Commun.*, 67:595.

Norton, D.P., Goyal, A., Budai, J.D., Christen, D.K., Kroeger, D.M., Specht, E.D., He, Q., Saffian, B., Paranthaman, M., Klabunde, C.E., Lee, D.F., Sales, B.C., and List, F.A., 1996, Epitaxial $YBa_2Cu_3O_7$ on biaxially textured nickel (001): An approach to superconducting tapes with high critical current density, *Science*, 274:755.

Ottosson, M., Anderson, T., Carlsson, J.-O., Harsta, A., Jansson, U., Norling, P., Niskanen, K., and Nordblad, P., 1989, Chemical vapor deposition of the superconducting $YBa_2Cu_3O_{7-x}$ phase using halides as metal sources, *Appl. Phys. Lett.*, 54:2476.

Paranthaman, M. and Beach, D.B., 1995, Growth of highly oriented $TlBa_2Ca_2Cu_3O_{9-y}$ superconducting films on Ag substrates using a dip-coated barium calcium copper oxide sol-gel precursor, *J. Am. Ceram. Soc.*, 78:2551.

Paranthaman, M., Goyal, A., List, F.A., Specht, E.D., Lee, D.F., Martin, P.M., He, Q., Christen, D.K., Norton, D.P., Budai, J.D., and Kroeger, D.M., 1997a, Growth of biaxially textured buffer layers on rolled Ni substrates by electron beam evaporation, *Physica C*, 275:266.

Paranthaman, M., Shoup, S.S., Beach, D.B., Williams, R.K., and Specht, E.D., 1997b, Epitaxial growth of $BaZrO_3$ films on single crystal oxide substrates using sol-gel alkoxide precursors, *Mater. Res. Bull.*, 32:1697.

Paranthaman, M., Shoup, S.S., Beach, D.B., Morrell, J.S., Goyal, A., Specht, E.D., Mathis, J.E., Verebelyi, D.T., and Christen, D.K., 1999, Growth of textured buffer layers and superconductors on rolled-Ni substrates using sol-gel alkoxide precursors, Symp. VI-Science and Engg. of HTC Superconducitivity, in: *Proc. of the 9th CIMTEC-World Ceramics Congress and Form on New Materials*, P. Vincenzini, ed., Tehna, Srl, p. 185.

Paranthaman, M., Feenstra, R., Lee, D.F., Beach, D.B., Morrell, J.S., Chirayil, T.G., Goyal, A., Cui, X., Verebelyi, D.T., Mathis, J.E., Martin, P.M., Norton, D.P., Specht, E.D., Christen, D.K., and Kroeger, D.M., 2000, Demonstration of high current density YBCO coated conductors on RE_2O_3-buffered Ni substrates with two new alternative architectures, in: *Advances in Cryogenic Eng. Mater.*, Vol. 46, U. Balachandran, D.U. Gubser, K.T. Hartwig, and V.A. Bardos, eds., Kluwer Academic/Plenum, New York, p. 879.

Paranthaman, M.P., Chirayil, T.G., List, F.A., Cui, X., Goyal, A., Lee, D.F., Specht, E.D., Martin, P.M., Williams, R.K., Kroeger, D.M., Morrell, J.S., Beach, D.B., Feenstra, R., and Christen, D.K., 2001a, Fabrication of long lengths of epitaxial buffer layers on biaxially textured nickel substrates using a continuous reel-to-reel dip-coating unit, *J. Am. Ceram. Soc.*, 84:273.

Paranthaman, M., Chirayil, T.G., Sathyamurthy, S., Beach, D.B., Goyal, A., List, F.A., Lee, D.F., Cui, X., Lu, S.W., Kang, B.W., Specht, E.D., Martin, P.M., Kroeger, D.M., Feenstra, R., Cantoni, C., and Christen, D.K., 2001b, Fabrication of long lengths of YBCO coated conductors using a continuous reel-to-reel dip-coating unit, *IEEE Trans. Appl. Supercond.*, 11:3146.

Rice, C.E., van Dover, R.B., and Fisanick, G.J., 1987, Preparation of superconducting thin films of $YBa_2Cu_3O_{7-\delta}$ by a novel spin-on pyrolysis technique, *Appl. Phys. Lett.*, 51:1842.

Risse, G., Schlobach, B., Habler, W., Stephan, D., Fahr, T., and Fischer, K., 1999, Textured YBCO-film formation by sol-gel process and post annealing, *J. European Ceram. Soc.*, 19:125.

Rosamilia, J.M. and Miller, B., 1988, Electrochemical behavior of several metal ion solutions at $YBa_2Cu_3O_7$ and CuO electrodes, *J. Electrochem. Soc.*, 135:3030.

Rosamilia, J.M. and Miller, B., 1989, Electrodeposition on cuprate-based superconductors in nonaqueous media, *J. Electrochem. Soc.*, 136:1053.

Rubin, H.D., Rosamilia, J.M., O'Bryan, H.M., Jr., and Miller, B., 1989, Properties of metal–$YBa_2Cu_3O_7$ composites formed by electrodeposition form nonaqueous solutions onto the superconductor, *Appl. Phys. Lett.*, 54:2151.

Rupich, M.W., Liu, Y.P., and Ibechem, J., 1992, Low-temperature formation of $YBa_2Cu_3O_{7-x}$ superconducting films from molecular Cu–Ba–Y precursors, *Appl. Phys. Lett.*, 60:1384.

Sathyamurthy, S. and Salama, K., 1998, Processing of $Y_1Ba_2Cu_3O_x$ films by solution techniques using metal–organic decomposition, *J. Supercond.*, 11:545.

Sathyamurthy, S., Paranthaman, M., Tolga, A., Kang, B.W., Martin, P.M., Goyal, A., Kroeger, D.M., and Christen, D.K., 2001, Chemical solution deposition of lanthanum zirconate buffer layers on biaxially textured Ni-3 at.% W alloy substrates for coated-conductor fabrication, *J. Mater. Res.* (submitted).

Schwartz, R.W., 1997, Chemical solution deposition of perovskite thin films, *Chem. Mater.*, 9:2325.

Shi, D., Xu, Y., Wang, S.X., Lian, J., Wang, L.M., McClellan, S.M., Buchanan, R., and Goretta, K.C., 2001, Deposition and interface structures of YBCO thin films via a non-fluorine sol-gel route, *Physica C* (in press).

Shibata, J., Yamagiwa, K., Hirabayashi, I., Ma, X., Yuan, J., Hirayama, T., and Ikuhara, Y., 1999, Effects of the initial heat-treatment conditions on microstructures of $YbBa_2Cu_3O_{7-\delta}$ superconducting final films deposited on STO (001) substrates by the dipping-pyrolysis process, *Jpn. J. Appl. Phys.*, 38:5050.

Shiohara Y. and Hobara, N., 2000, R&D of coated conductors in Japan, *Physica C*, 341–348:2521.

Shoup, S.S., Paranthaman, M., Beach, D.B., Specht, E.D., and Williams, R.K., 1997, Sol-gel synthesis of $LaAlO_3$: Epitaxial growth of $LaAlO_3$ thin films on $SrTiO_3$ (100), *J. Mater. Res.*, 12:1017.

Shoup, S.S., Paranthaman, M., Goyal, A., Specht, E.D., Lee, D.F., Kroeger, D.M., and Beach, D.B., 1998, Epitaxial thin film growth of lanthanum and neodymium aluminate films on roll-textured nickel using a sol-gel method, *J. Am. Ceram. Soc.*, 81:3019.

Suenaga, M., 2001, *DOE Superconductivity Annual Peer Review Meeting Presentations*, Washington, DC, August 1–3.

Tolga, A., Paranthaman, M., Sathyamurthy, S., Kang, B.W., Beach, D.B., Specht, E.D., Lee, D.F., Feenstra, R., Goyal, A., Kroeger, D.M., Leonard, K.J., Martin, P.M., and Christen, D.K., 2001, Continuous reel-to-reel chemical solution deposition of epitaxial Gd_2O_3 buffer layers on biaxially textured Ni-(3 at% W–1.7 at% Fe) alloy and Ni substrates for the fabrication of high-J_c $YBa_2Cu_3O_{7-\delta}$ coated conductors, *J. Am. Ceram. Soc.* (submitted).

Yamagiwa, K. and Hirabayashi, I., 1998, Structural and superconducting properties of biaxially aligned Yb123 films prepared by chemical solution deposition, *Physica C*, 304:12.

Yamagiwa, K. and Hirabayashi, I., 2000, Orientation behavior of $REBa_2Cu_3O_{7-y}$ (RE = Rare Earth and Y) films prepared by chemical solution deposition, *Jpn. J. Appl. Phys.*, 39:452.

Yamagiwa, K., Araki, T., Takahashi, Y., Hiei, H., Kim, S.B., Matsumoto, K., Shibata, J., Hirayama, T., Ikuta, H., Mizutani, U., and Hirabayashi, I., 2001, Epitaxial growth of $REBa_2Cu_3O_{7-y}$ films on various substrates by chemical solution deposition, *J. Crystal Growth*, 229:353.

Yokota, H. and Abell, J.S., 1994, Preparation of YBCO films on YSZ and silver substrates by the dipping pyrolysis process, *Physica C*, 235–240:629.

Chapter 13

JET VAPOR DEPOSITION FOR CONTINUOUS, LOW COST MANUFACTURE OF HIGH TEMPERATURE SUPERCONDUCTING TAPE

B.L. Halpern, T. Tamagawa, and Y. Di

Jet Process Corporation
24 Science Park
New Haven, CT 06511
USA

13.1 INTRODUCTION

Conducting tape is a thin film alternative to wire for HTS applications, and an economically attractive one (Goyal et al., 1999). But it has proven difficult to identify a thin film deposition method that can provide low cost, reliable, high throughput production of high quality oxide films. In this paper we show that the Jet Vapor Deposition™ (JVD™) process has excellent potential for manufacturing of HTS continuous conductors.

JVD uses inexpensive mechanical pumps and blowers to produce "supersonic jets in low vacuum" which propel a variety of film forming species to a wide range of substrates (Halpern, 1982; Schmitt and Halpern, 1988, 1998). Developed at Jet Process Corporation (JPC), the JVD process is remarkably versatile, depositing end product films of metals, oxides, nitrides, semiconductors, and organics in multicomponent and multilayer form, starting with incident species of atoms, molecules, and radicals, as well as clusters of controlled size. Film quality is equal or superior to conventional methods, and rates can be very high. JVD can operate in batch processes, or, more appropriate to HTS strips, as a reel-to-reel, continuous process; in all cases, the overall cost is low. We review the principles of JVD, and give a detailed discussion of the high rate "electron-jet, or e-jet," to highlight its potential in HTS manufacturing. We discuss a unique JVD stripcoater and air-to-vacuum seal. We summarize low rate JVD of multicomponent oxides, including ferrite and ferroelectric oxides, and CeO_2YSZ/CeO_2 buffer layers for HTS conductors. Lastly, we look at some of the challenges that JVD scaleup encounters.

13.2 PRINCIPLES OF JET VAPOR DEPOSITION

The basic JVD source is shown in Figure 13.1. The jet is formed by the flow of inert gas through a nozzle (often simply a hole in a plate), the flow driven by a mechanical pump and blower. The jet operates in "critical flow" conditions; when the upstream/downstream pressure ratio exceeds 2, carrier gas emerges at the speed of sound ($\sim 10^5$ cm/sec for He, 3×10^4 for Ar). JVD operates flexibly in a wide range of conditions: nozzle diameters from several mm to 15 mm, pumping speeds from 500 liters/min to 30,000 liters/min, nozzle pressures from 1 to 10 torr, and downstream pressures from 0.1 to 1 torr. Thus many disparate applications can be addressed (Halpern et al., 1992; Halpern and Schmitt, 1994; Zhang et al., 1997).

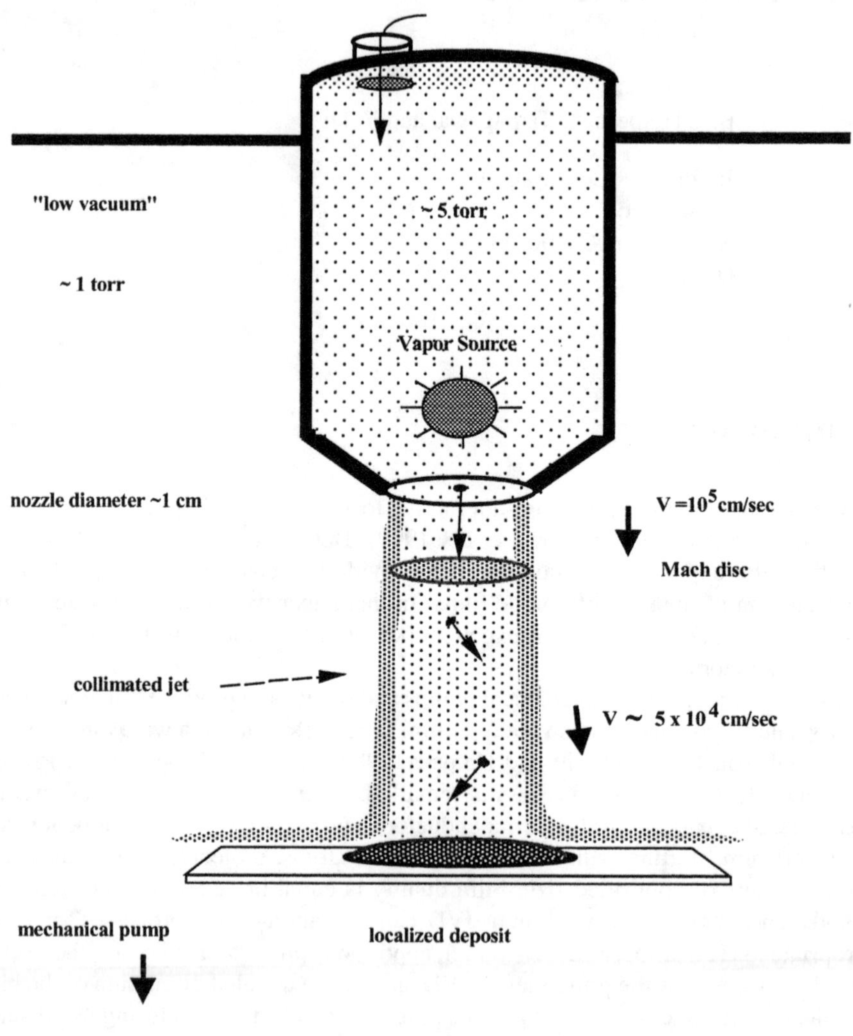

Figure 13.1. Basic Jet Vapor Deposition source showing typical operating conditions. At the Mach disc the jet becomes subsonic and its density rises. Vapor can be generated by thermal vaporization, sputtering and other techniques.

When a vapor source is placed upstream and close to the nozzle exit, liberated atoms or molecules are caught in the rapidly accelerating carrier gas flow and swept by the jet toward a substrate placed typically several cm downstream. Vapor can be generated by thermal evaporation, sputtering, microwave or DC plasma chemistry, or other means. Variations on this basic structure developed at JPC have lead to many JVD sources. A "hot filament-wirefeed" for Cu deposition will serve as simple illustration. Here, Cu wire of .25–.50 mm diameter is fed to a hot W filament placed several millimeters upstream of the nozzle. Wirefeeding is done via a knurled wheel, stepper motor driven mechanism under control of an intelligent driver or PC. When the Cu wire contacts the hot filament, a small length melts off, wets the filament, and vaporizes from it, almost as a point source. The vapor flux can thus be "pulsed" or held constant by maintaining a steady reservoir of molten Cu on the hot filament. Cu atoms vaporized and swept to the substrate downstream deposit efficiently, with few Cu atoms lost in the "wall" jet that flares radially at the substrate. The Cu deposit is localized, but diffusion of Cu atoms away from the jet axis gives a Gaussian deposit profile whose width at half maximum is comparable to the jet diameter.

Localized deposition means that uniform coverage of larger areas requires two degrees of relative jet-substrate motion. For batch processing, a number of relative motion schemes are used at JPC. Figure 13.2 shows one example, a "moving carousel"

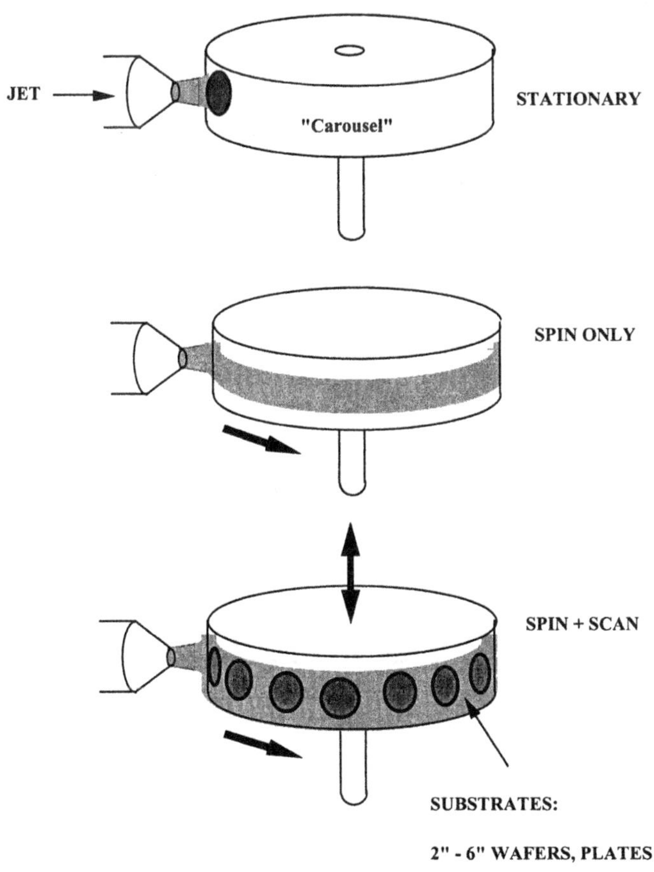

Figure 13.2. Uniform films over large areas: one example of relative jet-substrate motion in a batch process for coating discrete substrates.

which spins and oscillates along its axis. If the carousel is stationary, the jet deposits a spot; if it spins, a band; if it oscillates as well, a uniform layer over the entire carousel surface. We use the carousel mostly for coating small discrete substrates, but also for short strips of material wound around it. For continuous processing, JPC has developed a reel to reel stripcoater with a unique air-to-vacuum seal. This stripcoater will be discussed later.

Synthesis of complex multicomponents or multilayers is carried out by incorporating several metal vapor sources in a single nozzle, or by operating multiple jets either simultaneously or sequentially, as seen in Figure 13.3 (Halpern and Schmitt, 1994). Provided the critical flow condition is met, neither "information" nor material can propagate upstream against the sonic jet flow, so that multiple jets are independent of each other. The upstream conditions that determine the stability of vapor generation

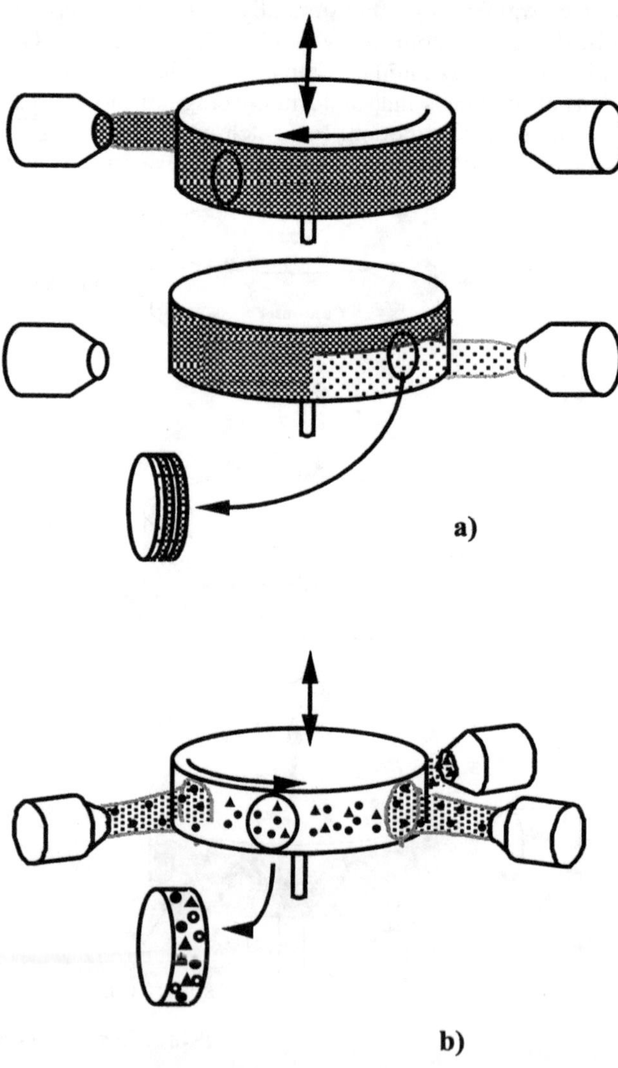

a)

b)

Figure 13.3. Multiple jets and moving substrates, for deposition of: (a) multilayers and (b) multicomponents, alloys and doped or graded films.

in one are unaffected by downstream changes due to another. This allows calibration of individual sources for precise control of deposition rate and thickness. It also permits "reactive" mode JVD; even the presence downstream of reactive atomic species such as H, O, or N cannot degrade a metal vapor source upstream in the nozzle. Thus, for example, a metal oxide can be deposited by bleeding O_2 into the downstream plasma of a metal vapor source such as the "electron jet" or "e-jet" described below. This is a general strategy in JVD: generate and deposit metal atomic vapor, and convert it rapidly to an oxide by reaction with a large flux of atomic oxygen at the growing film surface. In general, we do not use toxic metallorganic precursors; direct vaporization of metals in JVD is environmentally clean, a point that heavily favors a process intended for high throughput manufacture.

13.3 JVD SOURCES FOR HTS MATERIALS

HTS tape manufacture will require high rate, continuous deposition of multicomponent metal oxides. In JVD we generate metal vapor mainly in two ways: sputtering and thermal vaporization. Sputtering is more general; hollow cathode JVD sources can be used for virtually any metal; indeed, we used them to deposit CeO_2/YSZ/CeO_2 buffer layers. However, the throughput of sputter based JVD sources is orders of magnitude too low for future tape manufacturing needs. Thermal vaporization in, for example, a "hot filament-wirefeed" source is easily one hundred times faster. Unfortunately, most metals alloy with hot refractory filaments, and quickly destroy electrical continuity, so that the hot filament wirefeed is useful only for a limited number of metals, such as Au, Cu, Ag, Sn, Pb, that do not alloy with hot refractories. The "electron jet" or "e-jet," described below, was designed to overcome that limitation.

Among JVD sources, the e-jet offers the greatest power and versatility for making HTS tape, economically and of high quality. The e-jet combines high rate deposition with an extraordinarily high plasma ion density for ion bombardment of the growing film (Halpern et al., 1995; Halpern, 1996; Golz et al., 1997; Zhang et al., 1997). Because this chapter emphasizes potential manufacturability, we first describe in detail the workings and advantages of the e-jet, and then review slower JVD approaches.

13.4 THE E-JET JVD SOURCE

The e-jet, shown in Figure 13.4, has several distinguishing elements. A hot W filament, 0.5–1.0 mm diameter, supplies thermionic electrons to a low voltage, high current glow discharge. A "crucible," usually a 1–1.5 mm diameter W rod, serves as anode and collector in the discharge, and also, when hot, as the surface for vaporization of metal wire. A wirefeed mechanism, described previously, supplies wire; a radial gas injector allows input of O_2 or N_2 into the plasma for reactive deposition. Usually, Argon gas is supplied at nozzle pressures of 1–10 torr, but He, H_2, and N_2 can also be used as carriers or reactive components in the flow for specific applications.

In operation, the thermionic filament is heated to $>2000°C$ through a high current transformer, emitting electrons which are accelerated by the applied DC voltage. Avalanching and electron multiplication ignite and sustain a discharge at low voltages, typically $\sim$20 volts, with currents between 5 amps and 200 amps, depending on the design and purpose of the e-jet. The electrons arriving at the crucible deliver energies of 10–20 electron volts, sufficient to raise the crucible to the melting point

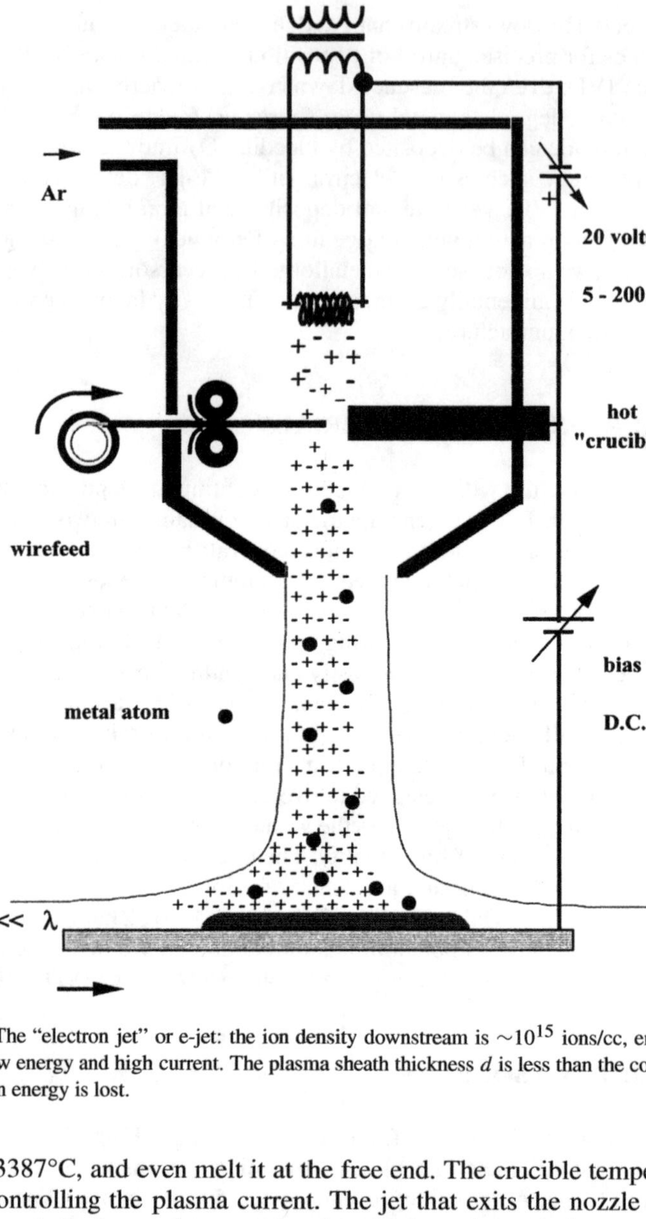

Figure 13.4. The "electron jet" or e-jet: the ion density downstream is $\sim 10^{15}$ ions/cc, enabling ion bombardment at low energy and high current. The plasma sheath thickness d is less than the collision mean free path l so no ion energy is lost.

of W, $T = 3387°C$, and even melt it at the free end. The crucible temperature is regulated by controlling the plasma current. The jet that exits the nozzle carries to the substrate a neutral, dense plasma; using Langmuir probe techniques, we have measured downstream Argon ion densities $>10^{15}$ ions/cc, and these ions are available for bombardment of the growing film, under DC or RF bias.

The hot crucible can be constructed in different forms to serve as a surface for vaporization. A simple rod can be used to vaporize wirefed metals at high rate and with excellent control. Among these are metals such as Cu, Au, and Ag, and others that do not alloy with hot refractories. In some cases such as Al, where alloying occurs, a protective sheath of TiB_2/BN suffices to prevent hot corrosion of the crucible. In other cases, such as Titanium, the crucible can be made from a rod of Tantalum; the Ti/Ta corrosion rate is low, so that a slow "crucible feed" can be used to replace the Ta distorted by alloying.

Low melting metals, or metals that vaporize at moderate temperature, such as Bi, Pb, Sn, and In, can be conveniently vaporized from a small "pot" shaped crucible which is heated by the e-jet plasma. The pot temperature is monitored by an embedded thermocouple, and the metal vapor flux is controlled via feedback to the plasma power supply. The source is calibrated by measurement of film thickness, or by loading a known weight of metal into the pot and evaporating it to completion. The pot can be made of refractory metal or, in some cases, ceramic sheathed in metal. For low vaporization temperatures, the pot can be sufficiently heated by surface catalyzed recombination of ions and electrons, so that no net current collection is required.

Metals that have high vapor pressure below their melting points can be vaporized in the e-jet by a rod feed mechanism. The tip of the rod is immersed in the plasma, and heated to the temperature necessary for vaporization. As the tip recedes due to evaporation, the rod is restored to the same position by a fiber optic position sensor, working in a closed loop with the rod feeding mechanism. In steady state the tip acquires a smooth taper, and the vaporization rate is stable. We make films of Cr, Mn, Mg and C in this way, as well as reactively deposited CrN.

Reactive deposition in the e-jet is straightforward. By means of a radial injector, O_2 is bled into the jet downstream of the nozzle where metal atoms are being vaporized. The high density plasma efficiently dissociates O_2 molecules to yield a high flux of O atoms at the growing film. Arriving metal atoms are rapidly converted to metal oxide, even at low substrate temperature.

Ion bombardment at high current and low voltage is a major advantage of the e-jet, a consequence of its high ion density. The plasma reaching the substrate can be biased by either a DC or RF voltage, for conducting or insulating substrates. The current of ions can "keep pace" with, and even outrun, a high flux of depositing neutral metal atoms, so that ion bombardment is effective even at high deposition rates. Ampere level ion bombardment currents are easily attained in the e-jet. This means that every depositing atom will be struck several times, while exposed on the growing film surface, by ions of an energy chosen in the range 20–60 eV, high enough to cause surface diffusion, but low enough to avoid damage such as resputtering or Ar^+ ion burial. Moreover, the high ion density of the e-jet plasma assures that the plasma sheath thickness (across which the bias voltage appears) is less than a collision mean free path, so that ions bombard the growing film with the intended energy. The e-jet feature of low energy, high flux bombardment confers powerful advantages for HTS production.

13.5 CONTINUOUS COATING: THE JVD STRIPCOATER

JPC has in current operation and development a high speed stripcoater for depositing metal or metal oxide/nitride layers on fast moving metal strip, which is supplied and taken up on reels outside a compact deposition chamber, shown schematically in Figure 13.5. The strip enters and leaves the chamber through thin, long slots, via a proprietary, unique air-to-vacuum seal. There is no need for load locks or differential pumping which add considerably to cost and upkeep. Although used mainly for single, multicomponent, or multilayer metal films, this stripcoater is easily adapted for reactive deposition of multicomponent oxides, and is potentially well suited to HTS strip production.

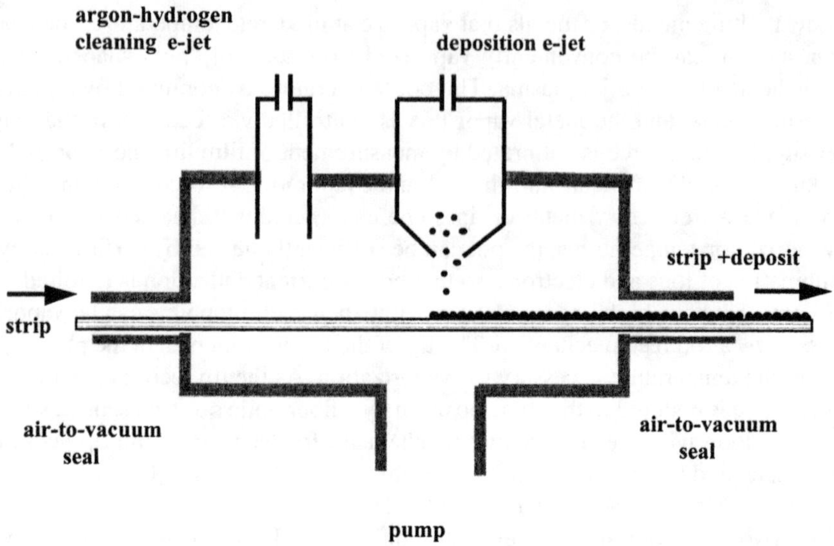

Figure 13.5. The JVD stripcoater: the proprietary air-to-vacuum seal simplifies entry and exit of strip; supply and takeup reels are not shown.

13.6 POTENTIAL ADVANTAGES OF JVD FOR HTS MANUFACTURE

The previous paragraphs suggest several advantages of JVD, the e-jet in particular, for production of HTS tape, summarized in the following paragraphs.

13.6.1 High Rate Deposition of a Variety of HTS Materials

One way to make HTS oxide films by JVD is to synthesize them from metal elements. The high rates of the e-jet are most easily delivered for elements that do not alloy with or degrade hot refractory crucibles, metal or ceramic. Metals of interest in HTS that fall into that category include Cu, Y, Ba, Bi, Ca, Sr and Pb (Bourdillon and Bourdillon, 1994). Many rare earths can replace Y; of these, La, Ce, Nd, Sm, Eu, Gd, Tb, Ho, Er, etc., are compatible with hot W or Ta. If wires are unavailable, rods, ingots, granules can be used with pots, either as single charges or with replenishment. In a stripcoater mode, vapor sources for all metal components need to be incorporated in a single nozzle; this can be done with many of the elements of interest, e.g., Y, Ba, and Cu. Injection of O_2 downstream completes the source.

13.6.2 Vaporization of Bulk Material with Easy Oxygen Replenishment

An alternative to synthesis from elements is vaporization and re-deposition of a solid of the correct composition. This can be done by the e-jet "rod feed" method described previously. For example, a $1/4''$ diameter rod of YBCO, several cm long, can be fed continuously into the e-jet plasma, as we do with Cr and C, so that its tip heats up and vaporizes; with a specific resistivity of magnitude milliohm-cm, a YBCO rod can easily sink the plasma current needed for vaporization. After steady state is achieved, the correct proportions of Y, Ba, and Cu leave the rod and deposit on the substrate.

If O atoms from the YBCO rod escape deposition, they can easily be replenished by injecting O_2 into the e-jet plasma downstream, where, as noted, dissociation to O atoms is extremely efficient. Even a few % of O atoms in the jet suffices to oxidize a high flux of co-depositing metal components.

With a rod fed e-jet we have easily deposited Cr metal at one micron thickness on a 1″ wide strip moving at a foot per minute. The front of the advancing rod becomes tapered and remains so in steady state. We expect a comparable throughput from a YBCO rod feed.

13.6.3 Ion Bombardment at High Current and Low Energy

Epitaxial growth of HTS films and buffer layers on biaxially textured substrates usually requires close control of temperature and deposition rate. But ion bombardment may provide an alternative to thermal energy for promoting the surface mobility that leads to correctly oriented grain growth. In the e-jet, ion bombardment currents can reach ampere levels, and even modest ion energies of, for example, 10–30 eV, provide on impact far more than the required diffusion activation energy without risk of damage. There is no point to using higher energies. Ample evidence exists showing that such ion bombardment enhances crystallization, at least in polycrystalline materials. Although it remains to be proven, it is reasonable to suggest that epitaxial growth can proceed at much higher rates, and possibly lower T, in the e-jet, than in other deposition methods that supply higher energy ions at far lower current.

13.6.4 Prevention of Oxidation of Ni Textured Substrates

Ni oxidizes easily under conditions of buffer layer and HTS oxide growth, and two methods are used to avoid this (Norton et al., 1996). In the first, a protective layer of Pt or Pd is deposited directly on the Ni. In the second, oxide deposition (e.g., CeO_2) is carried out under reducing conditions by introduction of H_2 during initial oxide growth; NiO will be reduced, whereas CeO_2 is stable in the presence of H_2 at elevated temperature.

In both cases, the e-jet would be of use in stripcoating. In method one, the noble metals are costly, and localized deposition by the e-jet would clearly minimize waste. (We typically capture 95% of vaporized Au in JVD). In method two, better control might result if deposition were carried out at low T. For example, we can clean the cold Ni surface with a brief exposure to atomic H, and then deposit a thin layer of Ce, which is then oxidized by exposure to a controlled flux of O atoms, again at low T. As usual, O and H are generated by bleeding the diatomics into the e-jet plasma. Without even introducing atomic H, NiO on the surface can be reduced with a few percent H_2 in the chamber ambient at temperatures around 600°C. Thicker CeO_2 is then built up by fast reactive deposition of Ce in the presence of O atoms. As described later, we used a very similar strategy at higher T with much slower sputter jets. With an e-jet, we expect these methods to be much faster in a stripcoating mode.

13.6.5 Simplified, Lower Cost Stripcoating Equipment

The stripcoater currently in operation at JPC is truly robust and low cost. It has a total length from reel to reel of ~8 feet. It runs on a moderate sized pump of speed ~10,000 liters/min, mounted on wheels and easily moved. It deposits micron thick films at speeds of feet per minute on a 1″ wide strip, and has proven effective for metals such as Au, Cu, Sn, Ni, Cr, and others. It currently has a dedicated "cleanup" e-jet which removes contaminants from incoming metal strip by means of ion etching or H atom attack. A unique, simple, proprietary air-to-vacuum seal eliminates need for differential pumping and associated vacuum equipment, reducing cost and assuring a small footprint. Refinements of the stripcoater now underway at JPC are enhancing its capability for complex multicomponents as well as its throughput.

13.7 HIGH RATE METAL OXIDE DEPOSITION IN THE E-JET: NICKEL FERRITE

Although we have not yet used the e-jet to deposit thick, multicomponent HTS oxides, the following example should suffice to show its capability. We deposited nickel ferrite (Dionne et al., 1995), $NiFe_2O_4$, by vaporizing NiFe alloy wire from the e-jet crucible, and oxidizing the growing film with O atoms from O_2 bled in downstream. The rate was high; we grew a ~25–100 micron film in several minutes on a ceramic substrate, $1'' \times 1''$. The substrate was moved in two ways. The first was to mount it on a small carousel with the usual spin and scan. In this case the e-jet power raised the substrate temperature to ~500°C; a Cu block substrate heater allowed heating to 650°C. The second method was to scan the substrate very slowly, both vertically and horizontally, but only through distances slightly greater than $1''$, just sufficient to assure deposit uniformity. In this case, because the substrate was always exposed to the energetic e-jet plasma, the substrate temperature could reach >850°C during deposition. This slow motion resembles that in strip coating, where a thick film builds up as a "moving cliff" on a substrate area whose temperature rises and falls as it enters and leaves the deposition zone. Substrate temperature was critical for $NiFe_2O_4$ quality; at 640°C the as-deposited saturation magnetization, $4\pi M_S$, was acceptably high, e.g., 2535 gauss, and after a long (4 h) anneal at 1000°C, it increased to the bulk crystalline value of 3000 gauss. But the process window was narrow, and $4\pi M_S$ decreased sharply as the deposition temperature dropped.

We wanted to learn if e-jet ion bombardment would increase magnetization during low T deposition. Although nickel ferrite might be sufficiently conductive at 900°C to permit DC bias bombardment, that is not the case at the lower temperatures of interest. We therefore investigated deposition with RF bias, in the frequency range 1.8 MHz to 13.6 MHz, and power inputs of 50–140 watts. At a temperature of 600°C, borderline for deposition of good quality ferrite, the as-deposited $4\pi M_S$ went from 840 gauss without RF ion bombardment to 2120 gauss with RF ion bombardment. This confirms that e-jet ion bombardment has a marked effect on film properties.

We continued deposition with RF bombardment down to temperatures $\ll$600°C, and found the as-deposited $4\pi M_S$ to be even lower. One film deposited at 530°C and 140 watts RF power had an as-deposited $4\pi M_S = 542$ gauss. But surprisingly, after a "flash" anneal at 850°C for 30 seconds, $4\pi M_S$ increased sharply to 3000 gauss. This contrasts with deposition in absence of RF ion bombardment, where a longer anneal at higher temperature (1000°C) was needed. It appears that e-jet ion bombardment makes the film more susceptible to crystallization by subsequent rapid thermal annealing; we believe that this result has the following interpretation.

Recall that the e-jet plasma has a high ion density (Halpern et al., 1995; Golz et al., 1997; Zhang et al., 1997), and can deliver high ion currents at small to moderate ion energies, in the range of 10–30 electron volts. When an ion collides with a group of surface atoms in the vicinity of a defect, i.e., an exposed atom or vacancy, the available kinetic energy is partitioned so that some atoms in the group receive enough energy to make them mobile, several eV, but not enough to sputter them, thresholds of ~20 eV. This low energy, high flux property of the e-jet is key: exposed surface atoms are hit with just the right amount of energy to mobilize them, but more importantly, they are hit many times.

Low energy ion bombardment is a two dimensional process that heals defects as the film grows, even at low T, creating small zones of crystallinity that can function later as nuclei for grain growth. Grain growth is a three-dimensional process that requires thermal energy, and can take place after deposition is complete. Because the e-jet bias currents are so large, many such potential nuclei are generated. This sets the stage for efficient growth induced by a subsequent rapid thermal anneal, and the magnetization $4\pi M_S$, which depends in part on film crystallinity, reaches the bulk limit quickly.

Although these exploratory results are not yet completely understood, they suggest that the e-jet's high current, low voltage ion bombardment capability will be important in deposition of HTS strip. Even if complete crystallization is not achieved in the deposition zone, a subsequent, in-line rapid anneal will be effective. As a further advantage, in contrast to the ferrites, many oxides of interest in HTS technology are good conductors; ion bombardment can be carried out under DC bias, which can be implemented more easily than RF ion bombardment.

13.8 METAL OXIDE AND HTC BARRIER LAYER VIA JVD SPUTTER JETS

Although much slower than the e-jet, JVD sputter sources can vaporize and deposit virtually any metal in the Periodic Table, and by means of multiple jet techniques, or using single alloy targets, synthesize multicomponent and/or multilayer oxides. Metals of interest for HTS materials include alkaline earths such as Ca, transition metals such as Co and Cr, and rare earths such as La and Ce. Typical metal sputtering rates per source lie in the range 0.1–1.0 milli-cc/minute, more than one hundred times slower than the e-jet. Accordingly, JVD sputter jet sources may be limited to deposition of very thin buffer layers, or order hundreds of angstroms, in HTS manufacture. But it is worth noting that concentrated high rate growth of epitaxial layers by any method has yet to be demonstrated, so that relatively slow growth using multiple sources may still prove necessary.

A brief review of selected examples of JVD sputter jet deposition is therefore useful. There exist many electronic applications where required layers are less than a micron thick, and for which low sputter jet rates are economical. There are other advantages. For very thin layers it is easier to control thickness at low rate, particularly in carousel batch processes, and there is better atomic level mixing when only submonolayers deposit per carousel rotation. Below we review deposition of two oxide materials with JVD sputter jets; single component CeO_2 for HTS buffer layers, and ferroelectric PZT (Hwang et al., 1991; Golz et al., 1993), as examples of complex multicomponent deposition. Both are batch processes using a rotating carousel with one or more metal atom jets aimed at the carousel. Both CeO_2 and PZT were deposited on discrete substrates, and CeO_2 was also deposited on a continuous Ni strip wrapped around the carousel.

Metal atoms are sputtered from a hollow cathode target inside the nozzle at typical pressures of ~1 torr, with downstream pressures in the range of tens to hundreds of millitorr, depending on the number of sources in operation. Conversion to oxide can take place in several ways. Oxygen molecules, 10% of the total flow or greater, can be bled into the neutral plasma of the jet downstream, where some are converted to O atoms. For many metals, e.g., transition metals such as Ti, reaction with O_2 molecules suffices to give the oxide, and O_2 can be injected anywhere in the chamber. For less reactive metals, such as Pb, oxidation is more difficult, and O atoms are needed; they can be gotten by injection of O_2 into the sputter jet plasma. Alternatively, O atoms can be generated and introduced via a separate microwave discharge jet, which operates at maximum efficiency in JVD conditions (1–2 torr). Use of atomic oxygen is a particularly useful approach in cases such as In–Sn oxide, and Al doped Zn oxide. Some gas phase reaction of metal atoms and O_2 may occur in the jet, but only infrequently; most oxidation occurs at the film surface, mainly during the short time that the substrate passes through the jet.

We are able to raise the temperature of substrates on the spinning carousel in several ways: (i) heat transfer from resistive heating elements (ii) radiant heat from quartz lamps and (iii) plasma heating from an auxiliary e-jet directed at the carousel.

13.8.1 Deposition of Lead Zirconate Titanate (PZT)

The lead zirconate titanate or PZT system is described as an example of a multicomponent oxide deposited from multiple JVD sources. The aim was to produce ferroelectric PZT films for memory applications (Hwang et al., 1991). Stoichiometric PZT has the formula $PbTi_xZr_{(1-x)}O_3$, and ferroelectric properties exist over a wide range of x. We obtained best control of composition and properties by using three separate JVD sputter jet sources of Pb, Zr, and Ti. We found that the simplest, most efficient conversion to the oxide was via O atoms, generated by injection of O_2 into the Pb source plasma downstream of the nozzle. Deposition of the perovskite phase was achieved at a substrate temperature of 550°C.

With these jet sources, PZT films were deposited on Si wafers pre-coated with a polycrystalline Pt film to serve as bottom electrode and diffusion barrier. The PZT films exhibited a preferential 100/001 orientation even though the polycrystalline Pt underlayer was strongly (111) oriented, as seen in Figure 13.6. This “orientation effect” is a unique but not yet understood feature of the JVD process. When ruthenium oxide was substituted for Pt, the PZT orientation was more random, as in Figure 13.7. Addition of other elements, La, for example, modifies PZT’s ferroelectric, pyroelectric, piezoelectric and electro-optic properties (Jaffe et al., 1971), and such addition is easily accomplished in JVD.

Table 13.1 gives some representative properties (dielectric constant, pyroelectric coefficient, and loss tangent) for several JVD oxide films:

Other PZT properties, such as remanent polarization, breakdown strength, and cycle lifetime were extensively characterized and found to be comparable to thin films done by other methods, and the as deposited films were always dense and crack free, even up to a thickness of more than 2 microns.

13.8.2 JVD of CeO_2 Buffer Layers

We used a sputter jet source of Ce to deposit CeO_2 on two types of Ni substrates: single crystal buttons, and rolling assisted biaxially textured strip (RaBiTs). Some preliminary work was also done on single crystal oriented lanthanum aluminate, $LaAlO_3$.

The substrate temperature was held at ~650°C. Deposition was divided into two stages. In the first stage, ~100–200 Å of Ce was deposited without any O_2 flow. At the

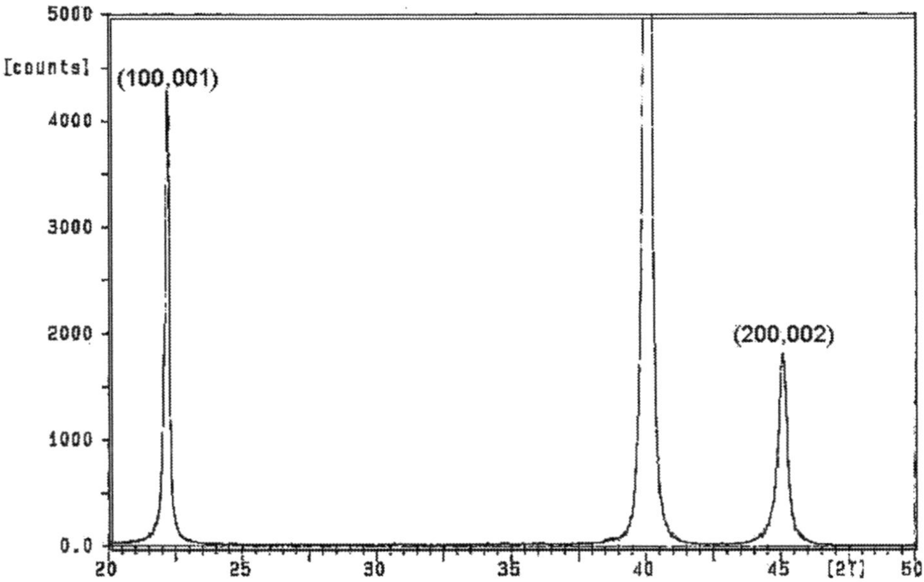

Figure 13.6. Theta–2-theta X-ray diffraction plot (CuK_a radiation) for 2500 Å PZT film deposited onto Pt-coated, oxidized Si substrate. (Peak near 40° due to (111) Pt signal.)

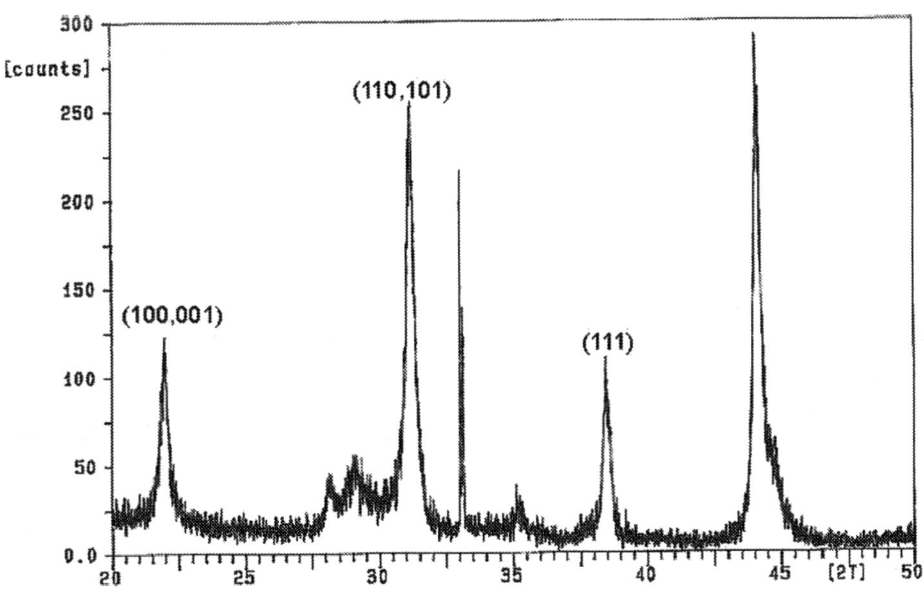

Figure 13.7. Theta–2-theta X-ray diffraction plot (CuK_a radiation) of 2500 Å PZT film deposited onto RuO_2 coated, oxidized Si substrate.

same time, we incorporated several % of H_2 either in the Argon flow to the Ce sputter jet, or via a separate inlet; this prevented oxidation of the Ni substrate which would have destroyed the 100 texturing. In the second stage, O_2 flow, via an inlet far from the Ce source, was begun only after the first 100–200 Å of Ce metal were deposited.

Table 13.1. Ferroelectric capacitor measurements on JVD thin film ferroelectrics Thicknessess: PZT: 0.7 micron, $PbTiO_3$: 0.3 micron, PLT: 0.25 micron

	PZT ($x = 0.5$)	$PbTiO_3$	PLT
ε	843	198	317
P	34 nC/cm^2 K	8	–
$\tan\delta$	0.05	0.1	0.033

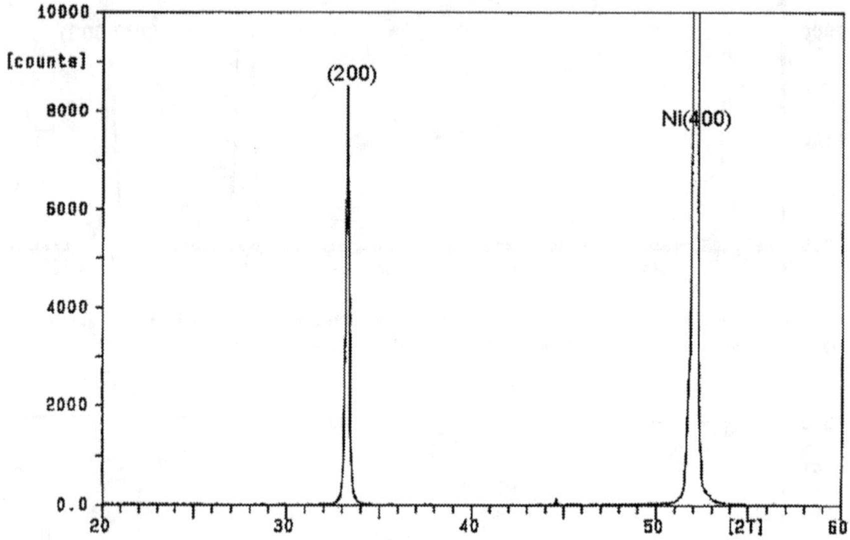

Figure 13.8. Theta–2-theta X-ray diffraction plot (CuK_a radiation) for 1500 Å CeO_2 film deposited onto biaxially textured nickel strip.

Deposition of ~500 Å of CeO_2 over 300 cm^2 took ~5–10 minutes. We do not know whether the initial, highly reactive Ce layer was converted to oxide by the low residual O_2 concentration in the chamber, or by the much larger O_2 flow in the second stage. XRD spectra show no presence of metallic Ce, so, although we cannot be sure of the oxidation state of the initial layers, nonetheless, oxidation must be complete.

This two stage process gave well textured CeO_2 films on both single crystal and RaBiTs substrates, Figures 13.8 and 13.9. The slight (111) nickel signal (near 28.6°) observed in the single crystal nickel sample may be due to inexactness in the nickel cutting and polishing process. It can be adapted to a continuous process utilizing two Ce jet sources and a Zr jet source on the JVD stripcoater; the economics remains to be judged.

13.8.3 Yttrium Oxide-Stabilized Zirconium Oxide (YSZ)

YSZ is used as a spacer in a YSZ/CeO_2/YSZ sandwich structure because thick layers of CeO_2 tend to crack. In JVD it is possible to use either separate Y and Zr sputter sources, but we found it simpler to use a single source with an alloy target of Y and Zr. Oxygen was injected downstream of this Y + Zr source. We deposited YSZ layers, 0.5–1 micron thick, on a previously deposited film of CeO_2 several hundred

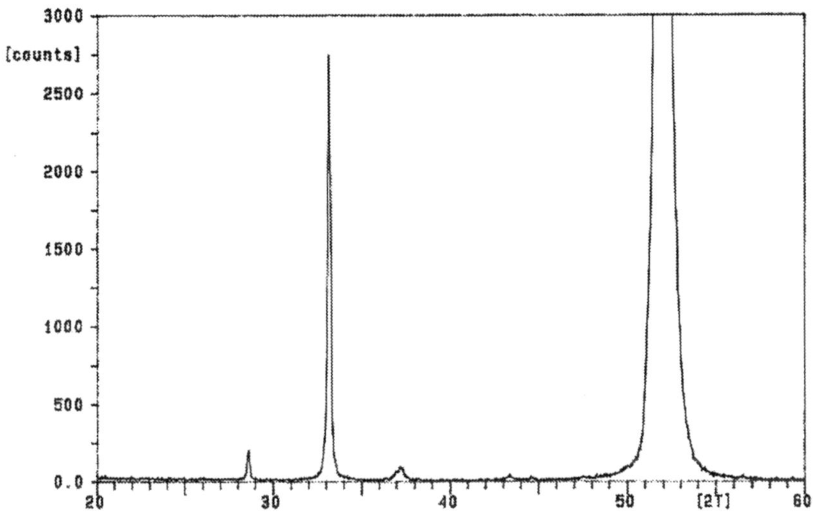

Figure 13.9. Theta–2-theta X-ray diffraction plot (CuK_a radiation) for 500 Å CeO_2 film deposited onto polished slice of single crystal Ni.

angstroms thick. Although better texturing is expected at higher temperatures, this work was limited to temperatures below 700°C.

Standard XRD measurements and pole figure measurements (not shown) indicate good texturing of YSZ on the CeO_2 surface. We deposited a YSZ/CeO_2/YSZ sandwich structure in which the YSZ layer was ~0.5 microns and the CeO_2 layers were a few hundred angstroms thick.

13.8.4 Status and Future of Barrier Layer Deposition by JVD

Although we need further characterization of YSZ and CeO_2 films made with JVD sputter jets, our exploratory work indicates that the properties of these films are suitable for HTS tape application. Scaleup to commercial production levels is a different issue. JVD sputter jets are comparable in rate to standard processes such as magnetron sputtering. In contrast to conventional sputtering processes however, JVD sources are readily operated in DC mode with metallic cathodes in oxidizing environments. Use of RF sputtering equipment and oxide cathodes is thereby avoided. Sputter jets may be adequate for continuous deposition of several hundred angstroms of CeO_2, but deposition of several thousands of angstroms of YSZ is problematic. We believe the best strategy, albeit a great challenge, is to adapt the e-jet for deposition of metals such as Ce and Zr. For example, Ce does not alloy with hot W or Ta, and should be vaporizable from refractory crucibles in an e-jet. Each metal presents a unique problem in e-jet source design, but ongoing development at Jet Process Corporation continues to extend the number of metal vapors the e-jet can deliver at high rate.

13.9 CHALLENGES IN HIGH RATE JVD STRIPCOATING

Continuous stripcoating offers production advantages over batch (carousel) processes that affects source design. In a carousel process, complex multicomponent oxides can be built up in many passes by means of single jets separated in space. In continuous coating, multicomponent layers must be built up in a single pass, by a single jet designed to vaporize and deposit several different metal vapors at one location. It is not difficult to do this with alloy targets in a sputter jet source. In the e-jet, we have also implemented simple cases. Nickel ferrite was deposited with an alloy wire of Fe and Ni. We deposit Cu and Au by vaporizing separate wires in a single jet, from a single hot filament, by using two independent feed mechanisms. But combinations of more than two metals, of which some may corrode hot refractories, can present a more intricate challenge than these simple cases.

However, it is worth noting that a single jet can carry metal atoms from several metal vapor sources just as easily as a single metal vapor. For example, if metal vapor is present at levels of 10^{-2} torr in a jet operating at 1 torr downstream, and travelling at 5×10^4 cm/second, the deposition rate would be $\sim$0.01 cc of metal per minute, a moderately high rate (1 micron/minute over 100 cm^2). If the mechanical and chemical problems of vaporization can be solved, it is easy to envision multiple vapor sources in a single e-jet, with O atoms generated downstream by O_2 injection, and suitable for stripcoat operation. Considerable effort at JPC is devoted to this end.

13.10 SUMMARY

Jet Vapor Deposition holds much promise for high rate, low cost manufacture of HTS tape. At low rates, JVD has proven its ability to deposit multicomponent oxides such as the ferrites and ferroelectrics, as well as HTS buffer layers and sandwiches of CeO_2 and YSZ. For more demanding future production, the versatile "e-jet" JVD source offers powerful advantages: high rate deposition of many HTS oxides, synthesis from elemental metals or bulk HTS oxides, easy replenishment of lost oxygen, control of crystallinity via low energy, extreme-high ion current bombardment, and control of surface reaction conditions on rolled, textured Ni substrates. Added to these advantages is the absence of environmental threat; there are no toxic precursors or products in JVD. For economic, continuous manufacture, the JVD stripcoater features a simple but unique air-to-vacuum seal that enables reel-to-reel production in a low cost, small footprint system. Accelerating work at Jet Process Corporation is aimed at realizing these advantages for HTS production.

REFERENCES

Bourdillon, A. and Bourdillon, N.X., 1994, *High Temperature Superconductors: Processing and Science*, Academic Press, Inc.

Dionne, G.F., Cui, G.-J., McAvoy, D.T., Halpern, B.L., and Schmitt, J.J., 1995, Magnetic and stress characterization of nickel ferrite ceramic films grown by Jet Vapor Deposition, *IEEE Transactions on Magnetics*, 31:3853.

Golz, J.W., Di, Y., Halpern, B.L., Schmitt, J.J., Cirino, P., and Bartlett, A., 1993, Jet Vapor Deposition of lead zirconate titanate (PZT) for thin film pyroelectric detectors, *Mat. Res. Soc. Symp. Proc.*, 284:541.

Golz, J., Zhang, J.Z., Han, H., Motherway, B., Srivatsa, A., Halpern, B.L., and Schmitt, J.J., 1997, New directions in the Jet Vapor Deposition process: Development and applications of the electron jet, in: *Advances in Coatings Technologies for Surface Engineering*, A.R. Srivatsa, C.R. Clayton, and J.K. Hirvonen, eds., The Minerals, Metals and Materials Society.

Goyal, A., Feenstra, R., List, F.A., Paranthaman, M., Lee, D.F., Kroeger, D.M., Beach, D.B., Morrell, J.S., Chirayil, T.G., Verebelyi, D.T., Cui, X., Specht, E.D., Christen, D.K., and Martin, P.M., 1999, Using RABiTS to fabricate high temperature superconducting wire, *Journal of Metals*, July 1999, p. 19.

Halpern, B.L., 1982, Fast flow deposition of metal atoms on liquid surfaces, *J. Colloid Interface Sci.*, 86:337.

Halpern, B.L., 1996, Electron Jet Vapor Deposition system, US Patent 5,571,332, 11/5/96.

Halpern, B.L., Schmitt, J.J., Di, Y., Golz, J.W., Johnson, D.L., McAvoy, Wang, D., and Zhang, J.Z., 1992, Jet Vapor Deposition of single and multicomponent thin films, *Metal Finishing*, December 1992, p. 37.

Halpern, B.L. and Schmitt, J.J., 1994, Multiple jets and moving substrates: Jet Vapor Deposition of multicomponent thin films, *J. Vac. Sci. Technol. A*, 12:1623.

Halpern, B.L. and Schmitt, J.J., 1994, Jet Vapor Deposition, in: *Handbook of Deposition Technologies for Films and Coatings*, 2nd edition, R.F. Bunshah, ed., Noyes Publications, Park Ridge, NJ.

Halpern, B.L., Golz, J.W., Zhang, J.Z., McAvoy, D.T., Srivatsa, A.R., and Schmitt, J.J., 1995, The "electron jet" in the Jet Vapor Deposition process: High rate film growth and low energy, high current ion bombardment, in: *Advances in Coatings Technologies for Corrosion and Wear Resistant Coatings*, A.R. Srivatsa, C.R. Clayton, and J.K. Hirvonen, eds., The Minerals, Metals and Materials Society.

Hwang, C.L., Chen, B.A., Ma, T.P., Golz, J.W., Di, Y., Halpern, B.L., and Schmitt, J.J., 1991, Ferroelectric $Pb(ZrTi)O_3$ thin films prepared by gas jet deposition, in: *Proceedings of the 3rd International Symposium of Integrated Electronics 1991*, p. 515.

Jaffe, B., Cook, W.R., and Jaffe, H., 1971, *Piezoelectric Ceramics*, Academic Press, London.

Norton, D.P., Goyal, A., Budai, J.D., Christen, D.K., Kroeger, D.M., Specht, E.D., He, Q., Saffian, B., Paranthaman, M., Klabunde, C.E., Lee, D.F., Sales, B.C., and List, F.A., 1996, Epitaxial $YBa_2Cu_3O_7$ on biaxially textured nickel(001): An approach to superconducting tapes with high critical current density, *Science V*, 274:755.

Schmitt, J.J. and Halpern, B.L., 1988, Method and apparatus for the deposition of solid films of a material from a jet stream entraining the gaseous phase of said material, US Patent 4,788,082, 11/29/88.

Schmitt, J.J. and Halpern, B.L., 1998, Apparatus for the high speed, low pressure gas jet deposition of conducting and dielectric thin films, US Patent 5,725,672, 3/10/98.

Zhang, J.Z., Golz, J.W., Gorski, M., Schmitt, J.J., and Halpern, B.L., 1997, Jet Vapor Deposition: A new, low cost metallization process, in: *1997 International Symposium on Microelectronics Proceedings*, p. 144.

Chapter 14

PROCESSING OF LONG-LENGTH TAPES OF HIGH-TEMPERATURE SUPERCONDUCTORS BY COMBUSTION CHEMICAL VAPOR DEPOSITION

Shara S. Shoup and Todd A. Polley

MicroCoating Technologies
5315 Peachtree Industrial Blvd.
Chamblee, GA 30341
USA

14.1 INTRODUCTION

Cost models by the Department of Energy (DOE) convincingly prove that for power transmission, high-temperature superconducting (HTS) wires and tapes can economically compete with copper and aluminum wires. Once long-length manufacturing capabilities for hts wire and tape have been established, manufacturing costs are expected to dramatically decrease. Users of HTS wire, such as manufacturers of transmission cables, motors, and transformers, need long lengths of wire that can carry at least 100 A at a cost no greater than $5–20 kA-m. A vacuum technique, Ion Beam Assisted Deposition (IBAD) (Foltyn et al., 1999), was the first to succeed in producing a meter length of tape with a critical current close to 100 A, but vacuum processes (e.g., pulsed laser deposition and electron beam evaporation) are not only expensive but impractical when addressing the needs for rapid, low-cost production of kilometer lengths of wire. MicroCoating Technologies is investigating the use of the Rolling Assisted Biaxially Textured Substrate (RABiTS™) (Goyal et al., 1996; Norton et al., 1996) process in combination with the low-cost, open atmosphere Combustion Chemical Vapor Deposition (CCVD) technique to manufacture second-generation HTS tape.

14.2 COMBUSTION CHEMICAL VAPOR DEPOSITION

The innovative Combustion Chemical Vapor Deposition (CCVD) process (Hunt et al., 1993, 1997; Hendrick et al., 1998), patented by the Georgia Institute of Technology and licensed exclusively to MicroCoating Technologies, Inc. (MCT, Atlanta,

GA) has demonstrated the ability to overcome many of the shortcomings of traditional vapor deposition techniques while yielding equal or better quality coatings at a lower cost. Developed in 1993, it can deposit thin films in the open atmosphere, requiring only inexpensive precursors. This obviates the need for costly furnaces, vacuum equipment, reaction chambers, and many post-deposition treatments.

Typically, precursors are dissolved in a combustible solvent which also acts as the fuel for the flame (Figure 14.1). This solution is atomized to form submicron droplets by means of a proprietary technology (Nanomiser™ device, patent pending). These droplets are then carried by an oxidizing gas stream to the flame in which they are combusted. Coating is accomplished either by drawing the combustion plasma containing the activated deposition species over the substrate's surface or moving the substrate across the plasma. The thermal energy from the flame provides the means to evaporate the droplets and for the precursors to react and deposit onto the substrate. The entire process for coupon samples generally takes no more than two hours from initial set-up to post-deposition cleaning. Multiple depositions utilizing the same solution require minimal additional time per run other than deposition time. Therefore, coatings and their properties can be optimized quickly in a systematic manner.

The CCVD process is often mistaken for plasma or thermal spray techniques. Though there are similarities to these processes as well as to traditional CVD, CCVD is a novel deposition process in itself as it does not deposit droplets (these evaporate in the flame environment) or powders as in traditional thermal spray processes. The CCVD technology is also drastically different from spray pyrolysis: In spray pyrolysis, a liquid mixture is sprayed onto a heated substrate, while CCVD atomizes a precursor solution into sub-micron droplets followed by vaporization of said droplets. The resulting coating capabilities and properties described hereafter qualifies CCVD as a true vapor deposition process. Substrate temperatures may be as low as 100°C, thus, enabling deposition onto a wide variety of materials including polymers. Physical structure and chemical composition of the deposited films can be tailored to the specific application requirements: This greatly facilitates the rational design of thin films.

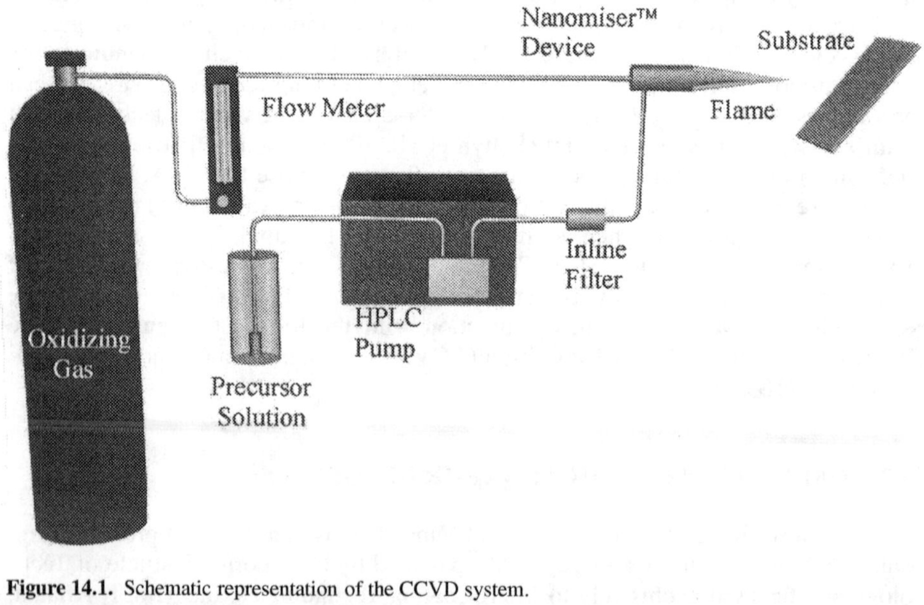

Figure 14.1. Schematic representation of the CCVD system.

In summary, the traditional ccvd process offers the following capabilities:

- **No Need for a Specialized Chamber.** Depositions are inexpensively performed at ambient conditions without the need for expensive, specialized equipment such as reaction furnaces and/or vacuum chambers, reducing repair times and costs.
- **Uses Inexpensive Precursors.** Soluble precursors, such as acetylacetonates or ethylhexanoates dissolved in alcohols, are used instead of expensive, high vapor pressure organometallics.
- **Excellent Composition Control.** Solution composition is adjustable to allow for great versatility in depositing a wide variety of complex multi-component compounds with targeted stoichiometries. Although the composition of the solutions may not always be that of the deposited film, the desired film stoichiometry is reproducible using a fixed solution and the same deposition parameters.
- **Suitable for Continuous Production.** The equipment enables a robust production system which has been shown to operate without failure around the clock for several days. Substrates that have been coated include fiber tows, pins, industrial rollers, wire, radiators, and roll sheet material.
- **Straightforward Integration into Existing Production Processes.** Advanced materials can be deposited onto large surface areas, including assembled parts that would be difficult to place inside a traditional coating chamber.
- **Ability to Deposit Multi-Layered Structures.** CCVD flames are regularly used in sequence to deposit multi-layered structures.
- **Wide Choice of Substrates.** The CCVD process allows deposition onto many different substrate materials, including plastics which can be incompatible with vacuum processes. Different sizes and shapes can be processed including non-planar substrates.
- **Precise Control of Coverage Area.** Coating can be limited to specific areas of a substrate by simply controlling the dwell time.
- **No Line-of-Sight Limit.** Operating at atmospheric pressure randomizes atom trajectories through diffusion. This permits infiltration into high aspect ratio holes and vias, as well as complete coverage around fibers and wires. This contrasts with conventional long mean-free-path PVD processes where the "shadowing" of molecular rays affects distribution of coatings over non-planar surfaces.
- **Outstanding Microstructure Control.** Microstructure of the deposited film is closely controlled, ranging from dense, *epitaxial* films to high surface area, nanoporous layers.
- **Accelerated Development Cycle for New Applications.** Development of coatings for specific applications is achieved more rapidly than with traditional technologies: a large number of samples are quickly prepared for testing and optimization.
- **Environmentally Friendly.** Relatively safe chemical precursors (non-toxic and halogen-free) are used, resulting in benign by-products, thus reducing environmental impact.

As a result, capital requirements and operating costs are reduced up to tenfold when compared to competing chamber- and vacuum-based technologies (e.g., sputtering and most CVD). The ability to deposit thin films in the open atmosphere enables continuous, production-line manufacturing. Consequently, throughput potential is far greater than with conventional thin-film technologies, most of which are generally restricted to batch processing. Thus far, CCVD has been used to prepare more than seventy distinct material compositions for a variety of applications (Table 14.1).

Table 14.1. Thin film materials deposited by CCVD

Metals	Ceramics	Others
Ag, Au, Cu, Ir, Ni, Rh, Pt.	Al_2O_3, doped-Al_2O_3, Al_2O_3•MgO, $3Al_2O_3$•$2SiO_2$, $BaCeO_3$, $BaTiO_3$, BST, CeO_2, Cr_2O_3, Cu_xO, $[La_{.95}Ca_{.05}]CrO_3$, Fe_2O_3, In_2O_3, ITO, $LaAlO_3$, LSC, LSM, MgO, Mn_2O_3, MoO_3, Nb_2O_5, NiO, NSM, $PbTiO_3$, PdO, PLZT, PMN, PMT, PNZT, PZT, RbO_x, RhO_x, RuO_2, SiO_2, Spinels (e.g., $NiAl_2O_4$, $NiCr_2O_4$), Silica Glasses, doped-SnO_2, $SrLaAlO_4$, $SrRuO_3$, $SrTiO_3$, Ta_2O_5, TiO_2, V_2O_5, WO_3, $YBa_2Cu_3O_x$, $YbBa_2Cu_3O_x$, YIG, YSZ, YSZ•Al_2O_3, YSZ-Ni, ZrO_2, fully and partially stabilized ZrO_2 (dopants: Y, Ce, Al).	$BaCO_3$, $LaPO_4$, $PbSO_4$.
Substrates Used		
Al, Brass, Ag, Cu, Pt, Ni, Stainless and C-Steel, Al_2O_3, Fiber Tows, Glass, Graphite, $LaAlO_3$, MgO, NAFION™, NiCr, Optical fibers, Polycarbonate, Silica, Si, Si–Ti/Pt wafers, SiC, Si_3N_4, Superalloys, Teflon™, Ti, TiAl alloy, YSZ.		
Possible Applications		
Architectural finishes, capacitors, catalytic applications, corrosion resistance, cutting tools, electronics, engines, ferroelectric materials, FGM, fuel cells, integrated circuits, optics, piezoelectrics, replacement for nickel plating, superconductors, thermal barrier, thermal control, wear resistance.		

14.3 DEPOSITION OF FUNCTIONAL MATERIALS

14.3.1 Buffer Layer

MicroCoating Technologies began investigating the deposition of buffer layers under a Small Business Innovation Research (SBIR) Phase I award granted by the Department of Energy in the fall of 1997. The intention was to combine the RABiTS™ (developed by the Oak Ridge National Laboratory, ORNL) and the CCVD processes for depositing the buffer layer. To achieve this goal, the CCVD process and equipment had to be modified: for example, in order to deposit on an oxidation-prone substrate (e.g., nickel), a localized reducing atmosphere surrounding the substrate was needed. This was accomplished by encasing the actual deposition (flame) zone and engineering dynamic seals to allow for continuous substrate passage. Further process modifications were needed to prevent carbon contamination under reducing process conditions. Using a solvent system with a lower carbon content and a hydrogen/oxygen flame alleviated these concerns.

Several advanced materials suitable for buffer layers have been deposited by CCVD. These include CeO_2, YSZ, $LaAlO_3$, $SrTiO_3$, Y_2O_3, and Yb_2O_3. These films were first deposited onto single-crystal substrates such as MgO, $SrTiO_3$, and $LaAlO_3$ to demonstrate process viability. For example, $SrTiO_3$ was grown on MgO with excellent full-width-at-half-maximum (FWHM) values of $2.020° \pm 0.004$ and $1.67° \pm 0.01$ for (110) and (200), respectively. Also, YBCO superconductors were deposited by pulsed laser deposition (PLD) at ORNL on buffer layers generated by CCVD. These results include $SrTiO_3$ on $LaAlO_3$ single crystal which enabled a superconductor with a critical current density (J_c) of 1.2 MA/cm^2 (Figure 14.2(left)) while a CeO_2 buffer layer on $LaAlO_3$ enabled a superconductor with a J_c of 3.5 MA/cm^2 (Figure 14.2(right)).

Following these early proof-of-concept studies, experiments have turned to depositions on commercially viable substrates such as textured nickel. MCT has part-

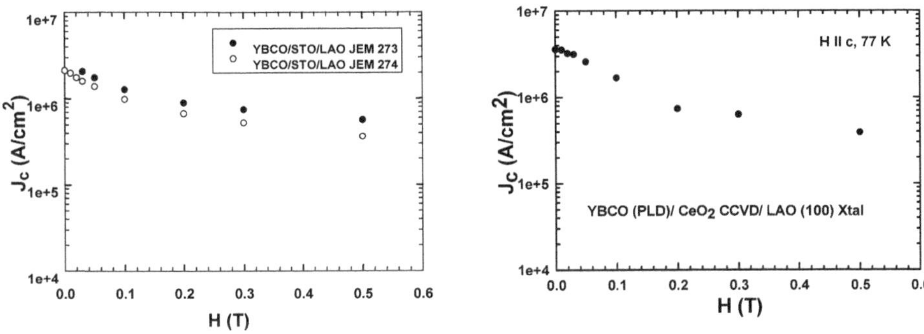

Figure 14.2. Critical current density vs. magnetic field of ybco deposited onto (left) $SrTiO_3$ on single crystal $LaAlO_3$ and (right) CeO_2 on single crystal $LaAlO_3$. This data shows that ccvd buffer layers enable high current density superconductors.

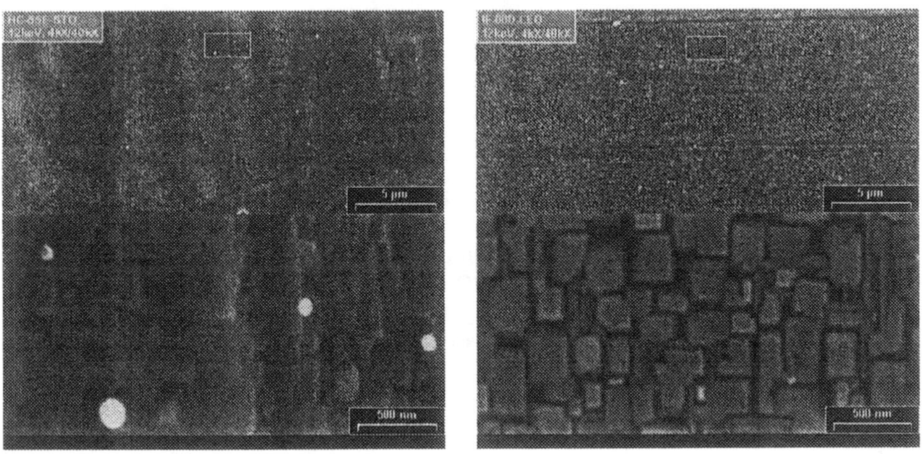

Figure 14.3. (Left) Buffer layers of $SrTiO_3$ on nickel and (right) architectures with a CeO_2 cap layer can exhibit a pronounced microstructure depending on the architecture and deposition temperature. Experiments are underway to determine the effect that different thin film surface characteristics have on the electrical performance of the superconducting film deposited. The lower portions of the figures are 10× magnifications of the boxed inset in the upper frame.

nered with Oxford Superconducting Technology (OST) in this effort with ost supplying many meters of textured nickel. Initially, epitaxial $SrTiO_3$ was deposited onto nickel. The FWHM values for $SrTiO_3$ on nickel closely match those of the nickel. For example, compare Ni (200) $\Delta\omega = 7.29°$ vs. $SrTiO_3$ (002) $\Delta\omega = 7.53°$ and Ni (111) $\Delta\phi = 8.36°$ vs. $SrTiO_3$ $\Delta\phi(111) = 8.89°$. Other buffer layers including $LaAlO_3$, CeO_2, and YSZ have also been deposited onto nickel with various architectures, but a base layer of $SrTiO_3$ must be used to transfer the epitaxial growth from the nickel to the other buffer layers. Phi scans performed on an architecture of $SrTiO_3$ with a CeO_2 cap layer yield a FWHM phi value of 9.73° and 6.67° for STO (110) and CeO_2 (111), respectively, while (002) omega values are 9.05° and 8.60°. The microstructure of solely $SrTiO_3$ on nickel (Figure 14.3(left)) typically exhibits small particles that are incorporated into the dense, continuous film while the microstructure of a film with a CeO_2 cap layer (Figure 14.3(right)) can look quite different with large block-shaped grains depending on the deposition temperature.

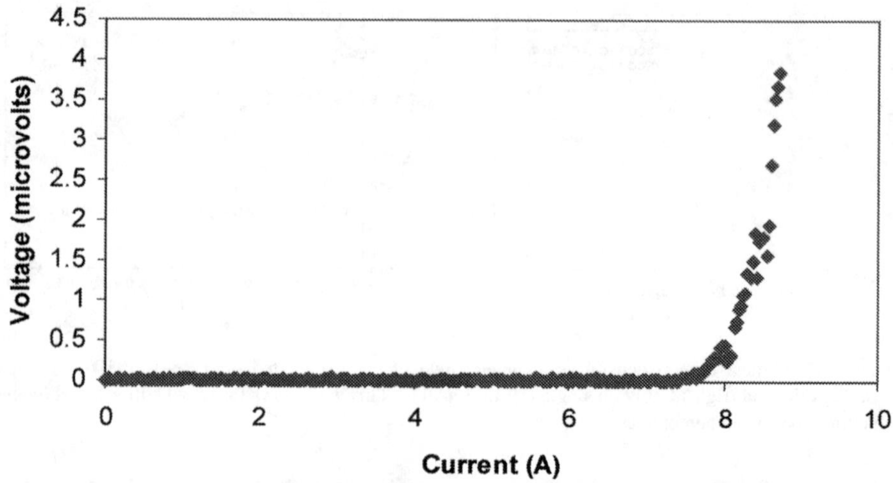

Figure 14.4. I_c as a function of applied voltage for PLD YBCO/CCVD Buffer/Ni: $I_c = 7.944$ A, 0.44 cm wide sample and 170 nm thick YBCO translates into a J_c of 1.06 MA/cm^2. Changes in architecture, thickness, and microstructure continue to be investigated to determine the result on the performance of YBCO.

The first research specimen for which YBCO was PLD-deposited by ORNL onto a CCVD-processed buffer layer of $SrTiO_3$ on a nickel substrate carried 91,000 A/cm^2. The buffer layer was quite thin and had a very rough microstructure. Specimens with $SrTiO_3$ and CeO_2 architectures having smoother $SrTiO_3$ microstructures and CeO_2 cap layers were also deposited with YBCO by PLD at the Los Alamos National Laboratory. These samples carried between 15 and 19 A with the best sample translating into 513,000 A/cm^2 (1 cm width and YBCO thickness of 370 nm). MCT has continued to refine the thickness and microstructure of the buffer layers and in conjunction with ORNL has demonstrated that PLD YBCO with a $J_c > 1$ MA/cm^2 can be deposited onto CCVD RABiTS™ (Figure 14.4). The total buffer layer thickness in such samples is estimated, using Rutherford Backscattering Spectroscopy at ORNL, at ~260 nm with the $SrTiO_3$ layer contributing approximately 250 nm and 10 nm from the CeO_2 cap. Experiments aimed at growing even smoother coatings of greater thickness (500–600 nm) continue.

The system used to deposit ccvd buffer layers on coupon size substrates was modified to enable reel-to-reel handling of tape. In addition, dynamic seals were added to allow passage of tape from the open atmosphere into the processing zone (reducing atmosphere) and back into the open atmosphere. The system is automated to control and monitor processing parameters such as gas flows, pressures, motion, temperature, etc. Thus far, tapes of 1 m and 10 m lengths with buffered $SrTiO_3$ on nickel have been fabricated using this system. A minor amount of secondary in-plane orientation was present at these lengths. The epitaxial quality along the length of these films is fairly uniform (Figure 14.5). The FWHM values for $SrTiO_3$ on a 1 m length of nickel are 9.455° (ϕ_{110}) and 9.233° (ω_{002}). Importantly, optimization of deposition parameters has resulted in films with 100% in-plane orientation.

14.3.2 Advanced Ceramics: YBCO

Initial experiments aimed at depositing YBCO by CCVD were funded through an SBIR Phase I grant. At the conclusion of the funding period, research work lay dormant for almost 24 months until new grant money was secured.

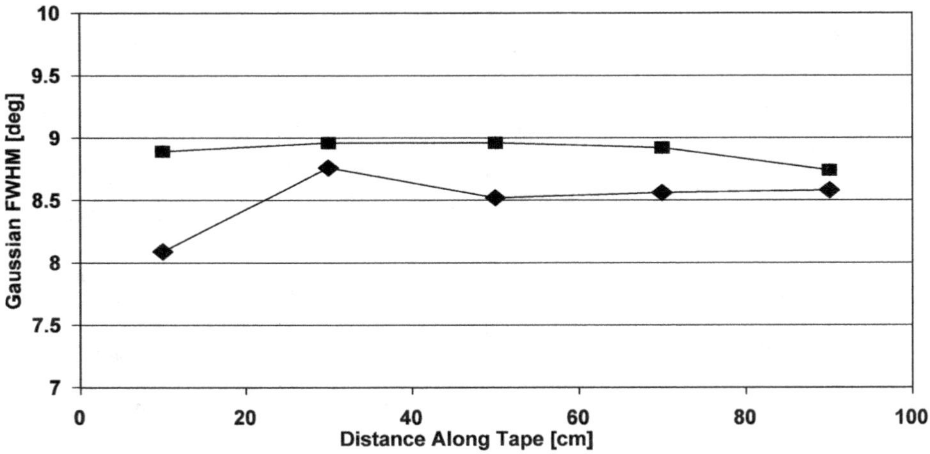

Figure 14.5. Texture data (■ = $SrTiO_3$ (111) ϕ-scan; ◆ = $SrTiO_3$ (002) ω-scan) for a 1 m-length nickel tape on which a buffer layer of $SrTiO_3$ has been deposited. Importantly, optimization of deposition parameters has resulted in films with 100% in-plane orientation.

One of the largest challenges in tailoring the ccvd system for deposition of YBCO ceramics was the removal of all carbon sources from the process. Once formed, barium carbonate is difficult to remove and detrimental to the performance of the superconductor coating. In the early stages of our research work, depositions were performed using combustible organic solvents. Albeit these depositions resulted in dense, epitaxial coatings, each coating had a measurable carbon content (as carbonate). Variations in deposition conditions did not eliminate the impurity and, therefore, the use of organic solutions was not deemed viable for superconductor depositions.

Hence, the process had to be modified to allow for use of aqueous solutions in the deposition of YBCO coatings. This decision created many challenges. For example, due to its surface tension, water is not the medium of choice for the generation of high-quality aerosols. Poor aerosol quality translates directly into poor coating microstructure and epitaxy. In early experiments, the difficulty in generating aerosols also led to system instability that resulted in system failure in generally less than an hour of deposition time. Thus, maintenance times were much higher than normal for the CCVD process. The poor coating quality and system instability made this approach unattractive for coating long lengths of conductor.

Eventually, it proved that the key to successful process redesign was the development of a new Nanomiser™ for the atomization of aqueous solutions and the use of a hydrox flame. Following these modifications, we have successfully deposited dense, epitaxial ybco coatings onto single crystal substrates. The coating exhibits cube texture with very good in-plane and out-of-plane epitaxy.

Remarkable progress has been made in a relatively short timespan in depositing high-quality superconductive films by CCVD. This is best evidenced by comparing micrographs of early and more recent samples (Figure 14.6). While the former exhibit a rough surface and poor epitaxy, the surface roughness and density are greatly improved in more recent samples.

Composition of the ybco coatings was assessed by energy-dispersive spectrometry (EDS). Though not an accurate measure of absolute composition, the EDS provides valuable qualitative information. A procedure was devised to quantify the deviation from a YBCO standard: For each cation the percent difference between measured

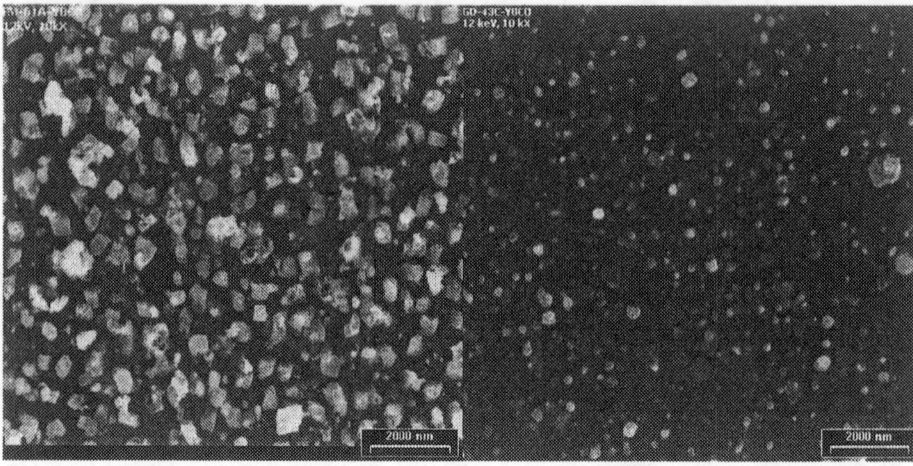

Figure 14.6. Micrographs of an early (left) and more recent (right) sample (10,000× magnification). While the former exhibit a rough surface and poor epitaxy, these properties are greatly improved in more recent samples.

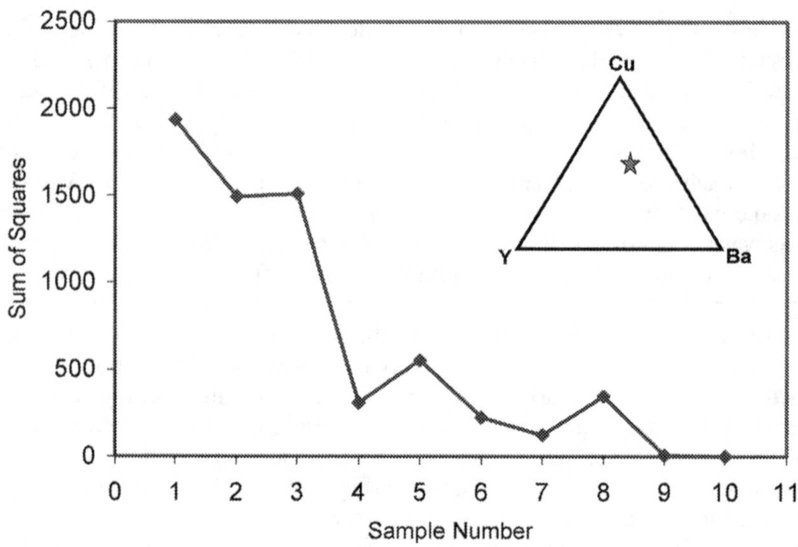

Figure 14.7. "Sum of squares" for the first ten depositions of YBCO using CCVD. This research took less than a week to complete: the ability to rapidly obtain the desired composition is a key to the success the CCVD process.

concentration and desired concentration is calculated. The sum-of-squares is the calculated according to $\sum_i \Delta^2_{\text{cat}(i)}$, where $i = \text{Y}$, Ba, and Cu. As the "sum of squares" converges towards zero, the target composition is reached. Figure 14.7 shows the "sum of squares" for the first ten depositions of ybco using CCVD. This research took less than a week to complete. The rapid ability to obtain the desired composition is a key to the success the CCVD process.

The YBCO coatings show excellent phase purity. An x-ray diffraction pattern of a typical YBCO coating is shown in Figure 14.8. The intense *c*-axis peaks are indicative of strong out-of-plane epitaxy. A pole figure shows cube texture with no apparent sec-

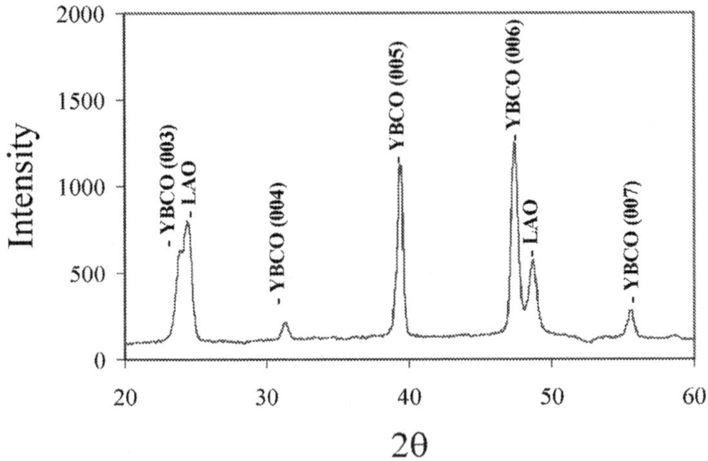

Figure 14.8. A typical x-ray diffraction pattern for ybco coating evidences excellent phase purity. The intense *c*-axis peaks indicate strong out-of-plane epitaxy.

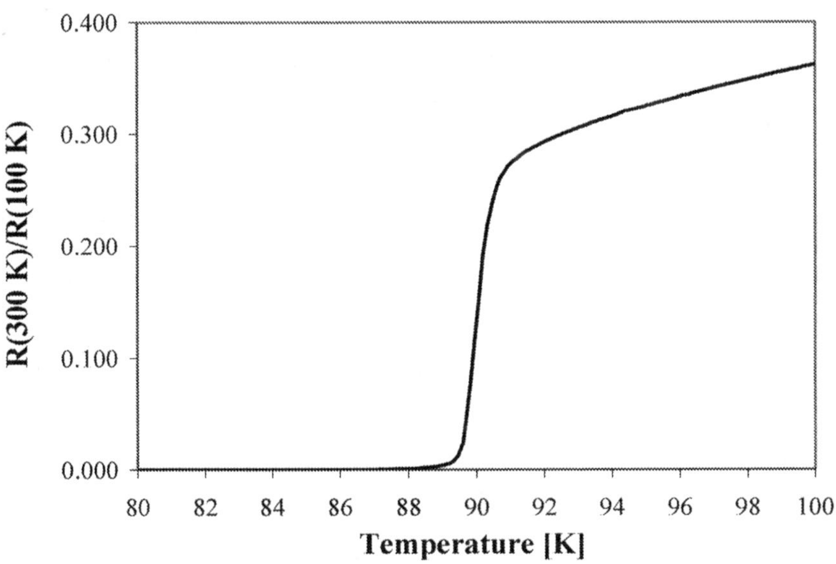

Figure 14.9. Resistance as a function of temperature for a YBCO coating deposited by CCVD.

ondary orientations. The quantitative measurement of in-plane and out-of-plane epitaxy were measured by phi and omega scans with FWHM values of 1.2° and 0.35°, respectively.

Most importantly, these films exhibit excellent superconducting properties. Microcracking and oxygen deficiency has a deleterious effect on performance of a superconductor thus an inline furnace helps control the substrate temperature and atmosphere before and after deposition. Such a furnace can also be easily incorporated into a system designed to coat practical lengths of coated superconductor. The critical temperatures are between 85–91 K (Figure 14.9). Critical current densities of these films on single crystal substrates are above 1.0 MA/cm^2 (Figure 14.10).

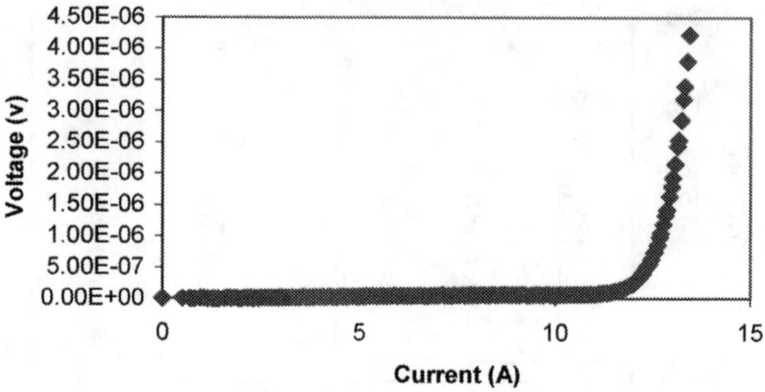

Figure 14.10. I_c as a function of applied voltage for CCVD YBCO/LAO single crystal: $I_c = 12.22$ A, 0.3 cm wide sample, and 380 nm thick YBCO translates into a J_c of 1.07 MA/cm^2.

14.4 DIRECTION OF FUTURE RESEARCH

Efforts are currently focused on increasing the rate at which buffer tape can be deposited using the CCVD process. This rate is determined by several parameters including solution concentration and the number of Nanomisers™ used. While a higher precursor concentration increases growth rate, this can also lead to film randomization. Similarly, microstructure is also affected by precursor concentration and growth rate: the growth rate must be slow enough to allow for deposition of epitaxial, dense, and continuous films but fast enough to be commercially viable. One approach to increase the overall growth rate while avoiding negatively impacting the epitaxy or microstructure of the film is the consecutive use of multiple Nanomiser™ devices. In a proof-of-concept study, a 30 cm long tape was coated with an epitaxial buffer layer under conditions that simulated 18 consecutive depositions. Conceptually, multiple layers of buffer and superconductor can be deposited in-line using a bank of flames with each bank depositing one component of the film architecture. Double-sided coatings are also feasible: not only can independent banks of flames point at the front and back of the tape, but ccvd growth can occur on the opposing side of the tape as it is not line-of-sight limited. We have experimentally observed epitaxial buffer layers deposited on the backside of the tape even though the Nanomiser™ device was directed at the front side of the tape only. A multi-Nanomiser™ system is now being constructed and tested.

Efforts are also underway to determine conditions necessary to deposit high quality YBCO onto CCVD RABiTS™ in order to fabricate a complete superconducting tape.

Other textured metal substrates besides nickel are also being investigated with Oxford Superconducting Technology. For example, alloys exhibit increased resistance to oxidation and improved mechanical strength.

14.5 CONCLUSIONS

The Combustion Chemical Vapor Deposition (CCVD) technique presents an attractive route for the deposition of low-cost, practical lengths of second-generation coated superconductors. CCVD-deposited buffer layers on textured nickel substrates have enabled high-performance superconductors of greater than 1 MA/cm^2. Thin films of YBCO deposited by CCVD exhibit excellent materials characteristics (epitaxy, composition, and microstructure) and electrical properties on single crystal substrates with critical current densities greater than 1 MA/cm^2. The CCVD technology is very amenable to scale-up as has been demonstrated by the fabrication of 1 and 10 m lengths of buffered nickel. The incorporation of ccvd superconductor depositions can also be easily realized. Because of the inexpensive chemical and capital equipment costs, CCVD is well-positioned to fabricate commercial quantities of YBCO-coated conductor that meet DOE's cost target of $10/kA-m.

ACKNOWLEDGMENTS

We gratefully thank the Department of Energy for its continuous support under a variety of contracts and research grants.* The progress described herein would also not have been possible without the fruitful collaboration of the Oak Ridge National Laboratory (Mariappan Paranthaman and team) and Los Alamos National Laboratory (Steve Foltyn).

Special recognition has to go to the hard-working and dedicated group of individuals at MCT, namely, Marvis White, Steve Krebs, Adam King, Yibin Xue, Dave Mattox, Guang-ji Cui, Ian Campbell, and Bert Bradley, that carried out the experimental work described herein. We also thank Andrew T. Hunt, Henry Luten, and S. Shanmugham also of MCT for valuable discussions and input. Lastly, we also acknowledge the excellent team at Oxford Superconducting Technology led by Ken Marken who has provided us with many meters of textured nickel and much discussion through a joint effort to advance CCVD RABiTS™ coated conductors beyond the laboratory scale.

REFERENCES

Foltyn, S.R., Arendt, P.N., Dowden, P.C., DePaula, R.F., Groves, J.R., Coulter, J.Y., Peterson, E.J., Maley, M.P., and Peterson, D.E., 1999, Critical issues in coated conductors: Progress at Los Alamos, in: *1999 Wire Development Workshop Proceedings*, January 12–13, Florida, pp. 215–228.

Goyal, A., Norton, D.P., Budai, J., Paranthaman, M., Specht, E.D., Kroeger, D.M., Christen, D.K., He, Q., Saffian, B., List, F.A., Lee, D.F., Martin, P.M., Klabunde, C.E., Hatfield, E., and Sikka, V.K., 1996, High critical current density superconducting tapes by epitaxial deposition of $YBa_2Cu_3O_x$ thick films on biaxially textured metals, *Appl. Phys. Lett.*, 69:1795.

Hendrick, M.R., Hampikian, J.M., and Carter, W.B., 1998, Combustion CVD-applied alumina coatings and their effects on the oxidation of a Ni-based chromia former, *J. Electrochem. Soc.*, 145:3986.

* The work described herein was completed under several federally-funded research grants and contracts. The DOE's support through the grant(s) does not constitute an endorsement by DOE of the views expressed in this chapter: (i) DOE SBIR Phase I grant: A.T. Hunt, "Buffer Layers on Textured Nickel Using Commercially Viable CCVD Processing," Contract # DE-FG02-97ER82345; (ii) DOE SBIR Phase I grant: S. Shanmugham, "Stoichiometric YBCO Epitaxial Coatings on RABiTS Using Low Cost CCVD Processing," Contract # DE-FG02-97ER82344; (iii) DOE SBIR Phase II grant: S. Shoup, "Buffer Layers on Textured Nickel Using Commercially Viable CCVD Processing," Contract # DE-FG02-97ER82345; (iv) DOD Air Force, SBIR Phase I: S. Shoup, "Low-Cost, High-Performance Superconducting Cable via CCVD," Contract # F33615-98-C-5418; (v) DOE SBIR Phase I grant: S. Shoup, "Non-Vacuum, Reel-to-Reel Processing of High Temperature Superconducting Coated Conductors," Contract # DE-FG02-99ER82834; (vi) DOE program, S. Shoup, "Non-Vacuum Continuous Processing for Low-Cost, High Performance RE-123 Coated Film Conductors on Textured Metallic Substrates," Contract # 4500011833; (vii) National Renewable Energy Laboratory: S. Shanmugham, "Advanced Buffer Layer for Superconductors on Metallic Substrates," Subcontract # ACQ-9-29612-01.

Hunt, A.T., Carter, W.B., and Cochran, J.K., 1997, *Method and Apparatus for the Combustion Chemical Vapor Deposition of Films and Coatings*, US Patent 5,652,021.

Hunt, A.T., Carter, W.B., and Cochran, J.K., 1993, Combustion chemical vapor deposition: A novel thin-film deposition technique, *Appl. Phys. Lett.*, 63:266.

Norton, D.P., Goyal, A., Budai, J.D., Christen, D.K., Kroeger, D.M., Specht, E.D., He, Q., Saffian, B., Paranthaman, M., Klabunde, C.E., Lee, D.F., Sales, B.C., and List, F.A., 1996, Epitaxial $YBa_2Cu_3O_7$ on biaxially textured nickel(001): An approach to superconducting tapes with high critical current density, *Science*, 274:755.

Chapter 15

MOCVD GROWTH OF YBCO FILMS FOR COATED CONDUCTOR APPLICATIONS

Alex Ignatiev

Space Vacuum Epitaxy Center
and Texas Center for Superconductivity
University of Houston
Houston, TX 77204-5507
USA

15.1 INTRODUCTION

Metal organic chemical vapor deposition (MOCVD) is one of the many currently available oxide thin film deposition techniques that can be applied to the growth of $YBa_2Cu_3O_{7-x}$ (YBCO) films. Physical vapor deposition (PVD) techniques such as laser ablation, evaporation, and magnetron sputtering, suffer from generally low growth rates, a requirement for high vacuum, continual source change-out, moderate area coverage, and a restriction to only line-of-sight deposition. Such restrictions, especially the low growth rates, are problematic for the commercialization of the YBCO film technology for HTS wires and tapes. MOCVD can overcome these drawbacks and produce high superconducting quality thin and thick YBCO films for coated conductor applications. MOCVD, first developed in the early 1970's is now a major thin film fabrication technique in the semiconductor-based microelectronics industry (Stringfellow, 1989). Given the industrial history for MOCVD, this technology has been directly transferred to YBCO film growth, and has shown the capability for fabrication of high quality YBCO samples. MOCVD growth of YBCO films has also demonstrated conditions that could directly translate to HTS coated conductor wire fabrication in industrial environments.

Initial work on the application of MOCVD to YBCO thin film growth was begun in the early period of high temperature superconductivity research (Berry et al., 1988; Yamane et al., 1988). Although the initial efforts yielded marginal material, a number of groups began the growth of YBCO films through the application of the, by then standard, microelectronics-developed MOCVD technique modified for higher temperatures, oxidizing atmospheres, and solid precursors (Dickenson et al., 1989; Panson et al., 1988; Zhang et al., 1989a, 1989b; Noh et al., 1989). The higher temperatures (more than 200 K higher than for semiconductor III–V compounds) required improved reactor designs and improved heaters, and the solid precursors required enhanced attention to precursor flow control and stability. The initial results were promising, and for films grown on single crystal oxide substrates $T_c > 90$ K and $J_c > 10^6$ A/cm^2 were realized (Yamane et al., 1989; Zhao et al., 1991; Schulte et al., 1991).

MOCVD was also applied to the growth of $Bi_2Sr_2Ca_2Cu_3O_x$ superconducting films, though with only marginal success, with T_c reaching ~110 K and J_c approaching 10^5 A/cm^2 (Sugimoto et al., 1991; Zhang et al., 1990; Yamasaki et al., 1992). Similar results were obtained for $Th_2Ba_2CaCu_2O_x$ superconducting films where $T_c \sim 110$ was reached and $J_c \sim 10^4$ A/cm^2 was obtained (Hamaguchi et al., 1991; Zhang et al., 1989a, 1989b). The poorer performance measures for the Bi and Th systems are principally due to the fact that the compositions of those systems are more complex that of YBCO, and the possibility of formation of different phases makes MOCVD deposition more involved.

With the success of generating YBCO films with high T_c (>90 K) and high J_c (>10^6 A/cm^2) by MOCVD on oxide substrates attention turned to two of the main challenges for the application of MOCVD to the large-scale growth of YBCO films: high growth temperatures (>800°C) and slow growth rates (~1 μm/h). Of the multiple parameters affecting the MOCVD growth of YBCO, principal focus was put on lowering growth temperatures by incorporating oxidizers such as N_2O and O_3, as well as plasma enhancement and photo enhancement of the reaction gases. N_2O is relatively inert to metalorganics, but highly reactive when dissociated. The use of N_2O as the oxidizer resulted in lower growth temperatures, but it also yielded lower growth rates which decreased in a linear manner with growth temperature (Tsuruoka et al., 1989; Zama and Oda, 1991; Zama et al., 1992a; Li et al., 1991; Chern et al., 1993). Additional work addressed the application of plasma enhancement to YBCO MOCVD growth. Plasma excitation was applied to the N_2O oxidizer prior to passage into the MOCVD reactor so as to enhance dissociation of the gas and thereby increase the presence of atomic and ionic oxygen (Zhao et al., 1991). High quality YBCO was obtained by plasma-enhanced MOCVD, but at growth rates no higher than ~2 μm/h. Plasma excitation was also applied to the metalorganic precursors to enhance their dissociation, and thus to reduce growth temperatures. Growth temperatures as low as 600°C were realized with good quality films, but again at growth rates of ~1 μm/h (Ebihara et al., 1993; Komatsu et al., 1999a, 1999b). Ozone was also used in attempts to enhance YBCO film quality but with only limited success (Endo et al., 1991).

Since the organometallic precursors are sensitive to photo-dissociation, photo-irradiation was also applied to MOCVD. Initial work focused on excimer laser irradiation of the reaction in attempts to increase the dissociation rate of the precursors (Ushida et al., 1991; Higashiyama et al., 1993; Mizushima and Hirabayashi, 1994). This yielded only moderate improvement in superconducting properties, but with some improvement in surface smoothness and in the generation of a-axis surface alignment. Irradiation of the surface of a growing YBCO sample by high flux visible/UV photons, however resulted in a significant increase in crystalline quality of the YBCO films, and more importantly, a large increase was realized in the growth rate of the films (Chou et al., 1995; Zhong et al., 1995). Growth rates of up to 1 μm/min were realized by photo-assisted MOCVD, thus allowing for future industrial viability of the MOCVD process for YBCO growth.

15.2 PHOTO-ASSISTED MOCVD

As noted, various attempts have been made to enhance MOCVD growth of YBCO. One such successful attempt has been the irradiation of the reactor with broad-band, high flux optical radiation to yield all of the energy required for the MOCVD reaction. This photo-assisted MOCVD (PhAMOCVD) technique has been pioneered at the University of Houston (Chou et al., 1994, 1995), and has shown extremely high growth rates for not only YBCO but for a variety of oxide thin film materials (Chen et al., 1998; Ritums et al., 1996; Ignatiev et al., 1998b). Photo-assisted MOCVD utilizes a bank of quartz-halogen lamps as the sole energy source for the reaction, which irradiate the surface of the substrate through a quartz window. This results in substrate surface temperatures that reach as high as 1000°C when using up to 10 kW of lamp power. Nominal temperatures for YBCO thin film growth are ~750–900°C. Therefore, the lamp output is controlled on both power output and sample surface temperature. Substrates are mounted in a vertical quartz reactor fed by organometallic precursors and oxidants, and the reactor is pumped to maintain a gas flow resulting in a nominal reactor pressure of ~1–5 Torr. Substrate temperature is monitored by a thermocouple imbedded into the succeptor (usually a Si wafer) onto which the substrates are mounted. The reactor has a provision for rotating the substrate for improved film uniformity, however, rotation becomes unworkable for the case of a continuously transported substrate as is required for the fabrication of long lengths of superconducting tape. Organometallic precursors are fed into the reactor by a carrier gas, and the residuals of the reaction are pumped away. A typical photo-assisted MOCVD reactor is shown in Figure 15.1.

The application of the photo-assisted MOCVD process to YBCO growth has resulted in the growth of high quality YBCO thin films with $T_c \sim 92$ K and $J_c \sim 4 \times 10^6$ A/cm^2 (Chou et al., 1994, 1995) as shown in Figures 15.2, 15.3, 15.4 and 15.5. YBCO films with thicknesses greater than 5 μm have been grown, and show excellent microstructural uniformity as can be seen in Figure 15.2. Enhanced crystalline quality was observed in the PhAMOCVD grown films as can be seen in Figures 15.3 and 15.4. It has been proposed that enhancement in crystalline quality is due to photo-enhanced diffusion of surface species leading to increased atomic ordering at the surface of the growing film (Zhong, 1996). Enhanced surface diffusion also leads to increased

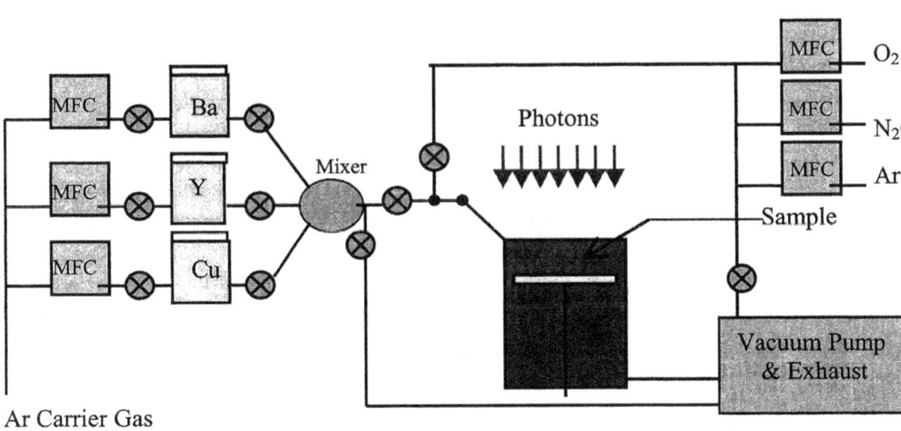

Figure 15.1. A schematic of a photo-assisted MOCVD vertical reactor.

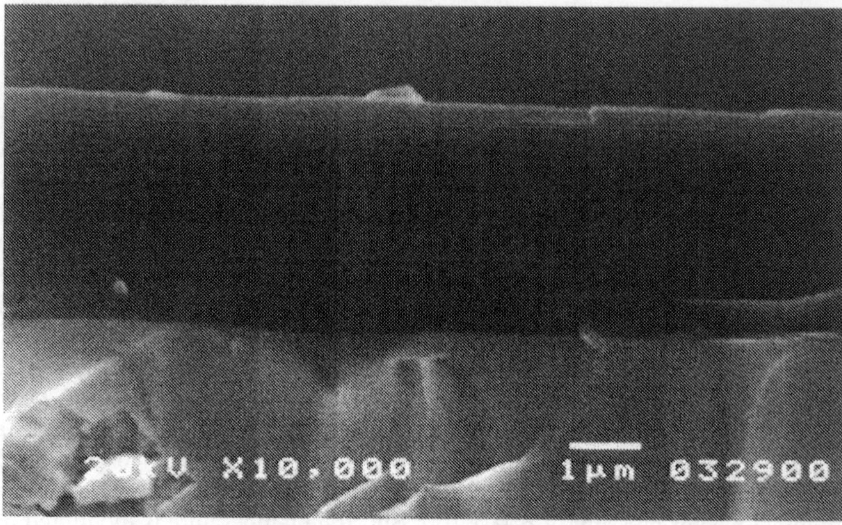

Figure 15.2. Scanning electron micrograph of a cross section of a 3 μm thick YBCO film grown on $LaAlO_3$ by PhAMOCVD. Note the homogeneity of the film cross section.

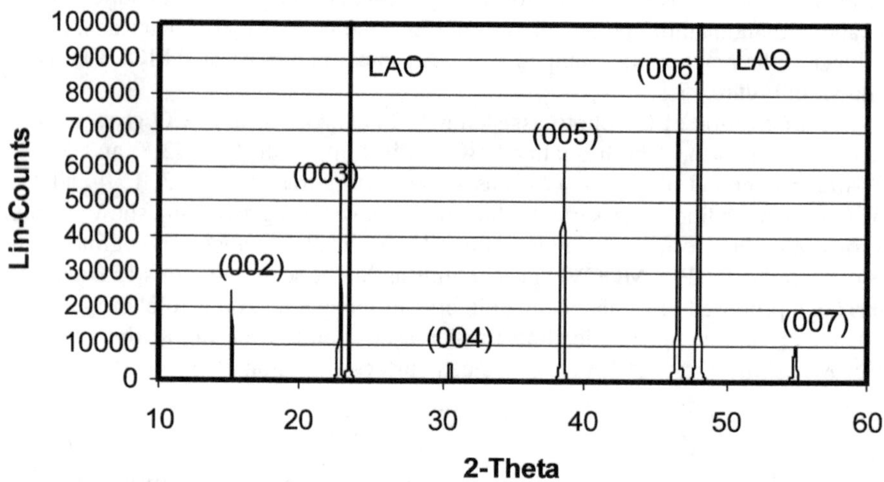

Figure 15.3. XRD θ–2θ scan of a 0.8 μm thick YBCO film grown on $LaAlO_3$ by PhAMOCVD showing only (00*l*) reflections for YBCO.

mass transport, which is the usual rate limiting step in MOCVD growth of thin films (Stringfellow, 1989; Weiss et al., 1997).

The increased mass transport due to photo-irradiation therefore, also allows for increased growth rates for YBCO films by PhAMOCVD. High growth rates were attained in the PhAMOCVD experiments (Zhong, 1996; Ignatiev et al., 1998a) where it was shown that growth rates higher than ~0.7 μm/min could still yield good quality (high T_c, and $J_c \sim 1 \times 10^6$ A/cm^2) YBCO films as seen in Figure 15.6. Best quality films ($J_c > 1 \times 10^6$ A/cm^2) were obtained for film growth rates of ~0.3 to 0.5 μm/min (Zhong et al., 1995; Zhong, 1996).

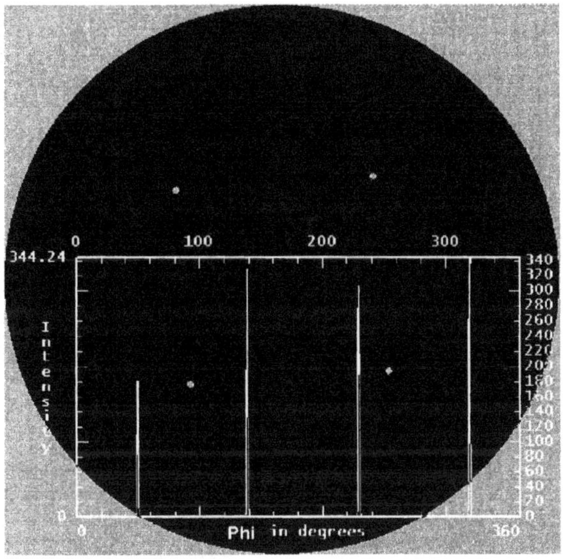

Figure 15.4. XRD pole figure for the 0.8 mm thick YBCO/LAO film at YBCO(103). FWHM of ϕ-scan peak is about 0.6°.

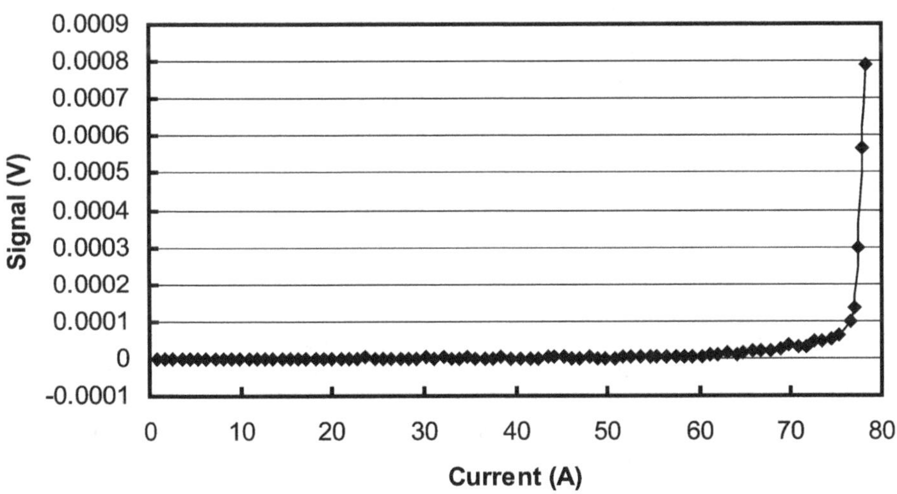

Figure 15.5. Current–voltage curve for an 0.8 μm thick × 2 mm wide YBCO film grown on $LaAlO_3$ by PhAMOCVD. The resultant critical current density of the YBCO film is ~4×10^6 A/cm^2.

Such rates, however are still from 10 to 100 times greater than those for thermal or plasma-assisted MOCVD (Busch et al., 1991; Chern et al., 1993; Ebihara et al., 1993; Komatsu et al., 1999a, 1999b), or physical deposition processes such as laser deposition (Eulenburg et al., 1999; Park et al., 1999), e-beam deposition/evaporation (Solovyov et al., 1998), or sputter deposition (Maraitakis et al., 1998; Goto et al., 1999). The extremely high growth rates for PhAMOCVD point to the possibility of its application in an industrial environment for the growth of YBCO films in commercial quantities.

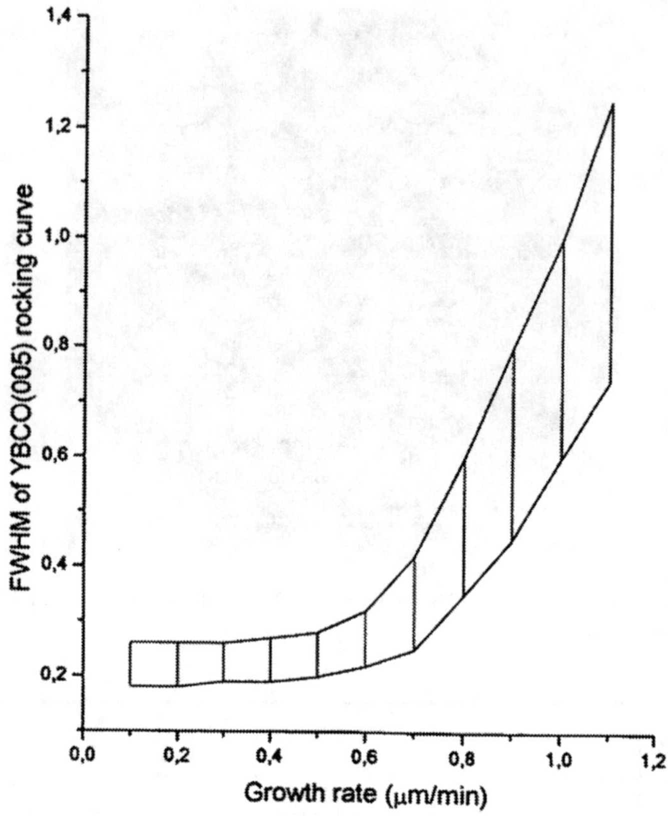

Figure 15.6. XRD rocking FWHM data for the (005) peak from YBCO films grown on $LaAlO_3$ as a function of growth rate up to $\sim$1.1 μm/min.

15.3 MOCVD PRECURSORS

One of the challenges in the application of MOCVD for YBCO growth is the quality and stability of the organometallic precursors used in the process. The original precursors used in the introduction of the MOCVD technique to YBCO thin film growth were the solid chelated compounds of the β-diketonates of Y, Ba and Cu, i.e., the 2,2,6,6-tetramethyl-3,5-heptanedionate (TMHD) of Y, Ba, and Cu: $Y(TMHD)_3$, $Ba(TMHD)_2$, and $Cu(TMHD)_2$. These precursors typically sublime at moderately high temperatures: between 100°C and 230°C; and since they are nominally solid at use temperatures, their transport and flow into the reactor is more complex than that for typical liquid precursors where transport by a carrier gas bubbled through the liquid suffices. The high sublimation temperatures also require heating of all reactor regions upstream of the precursor ovens. An additional and more critical difficulty is the instability of the precursors in air environments, with the $Ba(TMHD)_2$ having the lowest stability (Drozdov and Troyanov, 1995; Otway et al., 1997; Watson et al., 1994). The low concentration number of Ba in these compounds makes them sensitive to trace amounts of water or carbon dioxide, and the high chemical reactivity of these compounds is responsible for poor reproducibility of their volatility at elevated temperature. As a result, a number of approaches have been undertaken to mitigate these precursor problems. These include methods for enhanced volatilization, the development of new enhanced stability precursors, as well as new low-melting precursors.

Most attention has been given the Ba precursor system, which as noted is the most problematic, with major emphasis on fluorination of the β-diketonate complex (Richards et al., 1995a; Drozdov and Troyanov, 1995; Marks et al., 1997) yielding tetradecafluorononanedionates, hexafluoroacetylacetonates, and hexafluoropentanedionates of Ba. Enhanced stability of the Ba precursor was achieved with reduced melting temperatures, and higher volatility. In addition, good quality YBCO was grown with fluorinated Ba precursors typically yielding YBCO with $T_c \sim 90$ K, and $J_c > 1 \times 10^6$ A/cm^2. However, a reaction occurs between Ba and F to yield BaF_2. This needs to be removed by hydrolysis, and is best done by the introduction of H_2O into the gas phase mixture during the film growth phase. Such introduction of water vapor can affect YBCO properties including J_c and surface morphology (Zama et al., 1998), as well as reduce growth rates due to the need to reduce the BaF_2 (Richards et al., 1995a). As a result, other non-fluorinated Ba compounds have been studied with some success including Ba(TMHD)$_2$-2tetraen (Zama et al., 1998; Nagai et al., 1997). This system still has some higher temperature degradation (>140°C), possible dissociation on sublimation, and continued sensitivity to water vapor (Richards et al., 1995b). A mixture of Ba(THMD)$_2$ and Ba(tetramethyloctanedionato)$_2$ has also been studied with resulting lower melting temperature for the eutectic mixture and a higher volatility (Tasaki et al., 1998).

Other methods for improving precursor stability and delivery have focused on the combining of all three precursors prior to growth either through mixing of the precursor powders into one mixture that is then melted (Zhou et al., 1994; Lu et al., 1995), or through dissolving the precursors in a solvent and then injecting the liquid into a vaporizer connected to the growth reactor (Abrutis et al., 1998; Weiss et al., 1997; Takahishi et al., 2000; Senateur et al., 1997; Salazar et al., 1992). This liquid delivery process has used various vaporizers, which generally consist either of a thermal evaporator from which the vapor is extracted via carrier gas to the reactor, or an atomizer that creates an aerosol, which is then swept by carrier gas into the reactor. In both cases, the precursors are dissolved in a solvent (tetraglyme, diglyme, monoglyme, tetrahydrofuran, as well as other organic solvents). The solution is then either vaporized at nominal temperatures of ~200°C, or atomized. This solution approach results in enhanced stability of the TMHD precursors dissolved in the solvent, and uniform metering of the solution into the reactor thereby yielding more consistent composition of the YBCO films. The increased gas load of the solvent may however, result in increased CO_2 partial pressure in the growth zone, and may require additional pumping. To counteract this problem, a modified liquid delivery system has been used which initially deposits the precursor solution onto a metal tape or band (Klippe and Wahl, 1997; Senateur et al., 1997). The continuous band is then moved into a hot zone where the solvent is first evaporated. The precursors are then retained on the tape or band, and are evaporated in a subsequent hotter zone and then swept vie carrier gas into the reactor. The deposition rate under mixed oxide evaporation from a band seems to be defined by a single kinetics function and not a mixture of the single component kinetics thus yielding ease of operation. However, the system is mechanically additionally challenging and may not support industrial operation. Of the precursor stabilization approaches, liquid delivery accompanied by vaporization seems to be the most promising for industrial application.

15.4 INDUSTRIAL APPLICATION OF PHOTO-ASSISTED MOCVD

The principal driver for industrial application of photo-assisted MOCVD is the high growth rate reported for the technique. Although increases of growth rates have been seen in the continued development of MOCVD, the highest attained rates in any MOCVD approach other than photo-assisted MOCVD is ~3 μm/h with the nominal at ~1 μm/h, e.g., ~0.01 to 0.05 μm/min. Photo-assisted MOCVD, on the other hand, has shown rates for high quality YBCO growth as high as 0.7 μm/min with nominal rates at 0.4–0.5 μm/min. Such a significant multiplicative factor in growth rate indicates promise for industrial application of photo-assisted MOCVD.

Even though high growth rates have been realized in photo-assisted MOCVD, several challenges still remain before MOCVD can be applied at the industrial level for the fabrication of YBCO thin/thick film wires and tapes. These include: (i) continuous growth including buffer layer processing on flexible metal substrates; (ii) quality control and *in-situ* monitoring; (iii) further product yield enhancement as well as double-sided tape fabrication; and (iv) cost reduction of the precursors.

The requirement for buffer layers in the integration of YBCO with substrates other than perovskite oxides has been long recognized (Wu et al., 1991), and two leading buffer layer materials have emerged for use with YBCO: CeO_2 (Wu et al., 1991) and yttria stabilized zirconia (YSZ) (Fenner et al., 1991). These buffers have been used extensively in the development of YBCO films on metal substrates (Iijima et al., 1992; Yamaguchi et al., 1994; Krellmann et al., 1997; List et al., 1998; Ichinose et al., 1999; Wu et al., 1995).

The principle atomically ordered metal substrates used for with YBCO for superconducting wire applications are roll-textured nickel, and ion beam assisted deposition processes Hastelloy and Inconel. The application of CeO_2 and YSZ to roll-textured nickel has shown the ability for the buffer layers to support the atomic order of the metallic substrate and to prevent interdiffusion with YBCO (Goyal et al., 1996; List et al., 1998; Norton et al., 1996). Specifically, CeO_2 is grown initially on the nickel due to its lower growth temperature (~450°C) and resultant reduction of oxidation of the textured nickel substrate thus maintaining good atomic order in the buffer layer. The CeO_2 is followed by YSZ growth again maintaining good crystallographic order with respect to the CeO_2 and the substrate nickel. The YBCO is then grown on the YSZ buffer layer, or possibly on an additional CeO_2 layer grown on the YSZ. Ion beam assisted deposition (IBAD) has also been developed for the fabrication of atomically textured metal surfaces (Iijima et al., 1992; Wu et al., 1995; Yamaguchi et al., 1994; Thieme et al., 1999). The IBAD process generates YSZ or MgO textured layers on a variety of metal substrates, and has shown some benefit from the application of an additional CeO_2 buffer layer on the IBAD (YSZ or MgO) layer (Takahishi et al., 2000) since there is a better lattice match between CeO_2 and YBCO than either YSZ or MgO and YBCO ($a_{CeO_2} = 5.411$ Å; $a_{YSZ} = 5.139$ Å; $a_{YBCO} = 3.879$ Å; $a_{MgO} = 5.486$ Å).

For either of the two prevailing metallic substrates (roll-textured Ni or IBAD metal), buffer layers must be fabricated with high crystalline quality and at high rates. MOCVD can be applied to the fabrication of these buffers. CeO_2 and YSZ buffer layers have been fabricated by MOCVD with good results both in terms of crystallography, and prevention of interdiffusion (Frohlich et al., 1997; Komatsu et al., 1999b; Garcia et al., 1995; Becht and Morishita, 1997). Although the CeO_2 buffer is generally quite thin (a few to a few tens of nm), and therefore attainable by any of the MOCVD techniques. The YSZ layer is much thicker (0.5–1 μm), and thus requires the application of high growth rate photo-assisted MOCVD for successful industrial

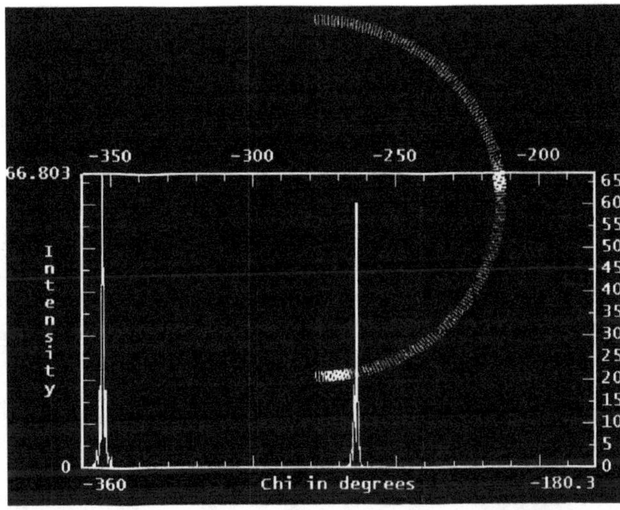

Figure 15.7. XRD pole figure for CeO_2 grown by PhAMOCVD on $LaAlO_3$. FWHM of the ϕ-scan from this sample yielded ~1.2°.

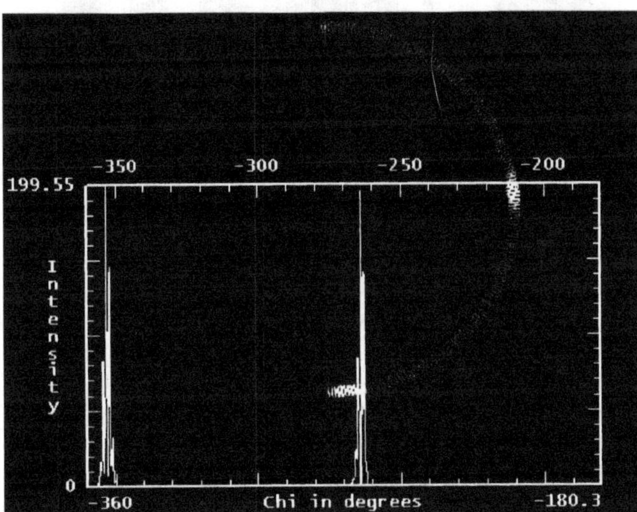

Figure 15.8. XRD pole figure for YSZ grown by PhAMOCVD on CeO_2 layer of Figure 15.7. FWHM of the ϕ-scan from this sample yielded ~1.5°.

utilization. The application of PhAMOCVD to both CeO_2 and YSZ growth has shown results similar to those of YBCO growth: excellent crystallography and high growth rates (Ignatiev et al., 1998b) as shown in Figures 15.7 and 15.8.

This now means that the complete heterostructure stack for HTS thin/thick film wire (CeO_2 buffer, YSZ buffer and YBCO film) can be grown by the same process, PhAMOCVD, and at equivalent rates, thereby resulting in an industrially viable process for YBCO wire production. Figure 15.9 shows an GADDS scan of a ~0.7 μm thick YBCO sample grown by PhAMOCVD in about 2 minutes time on an IBAD-prepared Hastelloy substrate (supplied by LANL). The scan shows good *a–b* align-

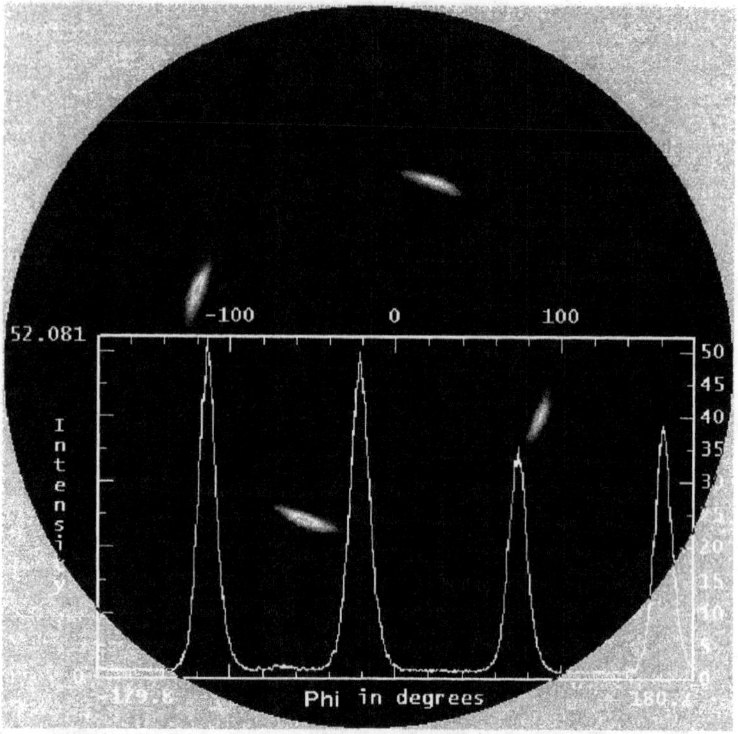

Figure 15.9. XRD pole figure for a 0.7 μm YBCO sample grown by PhAMOCVD on an IBAD processed Hastelloy substrate (courtesy of LANL) with FWHM of 12°, $T_c \sim 90$, and $J_c \sim 8 \times 10^5$ A/cm^2.

ment with 4-fold symmetry and FWHM of ~12°. The resultant critical current of the film is $J_c \sim 8 \times 10^6$ A/cm^2. An important point to note is that the high growth rates for PhAMOCVD can result in YBCO film thicknesses of more than 5 μm thereby effectively making the wires and tapes THICK film wires and tapes and thus enhancing the current carrying capacity of a YBCO wire. High performance YBCO films with thicknesses as high as 4.5 μm have been prepared by PhAMOCVD in less than ~10 minutes time.

Once thick film wire fabrication is initiated, quality control must become a concern under industrial production conditions. The MOCVD process is not exceptionally amenable to *in-situ* monitoring due to the high pressures of the reaction, and the high photon flux under photo-assisted MOCVD. There have, however, been previous attempts at *in-situ* monitoring including the measurement of optical reflectance oscillations (Zama et al., 1992a, 1992b), optical interference measurements (Higashiyama et al., 1992) from the growing films, and ultra-violet absorption spectroscopy (Musolf and Smith, 1999) and ultra-sonic measurement of the gas phase density (Mulsolf, 1997). The application of these techniques to photo-assisted MOCVD is not direct since most of them rely on optical measurements, and are thus complicated by the presence of high optical flux during growth. However, ultra-sonic measurement of gas density could be directly applied to the system to keep reactant flow constant. In addition, optical reflectance oscillation may be extracted from PhAMOCVD growth if a pulsed source is used and phase sensitive detection techniques are applied. The roughness of the growing surface complicates the collection of meaningful optical reflection oscillation data; hence the technique may not be directly applied. Similar pulsing and phase sensitive detection techniques could also be applied to optical interference measurements for the growing film.

It is well to note that in addition to monitoring growth rate and film thickness, the superconducting properties of the growing film would be appropriate to monitor. Contact measurements will probably not be appropriate under the need to maintain crystallographic integrity of all of the layers of the film heterostructure and the motion of the substrate through the reactor. Therefore, non-contact measurements in the normal state will need to be defined that have a one-to one correlation with critical current density. In any case, further work needs to be done to identify reliable quality control tools for YBCO thick film wire production.

The industrial viability of photo-assisted MOCVD can be further accentuated by the incorporation of double-sided growth of the buffer layers and YBCO film on a moving substrate, thus giving twice the current carrying capacity over a single side coated substrate. Past efforts in MOCVD growth have identified the possibility of double-sided growth (Lu et al., 1995; Ito et al., 1997). Both sequential and concurrent double-sided growths were undertaken with positive results with respect to materials quality (T_c and XRD measurements), but with the expected low growth rates from thermal MOCVD processing. Double-sided growths by photo-assisted MOCVD can alleviate the low growth rate problem, and result in enhanced current performance for YBCO coated conductor wires.

Finally, the present cost of precursors for YBCO and buffer layer growth needs to be significantly reduced. Current costs are at the >$15/gm level, and need to get to the ~$1/gm level to have MOCVD be cost effective when applied to YBCO thick film wire and tape production. It is well to note that prevailing precursor production is typically done in several 10's to several 100's gram batches. Economies of scale can greatly reduce these costs, and discussions with precursor suppliers have confirmed that increased demand to the several 10's of kg level could invariably reduce the cost of precursors by nearly a factor of 10. Such lower precursor costs coupled with costs of the metallic substrate at ~$0.2/m to $0.5/m should bring the cost of YBCO/buffer layer wire fabrication to the order of a dollar per meter. Such costs could make YBCO thick film wire produced by photo-assisted MOCVD extremely cost competitive with BSCCO powder-in-tube superconducting wire, and with the added benefit of much higher superconductor performance. Additional optimization of growth conditions and precursor use could even make the PhAMOCVD coated conductor wires cost competitive with copper wire.

Further additional work is required and is underway for advanced development of the photo-assisted MOCVD technique and it application to the fabrication of long lengths of thick film YBCO coated conductor wire. The promise however, of high quality YBCO superconducting films integrated with buffer layers and flexible metal substrates, and their production under industrially viable fabrication rates, assures the realization of high performance YBCO superconducting wire for electric power applications.

ACKNOWLEDGMENTS

The support of P.C. Chou, Y. Chen, X. Zhang, J. Zeng and Q. Zhong in the preparation and characterization of YBCO films by PhAMOCVD, and in the advancement of the PhAMOCVD technique is greatly acknowledged. Partial support for this work from the Texas Center for Superconductivity, from NASA through the Space Vacuum

Epitaxy Center, from the State of Texas through it Advanced Technology Development program, from Lockheed Martin/Oak Ridge National Laboratory through the Department of Energy, and from the R.A. Welch Foundation is gratefully acknowledged.

REFERENCES

Abrutis, A., Senateur, J.P., Weiss, F., Kubilius, V., Bigelyte, V., Saltyte, Z., Vengelis, B., and Juknas, A., 1998, Thin YBCO films on $NdGaO_3$(001) substrate grown by injection MOCVD, *Supercond. Sci. Technol.*, 10:959.

Becht, M. and Morishita, T., 1997, Preparation of buffer layers for HTS materials by MOCVD, *J. Alloys and Compounds*, 251:310.

Berry, A.D., Gaskill, D.K., Holm, R.T., Cukauskas, E.J., Kaplan, R., and Henry, R.L., 1988, Formation of high T_c superconducting films by organometallic chemical vapor deposition, *Appl. Phys. Lett.*, 52:1743.

Busch, H., Fink, A., and Muller, A., 1991, High-quality $YBa_2Cu_3O_{7-x}$ films on large area by chemical vapor deposition, *J. Appl. Phys.*, 79:2449.

Chen, Y.M., Ritums, D., Zhuang, W.W., Wu, N.J., and Ignatiev, A., 1998, Dielectric properties of (Ba,Sr)TiO_3 films deposited by Photo-Assisted Metal Organic Chemical Vapor Deposition, in: *Proc. of the 11th IEEE Intl. Symp. on Appl. of Ferroelectric*, 96CH35948:43.

Chern, C.S., Martens, J.S., Li, Y.Q., Galloios, B.M., Lu, P., and Kear, B.H., 1993, Metalorganic chemical vapor deposition of large area high quality $YBa_2Cu_3O_{7-x}$ films in a high speed rotating disc reactor, *Supercond. Sci. Technol.*, 6:460.

Chou, P.C., Zhong, Q., Li Q.L., and Ignatiev, A., 1994, Optimization of J_c for photo-assisted MOCVD prepared YBCO thin films by Robust Design, *Mat. Res. Symp. Proc.*, 335:279.

Chou, P.C., Zhong, Li, Q.L., Abazajian, K., Ignatiev, A., Wang, C.Y., Deal, E.E., and Chen, J.G., 1995, Optimization of J_c of YBCO thin films prepared by photo-assisted MOCVD through statistical Robust Design, *Physica C*, 254:93.

Dickenson, P.H., Geballe, T.H., Sanjurjo, A., Hildebrand, D., Craig, G., Zisk, M., Collman, J., Banning, S.A., and Sievers, R.E., 1989, Chemical vapor deposition of $YBa_2Cu_3O_{7-x}$ superconducting films, *J. Appl. Phys.*, 66:444.

Drozdov, A. and Troyanov, S., 1995, Barium diketonates as precursors for HTSC thin films: Structure and properties, *J. Physique, Colloque C5*, 5:373.

Ebihara, K., Fujishima, T., and Ikegami, T., 1993, Controlled superconducting transition of YBCO thin films prepared by plasma-enhanced MOCVD, *IEEE Trans. Appl. Supercond.*, 3:976.

Endo, K., Moriyasu, Y., Misawa, S., and Yoshida, S., 1991, MOMBE growth of superconducting oxide thin films, *Physica C*, 185:1995.

Eulenburg, A., Romans, E.J., and Pegrum, C.M, 1999, Pulsed laser deposition of YBCO and NBCO using Experimental Design, *Appl. Supercond.*, 9:2402.

Fenner, D.B., Viano, M., Fork, D.K., Connell, G.A.N., Boyce, J.B., Ponce, F.A., and Tramonatana, J.C., 1991, Reactions at the interface of thin films of Y–Ba–Cu and Zr-oxides with Si substrates, *J. Appl. Phys.*, 69:2176.

Frohlich, K., Souc, J., Rosova, A., Machajdik, D., Graboy, I.E., Svetchinkov, V.L., Figueras, A., and Weiss, F., 1997, Superconducting $YBa_2Cu_3O_{7-x}$ films prepared by aerosol metal organic chemical vapor deposition on Al_2O_3 substrates with CeO_2 buffer layer, *Supercond. Sci. Technol.*, 10:657.

Garcia, G., Casado, J., Llibre, J., Doudkowski, M., Santiso, J., Figueras, A., Schamm, S., Dorignac, D., Grigis, Ch., and Aguilo, M., 1995, Preparation of YBCO on YSZ layers deposited on silicon and sapphire by MOCVD, *J. Physique, Colloque 5*, 5:439.

Goto, T., Kuji, T., Jiang, Y.S., Moriya, M., Usami, K. and Kobayashi, T., 1999, Preparation of YBCO films on sapphire with CeO_2 depopsited by ion beam sputtering, *Appl. Supercond.*, 9:1653.

Goyal, A., Norton, D.P., Budai, J.D., Paranthaman, M., Specht, E.D., Kroeger, D.M., Christen, D.K., He, Q., Saffian, B., List, F.A., Lee, D.F., Martin, P.M., Klabunde, C.E., Hatfield, E., and Sikka, V.K., 1996,

High critical current density superconducting tapes by epitaxial deposition of $YBa_2Cu_3O_{7-x}$ thick films on biaxially textured metals, *Appl. Phys. Lett.*, 69:1795.

Hamaguchi, N., Gardiner, R., and Kirlin, P.S., 1991, MOCVD of TlBaCaCuO superconducting thin films: Structure–property–processing relationships, *Appl. Surface Sci.*, 48:441.

Higashiyama, K., Hirabayshi, I., and Tanaka, S., 1992, *In-situ* optical monitoring of growth rate and surface roughness of $YBa_2Cu_3O_{7-x}$ films in chemical vapor deposition, *Jpn. J. Appl. Phys.*, 31:L835.

Higashiyama, K., Ushida, T., Higa, H., Hirabayashi, I., and Tanaka, S., 1993, Growth mechanism of a-axis oriented $YBa_2Cu_3O_{7-x}$ thin film by laser-induced MOCVD, *Physica C*, 212:101.

Ichinose, A., Daniels, G., Yang, C.Y., Larbalestier, D., Kikuchi, A., Tachikawa, K., and Akita, S., 1999, Preparation and characterization of Y_2O_3 buffer layers and YBCO films on textured Ni tape, *IEEE Trans. Appl. Supercond.*, 9:2280.

Iijima, Y., Tanabe, N., Kohono, O., and Okeno, Y., 1992, In-plane aligned $YBa_2Cu_3O_{7-x}$ thin films deposited on polycrystalline metallic substrates, *Appl. Phys. Lett.*, 60:769.

Ignatiev, A., Zhong, Q., Chou, P.C., Zhang, X., Liu. J.R., and Chu, W.K., 1996, Large J_c enhancement by ion-irradiation for thick YBCO films prepared by photo-assisted MOCVD, *Appl. Phys. Lett.*, 70:1474.

Ignatiev, A., Chou, P.C., Zhong, Q., Zhang, X., and Chen, Y.M., 1998a, Photo-assisted MOCVD growth of YBCO thick films for wire applications, *Appl. Supercond.*, 4:455.

Ignatiev, A., Chou, P.C., Zhong, Q., Zhang, X., and Chen, Y.M., 1998b, Photo-assisted fabrication of YBCO thick films and buffer layers on flexible substrates for wire applications, *Intl. J. Mod. Phys. B*, 2:3162.

Ito, Y., Yoshida, Y., Iwata, M., Takai, Y., and Hirabayashi, I., 1997, Preparation of double-sided $YBa_2Cu_3O_{7-x}$ film by hot-wall type MOCVD, *Physica C*, 288:178.

Klippe, L. and Wahl, G., 1997, Single-source MOCVD of Y-, Ba- and Cu oxides from thd-precursors, *J. Alloys and Compounds*, 251:249.

Komatsu, Y., Sato, T., Ito, S., and Akashi, K., 1999a, Preparation of YBCO/ZrO_2 thin films on Si by MOCVD using a mode converting type of microwave plasma apparatus, *Thin Sol. Films*, 341:132.

Komatsu, Y., Sato, T., Ito, S., and Akashi, K., 1999b, Preparation of YBCO /ZrO_2 thin films on Si by MOCVD using a mode converting type of microwave plasma apparatus, *Thin Sol. Films*, 341:132.

Krellmann, M., Selbmann, D., Schmatz, U., and Weiss, F., 1997, YSZ buffer layers on technical substrates for HTSC current applications by AAMOCVD, *J. Alloys and Compounds*, 251:307.

Li, Y.Q., Zhao, J., Chern, C.S., Huang, W., Kulesha, G.A., Lu, P., Gallois, B., Norris, P., Kear, B., and Cosandey, F., 1991, High critical current densities in $YBa_2Cu_3O_{7-x}$ thin films formed by metalorganic chemical vapor deposition at 730°C, *Appl. Phys. Lett.*, 58:648.

List, F.A., Goyal, A., Paranthaman, M., Norton, D.P., Specht, E.D., Lee, D.F., and Kroger, D.M., 1998, High J_c YBCO films on biaxially textured Ni with oside buffer layers deposited using electron beam evaporation and sputtering, *Physica C*, 302:87.

Lu, Z., Truman, J.K., Johnasson, M.E., Zhang, D., Shih, C.F., and Liang, G.C., 1995, Large area double-sided $YBa_2Cu_3O_{7-x}$ films grown by single-source metal organic chemical vapor deposition, *Appl. Phys. Lett.*, 67:712.

Maraitakis, E., Anagnostou, M., Pissas, M., Psyharis, V., Niarchos, D., and Strantakos, G., 1998, Deposition of $YBa_2Cu_3O_{7-x}$ thin films over large areas with simple sputtering technique for microwave applications, *Supercond. Sci. Technol.*, 11:686.

Marks, T.J., Belot, J.A., Reedy, C.J., McNeely, R.J., Strudebaker, D.B., Neumayer, D.A., and Stern, C.L., 1997, MOCVD precursor desing issuess. Recent advances in the chemistry and vapor pressure characteristics of Ba(hexafluoroacetylacetonate)$_2$ •polyether complexes, *J. Alloys and Compounds*, 251:243.

Mizushima, Y. and Hirabayashi, I., 1994, Fabrication of $YBa_2Cu_3O_{7-x}$ film by laser chemical vapor deposition using Ba(hfa)$_2$•tetraglyme, *Physica C*, 235:577.

Nagai, H., Yoshida, Y., Ito, Y., Taniguchi, S., Hirabayashi, I., Matsunami, N., and Takai, Y., 1997, Preparation of $YBa_2Cu_3O_{7-x}$ films by metal organic chemical vapor deposition using liquid-state non-fluorinated sources, *Supercond. Sci. Technol.*, 10:213.

Noh, D.W., Gallois, B., Chern, C.S., Caracciolo, R., Kera, B.H., Zawadzki, P.A., Tompa, G.S., and Norris, P.E., 1989, Metalorganic chemical vapor deposition of superconducting $YBa_2Cu_3O_{7-x}$ in a high speed rotating disk reactor, *J. Appl. Phys.*, 66:5099.

Norton, D.P., Goyal, A., Budai, J.D., Christen, D.K., Kroeger, D.M., Specht, E.D., He, Q., Saffian, B., Paranthaman, M., Klabunde, C.E., Lee, D.F., Sales, B.C., and List, F.A., 1996, Epitaxial $YBa_2Cu_3O_{7-x}$ on biaxially textured nickel (001): An approach to superconducting tapes with high critical current density, *Science*, 274:755.

Otway, D.J., Obi, B., and Rees, W.S., 1997, Precursors for chemical vapor deposition of YBCO, *J. Alloys and Compounds*, 252:254.

Panson, A.J., Charles, R.G., Schmidt, D.N., Szendon, J.R., Machiko, G.J., and Braginski, A.I., 1988, Chemical vapor deposition of $YBa_2Cu_3O_{7-x}$ using metalorganic chelate precursors, *Appl. Phys. Lett.*, 53:1756.

Park, C., Norton, D.P., Christen, D.K., Verebelyi, D.T., Feentstra, R., Budai, J.D., Goyal, A., Lee, D.K., Specht, E.D., Kroger, D.M., and Paranthaman, M., 1999, Long length fabrication of YBCO on rolling assisted biaxially textured substrates (RABiTS) using pulsed laser deposition, *Appl. Supercond.*, 9:2276.

Richards, B.C., Cook, S.L., Pinch, D.L., Andrews, G.W., Lengeling, G., Schulte, B., Jurgensen, H., Shen, Y.Q., Vase, P., Freltoft, T., Spee, C.I.M.A., Linden, J.L., Hitchman, M.L., Shamlian, S.H. and Brown, A., 1995a, MOCVD of high quality $YBa_2Cu_3O_{7-x}$ thin films using a fluorinated barium precursor, *Physica C*, 252:229.

Richards, B.C., Cook, S.L., Pinch, D.L., and Andrews, G.W., 1995b, MOCVD of high quality $YBa_2Cu_3O_{7-x}$ thin films using novel fluorinated and non-fluorinated precrusors, *J. Physique, Colloque C5*, 5:407.

Ritums, D.L., Liu, D., Wu, N.J., Zhong, Q., Chen, Y.M., Zhang, X., Chou, P.C., and Ignatiev, A., 1996, Epitaxially deposited $SrVO_3$ conducting films by laser ablation and MOCVD, in: *Proc. of 10th IEEE Intl. Symp. on the Appl. Ferroelectrics*, 96CH35948:417.

Salazar, K.V., Ott, K.C., Dye, R.C., Hubbard, K.M., Peterson, E.J., and Coulter J.Y., 1992, Aerosol assisted chemical vapor deposition of superconducting $YBa_2Cu_3O_{7-x}$, *Physica C*, 98:303.

Schulte, B., Maul, M., Becker, W., Schlosser, E.G., Elschner, S., Haussler, P., and Adrian, H., 1991, Carrier gas-free chemical vapor deposition technique for in situ preparation of high quality $YBa_2Cu_3O_{7-x}$ thin films, *Appl. Phys. Lett.*, 59:869.

Senateur, J.P., Felten, F., Pignard, S., Weiss, F., Arbrutis, A., Bigelyte, V., Teiserskis, A., Saltyte, Z., and Vengalis, B., 1997, Synthesis and characterization of YBCO thin films grown by injection MOCVD, *J. Alloys and Compounds*, 251:288.

Stringfellow, G.B., 1989, *Organometallic Vapor-Phase Epitaxy: Theory and Practice*, Academic Press, San Diego.

Sugimoto, T., Yoshida, M., Yamaguchi, K., Yamada, Y., Sugawara, K., Shiohara, Y., and Tanaka, S., 1991, Fabrication and characterization of Bi–Sr–Ca–Cu–O MOCVD thin films, *J. Crystal Growth*, 107:692.

Takahishi, N., Koukitu, A., and Seki, H., 2000, Growth and characterization of $YBa_2Cu_3O_x$ and $NdBa_2Cu_3O_x$ superconducting thin films by mist microwave-plasma chemical vapor deposition using a CeO_2 buffer layer, *J. Mater. Sci.*, 35:1231.

Tasaki, Y., Yoshizawa, S., and Satoh, M., 1998, New method to increase solid precursor vaporization for metalorganic chemical vapor deposition, *Jpn. J. Appl. Phys.*, 37:649.

Thieme, C.L.H., Fleshier, S., Buczek, D.M., Jowett, M., Fritzmeier, L.G., Arendt, P.N., Foltyn, S.R., Coulter, J.Y., and Willis, J.O., 1999, Axial strain dependence at 77 K of the critical current of thick $YBa_2Cu_3O_{7-x}$ films on Ni-alloy substrtes with IBAD buffer layers, *IEEE Trans. Appl. Supercond.*, 9:1494.

Tsuruoka, T., Kawasaki, R., and Abe, H., 1989, Y–Ba–Cu–O film growth by OMCVD using N_2O, *Jpn. J. Appl. Phys.*, 28:1800.

Ushida, T., Higa, H., Higashiyama, K., Hirabayashi, I., and Tanaka, S., 1991, Preparation of a-axis oriented $YBa_2Cu_3O_{7-x}$ films by laser metalorganic chemical vapor deposition, *Appl. Phys. Lett.*, 59:860.

Watson, I.M., Atwood, M.P., and Haq, S., 1994, Investigation of barium β-diketonate complexes used in chemical vapor deposition of high T_c oxide films, *Supercond. Sci. Technol.*, 7:672.

Weiss, F., Schmatz, U., Pish, A., Felten, F., Pignard, S., Senateur, J.P., Frolich, K., Seldmann, D., and Klippe, L., 1997, HTS films by innovative MOCVD processes, *J. Alloys and Compounds*, 251:264.

Wu, X.D., Dye, R.C., Meunchausen, R.E., Flotyn, S.R., Maley, M., Rollet, A.D., Garcia, A.R., and Nogar, N.S., 1991, Epitaxial CeO_2 films as buffer layers for high temperature superconducting thin films, *Appl. Phys. Lett.*, 58:2165.

Wu, X.D., Foltyn, S.R., Arendt, P.N., Blumenthal, W.R., Campbell, H., Cotton, J.D., Coulter, J.Y., Hults, W.L., Maley, M.P., Safar, H.F., and Smith, J.L., 1995, Properties of $YBa_2Cu_3O_{7-x}$ thick films on flexible buffered metallic substrates, *Appl. Phys. Lett.*, 67:2397.

Yamane, H., Kurosawa, H., and Hirai, T., 1988, Preparation of $YBa_2Cu_3O_{7-x}$ films by chemical vapor deposition, *Chem. Lett.*, 939.

Yamane, H., Kurosawa, H., Hirai, T., Watanabe, K., Iwasaki, H., Kobayashi, N., and Muto, Y., 1989, High critical-current density of Y–Ba–Cu–O superconducting films prepared by CVD, *Supercond. Sci. Technol.*, 2:115.

Yamasaki, H., Endo, K., Nakagawa, Y., Umeda, M., Kosaka, S., Misawa, S., Yoshida, S., and Kajimura, K., 1992, Critical current density of high quality Bi–Sr–Ca–Cu–O thin films prepared by metalorganic chemical vapor deposition, *J. Appl. Phys.*, 72:2951.

Zama, H. and Oda, S., 1991, Low-temperature chemical vapor deposition of $YBa_2Cu_3O_{7-x}$ films, *Physica C*, 185:2103.

Zama, H., Miyake, T., Hattori, T., and Oda, S., 1992a, Preparation of YBCO superconducting films by low-temperature chemical vapor deposition using β-diketonate complex and N_2O, *J. Appl. Phys.*, 31:3839.

Zama, H., Sakai, K., and Oda, S., 1992b, *In-situ* monitoring of optical reflectance oscillation in layer-by-layer chemical vapor deposition of oxide superconductor films, *Jpn. J. Appl. Phys.*, 31:L1243.

Zama, H., Tanaka, N., and Morishita, T., 1998, Evaluation of a new Ba precursor, $Ba(DPM)_2$-2tetraen, for MOCVD of oxide superconductors, *Mat. Sci. and Engineering*, B54:104.

Zhang, K., Boyd, E.P., Kwak, B.S., Wright, A.C., and Erbil, A., 1989a, Metalorganic chemical vapor deposition of TlBaCaCuO superconducting thin films on sapphire, *Appl. Phys. Lett.*, 55:1258.

Zhang, K., Kwak, B.S., Boyd, E.P., Wright, A.C., and Erbil, A., 1989b, *C*-axis oriented $YBa_2Cu_3O_{7-x}$ superconducting films by metalorganic chemical vapor deposition, *Appl. Phys. Lett.*, 54:380.

Zhang, J.M., Wessels, B.W., Tonge, L.M., and Marks, T.J., 1990, Formation of oriented high T_c superconducting Bi–Sr–Ca–Cu–O thin films on silver substrates by organometallic chemical vapor deposition, *Appl. Phys. Lett.*, 56:976.

Zhao, J., Li, Y.Q., Chern, C.S., Lu, P., Norris, P., Gallios, B., Kear, B., Cosandey, F., Wu, X.D., Muenchausen, R.E., and Garrison, S.M., 1991, High-quality $YBa_2Cu_3O_{7-x}$ thin films by plasma-enhanced metalorganic chemical vapor deposition at low temperatures, *Appl. Phys. Lett.*, 59:1254.

Zhong, Q., 1996, High rate growth of YBCO films by photo-assisted metal organic chemical vapor deposition, Ph.D. thesis, University of Houston.

Zhong, Q., Chou, P.C., Li, Q.L., Taraldsen, G.S., and Ignatiev, A., 1995, High-rate growth of purely *a*-axis oriented YBCO high-T_c thin films by photo-assisted MOCVD, *Physica C*, 246:288.

Zhou, G., Meng, G., Schnider, R., Sarma, B., and Levy, M., 1994, Vaporization of a mixed precursor in chemical vapor deposition of YBCO films, *J. Superconductivity*, 7:235.

Chapter 16

LPE PROCESSING FOR COATED CONDUCTOR

Teruo Izumi and Yuh Shiohara

Superconductivity Research Laboratory
International Superconductivity Technology Center
10-13 Shinonome 1-chome, Koto-ku
Tokyo 135-0062
Japan

16.1 INTRODUCTION

Several kinds of processings have been tried to fabricate the coated conductors (Goyal et al., 1997, 1999; Holesinger et al., 2000; Thieme et al., 2000; Watanabe et al., 2001; Yoshino et al., 2001; Ohmatsu et al., 2001; Iijima et al, 2001). For tape application, long length stability not only on the superconducting characteristics but also on the mechanical property etc. is required. Therefore, the fabrication of the superconducting layer on the metal tape is considered to realize long tape application. On the other hand, not only high critical current density (J_c) but high J_e (engineering J_c) and high critical current (I_c) are required for real application. In order to realize high J_e and I_c, a thick superconducting layer with high J_c is expected. The Liquid Phase Epitaxy (LPE) process is one of the strong candidates to fabricate thick films with maintaining high J_c property (Miura et al., 1997). The LPE process for the growth of RE123 film was developed by modifying the SRL-CP method (Kitamura et al., 1995; Ishida et al., 1997; Yamada and Shiohara, 1993). The detail procedure can be found in the other article (Ishida et al., 1997). Briefly, Y_2BaCuO_5 (Y211) powder was placed at the bottom of the yttria crucible and the oxide powder of Ba–Cu–O was filled on the Y211 layer. The crucible was heated to obtain the complete melt of the Ba–Cu–O solvent. The temperature at the liquid surface is controlled to be lower than the equilibrium peritectic temperature (T_p), at which the $YBa_2Cu_3O_y$ (Y123) phase forms from Y211 and liquid, and the temperature difference along the vertical direction is applied to get the supersaturation for the growth of Y123 at the surface of the melt. The LPE film can be grown by the dipping the substrate into the liquid with a seed film, which is deposited by the vapor process. The LPE films even with 8 μm in thickness on MgO single crystal substrates revealed over 10^5 A/cm^2 of J_c at 77 K (Miura et al., 1997). In the LPE process, the crystallinity can be improved with increasing thickness, which is the main reason for maintaining high J_c even in thick film. Additionally, higher growth rate can be expected in the LPE process comparing with that of vapor process, since it is the thermodynamically equilibrated system at high temperature. Even in Y-system, which is a relatively low growth rate system, the order of μ/min can be easily obtained in the LPE method.

However, there are several problems to be solved for applying the LPE technique to fabricate a film on a long metal tape. The most serious problem is that the solvent in the LPE process for RE123 growth is very reactive with almost all metals except with silver. Then, the process to overcome this problem has to be developed at first. In order to prevent the reaction, a buffer layer growth on a metal tape from a solution saturated by the buffer layer materials, which is called "saturated system" is found to be effective (Kakimoto et al., 2000a, 2000b; Hobara et al., 2000). In this chapter, the recent progress of the development of LPE process using the saturated system for coated conductor is reviewed.

16.2 PREVENTATION OF REACTION

In order to avoid the reaction between metal tape and liquid phase, the buffer layer should be prepared between the metal tape substrate and the LPE layer. Several different kinds of oxide materials were considered for the buffer layer. Consequently, MgO was selected as a strong candidate, because it had been used as the substrate material for LPE growth of the Y123 phase, and the effective technique of ISD (Inclined Substrate Deposition) method to obtain an aligned MgO layer on a non-aligned metal tape had been already reported (Metzger et al., 2000; Hasegawa et al., 2001). The MgO buffer layer was deposited on a Hastelloy tape by the e-beam evaporation technique. The substrate with the buffer layer was dipped into the liquid for LPE process to confirm the ability for protection of Hastelloy from the liquid. The liquid used in this study was prepared by the procedure similar to that mentioned above. The oxide powder with its cationic ratio of Ba : Cu = 3 : 5 was filled on the Y211 layer. The surface and the bottom temperatures were controlled at 1000°C and 1010°C, respectively. This is a general growth condition for the Y123 growth. This liquid is called as an MgO-free liquid. After dipping of the sample, the dipped part disappeared as shown in Figure 16.1. In order to clarify the reason for the disappearance during dipping, the solubility of MgO to the liquid was measured. The excess amount of MgO powder was added into the solvent. Then, a small amount of the liquid was picked up at different temperatures to analyze the liquid compositions. The concentrations of Mg element in the samples were analyzed by the ICP (Inductively Coupled Plasma) method. Figure 16.2 shows the solubility of MgO in the liquid at different temperatures. A small but a finite value for solubility of MgO can be recognized in this figure. According to these results, the disappearance of the substrate can be explained by not melting but by dissolution of the MgO buffer layer. Therefore, it is expected that a saturation of liquid by MgO can prevent the dissolution. Then, the dipping of the Hastelloy with the MgO buffer layer into the MgO-saturated liquid was carried out. The MgO-saturated liquid was prepared by adding the excess amount of the MgO powder into the MgO-free liquid. In this case, the MgO buffer layer was survived and protects the Hastelloy tape. The cross-section of the dipped sample was shown in Figure 16.3. Through the investigation, it was clarified that the combination of MgO buffer layer and MgO-saturated liquid can prevent the reaction between liquid and Hastelloy tape.

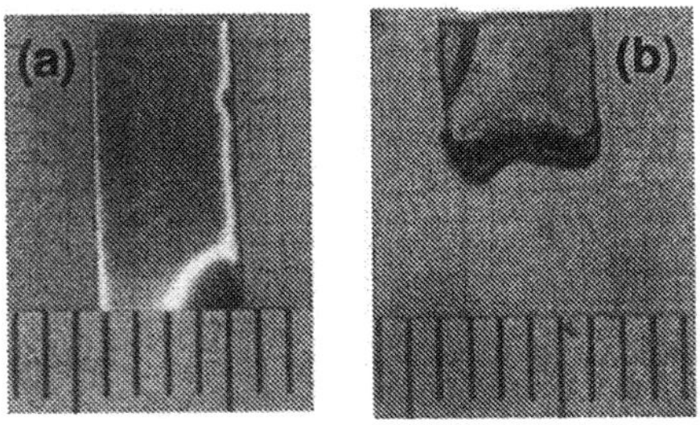

Figure 16.1. Outwards of the Hastelloy tapes covered with MgO buffer layer before/after dipping into Ba–Cu–O solution for LPE process: (a) before dipping; (b) after dipping.

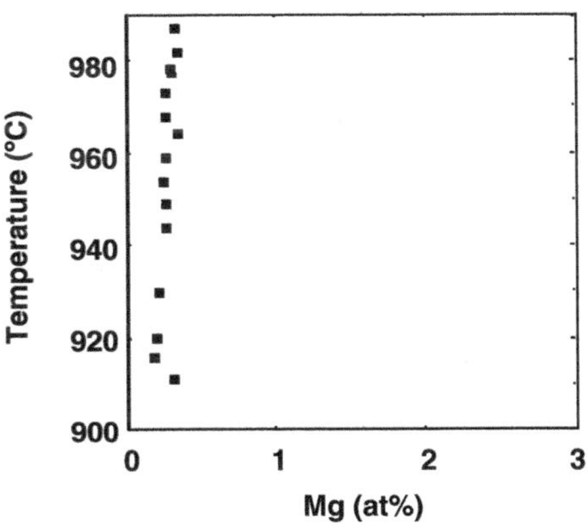

Figure 16.2. Mg-solubility in the $Ba_3Cu_5O_x$ melt.

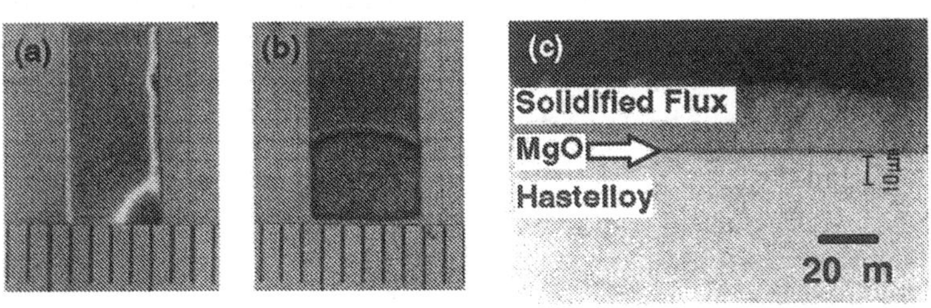

Figure 16.3. Outwards of the Hastelloy tapes covered with MgO buffer layer before/after dipping into Ba–Cu–O solution saturated with MgO: (a) before dipping and (b) after dipping; (c) cross-section of the Hastelloy tape with MgO after dippinginto the solution saturated with MgO.

On the other hand, NiO material is an another candidate for the buffer layer, because the SOE (Surface Oxidization Epitaxy) (Watanabe et al., 2001; Matsumoto et al., 2000) technique can be applied to obtain the aligned structure. Then, the similar saturated system was tried to apply for the case of the NiO buffer layer. The surface oxidized Ni tape was dipped into the liquid, which was saturated with NiO. The cross section of the dipped sample is shown in Figure 16.4. From this figure, it can be recognized that the NiO layer protected the Ni metal tape. This indicates that combination of NiO and NiO saturated liquid is also effective to prevent the reaction.

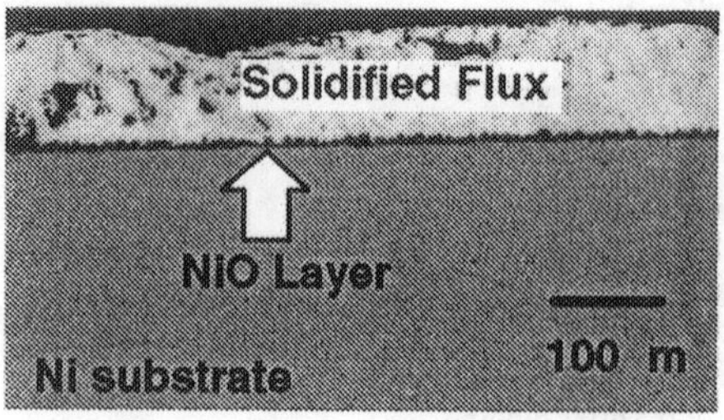

Figure 16.4. Cross-section of the Ni tape covered with NiO after dipping into the solution saturated with NiO observed by compositional images.

16.3 GROWTH IN MgO SATURATED SYSTEM

16.3.1 Influence of MgO Addition for Growth of Y123 Phase

In order to investigate influence of the MgO addition to the liquid for the growth of the Y123 phase, the solubility of Y_2O_3 in the Ba–Cu–O liquid was measured for two different kinds of liquids which were MgO-free and -saturated liquids. Then, a small amount of the liquid was taken for the samples at different temperatures. The concentrations of Y element in the samples were analyzed by the ICP method. Figure 16.5 shows the results of the solubility of Y in the two different liquids at different temperatures. In this figure, the negligible difference in the two solubility-curves was seen for the liquids with and without MgO. Additionally, the T_p in the both systems, which can be recognized from the bending point of two curves of high and low temperature regions, are almost the same. This result reveals that a similar supersaturation, which is a driving force for the Y123 growth, can be expected under the same growth conditions. The expectation could be confirmed by the actual growth on an MgO single crystal from MgO-saturated liquid under the same condition as that in the MgO-free system. An MgO single crystal substrate with a Y123 seed film, which is deposited by a pulsed laser deposition process, was dipped into the MgO-saturated liquid for 5 min. The average growth rate can be estimated as about 1 μm/min by the thickness (about 5 μm), which was obtained by SEM observation of the cross section of the sample, and the growth time (5 min). This value is almost the same as that in the MgO-free system.

16.3.2 Growth of Double Layered LPE Film

In the investigations as mentioned above, the growth of the Y123 phase from the MgO-saturated liquid was confirmed. However, the LPE film reveals relatively low T_c value of 40 K. This is caused by substitution of Mg for the Cu-site in the Y123 phase. In order to obtain higher T_c and J_c, a further LPE layer without MgO (2nd LPE layer) on the MgO-substituted layer (1st LPE layer) is required. Then, the 2nd LPE layer was grown on the 1st LPE layer. The growth condition was the same as that for the growth of the 1st LPE layer except that no MgO was added in the liquid. The double layered LPE films was grown on an MgO substrate as shown in Figure 16.6, which revealed a high T_c value of 90 K as shown in Figure 16.7. Furthermore, the double layered LPE construction was tried to grow on a metal tape. The MgO buffer layer was deposited on a Hastelloy tape by an e-beam evaporation technique. On the buffer layer, the thin film of the Y123 phase was deposited as a seed film. The substrate was dipped into the MgO-saturated liquid and successively dipped into the MgO-free liquid. The growth of the double layered LPE film on Hastelloy tape can be confirmed as shown in Figure 16.8.

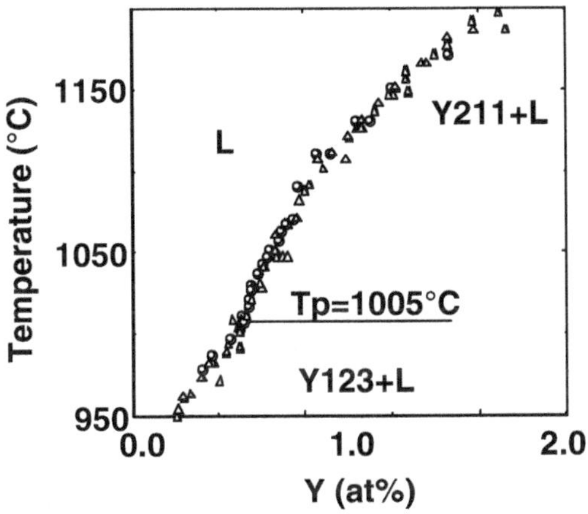

Figure 16.5. Y-solubility at different temperatures in the $Ba_3Cu_5O_x$ melt without MgO (○) and in that saturated with MgO (△).

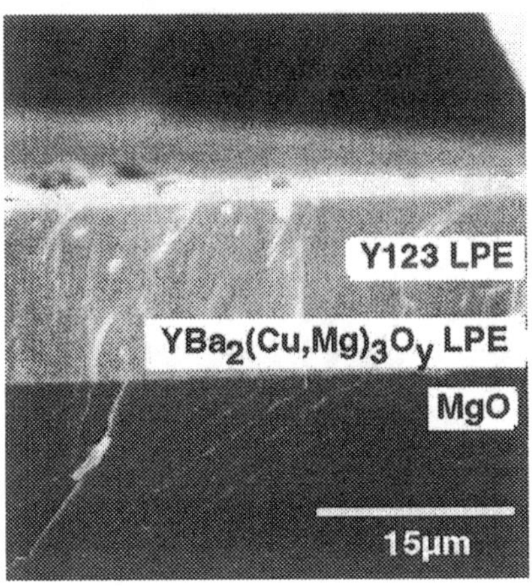

Figure 16.6. Cross-section views of the double layered LPE film grown on an MgO single crystal substrate. The 1st LPE layer on an MgO substrate is the $YBa_2(Cu,Mg)_3O_y$ phase and the 2nd LPE layer is the $YBa_2Cu_3O_y$ phase.

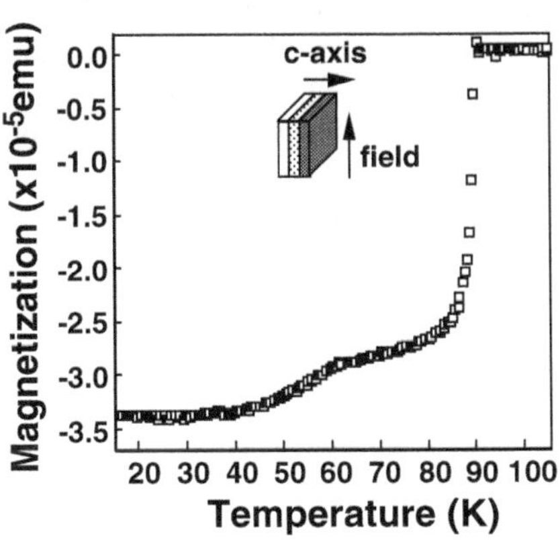

Figure 16.7. Magnetic measurement for superconductivity of the double layered LPE film grown on an MgO single crystal substrate. The direction of the magnetic field is in parallel with the film surface.

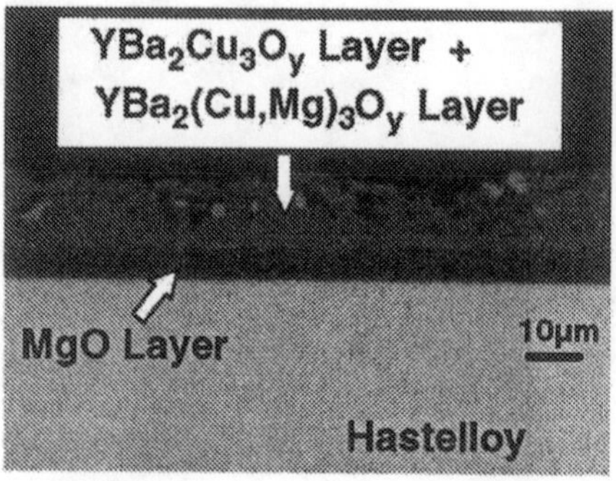

Figure 16.8. Cross-sectional view of the double layered LPE film grown on a Hastelloy tape substrate. The 1st LPE layer on the MgO buffer layer is the $YBa_2(Cu,Mg)_3O_y$ phase and the 2nd LPE layer is the $YBa_2Cu_3O_y$ phase.

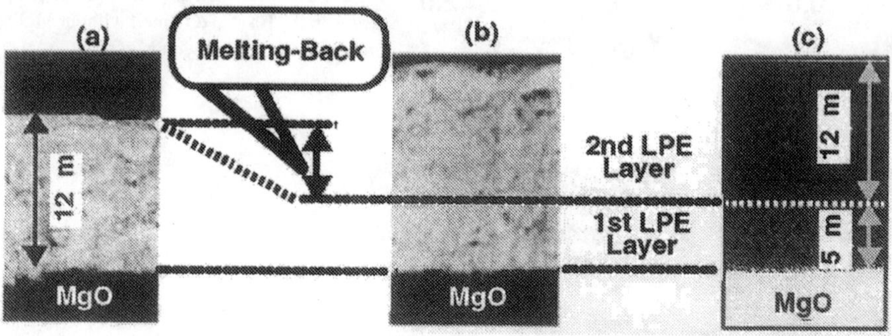

Figure 16.9. Cross-sectional views of the samples grown on an MgO single crystal substrate: (a) SEM observation of the single 1st layer of the $YBa_2(Cu,Mg)_3O_y$ phase grown from MgO-saturated liquid; (b) SEM obseravation of the double layered LPE film; (c) compositional mapping image of Mg element analysed by EPMA for the double layered film.

16.3.3 Melting-back of First Layer

Although the double layered construction was realized on the Hastelloy substrate as mentioned above, the samples sometimes disappeared after dipping for the growth of the 2nd layer. The reason for the disappearance phenomenon can be found in the compositional mapping analysis by Electron Probe Microanalysis (EPMA). Figure 16.9 shows the thickness change of the 1st layer due to the dipping for the growth of the 2nd layer. The thickness of the 1st layer in the double layered LPE sample is thinner than that of the sample before the 2nd dipping. This phenomenon can be understood by the dissolution of the 1st layer during the 2nd dipping.

The dissolution of the 123 solid solution (123ss) into the liquid can be discussed as,

$$123\text{ss(s)} = 123(1\text{ wt\%, liq.}). \tag{1}$$

123ss(s) and 123(1 wt%, liq.) represent a solid state of the Y123ss and a dilute solution state in liquid. The free energy change of this reaction (ΔG) was expressed by,

$$\Delta G = \Delta G^\circ + RT \ln\left(\frac{a_{123(1\text{ wt\%,liq.})}}{a_{123\text{ss(s)}}}\right), \tag{2}$$

where ΔG° is the standard free energy change for the reaction, R is gas constant, T is temperature, $a_{123\text{ss(s)}}$ and $a_{123(1\text{ wt\%,liq.})}$ are activities for the solid state of the 123ss and the dilute solution state in liquid, respectively. If the value of ΔG is negative, the dissolution of the 123ss proceeds. Here, ΔG° can be divided into two parts as,

$$\Delta G^\circ = \Delta G_{\mathrm{m}} + \Delta G_{\mathrm{d}} = (\Delta H_{\mathrm{m}} - T\Delta S_{\mathrm{m}}) + (RT \ln A\gamma^\circ), \quad (3)$$

where ΔG_{m} and ΔG_{d} are the free energy change for melting of the 123ss phase and the standard free energy change for dissolution from pure liquid to dilute one, respectively. ΔH_{m} and ΔS_{m} are the molar enthalpy and entropy changes for melting of the 123ss. Both should be positive in this situation. A is a constant for standard state change from pure materials (Raoult's law) to the 1 wt% dilute solution. γ° is an activity coefficient in Henry's law. In general, the products of $A\gamma^\circ$ is smaller than unity. Here, two approaches can be considered in order to suppress the dissolution of the 1st layer. One is the selection of lower growth temperature for the 2nd layer than that of the 1st layer. This leads to the decrease of the second and the third terms in Equation (3), which results in the increase of ΔG° and consequently ΔG should increase. The other approach is the choice of the 1st layer 123 phase with ΔH_{m} higher than that of the 2nd layer 123 phase. This makes the first term in Equation (3) increase, which leads to increase ΔG. For example, Nd- and Sm-system 123 have larger ΔH_{m} than Y123[19]. The above expectation was confirmed by using the combination of $Nd_{1+x}Ba_{2-x}(Cu,Mg)_3O_y$ (Nd123-Mg) material and Y123 for the 1st and the 2nd LPE layers respectively. The Nd123-Mg LPE layer was grown on an MgO single crystal substrate from MgO-saturated liquid. Then, Y123 layer without MgO was grown on it. The observations of the cross-section were shown in Figure 16.10. It can be recognized that the melting-back of the 1st layer was much suppressed and negligible. Additionally, the sample revealed a high T_{c} value over 90 K and a high J_{c} value of 10^5 A/cm^2 at 77 K as shown in Figure 16.11. Consequently, the suitable construction for the LPE process to the coated conductor in Mg-saturated system was developed as RE123(LPE)/RE123-Mg(LPE)/MgO/Hastelloy. Here, the T_{p} of the 2nd LPE layer should be lower than that of the 1st LPE layer.

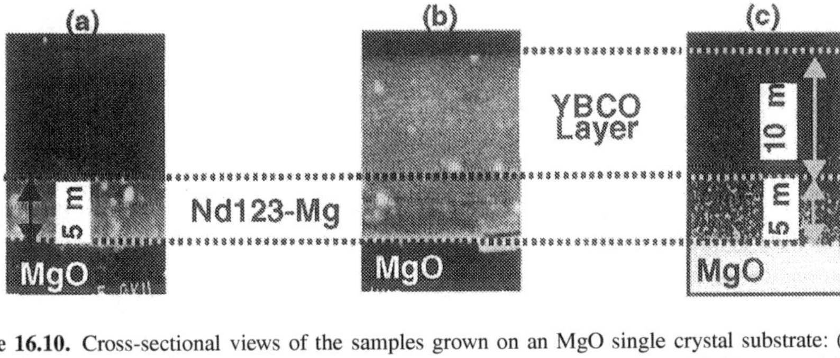

Figure 16.10. Cross-sectional views of the samples grown on an MgO single crystal substrate: (a) SEM observation of single 1st layer of the $NdBa_2(Cu,Mg)_3O_y$ phase grown from MgO-saturated liquid at around 1060°C; (b) SEM obseravation of double layered LPE film, where the 2nd LPE layer of the Y-123 phase was grown at around 990°C; (c) compositional mapping image of Mg element analysed by EPMA for the double layered film.

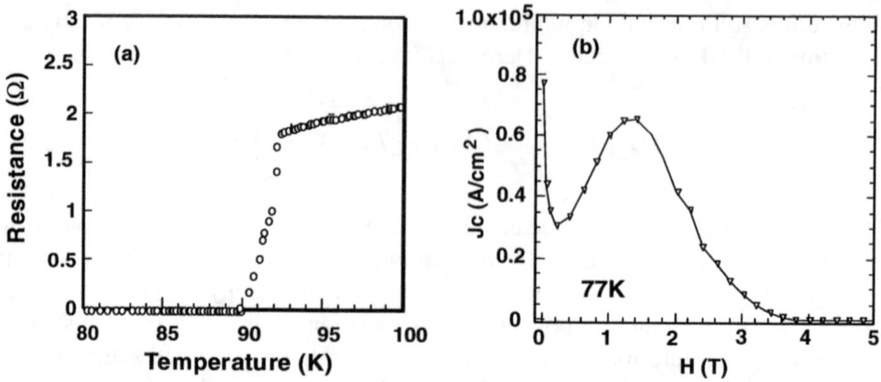

Figure 16.11. Superconducting characteristics of the double layered LPE film grown on an MgO single crystal substrate. The 1st LPE layer is the $NdBa_2(Cu,Mg)_3O_y$ phase and the 2nd LPE layer is Y-123 phase: (a) transport measurement by the four-probe method for T_c; (b) magnetic measurement for J_c-B property.

16.4 GROWTH IN NiO SATURATED SYSTEM

16.4.1 Influence of NiO Addition for Growth of Y123 Phase

According to DTA analysis of Y123 powders with different amounts of NiO, it was found that the T_p of the Y123 phase decreases with increasing NiO contents as shown in Figure 16.12, although there is little influence for T_p in the MgO-saturated system. Additionally, the crystallization temperature of the low temperature phase of Ba–Cu–Ni–O increases due to NiO-addition as shown in Figure 16.13. These results indicate that the possible temperature range for growth of the Y123 phase became small by NiO addition. Accordingly, a Y123 LPE film is difficult to grow from the NiO-saturated liquid. In order to solve this problem, the selection of RE of the RE123 phase, which has a higher T_p than that of Y123, for the 1st layer in NiO saturated system. It was considered to be effective, because the influence of the different kinds of RE elements might be little for the lower temperature phase crystallization temperature due to no RE in the phase. Then the efficiency was confirmed using the Nd-system as the high T_p phase by DTA analysis as shown in Figure 16.14. The clear temperature difference can be observed between two exothermic peaks, which correspond the crystallizing temperatures of the $NdBa_2(Cu,Ni)_3O_y$ (Nd123-Ni) and the Ba–Cu–Ni–O phases respectively. This represents a stable growth of Nd123-Ni can be expected. The similar temperature difference for the expected stable growth of Sm123-Ni was also confirmed in the Sm-system.

16.4.2 Growth of Double Layered LPE Films

Based on the experimental results as mentioned above, the Nd123-Ni phase layer was grown from NiO-saturated liquid on an MgO single crystal substrate by LPE process. Additionally, a Y123 LPE layer without NiO was also grown on a Nd123-Ni LPE layer in order to realize the high superconducting performance. Figure 16.15 shows the cross section of the double layer on the MgO single crystal substrate. T_h is sample revealed a T_c value of 90 K. Furthermore, the double layer LPE construction was also realized even on the metal tape of surface oxidized Ni as shown in Figure 16.16. The measurement of the temperature dependence of the resistivity revealed the T_{CO} value of 85 K as shown in Figure 16.17. Consequently, the suitable

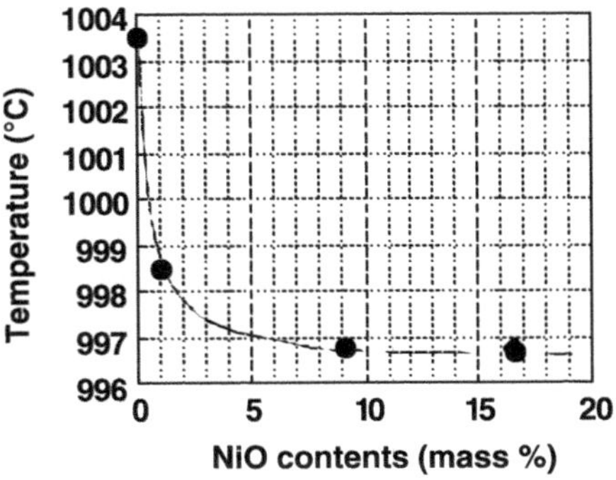

Figure 16.12. Added NiO amount dependence on the peritectic temperature of the $YBa_2Cu_3O_y$ phase measured by the DTA method.

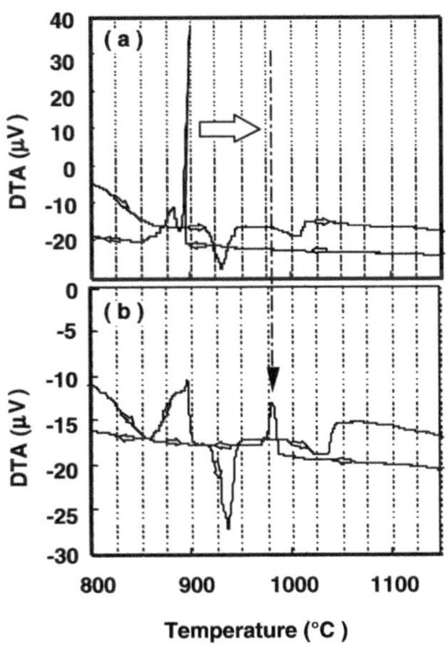

Figure 16.13. DTA data on the Ba–Cu–O powder with and without NiO: (a) without NiO and (b) with NiO. The crystallization temperature of Ba–Cu–Ni–O phase becomes higher due to NiO addition.

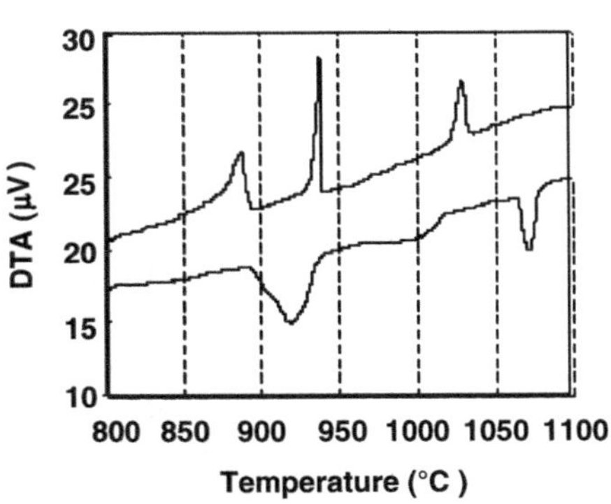

Figure 16.14. DTA data of the powder which consists of Sm123 : $3BaCuO_2 + 2CuO$: NiO = 1 : 1 : 1.5. The exothermic peaks of (α) and (β) correspond to the crystallizations of Ba–Cu–Ni–O and Sm-123 phases.

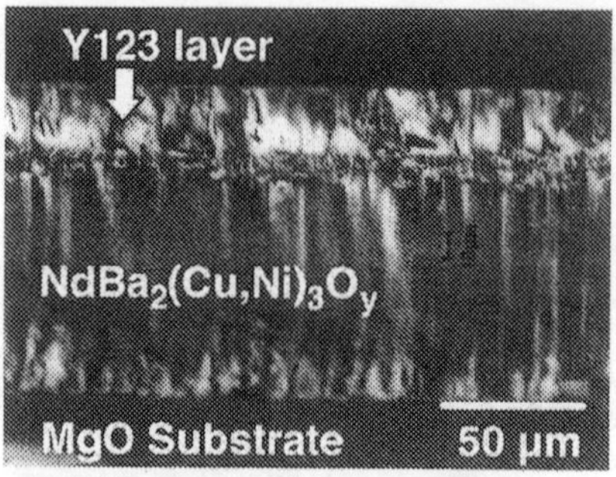

Figure 16.15. Cross-sectional view of the double layered LPE film grown on an MgO single crystal substrate. The 1st LPE layer on an MgO substrate is the $NdBa_2(Cu,Ni)_3O_y$ phase and the 2nd LPE layer is the $YBa_2Cu_3O_y$ phase.

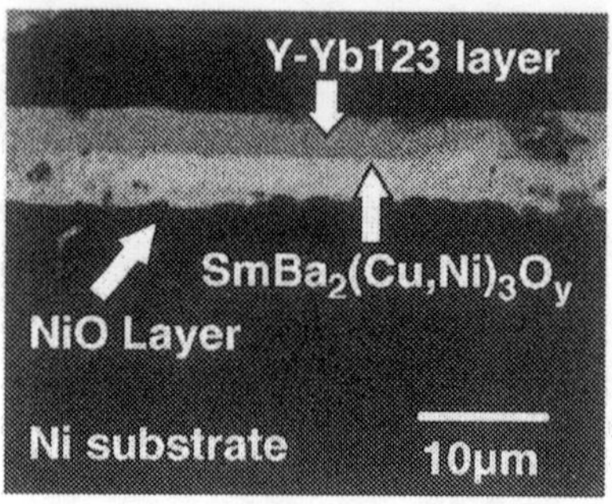

Figure 16.16. Cross-sectional view of the double layered LPE film grown on a Hastelloy tape substrate. The 1st LPE layer on the NiO buffer layer is the $SmBa_2(Cu,Ni)_3O_y$ phase and the 2nd LPE layer is the $(Y,Yb)Ba_2Cu_3O_y$ phase.

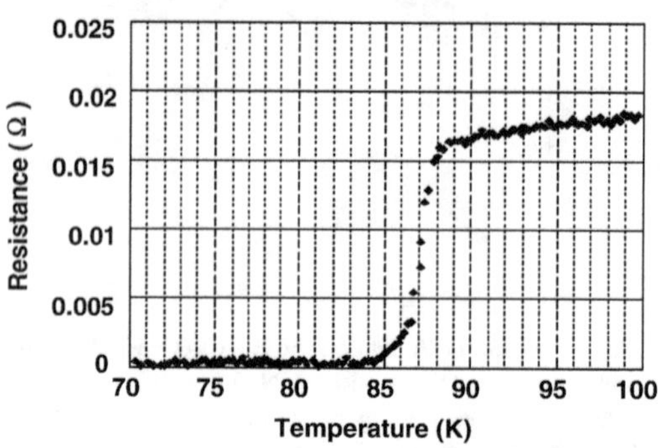

Figure 16.17. Transport superconducting characteristics of the double layered LPE film grown on Ni metal tape. The 1st LPE layer is the $SmBa_2(Cu,Ni)_3O_y$ phase and the 2nd LPE layer is the $YBa_2Cu_3O_y$ phase.

construction of the LPE coated conductor for the NiO-saturated system was confirmed as RE123(LPE)/RE (Nd or Sm)123-Ni(LPE)/NiO/Ni.

16.5 CONCLUSION

The recent progress on the development of LPE process for fabricating coated conductors was reviewed. In order to prevent the reaction between liquid and metal, it was clarified that the combination of MgO or NiO buffer layer and MgO- or NiO-saturated liquid are both effective. There is difference in the influence of buffer material-addition to liquid for growth of the 1st LPE layer. Although the MgO addition affect little to the Y-solubility and T_p of 123, NiO addition makes the able temperature range for Y123 growth decrease due to decrease of T_p and increase of crystallization temperature of the low temperature phase. Additionally, it was found that the suitable selection of RE for each layers to suppress dissolution of the 1st layer at dipping for the 2nd layer, which means stable growth of the 2nd layer. Consequently, the double layered LPE structure successfully realized on metal substrates, which reveal high superconducting performance. Then, the suitable constructions for the LPE coated conductors were confirmed as that lower growth temperature for the 2nd LPE layer should be selected than T_p of the 1st LPE layer material. Additionally, a higher T_p material than Y123 such as Nd- or Sm-system has to be selected for the 1st LPE layer in the NiO saturated system.

Next stage, the textured substrates such as the MgO deposited by using ISD technique or NiO fabricated by the SOE method will be applied to obtain high J_c and I_c. Furthermore, the process has to be extended to the long tape processing in parallel with improving J_c and I_c.

ACKNOWLEDGMENT

This work was supported by the New Energy and Industrial Technology Development Organization (NEDO) as Collaborative Research and Development of Fundamental Technologies for Superconductivity Applications.

REFERENCES

Goyal, A., Norton, D.P., Kroeger, D.M., Christen, D.K., Paranthaman, M., Specht, E.D., Budai, J.D., He, Q., Saffian, B., List, F.A., Lee, D.F., Hatfield, E., Martin, P.M., Klabunde, C.E., Mathis, J., and Park, C., 1997, Conductors with controlled grain boundaries: An approarch to the next generation, high temperature superconducting wire, *J. Mater. Res.*, 12:2924–2940.

Goyal, A., Ren, S.X., Specht, E.D., Kroeger, D.M., Feenstra, R., Norton, D.P., Paranthaman, M., Lee, D.F., and Christen, D.K., 1999, Texture formation and grain boundary networks in rolling assisted biaxially textured substrates and in epitaxial YBCO films on such substrates, *Micron*, 30:163–478.

Hasegawa, K., Nakamura, Y., Izumi, T., and Shiohara, Y., 2001, Investigation of texture development on MgO films prepared by inclined substrate deposition with electron-beam evaporation, in: *Proceedings of 13th International Symposium on Superconductivity*, October 14–16, 2000, Physica C, North-Holland.

Hobara, N., Kakimoto, K., Nakamura, Y., Izumi, T., Yuasa, T., Takahashi, Y., Fujino, K., Ohmatsu, K., and Shiohara, Y., 2000, Development of Y-system coated conductor on metal substrate by LPE method, in: *Proceedings of 12th International Symposium on Superconductivity*, October 17–19, 1999, Springer-Verlag, New York.

Holesinger, T.G., Foltyn, S.R., Arendt, P.N., Kung, H., Jia, Q.X., Dickerson, R.M., Dowden, P.C., DePaula, R.F., Groves, J.R., and Coulter, J.Y., 2000, The microstructure of continuously processed $YBa_2Cu_3O_y$ coated conductors with underlying CeO_2 and ion-beam-assisted yttria-stabilized zirconia buffer layers, *J. Mater. Res.*, 15:1110–1119.

Iijima, Y., Kakimoto, K., and Takeda, K., 2001, Long length IBAD templete film for Y-123 coated conductors, in: *Proceedings of 13th International Symposium on Superconductivity*, October 14–16, 2000, Physica C, North-Holland.

Ishida, Y., Kimura, T., Kakimoto, K., Yamada, Y., Nakagawa, Z., Shiohara, Y., and Sawaoka, A.B., 1997, Liquid phase epitaxy of $YBa_2Cu_3O_x$ on MgO substrates with seed films, *Physica C*, 292:264–272.

Kakimoto, K., Hobara, N., Krauns, C., Nakamura, Y., Izumi, T., Fujino, K., Ohmatsu, K., and Shiohar, Y., 2000a, Process and characteristics of $YBa_2Cu_3O_y$/$YBa_2(Cu,Mg)_3O_y$ films on MgO substrates by LPE, in: *Proceedings of 12th International Symposium on Superconductivity*, October 17–19, 1999, Springer-Verlag, New York.

Kakimoto, K., Hobara, N., Nakamura, Y., Izumi, T., Fujino, K., Ohmatsu, K., and Shiohara, Y., 2000b, Y-system coated conductor on metal substrate by LPE method, *Physica C*, 341–348:2489–2490.

Kitamura, T., Yoshida, M., Yamada, Y., Shiohara, Y., Hirabayashi, I., and Tanaka, S., 1995, Crystalline orientation of $YBa_2Cu_3O_{7-y}$ film prepared by liquid-phase epitaxial growth on $NdGaO_3$ substrate, *Appl. Phys. Lett.*, 65:1421.

Krauns, Ch., Sumida, M., Tagami, M., Yamada, Y., and Shiohara, Y., 1994, Solubility of RE elements into Ba–Cu–O melts and the enthalpy of dissolution, *Z. Phys. B: Condens. Matter.*, 96:207–212.

Matsumoto, K., Kim, S.B., Hirabayashi, I., Watanabe, T., Uno, N., and Ikeda, M., 2000, High critical current density $YBa_2Cu_3O_{7-d}$ tapes prepared by the surface-oxidationepitaxy method, *Physica C*, 330:150–154.

Metzger, R., Bauer, M., Numssen, K., Semerad, R., Berberich, P., and Kinder, H., 2000, Inclined substrate deposition of MgO buffer layers for YBCO coated conductors, in: *Proceedings of ASC 2000*, September 17–22, Virginia Beach, USA.

Miura, S., Hashimoto, K., Wang, F., Enomoto, Y., and Morishita, T., 1997, Structual and electrical properties of liquid phase epitaxially grown $Y_1Ba_2Cu_3O_x$ films, *Physica C*, 278:201–206.

Ohmatsu, K., Muranaka, K., Taneda, T., Fujino, K., Takei, H., Sato, Y., Matsuo, K., and Takahashi, Y., 2001, Development of in-plane aligned YBCO tapes fabricated by inclined substrate deposition, in: *Proceedings of 13th International Symposium on Superconductivity*, October 14–16, 2000, Physica C, North-Holland.

Thieme, C.L.H., Annavarapu, S., Zhang, W., Prunier, V., Fritzemeier, L., Li, Q., Schoop, U., Rupich, M.W., Gopal, M., Foltyn, S.R., and Holesinger, T., 2000, Non-magnetic substrates for low cost YBCO coated conductor, in: *Proceedings of ASC 2000*, September 17–22, Virginia Beach, USA.

Watanabe, T., Matsumoto, K., Maeda, T., Tanigawa, T., and Hirabayashi, I., 2001, Long length oxide templete for YBCO coated conductor prepared by surface-oxidation epitaxy method, in: *Proceedings of 13th International Symposium on Superconductivity*, October 14–16, 2000, *Physica C*, North-Holland.

Yamada, Y. and Shiohara, Y., 1993, Continuous crystal growth of $YBa_2Cu_3O_{7-x}$ by the modified topseeded crystal pulling method, *Physica C*, 217:182–188.

Yoshino, H., Ymazaki, M., Thanh, T., Kudo, Y., and Kubota, H., 2001, Preparation of Ag–Cu/Ni/Ag–Cu clad tapes for YBCO superconducting tape and its textured properties, in: *Proceedings of 13th International Symposium on Superconductivity*, October 14–16, 2000, Physica C, North-Holland.

Section C

Deposition of Other HTS Materials

Chapter 17

EX-SITU PROCESSING OF Tl-CONTAINING FILMS

J.Y. Lao [1], J.H. Wang [1], D.Z. Wang [2], S.X. Yang [2], and Z.F. Ren [2]

[1]Department of Chemistry
State University of New York
Buffalo, NY 14260
USA
[2]Department of Physics
Boston College
Chestnut Hill, MA 02460
USA

17.1 INTRODUCTION

Since the discovery of high temperature superconductivity by Bednorz and Müller (1986) on lanthanum barium copper oxide (cuprate), which becomes superconducting below 35 K in 1986, there has been a world-wide effort to develop large scale processes for fabricating long length, flexible polycrystalline conductors with high critical current density for a range of applications such as superconducting motors, generators, transformers, magnets, and transmission lines with great efficiency. However, this progress has been hindered by the intrinsic problems of HTS ceramic materials, such as weak link, flux creep and poor mechanical properties. The strong orientation dependence of critical current density is the major limiting factor for the fabrication of high J_c conductors (Dimos et al., 1988, Dimos and Chaudhari, 1990). For example, long length high J_c (up to 1×10^4 A/cm^2 at 77 K and zero field) conductors of (Tl,Pb)-1223 were fabricated by powder-in-tube (PIT) method by Ren et al. (1992b), but the tapes can not tolerate magnetic field, the J_c drops about 20 times with only a 0.1 T external magnetic field applied. Fortunately, thin film growth processes have been proven very successful to eliminate high angle grain boundaries. Epitaxial (Tl,Bi)-1223 and Tl-1223 thin films (Ren et al., 1994, 1996; Wang et al., 1995; Piehler et al., 1994b; Lee et al., 1994) were shown to have much higher critical current density (J_c) both in the absence and presence of an external magnetic field than thallium- or bismuth-containing tapes made by PIT method (Ren and Wang, 1992a, 1992b, 1993a, Ren et al., 1995) due to the highly biaxial alignment of grains (i.e. well-aligned along c-axis and a-, b-axes) as compared to those in PIT tapes. This better alignment minimises weak-link effects caused by high-angle grain boundaries. Doi et al. (1994) also found that the biaxially oriented (Tl,Pb)Sr-1223 film on $SrTiO_3$ showed more than two orders of magnitude improvement in J_c at 1 T, near 10^5 A/cm^2 at 77 K compared to the (Tl,Pb)Sr-1223 uniaxially textured film on Ag tape. It was concluded that, much like the oriented YBCO films, highly biaxially textured Tl-1223 films were needed to realise a practical Tl wire or tape with useful transport properties at high magnetic fields.

Among the four major groups of cuprates, Tl-based cuprates have higher T_c, better oxygen stability than YBCO, lower toxicity than Hg-based cuprates and smaller electronic anisotropy as compared to the BSCCO type of materials, which results in better flux pinning properties and field performance (Bonham et al., 1989; Gammel et al., 1988). In fact, the processing of Tl compound is not more dangerous than lead exposure and not less experienced than semiconductor industry which using high toxic gases such as phosphine, arsine, and germane routinely. The lethal exposure is 0.5–1.0 g for Tl and Tl toxic exposure can be easily detected in the body by urine or blood analysis. T_c value of 120 K in Tl–Ba–Ca–Cu–O system was first reported by Sheng and Hermann (1988) and was quickly reproduced by other groups (Parkin et al., 1988; Soeta et al., 1989; Inoue et al., 1989, 1990; Torri et al., 1989a; Okada et al., 1990). This Tl-based HTS materials have two types of generalised chemical formulas: $Tl_2Ba_2Ca_{n-1}Cu_nO_{2n+4}$ and $Tl_{1-x}M_xA_2Ca_{n-1}Cu_nO_{2n+3}$ (A $=$ Ba or Sr or a combination of both, M $=$ Pb or Bi or a combination of both) with $1 \leqslant n \leqslant 5$. The former is referred to as "thallium bilayers," including Tl-2201, Tl-2212, Tl-2223, and Tl-2234 were used with $n = 1, 2, 3$, and 4 respectively; and the latter is called as "thallium monolayers" (Torardi, 1992), including Tl-1201, Tl-1212, Tl-1223, and Tl-1234. Thallium monolayers have much stronger flux pinning property than the thallium bilayers due to the shorter insulating layer distance in the crystal lattice. Among all the thallium monolayer phases, Tl-1223 is most promising because it has high T_c (110–120 K), strong flux pinning, and has been successfully fabricated into single phase. The electrical transport T_c ($R = 0$) of 106–111 K and J_c of up to 3×10^6 A/cm^2 at 77 K and in zero magnetic field have been obtained in (Tl,Bi)-1223 films on the (001) surface of single crystal $LaAlO_3$ substrate by pulsed laser deposition (PLD) and post-annealed in a muffle furnace in stationary air (Ren et al., 1994, 1996; Wang et al., 1995). Since that, high J_c epitaxial (Tl,Bi)-1223 films were also fabricated on $LaAlO_3$, YSZ and CeO_2-capped YSZ substrate by pulsed laser deposition and *ex-situ* annealing in flowing argon (Wang et al., 1995; Guo et al., 1997; Ren et al., 1998b). Consequently, epitaxial (Tl,Bi)-1223 films (Ren et al., 1998a, 1999) were also synthesised on YSZ- and CeO_2-capped Rolling Assisted Biaxially Aligned Substrates (RABiTS) developed by Oak Ridge National Laboratory (ORNL) (Goyal et al., 1996; Norton et al., 1996). The films exhibit T_c value in the range of 105–110 K and a transport J_c value over 1×10^5 A/cm^2 for YSZ-topped RABiTS and 4×10^5 A/cm^2 for CeO_2-toped RABiTS. For large-scale production of HTS wire in electric conductor application, another important technical challenge that must be met is the successful demonstration of a low cost, reasonable field, and high current carrying wire with acceptable mechanical properties. Therefore, at the same time, other low cost, large scalable non-vapour transport methods, such as electrodeposition (Bhattacharya et al., 1998a, 1998b, 1999) and thermal spray pyrolysis (Mogro-Campero et al., 1995; Specht et al., 1996; Paranthaman et al., 1997; Li et al., 1999) method, were also successfully used to synthesise epitaxial high critical current density (Tl,Bi)-1223 films on LAO substrates. Also, recently, progress was made on the development of high performance Tl containing superconducting materials. Ihara et al. (1999) and Khan et al. (1999) demonstrated the high J_c, T_c and excellent magnetic field performance of TlCu-1234 film and $Cu_{1-x}Tl_x$-1223 films on $SrTiO_3$ substrate.

However, another single Tl layer superconducting material, Tl-1212 phase, has been paid attention due to its shortest inter Cu–O conducting layer distance among Tl-based superconductors with structural similarity to YBCO (Kim et al., 1991). This short distance could possibly lead to even stronger interlayer coupling, less severe thermally activated flux motion, higher critical current density, and better performance in magnetic field. Also, thermodynamically, Tl-1212 compound is more stable than Tl-1223 phase. Because of the high copper valence, reductive annealing (Siegal et al., 1995a) and elemental substitutions such as Pb, Bi for Tl (Subramanian et al., 1988; Haldar et al., 1988, Li and Greenblaat, 1989), rare earth and Y for Ca (Sheng et al., 1989; Liu et al., 1989; Myers et al., 1994), are widely used for Tl-1212 film synthesis. Two zone furnace method, which is large scalable compared to regular crucible method, has also been successfully used to synthesis Tl-1212 films (Siegal et al., 1995a, 1995b, 1997a, 1997b, 1998). However, previously, the synthesis of high J_c Tl-1212 film is not successful compared to Tl-1223 film, the J_c and T_c values of Tl-1212 film are much lower than that of Tl-1223 film. And our work on Pb, Bi substitution for Tl with Y substitution for Ca did not lead to any new promising results (Lao et al., 1998). Interestingly, the 3d element, Cr, was reported to be doped into the $TlSr_2CaCu_2O_7$ compound to produce Tl-1212 superconducting films by a two step process: laser ablation of a Tl-free target, followed by post-ablation annealing in air at 860°C to 870°C for 15 to 20 hours in the presence of $Tl_{1.3}Sr_2CaCr_{0.2}Cu_2O_z$ pellets (Sheng et al., 1991; Tang et al., 1993). The annealed films had T_c values in the range of 98–102 K and J_c approaching 1×10^6 A/cm^2 at 77 K, as measured by a self-inductance method.

Recently, we studied the growth of Tl-1212 films with high T_c and J_c as an alternative to Tl-1223 films. Epitaxial Cr-doped (Tl,Bi)-1212 films, with transport J_c of up to 1.5×10^6 A/cm^2 at 77 K, have been successfully synthesised on LAO substrate in static air with short annealing time (Lao et al., 2000a). Consequently, high quality Cr-doped (Tl,Bi)-1212 films were also deposited on LAO and CeO_2-capped YSZ substrate by PLD and *ex-situ* annealing in flowing argon (Lao et al., 2000b). Potential applications, such as transmission cables, are also very promising for this compound.

17.2 DEVELOPMENT OF THALLIUM-1223 FILMS FOR CONDUCTOR APPLICATIONS

17.2.1 Introduction

Tl-1223 system is a member of the series $TlBa_2Ca_{n-1}Cu_nO_{2n+3+\delta}$, which has single layers of octahedrally-coordinated Tl between the familiar $BaO/CuO_2/BaO$ modules (Parkin et al., 1988). Single octahedral layers do not require an offset of adjacent modules so this is primitive tetragonal, space group *P4/mmm*, with $3.8 \times 3.8 \times 15.9$ Å unit cells. In the structure Figure 17.1 below (Siegal et al., 1997b), the Tl atom is shown as open circle, with oxygen occupying in the centre of the Tl layer square and polyhedral elsewhere. Tl is in a flattened octahedron with four longer Tl–O bonds (about 2.7 Å) and two short Tl–O bonds parallel to *c*-axis (2.0 Å). Two Ba layers separate the Tl layers with the adjacent blocks of CuO_2 planes, and three superconducting CuO_2 layers are separated by two Ca layers, with oxygen occupying the corners of the square pyramids.

The Tl-1223 superconducting phase is metastable, since it is overdoped when prepared under ambient conditions. This metastable superconducting phase becomes

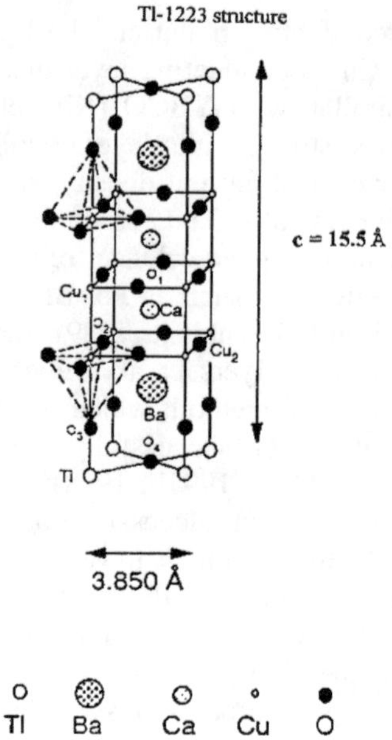

Figure 17.1. Structure of Tl-1223.

stable (Torri et al., 1989b; Kamo et al., 1991) when a partial substitution such as (Pb,Bi) for Tl, Ba for Sr, and (Y,La) for Ca is used. Compared to YBCO system, Tl-1223 has several advantages including higher transition temperature, better tolerance to intergranular J_c because of unique colony structure in the film growth morphology (Kroeger et al., 1994; Specht et al., 1994, 1995), and the successful synthesis of epitaxial films up to 10 μm thick by non-vacuum method (Lee et al., 1994). The magnetic field vs. temperature irreversibility line for Tl-1223 at 77 K compares favourably with the performance of YBCO (Nabatame et al., 1992). The $(Tl_{0.5}Pb_{0.5})Sr_2Ca_2Cu_3O_9$ phase has T_c as high as 122 K (Subramanian et al., 1988), while $(Tl_{0.8}Bi_{0.2})Sr_2Ca_2Cu_3O_9$ phase shows zero resistivity at 113 K. Pellet of composition $(Tl_{0.78}Bi_{0.22})(Sr_{1.6}Ba_{0.4})Ca_2Cu_3O_9$ was found to consist of a single 1223 phase according to the XRD patterns (Ren and Wang, 1992b) by the effect of partial substitution of Bi for Tl. High T_c $TlBa_2Ca_2Cu_3O_9$ thin films have been prepared by laser ablation in combination with thermal evaporation of thallium oxide (Piehler et al., 1993, 1994a). DC magnetisation measurements showed the onset of superconductivity at ~115 K. The J_c measured by magnetisation cycles were 9×10^5 A/cm^2 at 6 K and 2.5×10^5 A/cm^2 at 77 K. In a magnetic field to 1 T applied parallel to the c-axis the J_c were 3×10^5 A/cm^2 at 6 K and 3×10^3 A/cm^2 at 77 K. However, recent studies of the films with the composition of $(Tl_{0.78}Bi_{0.22})(Sr_{1.6}Ba_{0.4})Ca_2Cu_3O_9$ and $(Tl_{0.5}Pb_{0.5})Sr_{1.6}Ba_{0.4}Ca_2Cu_3O_9$ have shown the most promising results.

17.2.2 Synthesis of $(Tl_{0.78}Bi_{0.22})(Sr_{1.6}Ba_{0.4})Ca_2Cu_3O_9$[(Tl,Bi)-1223] Films by Vacuum Method

17.2.2.1 Synthesis of (Tl,Bi)-1223 Films on Single Crystal Substrates Although a number of single crystal substrates, such as MgO, $SrTiO_3$ and Al_2O_3 can be used as substrate for HTS film synthesis, $LaAlO_3$, YSZ and CeO_2-capped YSZ are most frequently used for HTS materials due to their good lattice match with HTS materials and chemical, thermal stability. Although these ceramic materials are expensive, brittle and not scalable for large amount production of long length, the work on these single crystal substrates can help to have better understanding of the epitaxial film synthesis and to explore and to demonstrate the possibility of Tl-based superconducting materials as a possible candidate for the next generation HTS wire. A number of techniques have been used to deposit precursor films with or without Tl. These methods include sputtering, e-beam evaporation, pulsed laser deposition, organometallic chemical vapour deposition, screen-printing mixed oxides, sol-gel methods, spray pyrolysis, electrodeposition and spin coating. In this section, pulsed laser deposition with the *ex-situ* annealing methods of the precursor films such as crucible method is used as research tool of deposition of amorphous precursor films of Tl-based HTS materials for demonstration purpose. The electrodeposition and thermal spray pyrolysis method, and *ex-situ* annealing two zone furnace annealing which are going to be discussed in following sections, are considered as low cost scalable method for future possible large sale production.

PLD has been widely used for the synthesis of superconducting materials (Liou et al., 1989; Holstein et al., 1993; Johs et al., 1989). The short wavelength radiation (193–351 nm) of excimer laser makes them an excellent noncontact tool for processing metals, plastics and ceramics. It has the advantage of high photon energy, congruent evaporation of target material that gives the stoichiometric deposition, and the low average power that makes the target thermally stable. Recently, excimer laser has been used to deposit various buffer materials, superconducting and high temperature superconducting thin films from bulk targets at lower power densities.

The $Tl_{0.78}Bi_{0.22}Sr_{1.6}Ba_{0.4}Ca_2Cu_3O_{9+\delta}$ superconducting thin films were prepared by pulsed laser ablation using a reacted superconducting source target, followed by post-annealing either in static air of muffle furnace or in a tube furnace with pure argon flowing through the tube all the time during annealing. To fabricate the reacted source target, a prepowder of $Sr_{1.6}Ba_{0.4}Ca_2Cu_3O_7$ was first prepared by grinding a stoichiometric mixture of $SrCO_3$, BaO_2, CaO and CuO, then the mixture was heated in an alumina crucible for 40 hours at 905–920°C with regrinding after each 10 hours of heating. Then a uniform mixture, with a stoichiometric composition corresponding to $Tl_{0.95}Bi_{0.22}Sr_{1.6}Ba_{0.4}Ca_{2.2}Cu_{3.3}O_{9+\delta}$, was prepared by grinding the mixture of one formula weight of the prepowder of $Sr_{1.6}Ba_{0.4}Ca_2Cu_3O_7$, 0.2 formula weight of CaO, 0.3 formula weight of CuO, 0.475 formula weight of Tl_2O_3 and 0.11 formula weight of Bi_2O_3. This mixture was subsequently compressed at a pressure of 1.75×10^8 Pa into a 1.9 cm diameter pellet, sandwiched between two gold plates, wrapped in silver foil and reacted at 885°C for 3.0–4.0 h in a muffle furnace with stationary air to make a superconducting material. The superconducting pellet was then pulverised, mixed with additional amount of Tl_2O_3 (0.475 formula weight), compressed at a pressure of 7.0×10^8 Pa into a 1.9 cm diameter pellet, again wrapped as described above, and heated at 850°C for 20 min the same way as in the previous step to become the superconducting source target for film fabrication.

The substrate was cleaned in methanol and acetone in ultrasonic bath for 5 min alternatively, and heated to 300°C on the heater in chamber before deposition to get rid of the moisture and/or hydrocarbon on the surface. Silver paste was used to provide the good thermal contact between the $LaAlO_3$ or YSZ substrate with heater. The chamber was evacuated to base pressure before deposition. Then, oxygen was introduced and the turbo pump was run at low speed mode. During the ablation, the oxygen pressure was maintained at 25 mTorr. The laser ablation was conducted at an energy range from 100 to 180 mJ/pulse, and 4 pulses per second repetition rate. The precursor films were deposited on the single crystal substrates at room temperature or 200°C. In fact, the

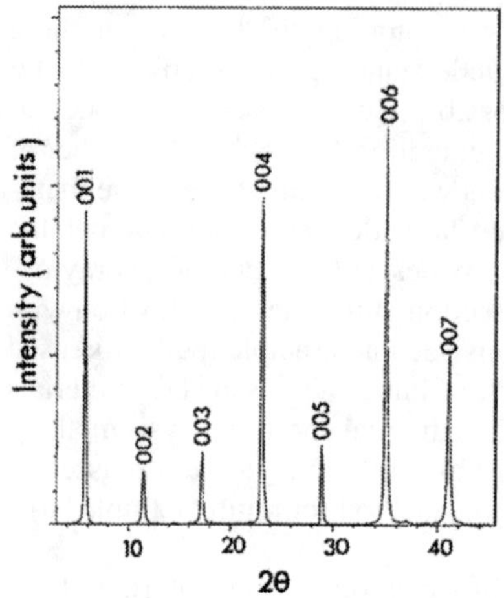

Figure 17.2. XRD θ–2θ scan of a typical (Tl,Bi)-1223 film annealed in static air.

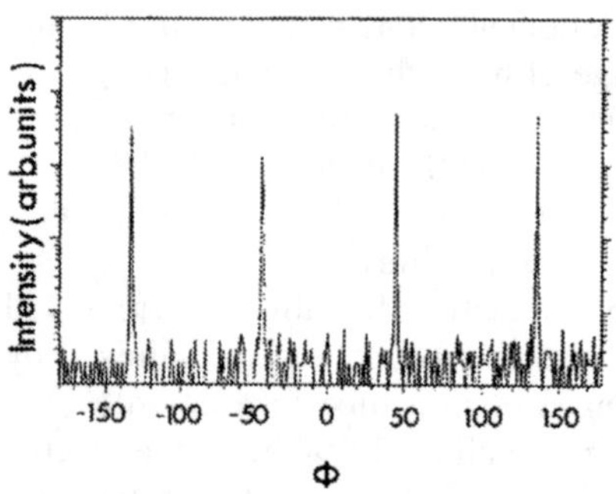

Figure 17.3. ϕ scan of the (Tl,Bi)-1223 film on LAO substrate.

substrate temperature has no observable effect on the film property as long as it is below 500°C. The as-deposited films are typically amorphous to nanocrystalline and are electrically insulating. The resulting precursor films were then processed *ex-situ* in a tube furnace by placing the sample on a gold plate situated between two (Tl,Bi)-1223 semicircular pellets for maintaining the partial pressure of Tl_2O. The assembly was wrapped in silver foil with adequate space for vapour diffusion and annealed at pre-set condition, resulting in a fully phase-developed (Tl,Bi)-1223 film.

Since $LaAlO_3$ substrate has excellent lattice match with HTS materials and can promote the phase formation of Tl-1223 phase, the synthesis of Tl-1223 films was started on $LaAlO_3$ substrate by the two-step procedure. The Tl-1223 phase was developed by heating the assembly in air at 840–870°C for 25–60 min or in argon at 750–780°C for 30 min. The annealing temperature and duration of annealing are the two key factors for the development of 1223 phase. It was concluded that that optimum annealing condition is 860°C, 60 min in air or 770°C, 30 min in argon. The dependence of Tl-1223 phase development on the film thickness was also observed. Film thickness over 1 μm is necessary to transform all the 1212 phase into 1223 phase. Figure 17.2 shows the XRD θ–2θ scan of the film annealed in static air, while the inset shows the rocking curve of the (006) reflection with FWHM of only 0.365°. The only existence of (00l) peaks in the figure indicate that c-axis is well aligned. The X-ray ϕ scan of (103) reflection of Tl-1223 phase which is shown in Figure 17.3 and (222) peak of $LaAlO_3$ substrate shows the films are epitaxial grown from the substrate. The phase development and epitaxial alignment of the films annealed in the argon is almost

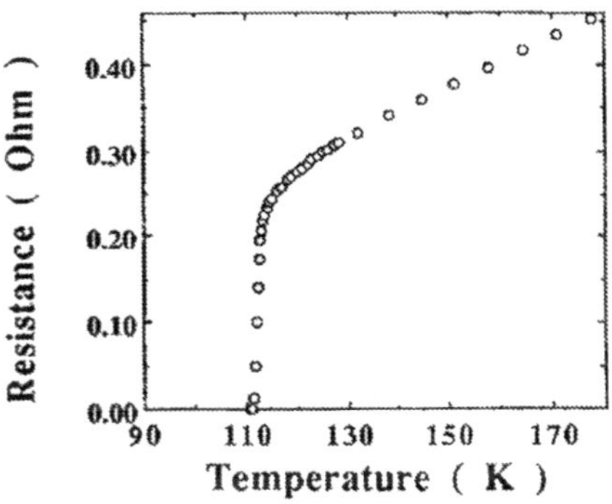

Figure 17.4. Temperature dependence of resistance of (Tl,Bi)-1223 film.

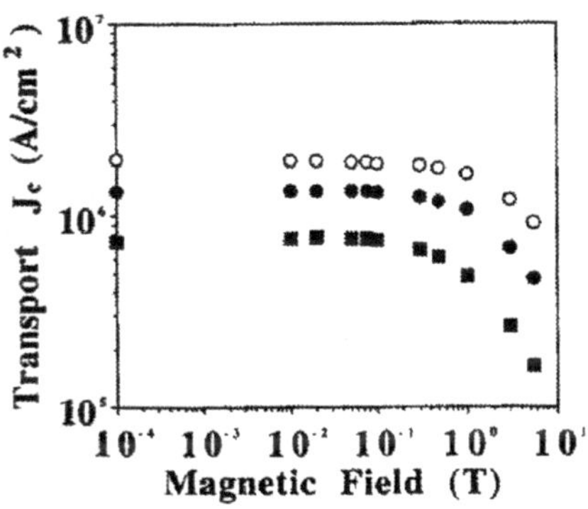

Figure 17.5. Magnetic dependence of J_c at different temperatures. Open circles; 67 K, filled circles: 77 K, and filled squares: 87 K.

as good as that of films annealed in air. The morphology of the as annealed films were studied by SEM. The dominant uniform round grains and a certain amount of a-axis oriented needle like grains were observed. TEM analysis also found fine a-axis oriented plates, dislocations, stacking faults, and other defects. Their fine scale and high density may partly be responsible for the good pinning properties of the (Tl,Bi)-1223 films. Figure 17.4 shows a typical temperature dependence of resistance of the Tl-1223 film. The as-grown films exhibited T_c's in the range of 105–111 K depending on the relative phase purity of 1223 to 1212. The best films prepared under the optimal conditions exhibited a transport critical current density close to 3×10^6 A/cm^2 at 77 K and zero field. Figure 17.5 shows both the temperature dependence and magnetic field dependence of the film transport J_c with field aligned parallel to the c-axis.

However, because of the associated difficulties of development of $LaAlO_3$ topped biaxially aligned metallic substrate (Parilla et al., 1997; Carlson et al., 1998), synthesis of epitaxial (Tl,Bi)-1223 films on YSZ and CeO_2-capped YSZ single crystal substrate are necessary as the first step for the possible electric conductor application. The processing of (Tl,Bi)-1223 films on YSZ and CeO_2-capped YSZ single crystal substrate was similar to that for films on $LaAlO_3$ substrate. The films were annealed at 780–810°C for 40–60 min with pure argon flowing through the tube all the time during the annealing, which is lower than that in static air because the argon annealing environment promotes the Tl_2O_3 dissociation. To deposit CeO_2 buffer layer, YSZ substrates were attached onto a stainless steel block by silver paste and heated to above 600°C by SiC heaters. The CeO_2 target was then laser ablated at energy of 100–120 mJ and repetition rate of 1–2 Hz. The substrate temperature was maintained

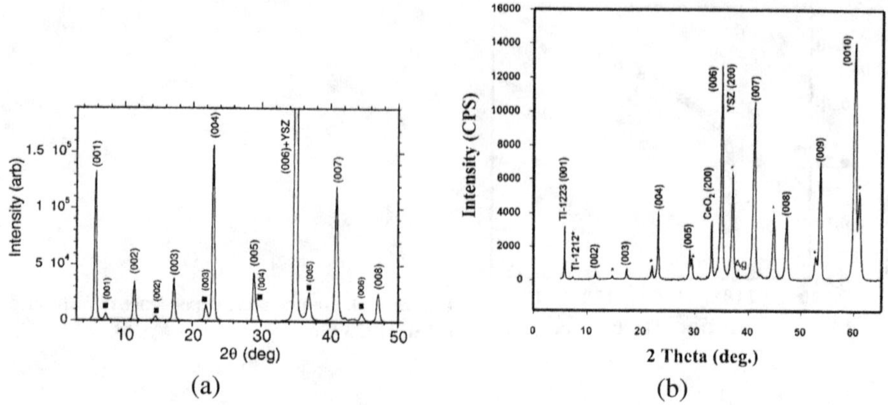

Figure 17.6. XRD θ–2θ scan of (Tl,Bi)-1223 film on: (a) YSZ substrate; (b) CeO_2-topped YSZ.

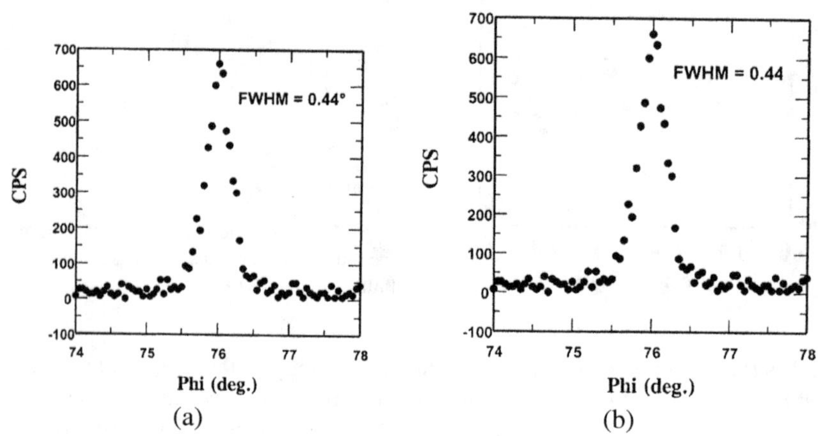

Figure 17.7. ϕ scan of (Tl,Bi)-1223 film on: (a) YSZ substrate; (b) CeO_2-topped YSZ.

at 600–650°C during deposition. The CeO_2 buffer thickness was controlled at ranges from 1000 to 2000 Å. After the deposition of the crystalline CeO_2, the substrates were then cooled inside the deposition chamber with controlled O_2 flow. When the temperature of the substrates cooled to below 100°C, the (Tl,Bi)-1223 target was switched into position for amorphous (Tl,Bi)-1223 deposition. The as annealed films, both on YSZ and CeO_2-capped YSZ substrates, contain some impure Tl-1212 phase as demonstrated by XRD θ–2θ scan in Figure 17.6. The ω scan of (007) peak of (Tl,Bi)-1223 phase shows FWHM of 1.1° for film on YSZ substrate, and 0.44° for film on CeO_2-capped YSZ, which demonstrated excellent c-axis alignment. Figure 17.7 shows the ϕ scan of films on YSZ surface and CeO_2-capped YSZ surface. The FWHM is 1.4° for the former and 0.44° for the latter. The better a-, b-axes alignments films on CeO_2-capped YSZ than that on YSZ surface is as expected since CeO_2 has better lattice alignment with Tl-1223 than YSZ. The zero resistivity T_c's of the films are in the range of 105–110 K. The zero field J_c (77 K) is around 0.6×10^6 A/cm^2 for the films on YSZ and 1.1×10^6 A/cm^2 for the films on CeO_2-capped YSZ. The magnetic field dependence of J_c at 77 K (H//c) are shown in Figure 17.8 with both films have irreversible line around 3 T.

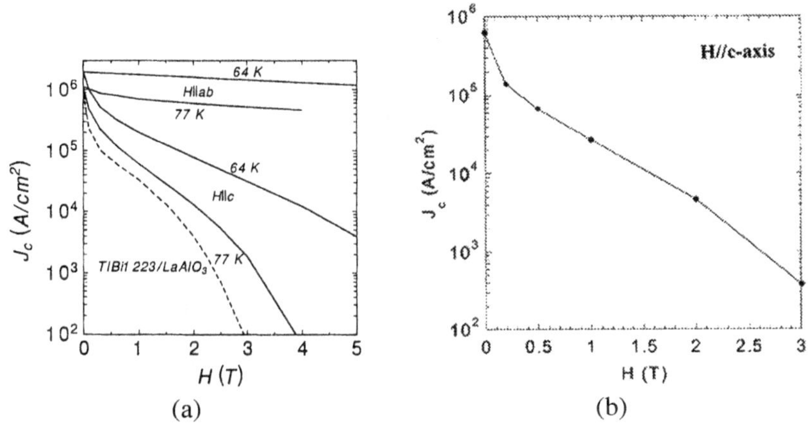

Figure 17.8. Magnetic field dependence of J_c of (Tl,Bi)-1223 film on: (a) YSZ substrate; (b) CeO_2-topped YSZ.

17.2.2.2 Synthesis of epitaxial (Tl,Bi)-1223 films on YSZ- and CeO_2-topped RABiTS With the successful growth of (Tl,Bi)-1223 film on single crystal YSZ and CeO_2-capped YSZ by laser ablation and post-deposition annealing in pure argon and the development of Rolling Assisted Biaxially Textured Substrates (RABiTS), the epitaxial growth of (Tl,Bi)-1223 film on RABiTS was successfully achieved for the first time by the authors' group (Ren et al., 1998a). The (Tl,Bi)-1223 superconducting films on RABiTS were prepared by pulsed laser ablation followed by post-deposition annealing in a tube furnace with flowing pure argon, which is similar to the previous synthesis of Tl-1223 films on single crystal substrate. During the deposition, RABiTS substrate was only loosely put on the heater because of the difficulty associated with handling the back silver-painted flexible thin RABiTS substrate. The RABiTS are provided by ORNL. For the YSZ-topped RABiTS, the base Ni tape is about 125 μm thick and the top buffer layer YSZ is about 0.5 μm thick, with a 40 nm thick CeO_2 in between. The structure for the CeO_2-capped RABiTS is CeO_2/YSZ/CeO_2/Ni, with actual thickness of 20 nm/0.7 μm/20 nm/120 μm correspondingly.

The XRD θ–2θ spectrum of a typical (Tl,Bi)-1223 superconducting thin film on a YSZ-topped RABiTS is shown in Figure 17.9. All the major reflections are indexed as either (00l) peaks of (Tl,Bi)-1223 phase or (200) of the RABiTS, with weaker peaks resulting from either (00l) of the (Tl,Bi)-1212 phase or NiO. The thin layer CeO_2 resulted in weaker intensity. Further optimisation of the deposition and annealing parameters might eliminate (Tl,Bi)-1212 phase and improve the physical properties of the films. The presence of only (00l) peaks shows that the films are strongly c-axis aligned. The degree of c-axis alignment has been determined by a ω-scan (rocking curve) of the (Tl,Bi)-1223 (007) peak. For this peak of the (Tl,Bi)-1223 phase, the full-width-at-half-maximum (FWHM) value determined from the rocking curve is about 6.8°, which is comparable with that of the RABiTS (Goyal et al., 1996). The out-of-plane FWHM of (005) peak of the minor (Tl,Bi)-1212 is about 11.3° which is much larger than the 6.8° of the major (Tl,Bi)-1223 phase. The elimination of the minor (Tl,Bi)-1212 phase will drastically improve the c-axis alignment of the major (Tl,Bi)-1223 phase. The in-plane (a- and b-axes) alignment was measured by XRD ϕ scans of the (Tl,Bi)-1223 (102) pole figure, as shown in Figure 17.10. The four well-developed diffraction spots with $\Delta\phi \sim 7.0°$ indicate that the a- and b-axes are aligned, with no indication of 45° misoriented domains which are frequently present in the YBCO films on YSZ, due

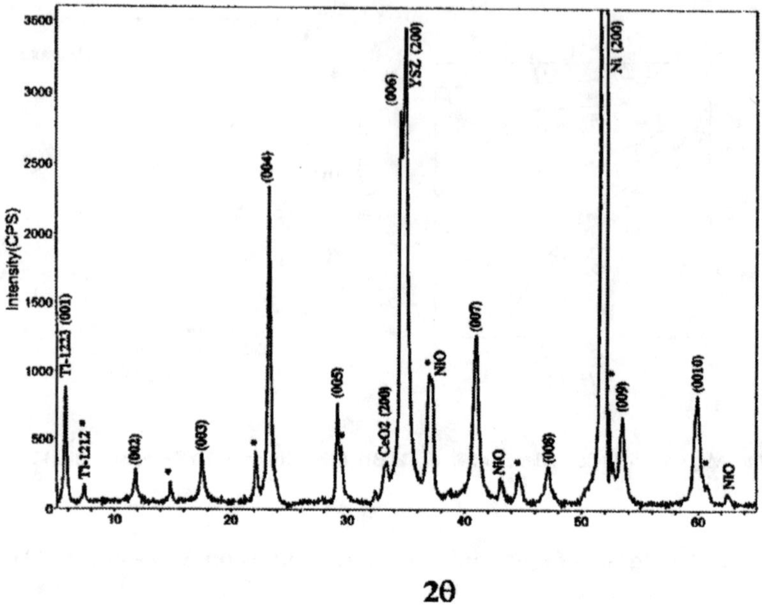

Figure 17.9. XRD θ–2θ scan of (Tl,Bi)-1223 film on YSZ-topped RABiTS.

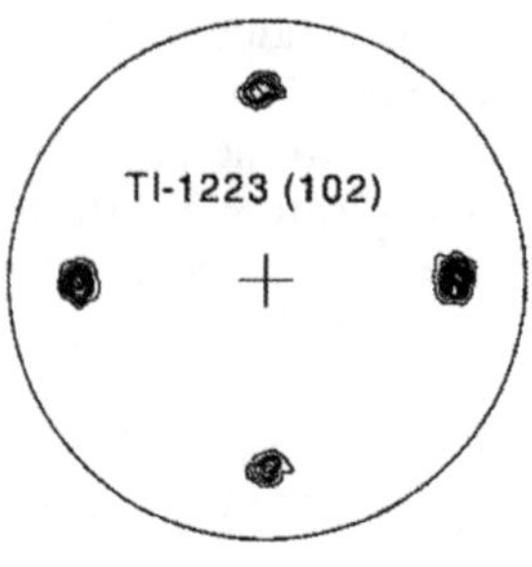

Figure 17.10. Pole figure of (Tl,Bi)-1223 film on YSZ-topped RABiTS.

to the lattice mismatch. Similar measurements carried out for the minor (Tl,Bi)-1212 phase showed that both a- and b-axes are also aligned with $\Delta\phi \sim 9.5°$. An epitaxy of $\langle 100 \rangle$ of (Tl,Bi)-1223 film on $\langle 110 \rangle$ of YSZ substrate was derived from the ϕ-scans of both the (Tl,Bi)-1223 film and the YSZ substrate. The magnetisation of the sample used for XRD θ–2θ diffraction shown in Figure 17.9 and pole figure shown in Figure 17.10 was measured in a SQUID magnetometer. The J_c calculated from Bean's model using the full width (3.5 mm) of the film as the appropriate lateral dimension was 2×10^5 A/cm^2 at 77 K and extrapolated to zero field. Transport measurements on another longer film which was processed a little differently showed that a J_c of 1×10^5 A/cm^2 was obtained at 77 K and zero field. The zero-resistance transition temperatures (T_c) of the films are in the range of 106 to 110 K. Figure 17.11 shows a typical temperature-dependent resistivity, with a zero-resistance T_c of 107 K for this particular film.

The XRD θ–2θ spectrum of a typical (Tl,Bi)-1223 superconducting thin film on CeO_2-capped RABiTS is shown in Figure 17.12. All the major reflections are indexed as (00l) peaks of either (Tl,Bi)-1223 phase or (Tl,Bi)-1212 phase, the peaks of the latter are marked by "*." The (200) peaks of YSZ and CeO_2 are also indexed. The pres-

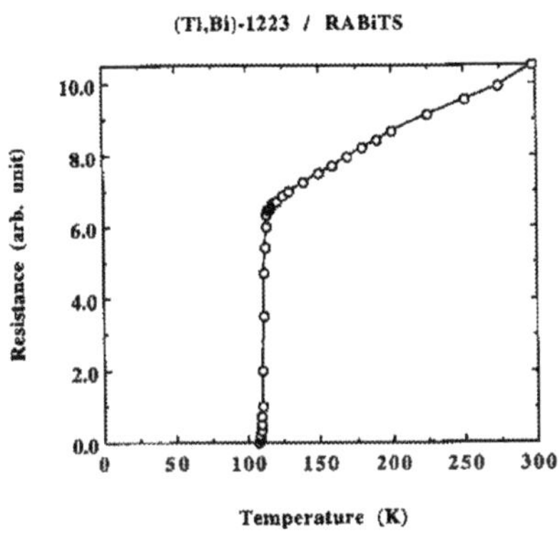

Figure 17.11. Temperature dependence of resistance of (Tl,Bi)-1223 film on YSZ-topped RABiTS.

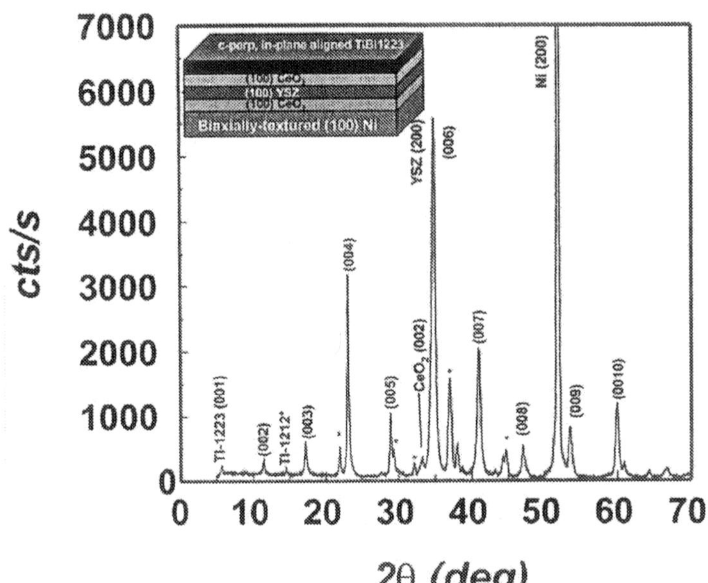

Figure 17.12. XRD θ–2θ scan of (Tl,Bi)-1223 film on CeO_2-capped RABiTS.

ence of only (00*l*) peaks shows that the films are strongly *c*-axis aligned. The FWHM values determined from the rocking curves are 8.99°, 9.7°, and 7.84° for (Tl,Bi)-1223 (007), (Tl,Bi)-1212 (005), and Ni (200) peaks, respectively. The in-plane (*a*- and *b*-axes) alignment was measured by XRD ϕ-scans of the (Tl,Bi)-1223 (102), (Tl,Bi)-1212 (102), YSZ (111), and Ni (111) as shown in Figure 17.13(a), (b), (c), and (d), respectively. The FWHM values are 7.22°, 8.86°, 7.91°, and 12.03° for (Tl,Bi)-1223 (102), (Tl,Bi)-1212 (102), YSZ (111), and Ni (111), respectively. An epitaxy of [100] of (Tl,Bi)-1223 film aligned with [110] of CeO_2 layer was derived from the ϕ-scans of both the (Tl,Bi)-1223 film and the CeO_2 cap layer. The FWHM values of ω-scans and ϕ-scans of (Tl,Bi)-1223 and (Tl,Bi)-1212 are just as narrow as that of the YSZ layer. The reason that FWHM values of YSZ instead of CeO_2 is used for comparison is that the signal from CeO_2 is very weak due to the limited thickness of CeO_2 (about 20 nm).

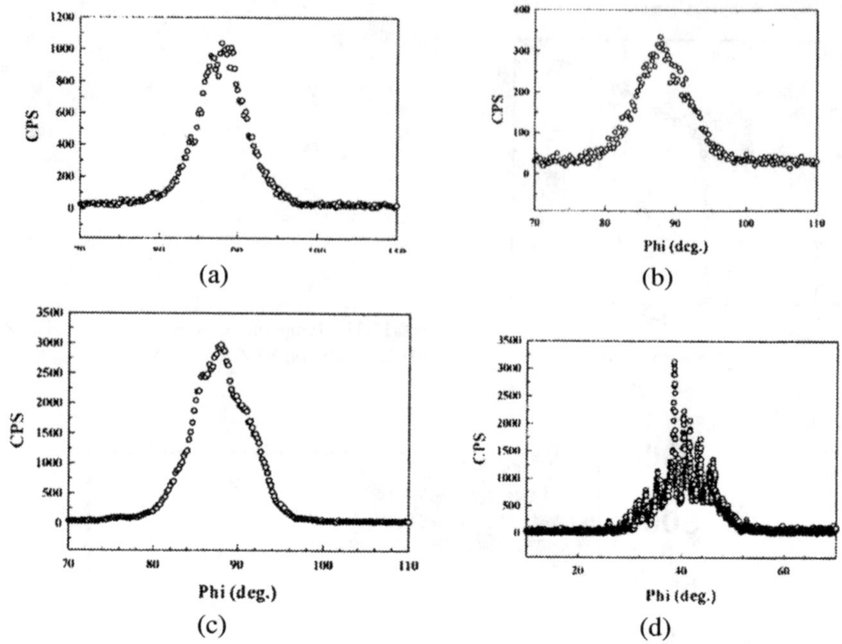

Figure 17.13. ϕ scan of: (a) Tl-1223 (102); (b) Tl-1212 (102); (c) YSZ (111), and (d) Ni (111).

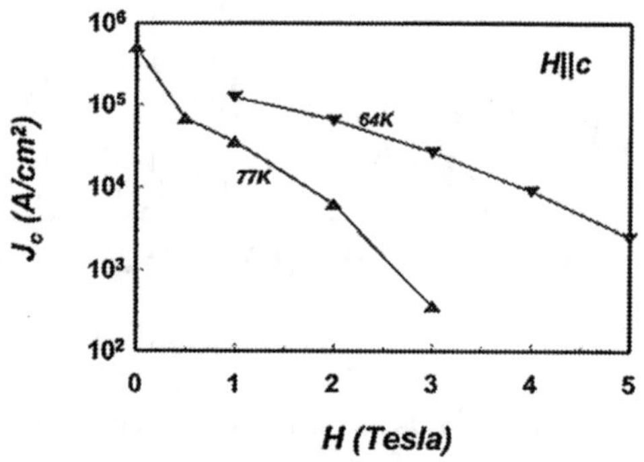

Figure 17.14. Magnetic field dependence of J_c of (Tl,Bi)-1223 film on CeO_2-capped RABiTS.

Transport T_c and J_c measurements have been carried out on typical (Tl,Bi)-1223 films on CeO_2-capped RABiTS. Typical resistivity vs. temperature measurements showed T_c's (zero-resistance) in the range of 105–110 K. A critical current (I_c) of 24.5 amperes was measured at 77 K and zero-field on a sample with a thickness of 1.8 μm and a width of 3.2 mm. This corresponds to a zero field J_c of 4.25×10^5 A/cm^2 at 77 K. Figure 17.14 shows the typical magnetic field dependence of transport J_c at 77 K and 64 K with the field applied along *c*-axis. These J_c vs. H curves show irreversibility fields reached beyond 3 T at 77 K and beyond 5 T at 64 K.

In conclusion of the results of epitaxial growth of (Tl,Bi)-1223 films on RABiTS by PLD and *ex-situ* annealing by crucible method, the potential of (Tl,Bi)-1223 films as the future electric conductor has been demonstrated.

17.2.3 Epitaxial Superconducting $Tl_{0.5}Pb_{0.5}Sr_{1.6}Ba_{0.4}Ca_2Cu_3O_9$[(Tl,Pb)-1223] Films on $LaAlO_3$ by Thermal Spray and Post-Spray Annealing

As described before, epitaxial superconducting films of (Tl,Bi)-1223 on $LaAlO_3$ (Ren et al., 1994, 1996; Wang et al., 1995), YSZ (Guo et al., 1997), CeO_2-coated YSZ (Ren et al., 1998b), and RABiTS (Ren et al., 1998a, 1999) have shown high J_c's at 77 K with good performance in an external magnetic field. However, all these high-J_c films were obtained by a very slow and expensive processing method: laser ablation and post-deposition annealing. This pulsed laser deposition technique is too costly for large-scale fabrication of long length wire. For scaled-up production of high J_c tapes, a simpler and faster method is necessary. Deluca et al. described a method for the preparation of superconducting $Tl_xCa_2Ba_2Cu_3O_9$ films by the reaction at 860°C of thallium oxide vapour with spray deposited Ca–Ba–Cu-oxide film containing silver on YSZ substrate (Deluca et al., 1993). Films are prepared routinely with zero resistance T_c of 104–107 K and zero J_c (77 K) in excess of 2×10^4 A/cm^2 with high value over 1.1×10^5 A/cm^2. Schulz described another route to synthesise thick (5–20 μm) highly c-axis textured, nearly phase pure superconducting $(Tl,Pb)(Ba,Sr)_2Ca_2Cu_3O_9/Ag_{0.37}$ tapes (Schulz et al., 1994). First, a Tl-free ink precursor powder in an ethanolic ethyl cellulose binder is sprayed onto a heated $LaAlO_3$ substrate. After an intermediate oxygen anneal to burn off the carbonaceous binder, the films were thallinated in a static two-zone furnace to get fully crystallised superconducting phase. Films exhibit excellent c-axis texturing with partial melting morphology as evidenced by SEM. Electrical characterisation of these films give zero resistance T_c of 99–101 K and transport J_c (77 K) up to 2.9×10^4 A/cm^2. Lee et al. reported the preparation of $Tl_{0.5}Pb_{0.5}Sr_{1.7}Ba_{0.3}Ca_2Cu_3O_x$ superconducting films by spin coating from metal acetate sol (Lee et al., 1994). Nearly phase pure films on MgO and $SrTiO_3$ substrates with preferred grain orientation were obtained. The T_c's of the films were above 120 K and J_c's are above 10^5 A/cm^2 at 77 K. He et al. reported the preparation of thick $TlBa_2Ca_2Cu_3O_y$ films (10 to 30 μm) with high J_c on polycrystalline Ag substrates by spin coating method followed by thallination in two-zone furnace (He et al., 1995). The films have highly textured c-orientation with FWHM of 4° for the (006) peak. SEM showed a dense, plate-like layered structure and almost no reaction between the film and the Ag substrate was found. The J_c at 77 K reaches up to 2.5×10^4 A/cm^2 in zero field and more than 1000 A/cm^2 in 1 T field with H//c.

Recently, big progress has been made on electrochemical (Bhattacharya et al., 1998a, 1998b, 1999, 2000) and thermal spray pyrolysis (Mogro-Campero et al., 1995; Specht et al., 1996; Paranthaman et al., 1997; Li et al., 1999) depositions of high quality Tl-1223 films. Here, we describe the recent development of fabricating epitaxial high-J_c (Tl,Pb)-1223 films on single crystal $LaAlO_3$ substrates by thermal spray and post-spray annealing (Li et al., 1999). The results on single-crystal will establish a base-line reference and proof of principle for transition to a suitably buffered metallic substrate such as RABiTS. The overall goal is to develop a Tl-based superconducting HTS wire with $J_c > 10^5$ A/cm^2 at 77 K and zero field with a high degree of biaxial texture on an oriented metallic substrate.

The spray solution was formed by dissolving $Tl(NO_3)_3$, $Pb(NO_3)_2$, $Sr(NO_3)_2$, $Ba(NO_3)_2$, $Ca(NO_3)_2$, and $Cu(NO_3)_2$ in distilled water according to a stoichiometric formula $Tl_{0.5}Pb_{0.5}Sr_{1.6}Ba_{0.4}Ca_2Cu_3O_9$. The transparent solution with a total concentration of 0.64 M was light blue. Before spraying, the solution was warmed to about 85°C, and the substrates were attached to a heater by Ag-paint for good thermal contact. The temperature of the heater was monitored by a thermocouple embedded inside

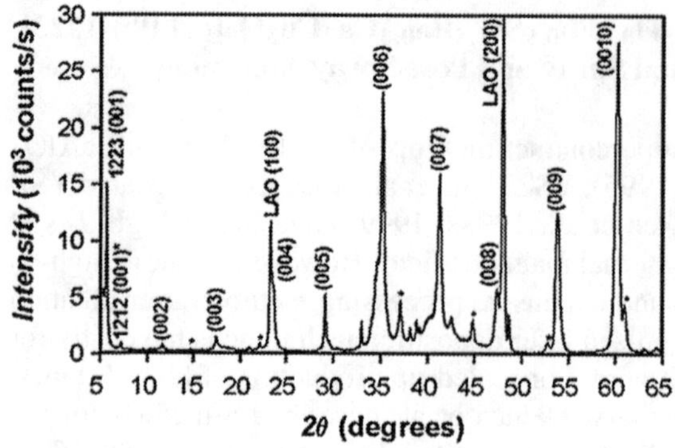

Figure 17.15. XRD θ–2θ scan of (Tl,Pb)-1223 film made by thermal spray and pyrolysis.

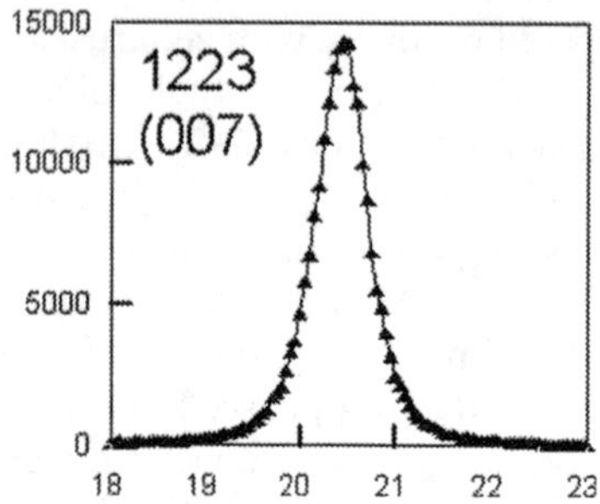

Figure 17.16. ω scan of (007) peak of (Tl,Pb)-1223.

the heater. When a stable temperature of about 490°C was reached, the spray process was started. The sprayer was pressurised by compressed air to a pressure of 3.0 psi. A total spray-deposition time was controlled from 30 to 60 seconds, depending on the final film thickness requirement. This deposition rate is at least 100 times faster than most vacuum deposition techniques. As soon as the spray starts, the temperature begins to decrease. A decrease of 60–110°C was normally observed. During the spray deposition, the surface of the substrates changes from shiny to brown, and eventually to black. The as-sprayed films were rough and amorphous.

After spray deposition the film was wrapped in silver foil along with two-semicircular unfired pellets having the composition of $Tl_{0.5}Pb_{0.5}Sr_{1.6}Ba_{0.4}Ca_2Cu_3O_9$. The wrapped package was placed in a furnace and annealed in air in a two-step program. The temperature was raised from room temperature to 650°C at a rate of about 10°C/min, held at 650°C for one hour. The purpose of this initial temperature soak is to decompose all the nitrates into oxides. The temperature was then increased to 870°C at the same rate and held for another 40 minutes so that the amorphous film could fully react and crystallise. After the annealing, the films were uniform and shiny.

The XRD $\theta/2\theta$ spectrum of a typical (Tl,Pb)-1223 superconducting thin film on $LaAlO_3$ is shown in Figure 17.15. All the major reflections are indexed as either (00l) peaks of (Tl,Pb)-1223 phase or those of (Tl,Pb)-1212 phase, which are marked by "$*$." The presence of only (00l) peaks shows that the films are strongly c-axis aligned. The degree of c-axis alignment has been determined by θ-scans (rocking curve) of the (Tl,Pb)-1223 (007) peak, as shown in Figure 17.16, with a full width at half maximum (FWHM) value of 0.79°. The θ-scans of (Tl,Pb)-1212 (005) shows a FWHM value of 0.73°. From the FWHM values, it is seen that the out-of-plane alignment of (Tl,Pb)-1223 and (Tl,Pb)-1212 are the same, which is understandable since they are intergrown phases. The in-plane (a- and b-axes) alignment was measured by XRD ϕ-scans of the

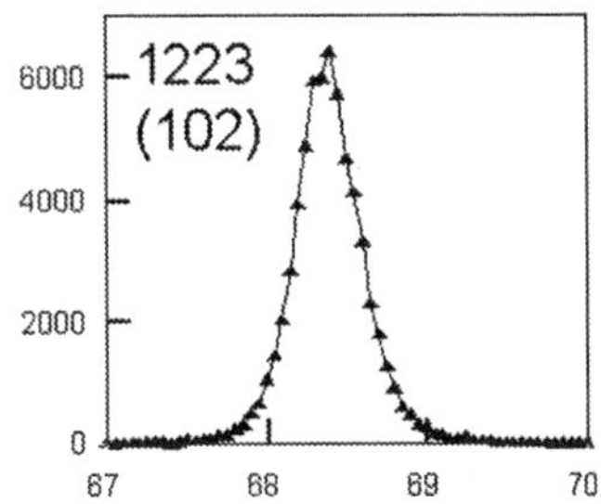

Figure 17.17. ϕ scan of (Tl,Pb)-1223 (102) peak.

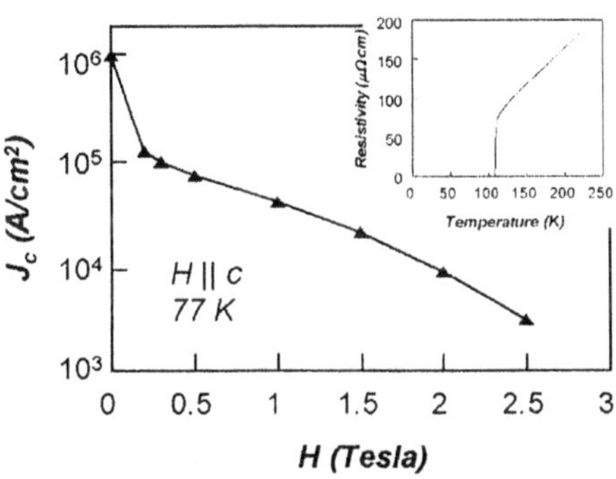

Figure 17.18. Magnetic field dependence of J_c of (Tl,Pb)-1223 film.

(Tl,Pb)-1223 (102) peak, as shown in Figure 17.17. The FWHM value of this peak is only 0.50°, indicating very good in-plane alignment. The ϕ-scans of the (Tl,Pb)-1212 (102) shows a FWHM value of 0.6°. The FWHM values of the superconducting-phase ϕ-scans are comparable to that of the $LaAlO_3$ substrates, confirming that full epitaxial alignment was obtained. From the intensity of the θ- and ϕ-scans, it was concluded that the (Tl,Pb)-1212 phase is indeed much less than the (Tl,Pb)-1223 phase, which is in very good agreement with the $\theta/2\theta$ spectrum shown in Figure 17.15. With further optimisation of both the spray deposition starting composition and annealing conditions, the amount of (Tl,Pb)-1212 phase could be greatly reduced.

Transport T_c and J_c measurements were carried out on one of the typical (Tl,Pb)-1223 samples. Figure 17.18 shows the magnetic field dependence of transport J_c at 77 K, with an inset showing the relationship of resistivity vs. temperature. The zero-resistance transition temperature is determined to be 108 K. A transport I_c of 51 A was obtained at 77 K in zero applied field. For this sample of 4.0 mm width and 1.2 μm thickness, a self-field J_c of 1.1×10^6 A/cm^2 was realised. The measured irreversibility field at 77 K was about 3 T, defined by a quadratic power-law dependence of voltage on current. With better control on both the spray deposition and annealing process parameters, the (Tl,Pb)-1212 phase could be eliminated, and even higher J_c could be obtained.

The SEM micrograph in Figure 17.19, taken at an inclined angle to the surface, shows that the surface is relatively dense and smooth, although there are a number of voids and particles in the size range of 1–6 μm. This observation is consistent with transmission optical microscopy, which shows light and dark areas. If the voids can be reduced to a very low level or even eliminated, a significantly higher J_c is expected. Probably the most important factor contributing to the formation of voids and particles is the surface roughness after spraying. So developing a smoother initial surface will be interesting.

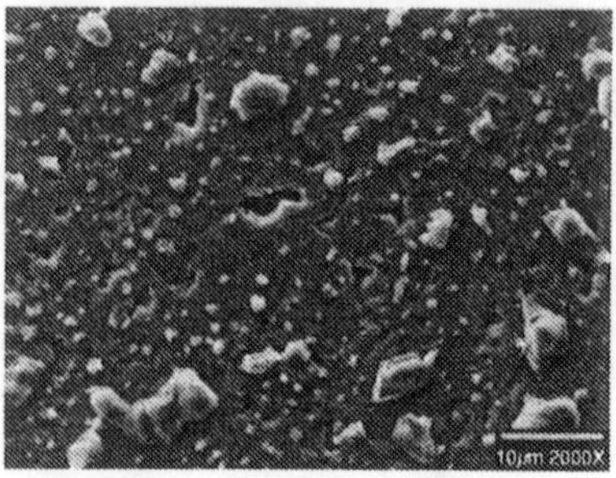

Figure 17.19. SEM micrograph of surface of (Tl,Pb)-1223 film.

17.2.4 Electrodeposition of Epitaxial (Tl,Bi)-1223 Films on Single Crystal Substrate (Bhattacharya et al., 1998a, 1998b, 1999, 2000)

The electrodeposited precursor films were obtained by coelectrodeposition of the constituent metals using nitrate salts dissolved in dimethyl sulfoxide (DMSO) solvent. The electrodeposition was performed in a closed-cell configuration at 24°C. A number of electrodeposition runs were performed with different electrolyte compositions, and the precursor films were analysed by inductively coupled plasma (ICP) spectrometry to establish the stoichiometric ratios of the deposited elements. The cation ratios of the electrodeposition bath were adjusted systematically to obtain $Ba_2Ca_2Cu_3Ag_{0.2}$(BCCO-Ag) and $(Tl,Bi)_{0.9}Sr_{1.6}Ba_{0.4}Ca_2Cu_3Ag_{0.2}$(TBSBCCO-Ag) precursor compositions. A typical electrolyte bath composition for depositing BCCO-Ag films consisted of 57.56 at% Ba [$Ba(NO_3)_2$], 27.41 at% Ca [$Ca(NO_3)_2{\cdot}4H_2O$], 15.03 at% Cu [$Cu(NO_3)_2{\cdot}6H_2O$], and 0.9 at% Ag [$AgNO_3$] dissolved in DMSO solvent. A typical electrolyte bath composition for the TBSBCCO-Ag films consisted of 2.7 at% Tl [$TlNO_3$], 1.5 at% Bi [$Bi(NO_3)_3{\cdot}5H_2O$}, 43.3 at% Sr [$Sr(NO_3)_2$]$\cdot4H_2O$], 11.3 at% Cu [$Cu(NO_3)_2{\cdot}6H_2O$], and 0.9 at% Ag [$AgNO_3$] dissolved in DMSO solvent. The substrates were single crystal $LaAlO_3$ (LAO) coated with 300 Å Ag, and commercial-grade flexible 0.125 mm thick Ag foils (99.9% pure). The films were electroplated by using a pulse-potential cycle of 10 s at −4 V followed by 10 s at −1 V and also at −3 V constant potential. All samples were electrodeposited in a 'vertical cell,' where the electrodes (working, counter, and reference) were suspended vertically from the top of the cell. All chemicals were of Analar or Puratronic grade purity and were used as received. The reference electrode was Ag (pseudo-reference) and the counter electrode was a Pt gauze. A Princeton Applied Research potentiostat/galvanostat Model 273A with an IBM PC AT computer interface was used for controlling the pulsed-potential electrolysis and to monitor the current and voltage profiles.

A two-zone thallination process was used to react the electrodeposited BCCO-Ag films on Ag foils or Ag coated LAO. The reaction consists of inserting the sample and heating to 860°C while a separate Tl source was kept initially at a low temperature of 685°C for 24 min and then increased to 728°C and held for 34 min. The thallination was carried out using a flowing O_2 ambient at 1 atm. The processed films on Ag foil show phase pure, *c*-axis oriented Tl-1223 phase, but not biaxially aligned as indicated by pole figure. The transport J_c at 77 K and zero field is 7.0×10^4 A/cm^2 for a 1 μm thick film and 2×10^4 A/cm^2 for a 10 μm thick film. The processed eletrodeposited BCCO-Ag precursor films on LAO substrates are between 1 and 5 μm thick. A representative X-ray diffraction of the annealed electrodeposited BCCO-Ag film on 300 Å Ag/LAO shown in Figure 17.20 indicates a highly phase-pure, *c*-axis-oriented 1223

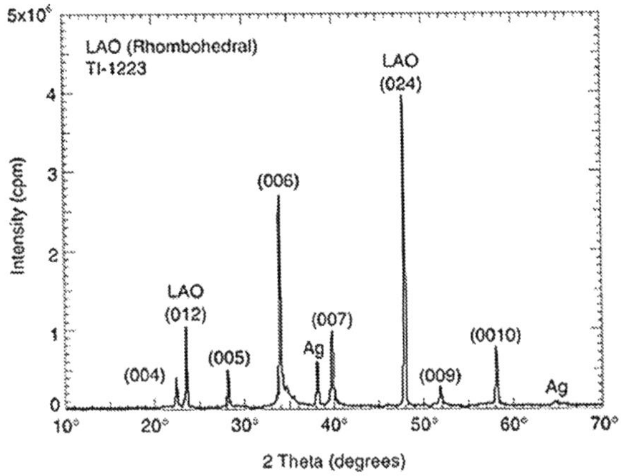

Figure 17.20. XRD θ–2θ scan of ED-BCCO film on Ag foil.

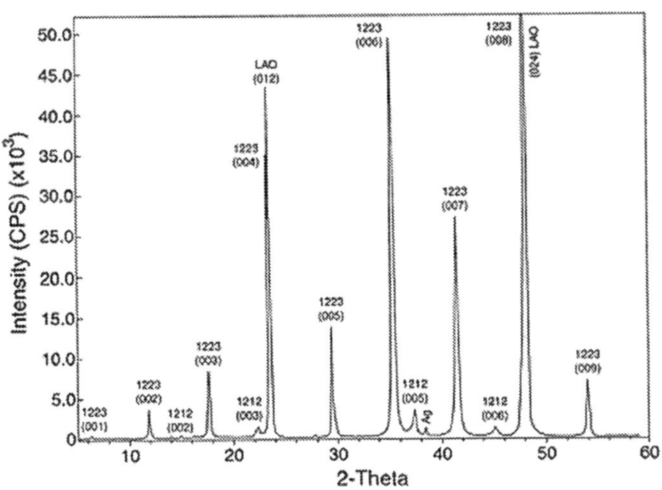

Figure 17.21. The X-ray diffraction of ED-TBSBCCO-Ag film on 300 Å Ag/LAO.

TBCCO phase, The pole figure measurements of the (103) *hkl* peak for all electrodeposited films on 300 Å Ag/LAO, in the range between 1 and 5 μm thick, show biaxial texturing. The phi scan indicates the full width at half maximum (FHM) value of only 0.7°, which shows a very high quality film. SEM analysis of the film showed plate-like structure development. The T_c measurement of a 1 μm thick electrodeposited film is over 105 K. At 76 K and zero magnetic field, the transport critical current density was $\geqslant 1.6 \times 10^5$ A/cm^2.

An electrodeposited TBSBCCO-Ag precursor film on 300 Å Ag/LAO, annealed in air at 870°C in the presence of a TBSBCCO pellet, shows major Tl-1223 phase development, with 1212 as a minor phase, as demonstrated in Figure 17.21. The pole-figure measurements of the (105) *hkl* peak show biaxial texturing (Figure 17.22) indicate the full width at half maximum (FWHM) of only 0.9° and 1.2°, respectively, which indicate a very high-quality film. The SEM analyses of the presently annealed film show dense and melted plate-like structure development (Figure 17.23). The SEM analyses

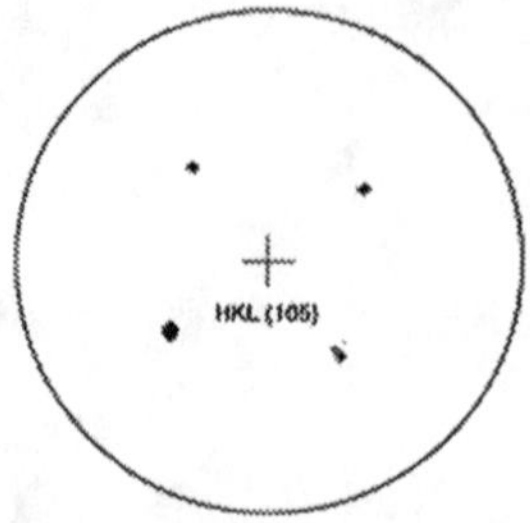

Figure 17.22. Pole figure of ED-TBSBCCO-Ag film on 300 Å Ag/LAO.

Figure 17.23. Surface feature of ED-TBSBCCO-Ag film on LAO.

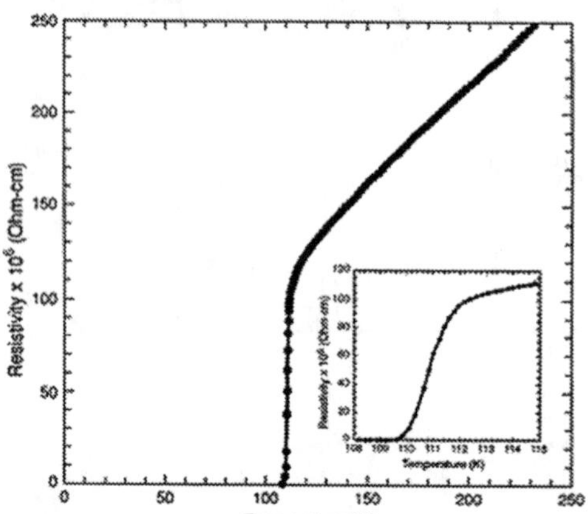

Figure 17.24. Temperature dependence of resistivity of as-annealed ED-TBSBCCO-AB on LAO.

of the annealed film also indicate about 30% void in the film caused pinhole formation. The thickness of the film varied from 0.6 to 0.9 μm. Figure 17.24 shows the temperature dependence of resistivity of this film, with superconductive transition temperature determined to be around 110 K. At 77 K and no magnetic field, the transport current of TBSBCCO on 300 Å Ag/LAO was 24.2 A, which corresponds to 7.84×10^5 A/cm^2 critical current density.

However, from the cost perspective, thicker films with good total current capacities are required. The film thickness indeed can be increased by using longer deposition time, but the film morphology was poor. A two-layer technique was then tried which used two layers of electrodeposited TBSBCCO films with an intermediate layer of Ag to improve the film uniformity. Better film quality was observed in presence of dissolved oxygen and with excess copper deposited in the precursor film.

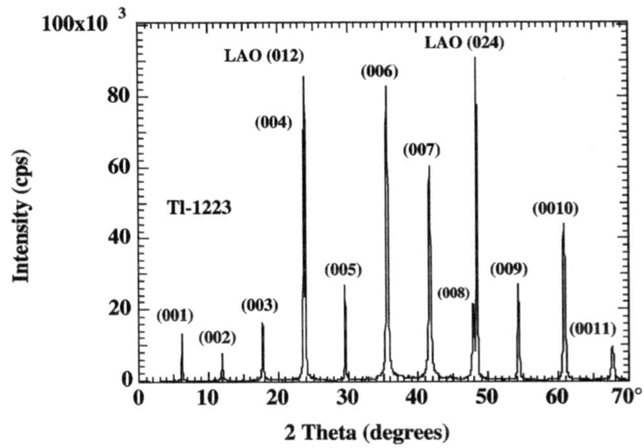

Figure 17.25. XRD θ–2θ scan of a two layer ED-TBSBCCO-Ag film on LAO.

The cation ratios of the electrodeposition bath were adjusted systematically to obtain $(Tl,Bi)_{1.1}Sr_{1.6}Ba_{0.4}Ca_2Cu_4$(TBSBCCO) precursor compositions. The increase of Cu content in the film from 3 to 4 can help to obtain good quality films reproducibly. The deposition process of two-layer technique for TBSBCCO films with an intermediate layer of Ag is as follows: (a) Single crystal substrates are coated with 300 Å Ag; (b) TBSBCCO films (0.8 μm to 1.3 μm) are prepared by electrodeposition (ED) on Ag/LAO; (c) 300 Å Ag is deposited on ED-TBSBCCO/Ag/LAO; (d) second layer of TBSBCCO is electrodeposited (0.8 μm to 1.3 μm) on Ag/ED-TBSBCCO/Ag/LAO and the complete two layer system is reacted.

To determine the effect of dissolved oxygen on deposition potential, a cyclic voltammogram experiment was performed on a solution mixture containing $Bi(NO_3)_3{\cdot}5H_2O$, $Ba(NO_3)_2$, $Ca(NO_3)_2{\cdot}4H_2O$, and $Cu(NO_3)_2{\cdot}6H_2O$ dissolved in DMSO solvent with and without bubbled oxygen. The reduction peaks of the corresponding Bi, Ba, Ca, and Cu were shifted towards the favourable positive direction in presence of oxygen. The deposited materials were more rigid in presence of oxygen and were not stripped significantly from the electrode surface on the positive-going scan. This behaviour is most likely due to the deposition of BiBaCaCu-oxide precursor as described by the following reactions

$$M^{n+} + \frac{n}{2}O_2 + ne^- \rightarrow MO \quad E_o < E_o\ (\text{Std}). \tag{1}$$

An electrodeposited TBSBCCO precursor film on 300 Å Ag/LAO, annealed in air at 870°C in the presence of a TBSBCCO pellet, shows Tl-1223 phase development, as demonstrated by XRD scan of Figure 17.25. The pole-figure measurement of the (105) *hkl* peak shows biaxial texture (Figure 17.26). The omega scan and phi scan indicate the full width at half maximum (FWHM) of only 0.92° and 0.6°, respectively, which indicate a very high-quality film. The SEM analyses of the presently annealed two-layer film show dense and melted plate-like structure development without any voids (Figure 17.27), compared with the previous one layer annealed film with void. The thickness of the annealed two-layer film varied from 0.8 μm to 2.6 μm. The superconductive transition temperature of this film determined resistively is about 110 K. Figure 17.28 shows the critical current density vs. magnetic field values at 77 K of

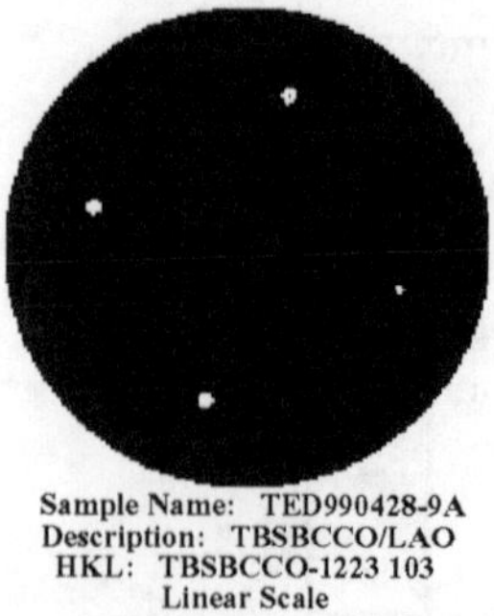

Figure 17.26. Pole figure of two layer ED-TBSBCCO-Ag film on LAO.

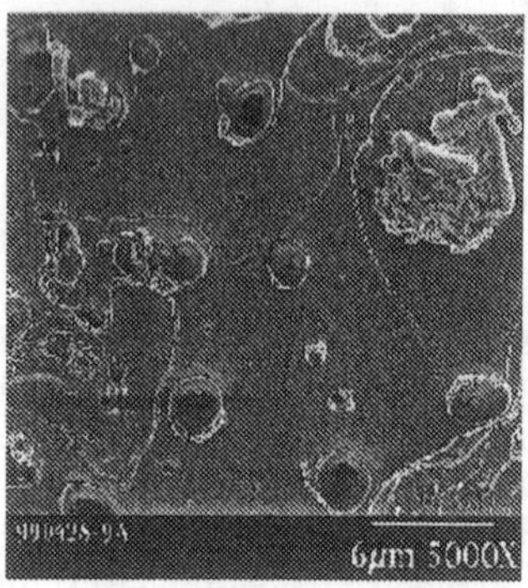

Figure 17.27. SEM micrograph of two layer ED-TBSBCCO-Ag film on LAO.

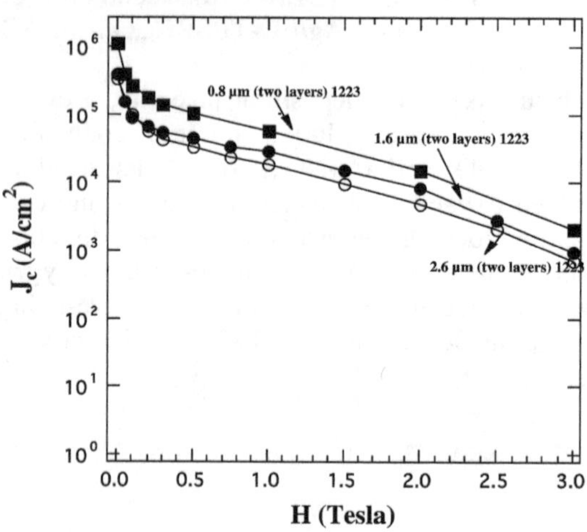

Figure 17.28. The magnetic field dependencies of transport J_c at 77 K (H//c) for two-layer 0.8 μm, 1.6 μm, and 2.6-μm ED-TBSBCCO/Ag/ED-TBSBCCO/Ag/LAO film.

0.8 μm, 1.6 μm, and 2.6 μm two layer films. At 77 K and no magnetic field, critical current density value of a two-layer 0.8-μm thick film is 1.1×10^6 A/cm^2. The critical current density of the film is calculated using the full cross-section of the sample (3.7 mm $\times$ 0.8 μm). The two layer 2.6 μm (width = 3.2 mm) thick film TBSBCCO film prepared by the electrodeposition process showed I_c of 28.24 A at 77 K (Normalised $I_c = 88.25$ A for 1 cm wide samples). The critical current density values vs. magnetic field measured at 40 K, 64 K and 77 K temperatures for two layers 1.6 μm and 2.6 μm thick films are shown in Figure 17.29 and Figure 17.30, respectively. These values for current density for ED Tl-1223 films are among the highest ever reported

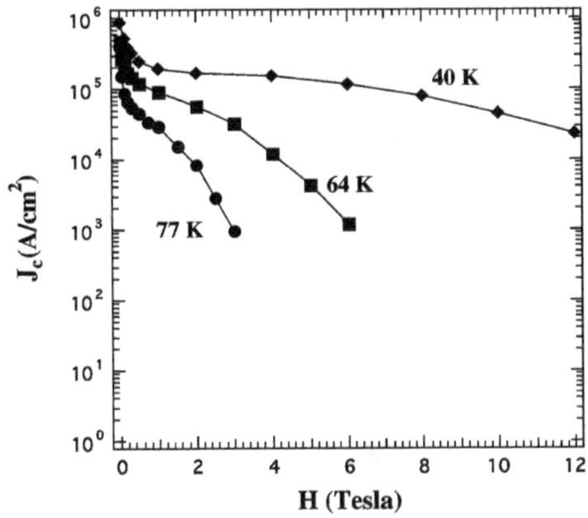

Figure 17.29. The magnetic-field dependence of transport J_C at 40 K, 64 K, and 77 K (H//c) for two-layer 1.6 μm ED-TBSBCCO/Ag/ED-TBSBCCO/Ag/LAO film.

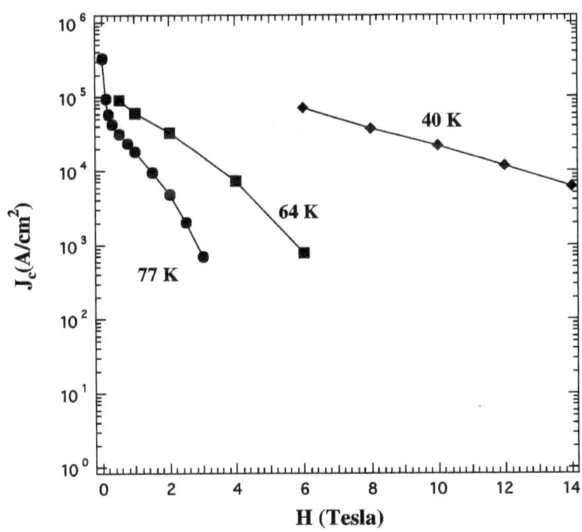

Figure 17.30. The magnetic-field dependence of transport J_C at 40 K, 64 K, and 77 K (H//c) for two-layer 2.6 μm ED-TBSBCCO/Ag/ED-TBSBCCO/Ag/LAO film.

for a processing technique that does not involve a vapour transport method such as PLD, sputtering, or e-beam. These latter methods have demonstrated typical current densities for Tl-1223 epitaxial films on single-crystal substrates (LAO, YSZ, etc.) of around 10^6 A/cm^2 at 77 K in zero field, which represents the highest value that is also obtained on ED-TBSBSCCO film.

Films were also tried to grow on 300 Å Ag/CeO_2/YSZ/CeO_2/Ni substrate by electrodeposition method. The XRD scan found dominant 1212 phase. The pole figure of (103) peak shows the biaxial texture. The FWHM of a ϕ scan peak is around 12°. The transport J_c of the film is 1.36×10^5 A/cm^2 at 77 K and zero field. This result demonstrates the ability to grow textured Tl-based films on RABiTS. With the further improvement, the J_c values on RABiTS could be comparable to that on single crystals.

17.2.5 Recent Development of High J_c $Cu_{1-x}Tl_x$-1223 and TlCu-1234 Thin Films

Recently, a new bulk superconducting compound $CuBa_2Ca_3Cu_4O_{12-y}$ with T_c up to 118 K has been prepared under high pressure (Ihara et al., 1994, Ihara, 1995). This is a very potential important compound for electric conductor and magnetic applications because of its low superconducting anisotropy and long coherence length. Because the Cu-1234 compound is a high-pressure phase, it is difficult to synthesise its thin films under ambient pressure. Thallium was found to have marvellous chemical effects acting as reaction accelerator, structure stabiliser; charge-reservoir-layer component and T_c enhancement element (Siegal et al., 1997b; Ihara et al., 1997). The derivatives of this phase in the form of CuTl-1223 phase can been synthesised at ambient and high quality CuTl-1223 thin films have been successfully synthesised on $SrTiO_3$ substrate (Ihara et al., 1999).

The $Cu_{1-x}Tl_x$-1223 films were grown by the combination of the amorphous phase epitaxial (APE) technique and the Tl effect in the annealing process. This technique can transform amorphous precursor films into crystalline superconducting phases by using the substrate effect on epitaxy growth and grain boundary diffusion effect. First, amorphous $Cu_{1-x}Tl_x$-1223 phase was deposited on the $SrTiO_3(100)$ single crystal substrates by sputtering. Then, the films were treated by enclosing them in an Ag capsule with $Cu_{1-x}Tl_x$-1223 pellets ($x = 0.25{\sim}0.5$) at 855~890°C for 30~60 min. A nearly equilibrium condition between films and $Cu_{1-x}Tl_x$-1223 pellets is important to get homogeneous and high T_c, J_c thin films.

The XRD of the as annealed samples is shown in Figure 17.31. All the peaks are assigned as $(00l)$ plane of substrate, $Cu_{1-x}Tl_x$-1223, -1234, -1212 phases. The $Cu_{1-x}Tl_x$-1223 phase is an over 80% major phase. The lattice parameter of the c-axis is 15.45 Å, which is between the value of 14.79 Å for Cu-1223 and 15.93 Å for Tl-1223. The pole figure and ϕ scan found the excellent biaxially alignment of the film with the substrate. The FWHM of the (102) peak is only $\Delta\phi = 0.7°$. The Tl content reduces with the increase of temperature during 50 min and approaches that of pellet content ($x = 0.25$) during annealing, as measured by EDX. The optimum value of Cu occupation for the high J_c was $1 - x = 0.5 + 0.1$ for the $Cu_{1-x}Tl_x$-1223 thin films. The T_c's of the films are in the range of 100 to 113 K, while Figure 17.32

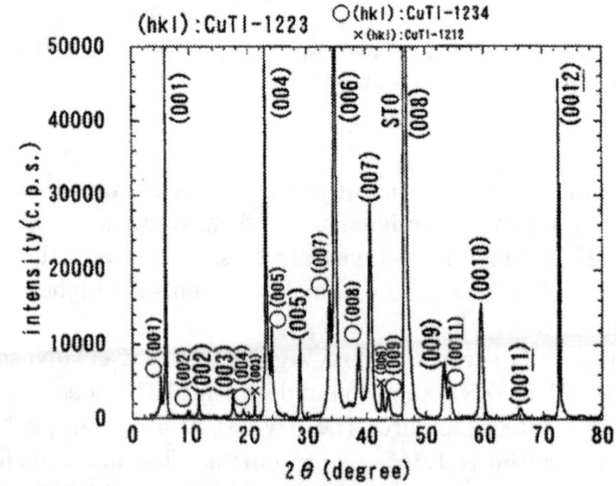

Figure 17.31. XRD diffraction data of $Cu_{1-x}Tl_x$-1223 film.

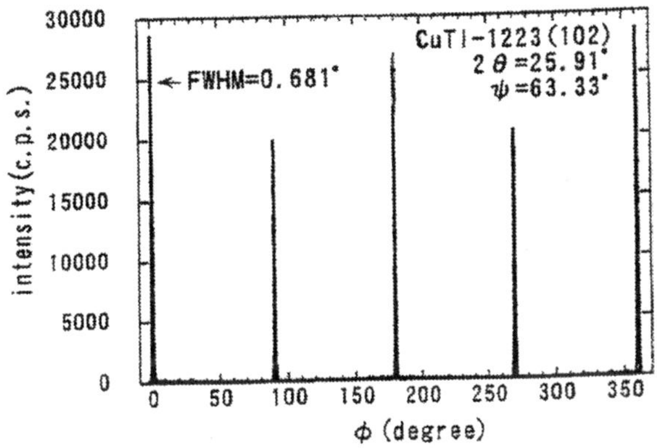

Figure 17.32. ϕ scan of $Cu_{1-x}Tl_x$-1223 film.

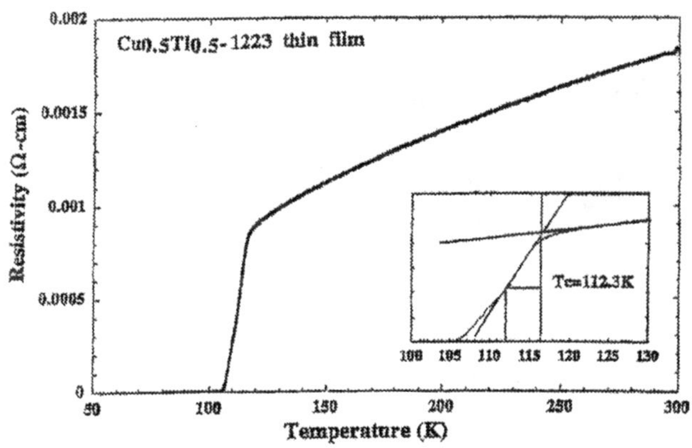

Figure 17.33. The transition of resistivity of $Cu_{1-x}Tl_x$-1223 film.

showing temperature dependence of a $Cu_{0.5}Tl_{0.5}$-1223 thin films with T_c 112 K. The transport J_c of the films with thickness of 0.2∼1.0 μm are 1 to 5 × 10^6 A/cm^2 at 77 K and zero field with high value of 2 × 10^7 A/cm^2. Figure 17.33 shows the magnetic field dependence of J_c. The maximum J_c values of the $Cu_{1-x}Tl_x$-1223 thin films are 1.2 × 10^6 and 4 × 10^5 A/cm^2 for 6 and 10 T at 77 K, which is twice the values of YBCO thin films. This value is also comparable to the highest J_c values of low temperature conductor of Nb_3Al at 4.2 K and very promising for the future wire application.

Since the anisotropy (Ihara et al., 1996a, 1996b) of these compound is found to decrease with the increase in number of Cu–O planes, which means that the anisotropy of TlCu-1234 is lower than that of CuTl-1223 and hence it is capable of carrying higher current, it is interesting to synthesise Cu-1234 superconducting thin films for the future HTS wire in magnetic applications. Although Cu-1234 is a high-pressure phase, the existence of Tl can facilitate its phase formation. Epitaxial TlCu-1234 films have been successfully synthesised on $SrTiO_3$ substrate by RF sputtering of $CuBa_2Ca_3Cu_4O_x$ target and the following APE method (Khan et al., 1999). The APE process is performed by treating amorphous thin film in an Au capsule containing pellets of the composition $Cu_{0.5}Tl_{0.5}Ba_3Cu_4O_x$. The gold capsule is heated at 920°C for 60 min followed by quenching to room temperature after the heat treatment. The SEM of the surface of the film is shown in Figure 17.35. The as-annealed films have typical grain size

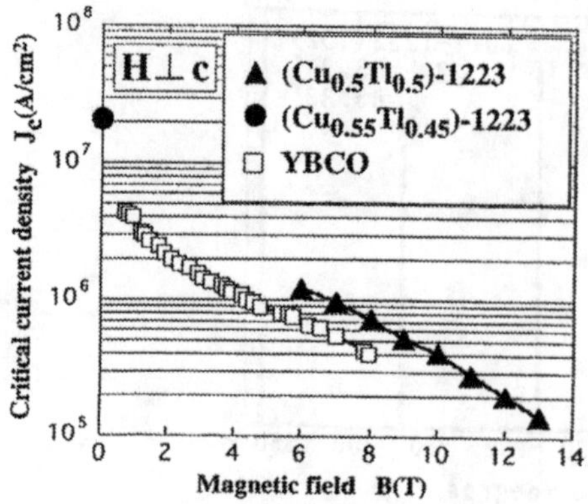

Figure 17.34. The magnetic field dependence of J_c of $Cu_{1-x}Tl_x$-1223 film.

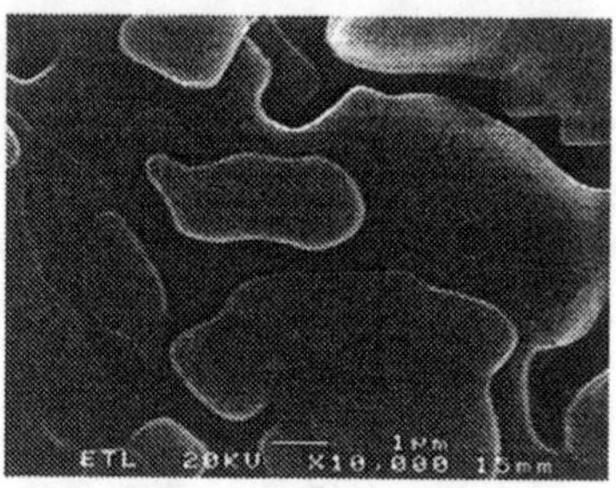

Figure 17.35. SEM micrograph of $Tl_{0.8}Cu_{0.2}Ba_2Ca_3Cu_4O_{12-y}$ thin film.

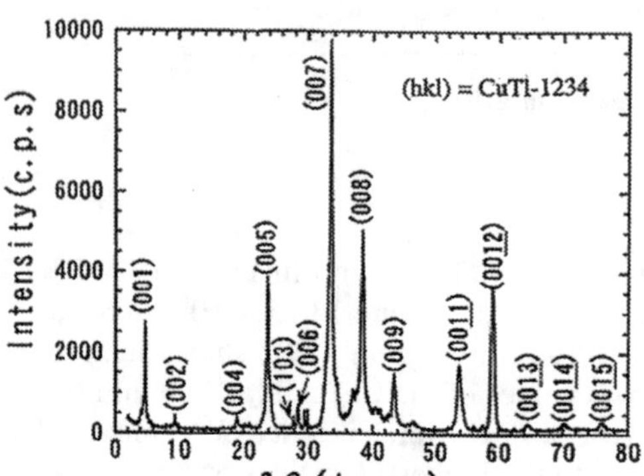

Figure 17.36. XRD θ–2θ scan of $Tl_{0.8}Cu_{0.2}Ba_2Ca_3Cu_4O_{12-y}$ thin film.

in few microns and these grains are well connected. The surface roughness is less than 0.2 μm for the 1 μm thick film. The XRD reflection is shown in Figure 17.36, which demonstrated a predominant single phase with the *c*-axis lattice constant of 18.74 Å. This lattice value is between the 17.99 Å for Cu-1234 and 19.11 Å for Tl-1234. The ϕ scan measurement of (103) peak of Figure 17.37 showing the in plane aligned film with FWHM of 0.8°. The composition of the films is $Tl_{0.8}Cu_{0.2}Ba_2Ca_3Cu_4O_{12-y}$ as measured by EDX. The T_c of the film is 113 K as shown in Figure 17.38 and the transport J_c is 3.3×10^5 A/cm^2 at 77 K and in zero field. The J_c is improved to 1.0×10^6 A/cm^2 after annealed in oxygen at 450°C for 20 h. Further optimisation is necessary to improve the superconducting properties of the CuTl-1234 films.

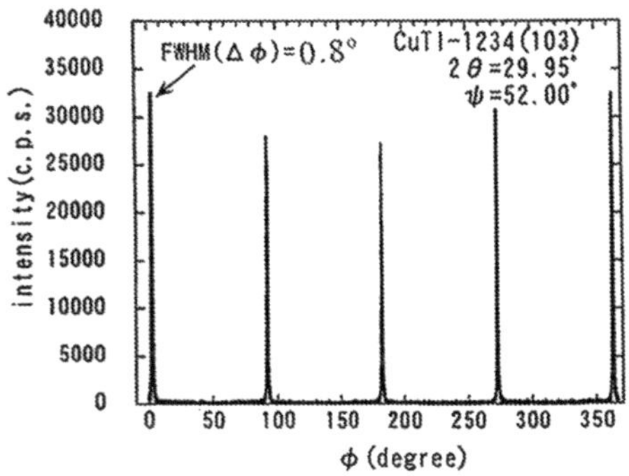

Figure 17.37. ϕ scan of (103) of $Tl_{0.8}Cu_{0.2}Ba_2Ca_3Cu_4O_{12-y}$ thin film.

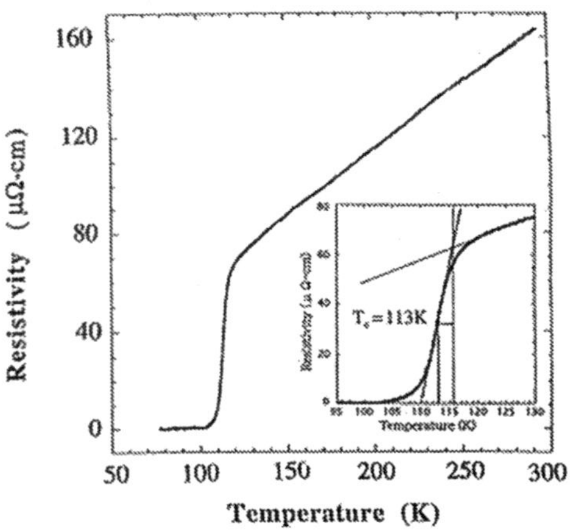

Figure 17.38. Temperature dependence of resistivity of $Tl_{0.8}Cu_{0.2}Ba_2Ca_3Cu_4O_{12-y}$ thin film.

17.3 DEVELOPMENT OF THALLIUM-1212 MATERIALS AS A POSSIBLE ALTERNATE FOR THE NEXT GENERATION OF HTS WIRES

17.3.1 Introduction

Tl-1212 has tetragonal structure, which resembles YBCO structure. Figure 17.35 shows the Tl-1212 structure. The O(2) and O(3) ions are strongly drawn toward Tl and Ca ions, respectively. The CuO_2 planes are not flat. As shown in Figure 17.36, the only structural difference between YBCO and Tl-1212 is that the Cu–O chain in the YBCO structure is replaced by the TlO plane and the Y atom in the YBCO structure is replaced by Ca atom of 1212 structure. This similarity can be further clarified if the $TlBa_2CaCu_2O_7$ structure is rewritten as $CaBa_2TlCu_2O_7$. Also, Tl-1212 system has the shortest insulating distance in the unit cell among all the single and double Tl system superconductors. This short insulating layer distance, and the structure similarity, could lead to its superior flux pinning property (Kim et al., 1991). Indeed, better intrinsic flux pinning properties has been observed by comparing normalised J_c of silver-sheathed tapes of $Tl_{0.5}Pb_{0.5}Sr_{1.6}Ba_{0.4}Ca_{0.8}Y_{0.2}Cu_2O_y$ and $Tl_{0.5}Pb_{0.5}Sr_{1.6}Ba_{0.4}Ca_2Cu_3O_y$ at 77 K under magnetic field (Ren and Wang, 1993b).

However, the synthesis of high J_c Tl-1212 film has not been as successful and film has much lower J_c and T_c values as compared to Tl-1223 film. While Tl-1223 single crystals have been grown with T_c around 105 K (Morosin et al., 1990), thin film samples typically have T_c ranging from 65 to 85 K. Mixed valence theory has been used to

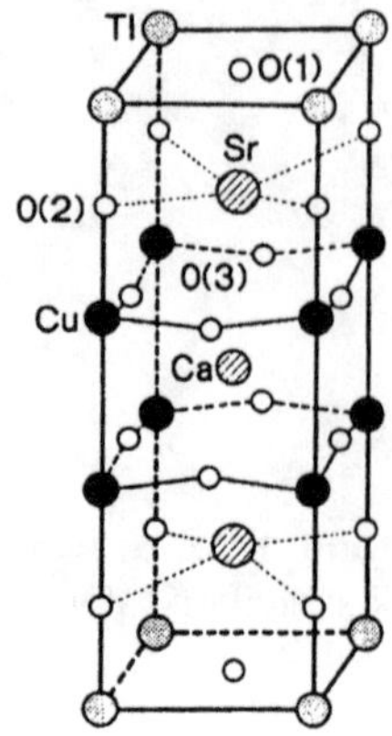

Figure 17.39. Structure of Tl-1212.

explain this phenomena (Presland and Tallon, 1991; Xia et al., 1989). According to this theory, the average copper valence is +2.5 for an ideal Tl-1212 structure if we assume +3 valence for Tl, +2 for Ba, Sr and Ca, and −2 for oxygen. This copper valence is higher than the generally accepted +2.2 optimal copper valence. The assumption that Tl is at +3 valence is reasonable since studies (Suzuki et al., 1989, 1994) have shown that Tl valence should be +3 for undoped single Tl–O layer compound since interstitial oxygen can only exist in double Tl–O layer (Tallon et al., 1989; Presland et al., 1991). Other study (Suzuki et al., 1989) also found that Tl in $TlBa_2CaCu_2O_{7-\delta}$ is trivalent compared to the valence between +3 and +1 of double layer Tl system. The T_c of Tl-1212 phase is also affected by other factors, such as Cu–O distance that is closely correlated to the copper valence. There are two ways to lower the Cu valence, one is the reductive annealing, and the other is the elemental substitution. Face and Nestlerode reported the first promising results for $TlBa_2CaCu_2O_7$ films for *in-situ* growth with *ex-situ* high temperature anneals in the presence of both oxygen and Tl-oxide to achieve T_c of 97 K (Face and Nestlerode, 1992). With a partial substitution of yttrium for calcium performed (Face and Nestlerode, 1993), the same high T_c value was achieved with only a post-deposition oxygen anneal. $TlBa_2CaCu_2O_7$ superconducting films were also successfully grown on $LaAlO_3$ substrates in a two-zone thallination furnace followed by reductive annealing (Siegal et al., 1995a). The resulting film has T_c round 100 K and J_c over 1×10^5 A/cm^2 at 77 K. Another type of Tl-1212 system, $TlSr_2CaCu_2O_7$, was reported to be superconducting at 70–80 K (Maysuda et al., 1988). However, this structure is not as stable as $TlBa_2CaCu_2O_7$ because Sr^{2+} is much smaller than Ba^{2+}. With elemental substitutions, such as (Pb,Bi) for Tl (Subramanian et al., 1988; Haldar et al., 1988; Li and Greenblaat, 1989), rare earth and Y for Ca (Sheng et al., 1989; Liu et al., 1989), the pure 1212 phase can be formed with T_c in the range of 75–90 K. One of the major reasons of the success of these substitutions is that Pb^{+4}, $Bi^{+3/+5}$, RE^{+3} and Y^{+3} reduce the copper valence from +2.5 to +2.2. A combination of both substitutions (Liu et al., 1989; Liang et al., 1990), has raised T_c value to above 105 K. Kountz et al. (1993) grew *ex-situ* (Tl,Pb)-1212 films with T_c around 88 K by Sr substituting for Ba using a standard closed crucible for thallination. Myers et al. reported the growth of highly epitaxial $(Tl,Pb)Sr_2Ca_{0.8}Y_{0.2}Cu_2O_7$ thin film by off-axis magnetron sputtering in the presence of Tl vapour on $LaAlO_3$ and $NdGaO_3(001)$ substrates with T_c 93 K (Myers et al., 1994). However, these *in-situ* growth methods are both cumbersome and difficult to scale for production. The volatility of Tl-oxides at high temperature makes it difficult to control the formation of single phase Tl-superconductors (Aselage et al., 1994).

17.3.2 Development of Two-Zone Furnace

As mentioned before, two-zone furnace allows the large-scale production of Tl-based HTS films. Here we describe the development of two-zone furnace method for the synthesis of Tl-1212 films (Siegal et al., 1995a, 1995b, 1997a, 1997b, 1998) made by group of scientist at Sandia National Laboratory. The furnace contains a high temperature and a low temperature zone with independent temperature control. The amorphous precursor films are placed in the high temperature zone of the furnace, while Tl_2O_3 is placed in the low temperature zone. The gas ambient is controlled by mass flow controllers. *This design allows independent control of the substrate temperature, oxygen partial pressure, and Tl-oxide partial pressure, each of which is a critical variable for the formation of a given Tl-superconducting phase* (Siegal et al., 1995a). First, precursor 0212 oxide films with thickness around 6000 Å are deposited onto $LaAlO_3$(100) substrate by off-axis sputtering from a Tl free BaCaCuO target. Then the films are put in a static two-zone furnace for thallination and crystallisation into superconducting phases. The substrate temperatures are kept in the range of 800 to 825°C during the annealing. The annealing time is 30 min and the annealing ambient is 0.8 atom of pure oxygen. Tl_2O_3 powder is used for the Tl-oxide source and preconditioned at 710°C (which is a higher operating temperature than that of typically used during film growth) for several hours, then the Tl_2O_3 powder is used repeatedly throughout the experiment. The optimum Tl-oxide source temperature for Tl-1212 phase formation is found to be in the range of 670 to 700°C, respectively. Certain match between substrate temperature and Tl-oxide source temperature is required for control purpose of Tl partial vapour pressure. To grow 1212 films at 825°C, the optimum Tl-oxide source temperature is 700°C. The T_c of these as-grown Tl-1212 films is 70 K. The superconducting properties of these films can be greatly improved by annealing in inert ambient such as nitrogen at temperatures ranging from 250 to 600°C. After annealing at 250°C for 1 h, Tl-1212 films can have T_c around 100 K, J_{cm} (5 K) $> 10^7$ A/cm^2 and J_{cm} (77 K) $> 10^5$ A/cm^2, as demonstrated in Figure 17.40 and Figure 17.41, respectively.

The best films reported to date grown in a two-zone furnace are not as god as those grown in crucibles. Film quality is dependent upon processing conditions that drift with usage. The variability in properties correlates with the inhomogeneity of thallination. The simple two-zone furnace fails because it does not take into account the kinetics of Tl_2O vapour transport from one end of the furnace to the other, i.e. How fast does the $p(Tl_2O)$ build-up at the film surface? This is partly dependent on the Tl-oxide source-to-film distances and on the total volume (within the furnace) that Tl_2O vapour must expand into. As the furnace size is increased (for growth of large-area films), these factors become significant. In addition, use of pure Tl_2O_3 powder as the

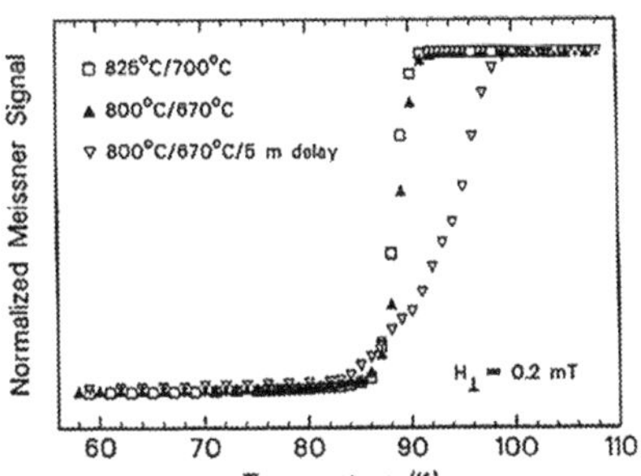

Figure 17.40. Meissner transition for Tl-1212 films grow at: filled triangle, 800°C; open square, 825°C; open triangle, Tl-1212 film grown with less Tl-oxide during the early stage of growth.

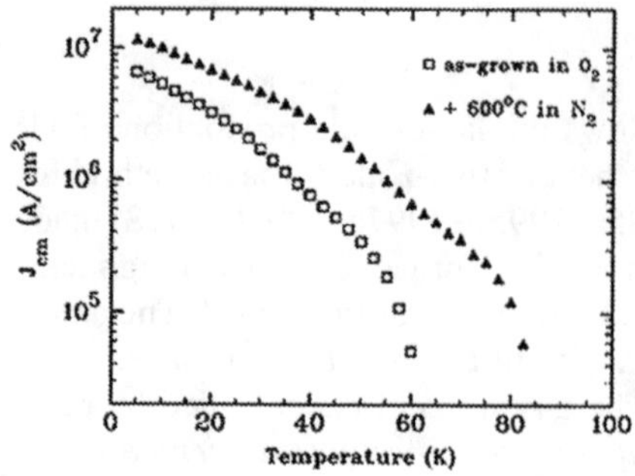

Figure 17.41. J_{cm} in zero field vs. temperature for Tl-1212 films as-grown in O_2 at 800°C and following a 1 h anneal in N_2 at 600°C.

source requires a detailed knowledge of the equilibrium Tl_2O partial pressure not just at the growth temperature used, but at all temperatures below the growth temperature where the film is reactive; i.e. $p(Tl_2O)$ at the film surface must be carefully controlled during the rise, hold and cooling cycles of the anneal (Siegal et al., 1998).

Therefore, a new concept two-zone furnace has been developed by studying the advantages of simple crucible method (Siegal et al., 1998). In the crucible process, the source of Tl_2O vapour is placed immediately below or besides the substrate so vapour transport is not a disturb issue. Also, Tl_2O vapour is supplied by a Tl–Ba–Ca–Cu–O mixture source rather than pure Tl_2O_3, which can control the thermodynamics and kinetics of release of Tl vapour from its source material in a manner similar to the uptake of Tl_2O vapour in a similar composition thin films. The new concept two-zone furnace was designed to achieve the attributes of the crucible process while maintaining the Tl-oxide content in a source material. The mixed Tl–Ba–Ca–Cu–O powder or pellet was put into the high temperature zone of the furnace as the main Tl_2O vapour source for the film growth. The Tl vapour pressure is controlled by the Tl-oxide content in the mixture, the source temperature and the oxygen partial pressure. This design simulates the crucible method by controlling Tl_2O vapour pressure at the film surface throughout the whole process. The Tl_2O_3 powder is placed at low temperature zone to maintain the Tl-oxide content in the mixture in the high temperature zone and fine tune the process for optimisation. The initial work of this new concept two-zone furnace appears to be very promising. Single-phase, highly *c*-axis aligned Tl-2212 thin films have been synthesised with smooth morphology, $T_c > 101$ K, and J_c around 1×10^7 A/cm^2 at 5 K and zero field for as many as twenty consecutive runs without having to change or add to the Tl-oxide source.

17.3.3 Recent Development of High Current Density Cr-Doped (Tl,Bi)SCCO Films on Single Crystal Substrates by Vacuum Method

Recently, the growth of Tl-1212 with high T_c and J_c as an alternative to Tl-1223 was studied by the authors' group. Among all the possible doping choice, a combination of Cr and Bi is found to be the best. Cr-doped (Tl,Bi)-1212 film, with transport J_c of up to 1.5×10^6 A/cm^2 at 77 K and self-field, have been successfully synthesised. The total annealing time in static air was less than an hour, which is ten times shorter than the previous reported times (Tang et al., 1993). More importantly, the annealing temperature window of 875–925°C is much larger than that of (Tl,Bi)-1223, 865–875°C. As far as to the authors' knowledge, these J_c values are the highest ever reported for Tl-1212 films. Potential applications, such as transmission cables, are very promising.

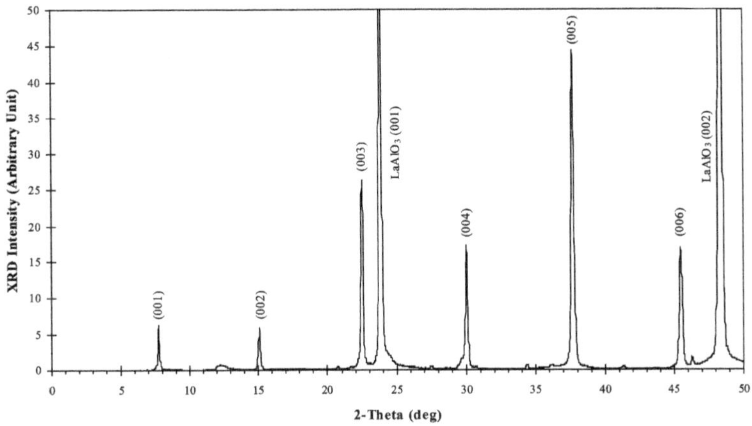

Figure 17.42. XRD θ–2θ pattern of Cr-doped (Tl,Bi)-1212 film on $LaAlO_3$ substrate.

Superconducting (Tl,Bi)$Sr_{2.0}Ca_{0.85}Cr_{0.15}Cu_2O_7$ films were prepared by pulsed laser ablation method using a reacted superconducting source target, followed by post-deposition annealing in muffle furnace in static air. In fabricating the reacted source target, a prepowder of $Sr_{2.0}Ca_{0.85}Cr_{0.15}Cu_2O_7$ was first prepared by grinding a stoichiometric mixture of $SrCO_3$, Cr_2O_3, CaO and CuO. The other procedures of target preparation are similar to that of Tl-1223 target except that amount of Tl in the target was maintained at 1 atom per unit. The films were deposited on $LaAlO_3$ (001) single crystals by an ArF 193 nm excimer laser, with the energy range from 90 to 120 mJ/pulse at the laser repeat rate of 4 Hz for 60 min with oxygen pressure around 25 mTorr. The resulting precursor film was then processed *ex-situ* in a muffle furnace at 885–905°C for 35 to 45 min in static air with the presence of two $(Tl_{0.85}Bi_{0.3})Sr_{2.0}Ca_{0.85}Cr_{0.15}Cu_2O_7$ semi-circular pellets for maintaining the partial pressure of Tl_2O to result in superconductive (Tl,Bi)$Sr_{2.0}Ca_{0.85}Cr_{0.15}Cu_2O_7$ film. The as-annealed film thickness is between 0.4 to 0.7 μm. The ICP emission spectroscopy measurements showed that the as-deposited film has only 0.45–0.55 atomic percent of Tl because of the high volatility of thallium during pulsed laser deposition. The as-annealed films have good superconductive properties although they are Tl deficient with composition of 0.5 atomic percentage. There is about 0.15 atomic percent of Bi in the as-annealed films, which transferred from the pellet. Reproducibility over more than 20 samples was very good.

The XRD θ–2θ diffraction spectrum of a good quality Cr-doped (Tl,Bi)-1212 film on $LaAlO_3$ substrate is shown in Figure 17.42. All the major reflections are indexed as (00l) peaks of (Tl,Bi)-1212 phase and (001), (002) peaks of the $LaAlO_3$ substrate. The strong (00l) peaks of the (Tl,Bi)-1212 phase indicate 1212 phase to be dominant with a large degree of uniaxial alignment of the c-axis normal to the substrate. Some weak minor impurity peaks are also found in the spectrum. The XRD ω scan of (005) peak of (Tl,Bi)-1212 phase showed a FWHM of only 0.58°, which indicates the good out-of-plane alignment of the (Tl,Bi)-1212 phase. The in-plane alignment was measured by a ϕ scan of the (103) peak of (Tl,Bi)-1212, as shown in Figure 17.43. The four strong equally separated peaks, with a FWHM value of 0.6°, indicate the excellent a- and b-axes alignment of the (Tl,Bi)-1212 phase.

Transport measurement of the (Tl,Bi)-1212 film showed that the zero-resistance T_c of the films is in the range of 94–100 K. Figure 17.44 shows the typical transport temperature-dependent resistivity curve of a (Tl, Bi)-1212 film with zero-resistance

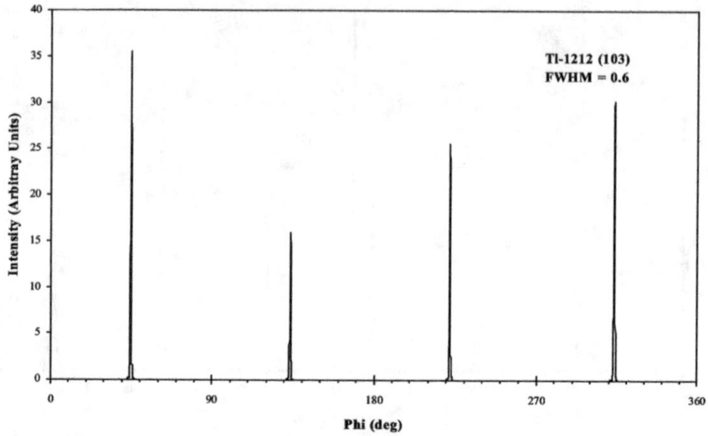

Figure 17.43. ϕ scan of Cr-doped (Tl,Bi)-1212 film on $LaAlO_3$ substrate.

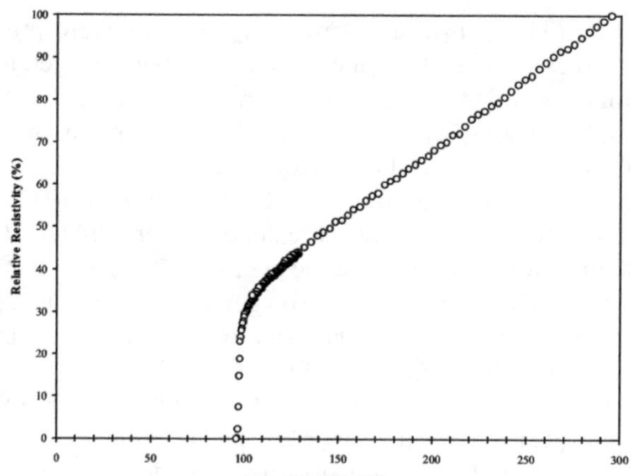

Figure 17.44. The temperature dependence of the resistivity, showing the zero-resistance superconducting transition temperature of 96.5 K.

T_c of 96.5 K. The highest transport J_c at self-field and 77 K was 1.5×10^6 A/cm^2, with reproducible J_c values of over 1.0×10^6 A/cm^2. Figure 17.45 shows the typical magnetic field dependence of the transport J_c at 77 K with the field applied parallel to c-axis. This curve shows the irreversibility line is $\sim$1.6 T, which is smaller than the 2.5 T of the (Tl,Bi)-1223 film grown on the $LaAlO_3$ substrate.

Figure 17.46 shows the surface feature of the as-annealed films. The film is smooth, well connected and plate like. Some pinholes can be found on the film surface. The small rods on the surface could be the minor-phase impurities shown in the XRD θ–2θ scan of Figure 17.42. Also, many small balls can be seen on the surface. The size and distribution of these corresponds to the small balls on the as-deposited film surface, and this feature is typical of PLD films. The superconducting properties should be further improved by optimising the stoichiometry, eliminating features such as pinholes, rods and balls.

A section of the film was analysed under cross-sectional TEM. Figure 17.47 shows the typical bright field TEM image of the cross section. The black spots in

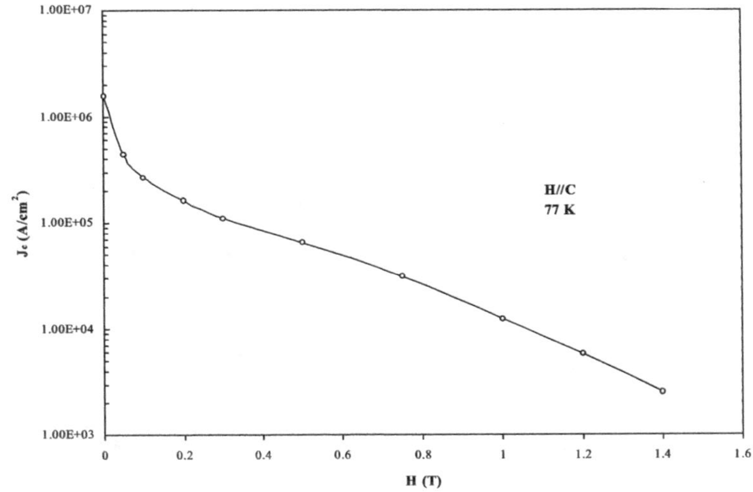

Figure 17.45. The magnetic field dependence of J_c for the field oriented perpendicular to the film plane (H//c) at 77 K.

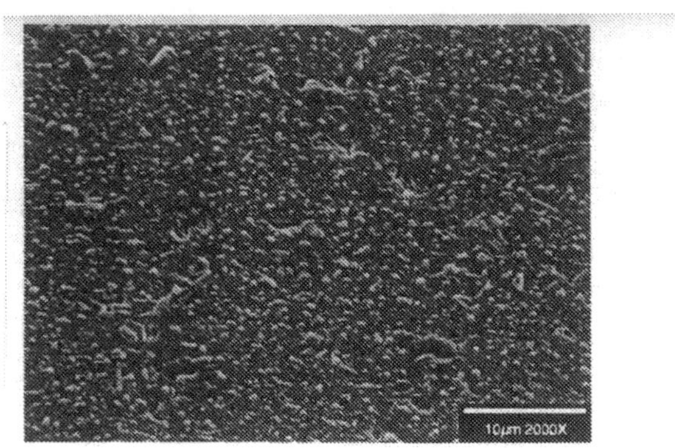

ıre 17.46. Surface microstructure of as- ꞔaled Tl-1212 film.

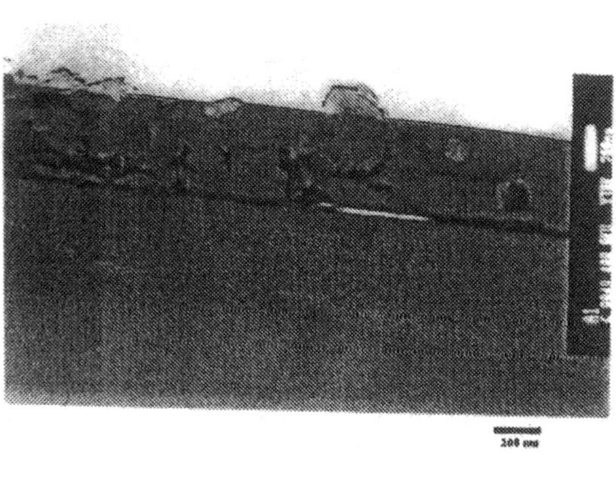

Figure 17.47. Typical bright field TEM image of Tl-1212 film on $LaAlO_3$ substrate.

the film are possibly impurity precipitates. This is not surprising considering the multi-element nature of the system, high volatility of thallium and particulates in the films due to the PLD method. For some areas, a very thin amorphous layer was found at the interface between the substrate and the film. However, the film on top

Figure 17.48. High resolution electron microscopy image of Tl-1212 film on $LaAlO_3$ substrate.

of this amorphous retains good epitaxy. The film thickness measured at this cross section is about 0.4 μm, which is in good agreement with surface profile measurement. Figure 17.48 shows the high-resolution electron microscopy (HREM) image of the interface between the film and substrate. The film is highly epitaxial with substrate. This is expected since the lattice parameters of $LaAlO_3$ substrate and high T_c superconducting materials are well matched. However, three layers of intergrown 1201 phase were found at the interface between the film and substrate. This kind of phases with less n of the $Ca_{n-1}Cu_2O_{2n}$ perovskite units are observed at the interface of film/substrate and bulk (such as 2212)/Ag tape (Feng et al., 1992; Wen et al., 1995). The measured c-axis of the (Tl,Bi)-1212 phase is between 1.190–1.195 nm, and the a-axis is 0.382 ± 0.0005 nm according to Figure 17.48. Compared to the data from Sheng et al. (1991), the a-axis of the film is slightly longer, and c-axis is slightly shorter. This difference could be due to the Bi addition or film strain, or both.

In fact, the Cr-doped (Tl,Bi)-1212 system has a potential advantage over other systems for applications as coatings on buffered metallic substrates due to a resistance to Cr contamination. For example, the RABiTS used for (Tl,Bi)-1223 synthesis were made from pure Ni (Ren et al., 1998a), which is ferromagnetic, and problematic for *ac* applications due to hysteresis losses. The Cr-alloyed Ni substrates (such as Hastelloy C, Inconel, etc.) have near-zero magnetism at 77 K, better oxidation resistance, better mechanical strength and better thermal expansion match with buffer layer materials. However, the superconductive properties of the HTS film could be seriously suppressed by Cr diffusion into the YBCO or Tl-1223 through the buffer layer. For this reason, the Cr-doped Tl-1212 film should have much better tolerance to Cr contamination and could even reduce the number of required buffer layers on the metallic substrate. The developed texture in Ni–Cr alloys has been found to be comparable to that of pure nickel. Although secondary recrystallisation can occur if the alloy is taken to too high temperature, at a low enough temperature the buffer-layer epitaxy can be transferred and should remain, even if the alloy later recrystallises during HTS processing. A possible concern is a change in physical properties (such as dimension, thermal expansion coefficient) associated with grain growth and secondary recrystallisation, which could damage or crack the buffer layer. Since we don't know if any

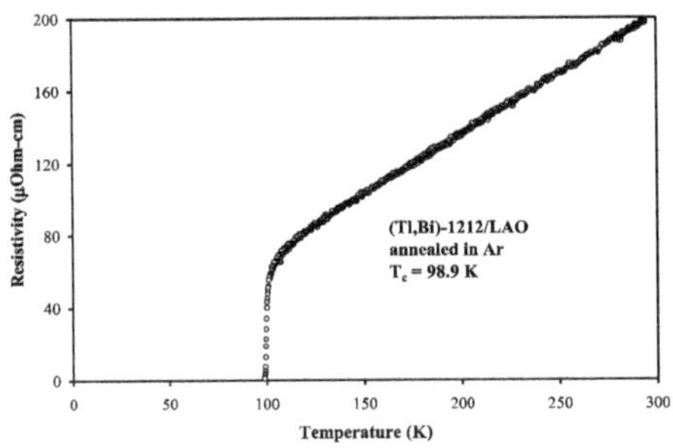

Figure 17.49. The temperature dependence of the linear resistivity of a Cr-doped (Tl,Bi)-1212 film on $LaAlO_3$ substrate annealed in flowing argon, showing the zero-resistance transition temperature of 98.9 K.

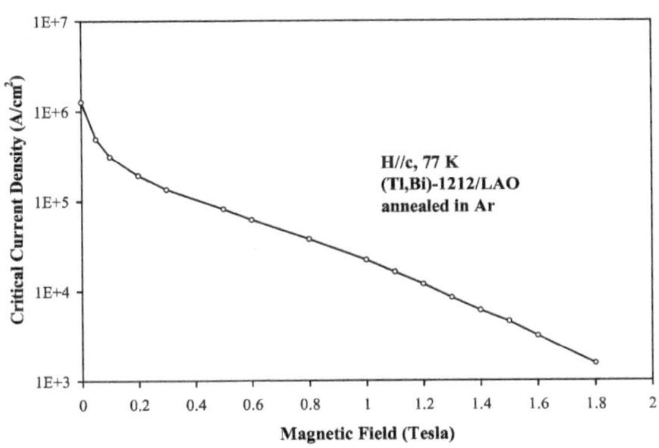

Figure 17.50. The magnetic field dependence of J_c of Cr-doped (Tl,Bi)-1212 film on $LaAlO_3$ substrate annealed in flowing argon for the field oriented perpendicular to the film plane (H//c) at 77 K.

of these things will happen, as a feasibility test of these issues, a study of Cr-doped (Tl,Bi)-1212 superconducting coatings on such alloy tapes will be interesting.

Since the aerobic environment used to post-anneal the films at high temperature is deleterious to the flexible metallic substrate, an oxygen-free argon annealing procedure is necessary. To avoid the possible sensitivity of the present Tl-1212 system to oxygen content, such as seen for $TlBa_2CaCuO_7$ films (Siegal et al., 1995a), $LaAlO_3$ single crystals were used as substrates. This approach anticipates the future availability of $LaAlO_3$-capped flexible metallic substrates (Parilla et al., 1997; Carlson et al., 1998). During the annealing, excess thallium was added to the pellet to compensate for thallium loss in the argon ambient. However, ICP measurements show that the as-annealed films have thallium content of over 1.0 atom per formula unit, and Bi content over of 0.15 atom per formula unit, gained from the source pellet. The as-annealed film thickness was in the range of 500 nm to 1500 nm.

The T_c values of the as-annealed films on $LaAlO_3$ substrates are in the range of 94–100 K, determined by electrical transport. Figure 17.49 shows the typical temperature dependent resistivity as measured by the four-probe method with zero resistance T_c of 98.9 K. Figure 17.50 shows the field dependence of transport J_c with magnetic fields applied parallel to the crystal c-axis. The film has a zero field J_c value of 1.2×10^6 A/cm^2, and an irreversible field of 2.0 T. The latter is greater than that of films annealed in air. However, this value is somewhat lower than that of (Tl,Bi)-1223 films on $LaAlO_3$. According to present understanding (Kim et al., 1991), the

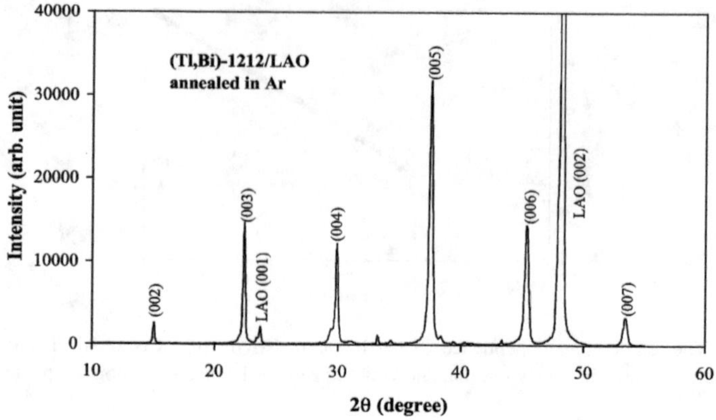

Figure 17.51. XRD θ–2θ scan of Cr-doped (Tl,Bi)-1212 film on $LaAlO_3$ substrate annealed in flowing argon.

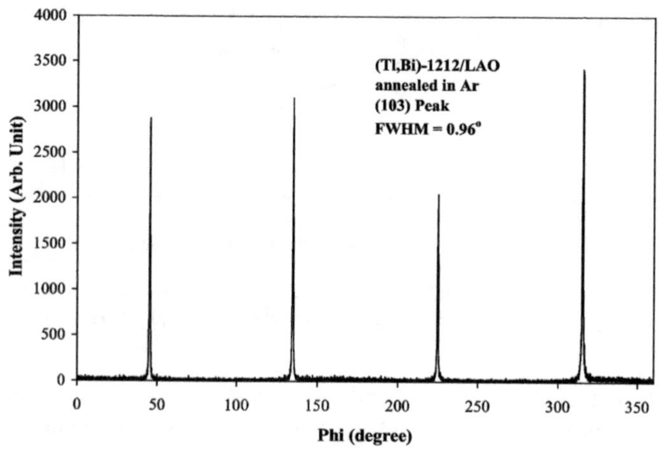

Figure 17.52. Phi scan of (103) peak of Cr-doped (Tl,Bi)-1212 film on $LaAlO_3$ substrate annealed in flowing argon. The peak FWHM is 0.96^0.

Tl-1212 system should have equal or better magnetic field performance compared to the Tl-1223 system, due to the short insulating layer distance of the Tl-1212 structure. A possible explanation is the reduced $T_c \approx 99$ K, compared to $T_c \approx 107$ K for the Tl-1223 system.

The XRD θ–2θ scan of the film is shown in Figure 17.51. All the major reflections are indexed as (00l) peaks of the (Tl,Bi)-1212 phase, and (001), (002) peaks of the $LaAlO_3$ substrate. The strong (Tl,Bi)-1212 (00l) peaks indicate that the 1212 phase is dominant, with a high degree of uniaxial alignment of the c-axis normal to the substrate surface. Some weak impurity peaks are also found in the spectrum. The film out-of-plane alignment was measured by an XRD ω scan of the (Tl,Bi)-1212 (005) peak with a FWHM of 0.93°, which is larger than the 0.58° of the films annealed in air (Lao et al., 2000a). Figure 17.52 shows the in-plane, ϕ scan of the (Tl,Bi)-1212 (103) peak. The four strong equally separated peaks, with a FWHM value of 0.96°, manifest the excellent alignment of a- and b-axes of the (Tl,Bi)-1212 phase.

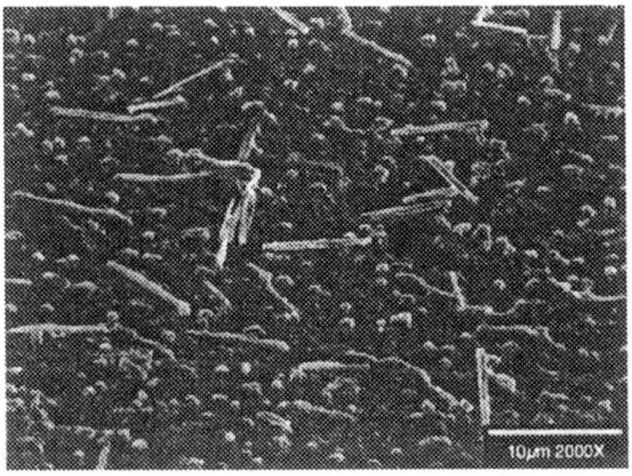

Figure 17.53. SEM picture of Cr-doped (Tl,Bi)-1212 film on $LaAlO_3$ substrate annealed in flowing argon.

Figure 17.53 is SEM micrograph that shows the surface features of as-annealed films. The film appears smooth and well connected, with small rods on the surface that could be the minor-phase impurities evidenced in the XRD of Figure 17.51. Very few pores, cracks, or pinholes were found on the film surface. More pinholes were observed in the films annealed in air; these differences may be related to the excess thallium observed by ICP in the films annealed in argon.

However, previously, only two kinds of RABiTS architectures, YSZ/CeO_2/Ni and CeO_2/YSZ/CeO_2/Ni, were used to produce superconducting Tl-1223 films. Partial success in producing $LaAlO_3$ epitaxial buffer films with the in-plane [100]//CeO_2/YSZ [110] has been reported (Parilla et al., 1997). Apparently, high-quality $LaAlO_3$ films with the desired texture can not be produced routinely directly on YSZ. Good bi-axially texture has been achieved for the configurations LAO/STO/CeO_2/YSZ and LAO/STO/CeO_2/Ni, but the out-of-plane texture has not been consistent (Carlson et al., 1998). Therefore, the research of the synthesis of Cr-doped (Tl,Bi)-1212 films on textured YSZ and CeO_2 surfaces will be interesting. For control studies, YSZ single crystal substrates were substituted for the textured YSZ/CeO_2 buffered RABiTS. Based on the experience of Tl-1223 film synthesis, the synthesis of superconducting films on YSZ or CeO_2-buffered YSZ single-crystal substrates can be successfully transferred to the synthesis of films on RABiTS.

However, the synthesis of Cr-doped 1212 films directly on YSZ single crystal substrates was not successful. The as-annealed 1212 films on YSZ have very high room temperature resistance, T_c values of $\sim$94 K, and very low J_c at 77 K, apparently due to a substrate-film reaction. In this regard, it is interesting to note that Y has often been doped into the Tl-1212 phase to improve its T_c and J_c values (Liu et al., 1989; Hong and Wang, 1993). Compared to the common Pb, Bi and rare-earth doped $TlBa_2CaCu_2O_7$ and $TlSr_2CaCu_2O_7$ systems, it seems that Cr doping may significantly change the chemistry of this Tl-1212 system. The results of Cr-doped Tl-1212 films on CeO_2 surfaces are very promising. In fact, compared to YSZ, the CeO_2/YSZ substrates have certain advantages for synthesis of Cr-doped (Tl,Bi)-1212 films in argon. First, CeO_2 is more chemically and thermally stable than YSZ in contact with the thallium based superconducting films. Second, the CeO_2 crystalline layer has better lattice match with the superconducting film than the YSZ (001) surface. During the synthesis, the Tl-1212 film grows its with [100] axis parallel to the [110] axis of either YSZ or CeO_2. The half-length of the lattice diagonal of CeO_2 ($a/\sqrt{2} = 0.383$ nm) has about 0.7% mismatch with the 0.38 nm of Tl-1212, while YSZ ($a/\sqrt{2} = 0.363$ nm) has a 4.4% mismatch.

CeO_2-capped YSZ single crystals were used as prototypes for CeO_2-capped RA-BiT substrates. The epitaxial CeO_2 buffer layer was deposited on YSZ by PLD at a

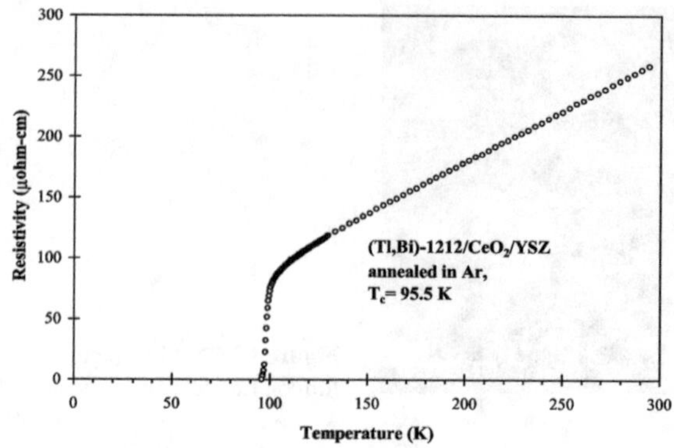

Figure 17.54. Temperature dependence of Cr-doped (Tl,Bi)-1212 film on CeO_2/YSZ substrate annealed in flowing argon.

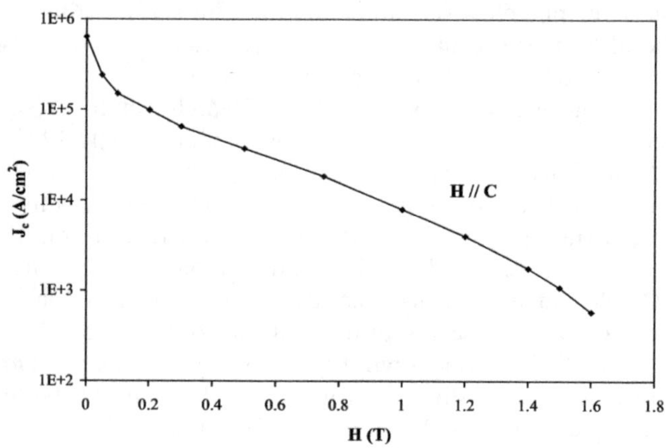

Figure 17.55. Magnetic field dependence critical density of Cr-doped (Tl,Bi)-1212 film on CeO_2/YSZ substrate annealed in flowing argon.

laser energy of 90 to 120 mJ and repetition rate of 1 Hz. The oxygen partial pressure was maintained at 25 mTorr. The YSZ substrates were affixed to a superalloy heater by silver paint. The heater temperature was maintained at 600 to 620°C during the deposition of the CeO_2 layer. The thickness of the deposited CeO_2 buffer was about 100 to 200 nm. After deposition of the CeO_2 cap layer, the heater was cooled to 200°C and the deposition of the Tl-1212 precursor began, using the same procedure as that of the Tl-1212 films on $LaAlO_3$ substrates. The as-deposited film was then furnace annealed at 850°C to 870°C for 35 to 45 minutes in flowing argon, along with pellets of composition $Tl_{1.05-1.25}Bi_{0.3}Sr_{2.0}Ca_{0.85}Cr_{0.15}Cu_2O_7$, to maintain the thallium vapour pressure in the assembly.

Figure 17.54 shows the temperature dependent resistivity of a Cr-doped (Tl,Bi)-1212 film on CeO_2/YSZ substrate. The T_c value is 95.5 K, and the zero field J_c (77 K) of the film is 6.0×10^5 A/cm^2. The magnetic field dependence is shown in Figure 17.55 where the irreversibility field is found to be 1.5 T.

Figure 17.56 shows the XRD θ–2θ scan of the film. All the major reflections are indexed as (00l) of the Tl-1212 phase and (002) peak of CeO_2 buffer layer. Very few minor-phase peaks exist. The FWHM of ω scan is only 0.73° for (005) peak of Tl-

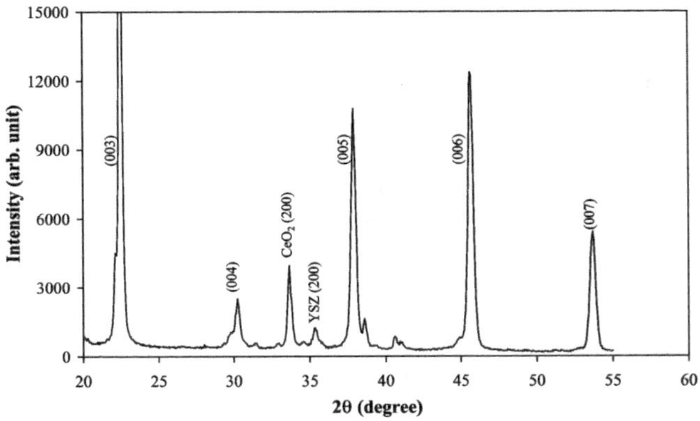

Figure 17.56. XRD θ–2θ scan of Cr-doped (Tl,Bi)-1212 film on CeO_2/YSZ substrate annealed in flowing argon.

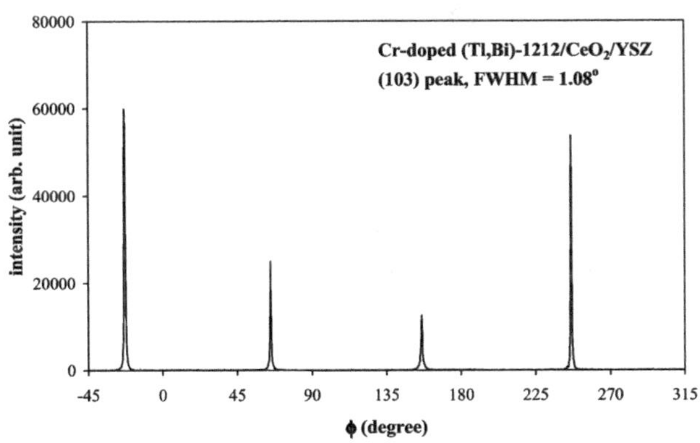

(a)

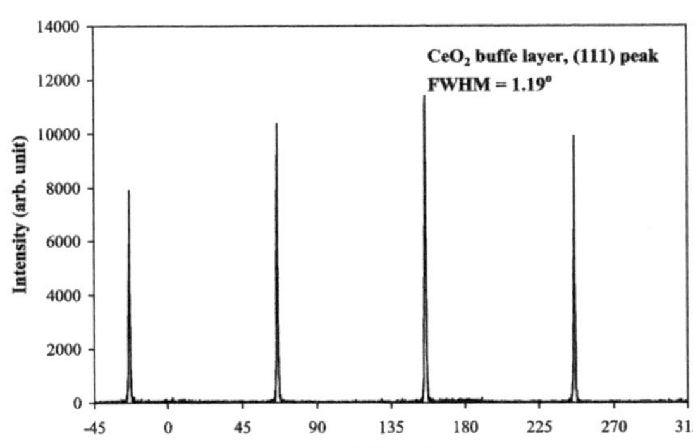

(b)

Figure 17.57. Phi scan of: (a) Cr-doped (Tl,Bi)-1212 (103); (b) CeO_2 (111).

1212 phase and 0.62° for (200) plane of the CeO_2 buffer layer. The ϕ scans of both the Cr-doped (Tl,Bi)-1212 film and the CeO_2 buffer layer were taken to examine the in-plane alignment of the films. Figure 17.57(a) and (b) show the ϕ scan of the Cr-doped (Tl,Bi)-1212 (103) plane and the CeO_2(111) plane, respectively. In each case,

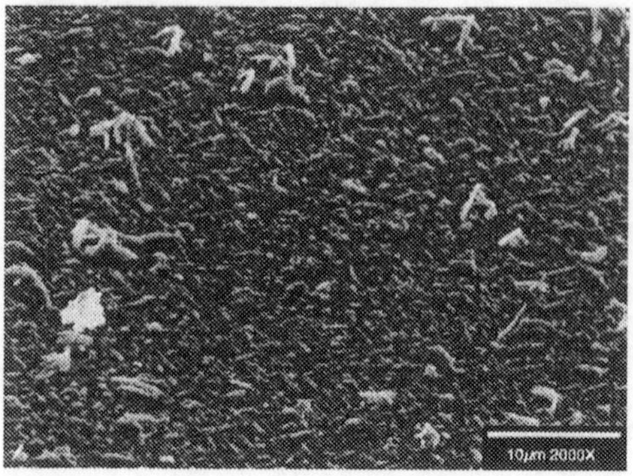

Figure 17.58. SEM picture of Cr-doped (Tl,Bi)-1212 film on CeO_2/YSZ substrate annealed in flowing argon.

the four strong peaks, spaced 90 degrees apart, demonstrate the excellent epitaxy of buffer layer and film. The FWHM of the ϕ scan is 1.08° for the 1212 film and 1.19° for the buffer layer.

Figure 17.58 shows the SEM surface features of the as-annealed films on CeO_2/YSZ substrates. The film surface is plate like, but not as smooth as that on the $LaAlO_3$ substrates. Further works will optimise the superconducting properties of the Cr-doped (Tl,Bi)-1212 films for potential electric conductor and/or electronic applications.

The electrodeposition of epitaxial Cr-doped (Tl,Bi)SCCO films up to 2.5 μm thick on $LaAlO_3$ substrate is still in development (Blaugher et al., 1999). XRD θ–2θ scan and pole figure of (103) peak have demonstrated excellent phase development and in-plane, out-of-plane alignment. The preliminary result shows a J_c value over 10^5 A/cm^2 at 77 K and zero field. Further development is still in progress.

17.4 CONCLUSIONS

The great potential of Tl-1223 films for conductor application has been demonstrated by the continuous research and improvements. With the successful synthesis of high J_c (Tl,Bi)-1223 films on $LaAlO_3$, YSZ and CeO_2-capped YSZ single crystal substrate by pulsed laser deposition method and post annealing in argon environment, epitaxial (Tl,Bi)-1223 films were also successfully grown on both YSZ and CeO_2-topped RABiTS. The films are biaxially aligned, with T_c in the range of 105–110 K and J_c (77 K) in the range of 1–5 $\times$ 10^5 A/cm^2. This J_c value could be further improved by reducing the amount of the intergrown Tl-1212 phase. Progress has also been made on CuTl-1233 films and CuTl-1234 films.

On the other hand, low cost upscalable precursor film deposition method, such as thermal spray pyrolysis and electrodeposition method, were also used to develop high quality (Tl,Bi)-1223 films on LAO single crystal substrates as the pre-step for conductor application developments on RABiTS. Epitaxial (Tl,Bi)-1223 films, with thickness in the range of 1–5 μm, have been successfully grown. The films exhibit T_c over 105 K and high J_c value over 1 $\times$ 10^6 A/cm^2. The growth of high quality epitaxial films on RABiTS is still in developing.

Another single layer Tl based HTS system-Tl-1212 system, has also been studied as the candidate for the conductor application. The successful development of two-zone furnace has provided an excellent practical *ex-situ* annealing method for Tl-based HTS films. Although Y and RE doping and reductive annealing are widely used for this system, but up to now, only the Cr-doped (Tl,Bi)SCCO film shows competitive high critical current density for conductor application. Epitaxial films on LAO and CeO_2-capped YSZ single crystal substrate, grown by PLD and post-annealing in argon environment, have T_c in the range of 94–100 K and J_c value over 5×10^5 A/cm^2, which is practical for conductor application. However, this system has not fully understood and further developments are necessary.

ACKNOWLEDGMENTS

The authors would like to express their sincere appreciation to many collaborators and advisors involved in the work described in this chapter. Especially we would like to thank D.K. Christen, R. Hawsey, M. Paranthaman, D.T. Verebelyi, and A. Goyal at Oak Ridge National Laboratory; D.J. Miller at Argonne National Laboratory; Drs. R.N. Bhattacharya and D. Blaugher at National Renewable Laboratory; Prof. M.J. Naughton at Boston College. The work was sponsored in part by National Science Foundation (NSF) under grant DMR-9996289, Department of Energy (DOE) under grant DEFG0298ER45719, Oak Ridge National Laboratory (ORNL), and National Renewable Energy Laboratory (NREL).

REFERENCES

Aselage, T.L., Venturini, E.L., and Van Deusen, S.B., 1994, *J. Appl. Phys.*, 75:1023.

Bednorz, J.G. and Muller, K.A., 1986, *Z. Phys. B*, 64:189.

Bhattacharya, R.N., Blaugher, R.D., Ren, Z.F., Li, W., Wang, J.H., Paranthaman, M., Verebelyi, D.T., and Christen, D.K., 1998a, *Electrochemical and Solid-State Letters*, 1(4):165.

Bhattacharya, R.N., Blaugher, R.D., Ren, Z.F., Li, W., Wang, J.H., Paranthaman, M., Verebelyi, D.T., and Christen, D.K., 1998b, *Physica C*, 304:55–56.

Bhattacharya, R.N., Parilla, P.A., Blaugher, R.D., Ren, Z.F., Li, W., Wang, J.H., Wang, Y.T., Hermann, A.M., Paranthaman, M., Goyal, A., Verebelyi, D.T., and Christen, D.K., 1999, *IEEE Trans. Appl. Supercond.*, 9:1681.

Blaugher, D., Bhattacharya, R., Parilla, P., Ginley, D., Alleman, J., Duda, A., Carlson, C., Wang, Y.T., Wu, H., Chrisen, D., Paranthaman, M., Verebelyi, D.T., Tu, Y., Siegal, M., Clem, P., Ren, Z.F., Lao, J.Y., Hermann, A., Price, J., Shoup, S., Shanmugham, S., and Tomsic, M., 1999, *National Renewable Energy Laboratory Superconductivity Program for Electric Systems Annual Peer Review*, July 27.

Bhattacharya, R.N., Wu, H.L., Wang, Y.-T., Blaugher, R.D., Yang, S.X., Wang, D.Z., Ren, Z.F., Tu, Y., and Christen, D.K., 2000, *Physica C* (submitted).

Bonham, S., Foster, J., Chang, Y.C., Jiang, P.Z., Vandervoot, K.G., Lam, D.J., and Kogan, V.G., 1989, *Phys. Rev. Lett.*, 63:782.

Carlson, C.M., Price, J.C., Parilla, P.A., Ginley, D.S., Niles, D., Blaugher, R.D., Goyal, A., Paranthaman, M., Kroeger, D.M., and Christen, D.K., 1998, *Physica C*, 304:82.

Deluca, J.A., Karas, P.L., Tkaczyk, J.E., Bednarczyk, P.J., Garbauskas, M.F., Briant, C.L., and Sorensen, D.B., 1993, *Physica C*, 205:21.

Dimos, D., Chaudhari, P., Mannhart, J., and LeGoues, F.K., 1988, *Phys. Rev. Lett.*, 61:219.

Dimos, D. and Chaudhari, P., 1990, *Phys. Rev. B*, 41:4038.

Doi, T.J., Yuasa, T., Ozawa, T., and Higashiyama, K., 1994, *Jpn. J. Appl. Phys.*, 33:5692.

Face, D.W. and Nestlerode, J.P., 1992, *Appl. Phys. Lett.*, 61:1838.

Face, D.W. and Nestlerode, J.P., 1993, *IEEE Trans. Appl. Supercond.*, 3:1516.

Feng, Y., Larbalestier, D.C., Babcock, S.E., and Vander Sande, J.B., 1992, *Appl. Phys. Lett.*, 61(10):1234.

Gammel, P.L., Schneemeyer, L.F., Waszczak, J.V., and Bishop, D.J., 1988, *Phys. Rev. Lett.*, 61:1666.

Goyal, A., Norton, D.P., Budai, J.D., Paranthaman, M., Specht, E.D., Kroeger, D.M., Christen, D.K., He, Q., Saffian, B., List, F.A., Lee, D.F., Martin, P.M., Klabunde, C.E., Hatfield, E., and Sikka, V.K., 1996, *Appl. Phys. Lett.*, 69:1795.

Guo, L.P., Ren, Z.F., Lao, J.Y., Wang, J.H., Christen, D.K., Klabunde, C.E., and Budai, J.D., 1997, *Physica C*, 277:13.

Haldar, P., Sridhar, S., Roig-Janiki, A., Kennedy, W., Wu, D.H., Zahopoulos, C., and Giessen, B.C., 1988, *J. Superconductivity*, 1:211.

He, Q., Christen, D.K., Klabunde, C.E., Tkaczky, J.E., Lay, K.W., Paranthaman, M., Thompson, J.R., Goyal, A., Pedraza, A.J., and Kroeger, D.M., 1995, *Appl. Phys. Lett.*, 67(2):294.

Holstein, W.L., Wilker, C., Laubacher, D.B., Face, D.W., Pang, P., Warrington, M.S., Carter, C.F., and Parisi, L.A., 1993, *J. Appl. Phys.*, 74:1426.

Hong, Q. and Wang, J.H., 1993, *Physica C*, 214:286.

Ihara, H., Tokiwa, K., Ozawa, H., Hirabayashi, M., Negishi, A., Matuhata, M., and Song, Y.S., 1994, *Jpn. J. Appl. Phys.*, 33:503.

Ihara, H., 1995, *Advances in Superconductivity*, VII:255.

Ihara, H., Iyo, A., Tokiwa, K., Terada, N., Tokumoto, M., and Umeda, M., 1996a, *Advances in Superconductivity*, VIII:247.

Ihara, H., Iyo, A., Tokiwa, K., Tokamoto, T., Terada, N., Tokumoto, M., and Umeda, M., 1996b, *Czechoslovak J. Phys.*, 46:3185.

Ihara, H., Tokiwa, K., Tanaka, K., Tsukamoto, T., Watanabe, T., Yamamoto, H., Iyo, A., Tokumoto, M., and Umeda, M., 1997, *Physica C*, 282–287:957.

Ihara, H., Sekita, Y., Tateai, H., Khan, N.A., Ishida, K., Harashima, E., Kojima, T., Yamamoto, H., Tanaka, K., Terada, N., and Obara, H., 1999, *IEEE Trans. Appl. Supercond.*, 9(2):1551.

Inoue, O., Adachi, S., and Kawashima, S., 1989, *Jpn. J. Appl. Phys.*, 28, L1187.

Inoue, O., Adachi, S., and Kawashima, S., 1990, *Jpn. J. Appl. Phys.*, 29, L763.

Johs, B., Thompson, D., Ianno, N.J., Woolam, J.A., Liou, S.H., Hermann, A.M., Sheng, Z.Z., Kiehl, W., Shams, Q., Fei, X., Sheng, L., and Liu, Y.H., 1989, *Appl. Phys. Lett.*, 54:1810.

Kamo, T., Toi, T., Soeta, A., Yuasa, T., Inoue, N., Aihara, K., and Matsuda, S., 1991, *Appl. Phys. Lett.*, 59:3186.

Khan, N.A., Sekita, Y., Tateai, F., Kojima, T., Ishida, K., Terada, N., and Ihara, H., 1999, *Physica C*, 320:39.

Kim, D.H., Gray, K.E., Kampwirth, R.T., Smith, J.C., Richeson, D.S., Marks, T.J., Kang, J.H., Talvacchio, J., and Eddy, M., 1991, *Physica C*, 177:431.

Kountz, D.J., Gai, P.L., Wilker, C., Hoilstein, W.L., Pellicone, F.M., and Brainard, R.J., 1993, *IEEE Trans. Appl. Supercond.*, 3:1222.

Kroeger, D.M., Goyal, A., Specht, E.D., Wang, Z.L., Tkaczyk, J.E., Sutliff, J.A., and Deluca, J.A., 1994, *Appl. Phys. Lett.*, 64(1):106.

Lao, J.Y., Ren, Z.F., Wang, D.Z., Yang, S.X., and Wang, J.H., 1998 (unpublished).

Lao, J.Y., Wang, J.H., Wang, D.Z., Yang, S.X., Tu, Y., Wen, J.G., Wu, H.L., Ren, Z.F., Verebelyi, D.T., Paranthaman, M., Aytug, T., Christen, D.K., Bhattacharya, R.N., and Blaugher, R.D., 2000a, *Supercond. Sci. Technol.* (accepted).

Lao, J.Y., Wang, J.H., Wang, D.Z., Tu, Y., Yang, S.X., Wu, H.L., Ren, Z.F., Verebelyi, D.T., Paranthaman, M., Aytug, T., Christen, D.K., Bhattacharya, R.N., and Blaugher, R.D., 2000b, *Physica C* (submitted).

Lee, B.I., Doi, T.J., and Yuasa, T., 1994, *Physica C*, 226:377.

Li, S. and Greenblaat, M., 1989, *Physica C*, 157:365.

Li, W., Wang, D.Z., Lao, J.Y., Ren, Z.F., Wang, J.H., Paranthaman, M., Verebelyi, D.T., and Christen, D.K., 1999, *Supercond. Sci. Technol.*, 12:L1.

Liang, J.M., Liu, R.S., Huang, Y.T., Wu, S.F., and Wu, P.T., 1990, *Physica C*, 165:347.

Liou, S.H., Aylesworth, K.D., Ianno, N.J., Thompson, D., Meyer, D., Wollam, J.A., and Barry, C., 1989, *Appl. Phys. Lett.*, 54:760.

Liu, R.S., Liang, J.M., Huang, Y.T., Wang, W.N., Wu, S.F., Koo, H.S., Wu, P.T., and Chen, L.J., 1989, *Physica C*, 162–164:869.

Maysuda, S., Takeuchi, S., Soeta, A., Suzuki, T., Alihara, K., and Kamo, T., 1988, *Jpn. J. Appl. Phys.*, 27:2062.

Mogro-Campero, A., Bednarczyk, P.J., Tkaczyk, J.E., and DeLuca, J.A., 1995, *Physica C*, 247:239.

Morosin, B., Baughman, R.J., Ginley, D.S., Schirber, J.E., and Venturini, E.L., 1990, *Physica C*, 161:115.

Myers, K.E., Face, D.W., Kountz, D.J., and Nestleerode, J.P., 1994, *Appl. Phys. Lett.*, 65(4):490.

Nabatame, T., Sato, J., Aihara, K., Kamo, T., and Matsuda, S., 1992, *Physica C*, 193:390.

Norton, D.P., Goyal, A., Budai, J.D., Christen, D.K., Kroeger, D.M., Specht, E.D., He, Q., Saffian, B., Paranthaman, M., Klabunde, C.E., Lee, D.F., Sales, B.C., and List, F.A., 1996, *Science*, 274:755.

Okada, M., Yuiasa, T., Matsumoto, T., Aihara, K., Seido, M., and Matuda, S.P., 1990, *Jpn. J. Appl. Phys.*, 29:L2732.

Paranthaman, M., List, F.A., Goyal, A., Specht, E.D., Vallet, C.E., Kroeger, D.M., and Christen, D.K., 1997, *J. Mater. Res.*, 12:619.

Parilla, P.A., Mcgraw, J.M., Schulz, D.L., Wendelin, J., Bhaatacharya, R.N., Blaugher, R.D., and Ginley, D.S., 1997, *IEEE Trans. Appl. Supercond.*, 7(2):1051.

Parkin, S.S.P., Lee, V.Y., Nazzal, A.I., Savoy, R., Beyers, R., and Placa, S.J.L., 1988, *Phys. Rev. B*, 38:6531.

Piehler, A., Low, R., Betz, J., Schonberger, R., and Renk, K.F., 1993, *J. Appl. Phys.*, 74(10):6437.

Piehler, A., Strobel, J.P., Reschauer, N., Low, R., Schonberger, R., Renk, K.F., Kraus, M., Daniel, J., and Saemann-Ischenko, G., 1994a, *Physica C*, 223:391.

Piehler, A., Reschauer, N., Spreitzer, U., Strobel, J.P., Schonberger, R., and Renk, K.F., 1994b, *Appl. Phys. Lett.*, 65:1451.

Presland, M.R. and Tallon, J.L., 1991, *Physica C*, 177:1.

Presland, M.R., Tallon, J.L., Buckley, R.G., Liu, R.S., and Flower, N.E., 1991, *Physica C*, 176:95.

Ren, Z.F. and Wang, J.H., 1992a, *Appl. Phys. Lett.*, 61:1715.

Ren, Z.F. and Wang, J.H., 1992b, *Physica C*, 192:55.

Ren, Z.F. and Wang, J.H., 1993a, *Physica C*, 216:199.

Ren, Z.F. and Wang, J.H., 1993b, *Appl. Phys. Lett.*, 62(23):3025.

Ren, Z.F., Wang, C.A., and Wang, J.H., 1994, *Appl. Phys. Lett.*, 65:237.

Ren, Z.F., Wang, C.A., Wang, J.H., Miller, D.J., and Goretta, K.C., 1995, *Physica C*, 247:163.

Ren, Z.F., Wang, C.A., Wang, J.H., Miller, D.J., Christen, D.K., Hettinger, J.D., and Gray, K.E., 1996, *Physica C*, 258:129.

Ren, Z.F., Lao, J.Y., Guo, L.P., Wang, J.H., Budai, J.D., Christen, D.K., Goyal, A., Paranthaman, M., Specht, E.D., and Thompson, J.R., 1998a, *J. Superconductivity*, 11(1):159.

Ren, Z.F., Li, W., Wang, D.Z., Lao, J.Y., Wang, J.H., Paranthaman, M., Verebelyi, D.T., and Christen, D.K., 1998b, *Physica C*, 306:149.

Ren, Z.F., Li, W., Wang, D.Z., Lao, J.Y., Wang, J.H., Paranthaman, M., Verebelyi, D.T., Christen, D.K., Lee, D.F., Goyal, A., and Kroeger, D.M., 1999, *Physica C*, 313:241.

Schulz, D.L., Parilla, P.A., and Ginley, D.S., 1994, *Appl. Phys. Lett.*, 65(19):2472.

Sheng, Z.Z. and Hermann, A.M., 1988, *Nature*, 332:138.

Sheng, Z.Z., Sheng, L., Fei, X., and Hermann, A.M., 1989, *Phys. Rev. B*, 39:2918n.

Sheng, Z.Z., Gu, D.X., Xin, Y., Pederson, D.O., Finger, L.W., Hadidiacos, C.G., and Hazen, R.M., 1991, *Mod. Phys. Lett. B*, 5:635.

Siegal, M.P., Venturini, E.L., Newcomer, P.P., Overmyer, D.L., Dominguez, F., and Dunn, R., 1995a, *J. Appl. Phys.*, 78(12):7186.

Siegal, M.P., Venturini, E.L., Newcomer, P.P., Morosin, B., Overmyer, D.L., Dominguez, F., and Dunn, R., 1995b, *Appl. Phys. Lett.*, 67(26):3966.

Siegal, M.P., Overmyer, D.L., Venturini, E.L., Newcomer, P.P., Dunn, R., Dominguez, F., Padilla, R.R., and Sokolowski, S.S., 1997a, *IEEE Trans. Supercond.*, 7:1881.

Siegal, M.P., Venturini, E.L., Morosin, B., and Aselage, T.L., 1997b, *J. Mater. Res.*, 12(11):2825.

Siegal, M.P., Overmyer, D.L., Venturini, E.L., Dominguez, F., and Padilla, R.R., 1998, *J. Mater. Res.*, 13(12):3349.

Soeta, A., Suzuki, T., Takeuchi, S., Kamo, T., Usami, K., and Matsuda, S.P., 1989, *Jpn. J. Appl. Phys.*, 28:L1186.

Specht, E.D., Goyal, A., Kroeger, D.M., Deluca, J.A., Tkaczyk, J.E., Briant, C.L., and Sutliff, J.A., 1994, *Physica C*, 226:76.

Specht, E.D., Goyal, A., Kroeger, D.M., Deluca, J.A., Tkaczyk, J.E., Briant, C.L., and Sutliff, J.A., 1995, *Physica C*, 242:164.

Specht, E.D., Goyal, A., Kroeger, D.M., Mogro-Campero, A., Bednarczyk, P.J., Tkaczyk, J.E., and DeLuca, J.A., 1996, *Physica C*, 270:91.

Subramanian, M.A., Toradi, C.C., Gopalakrishnan, J., Gai, P.L., Calabrese, J.C., Askew, T.R., Flippen, R.B., and Sleight, A.W., 1988, *Science*, 242:249.

Suzuki, T., Nagoshi, M., Fukuda, Y., Nakajima, S., KiKuchi, M., Syono, Y., and Tachiki, M., 1989, *Physica C*, 162–164:1387.

Suzuki, T., Nagoshi, M., Fukuda, Y., Nakajima, S., KiKuchi, M., Syono, Y., and Tachiki, M., 1994, *Supercond. Sci. Technol.*, 7:817.

Tallon, J.L., Liu, R.S., Wu, K.L., Blunt, F.J., and Wu, P.T., 1989, *Physica C*, 161:523.

Tang, Y.Q., Chen, K.Y., Chan, I.N., Chen, Z.Y., Shi, Y.J., Salamo, G.J., Chan, F.T., and Sheng, Z.Z., 1993, *J. Appl. Phys.*, 74(6):4259.

Torardi, C.C., 1992, *Chemistry of Superconductor Materials*, T.A. Vanderah, ed., Noyes Publications.

Torri, Y., Takei, H., and Tada, K., 1989a, *Jpn. J. Appl. Phys.*, 28:L2192.

Torri, Y., Kotani, T., Takei, H., and Tada, K., 1989b, *Jpn. Appl. Phys.* 28:L1190.

Wang, C.A., Ren, Z.F., Wang, J.H., and Miller, D.J., 1995, *Physica C*, 245:171.

Wen, J.G., Morishita, T., Koshizuka, N., Traeholt, C., and Zandbergen, H.W., 1995, *Appl. Phys. Lett.*, 66(14):1830.

Xia, J.S., Sun, S.F., Zhang, T., Cao, L.Z., Zhang, Q.R., Chen, J., and Chen, Z.Y., 1989, *Physica C*, 158:477.

Chapter 18

EPITAXY OF Hg-BASED HIGH-T_c SUPERCONDUCTING THIN FILMS

Judy Wu

Department of Physics and Astronomy
University of Kansas
Lawrence, KS 66045
USA

18.1 INTRODUCTION

The superconductivity above 130 K in Hg-based high-temperature superconductors (Hg-HTS's: $HgBa_2Ca_{n-1}Cu_nO_{2n+2}$, $n = 1, 2, 3, \ldots$) has generated much excitement since it was discovered in 1993 (Putilin et al., 1993a, 1993b; Schilling et al., 1993; Antipov et al., 1993; Capponi et al., 1996). Several members in the Hg-HTS's family have their superconducting transition temperatures (T_c) above 100 K with the highest ambient zero-resistance T_c of 135 K in $HgBa_2Ca_2Cu_3O_{8+\delta}$ (Hg-1223), over 40 K higher than that of $YBa_2Cu_3O_{7-\delta}$ (YBCO). Under a hydrostatic pressure, Hg-HTS's seem to be much more responsive to the applied pressure. Although the mechanism of this pressure induced T_c enhancement is still under investigation, the onset superconducting transition temperature of Hg-1223 could be driven to above 160 K at hydrostatic pressures of 25–30 GPa (Chu et al., 1993). It should be realized that the pressure applied to the sample is uniaxial in the case of hydrostatic pressure. If the pressure derivatives of T_c have different signs along the ab-plane and c-axis, respectively, as is the case for La–Sr–Cu–O system (Locquet et al., 1998), the T_c enhancement in Hg-HTS's could be much higher when the pressure is compressive in the ab-plane and tensile along the c-axis. Hg-HTS's present one of the most interesting systems for fundamental studies of high-temperature superconductivity. Moreover, they are also extremely attractive for numerous superconducting device applications due to the promise of higher operation temperatures that imply lower cost and better performance.

It should be realized that many superconductor-related applications require high critical current density (J_c), especially in the presence of a magnetic field. In another word, a high irreversibility field (H_{irr}) is necessary, in addition to the high T_c's. Fortunately, Hg-HTS's were found (Welp et al., 1993; Huang et al., 1994) to have a moderately high H_{irr}, somewhere between that of less anisotropic YBCO and more anisotropic Bi-based high-temperature superconductors (Bi-HTS's) and Tl-based high-temperature superconductors of two Tl–O planes in a unit cell (double-layered Tl-HTS's). On the reduced temperature (T/T_c) scale, the H_{irr} depends on temperature (T) approximately via a simple power law: $H_{irr} \sim (1 - T/T_c)^n$ for many HTS's. The exponent "n" was reported to be 3/2, 5/2, and 11/2, respectively, for YBCO, Hg-HTS's, and Bi-HTSs or double-layered Tl-HTS's (Huang et al., 1994). This not only suggests that Hg-HTS's may be one of few alternatives for YBCO with comparable performance at 77 K, but also means that Hg-HTS's would be a unique choice for many applications at temperatures above 77 K.

Given such a motivation, many efforts have been put into development of Hg-HTS's in the form of both bulks and films immediately after Hg-HTS's were discovered. In particular, since epitaxial films (which are going to be the main topic of this chapter) are essential for both microelectronic applications and power-related electrical applications, growth of high-quality thin and thick films of Hg-HTS's has been a major focus of many groups in the world. It was soon realized that, despite the excellent physical properties Mother Nature gives to Hg-HTS's, synthesis and epitaxy of these materials presents one of the toughest challenges so far in the research of HTS materials due to a highly volatile nature of Hg-based compounds. Although some successes were reported shortly after the discovery of the Hg-HTS's, most early works suffered from problems of poor sample quality and reproducibility (see, for example, review articles by Wu and Tidrow (1999) and many references therein, and by Schwartz (2001)). T had been doubtful whether the Hg-HTS's could ever make the applications since most of the technical problems encountered in Hg-HTS thin film epitaxy, which we will discuss in this chapter, seem to be unavoidable to the Hg-HTS's.

Despite the difficulties associated to epitaxy of Hg-HTS thin and thick films, exciting progress has been achieved recently through development of new fabrication processes. High-quality Hg-1223 and $HgBa_2CaCu_2O_{6+\delta}$ (Hg-1212, with ambient zero-resistance T_c of 125 K) thin films have been obtained by several groups and can be fabricated routinely now in those laboratories. Many promising results have been obtained. Remarkably, a J_c in exceeding 1 MA/cm^2 can be still maintained in these films grown on both single-crystal (Yan et al., 1998; Kang et al., 1999) as well as metal substrates and (Xie et al., 2000) at temperatures YBCO and many other HTS's become non-superconducting. Low microwave surface resistance, with the value similar to that of other HTS's at 77 K, was observed on Hg-HTS films at temperatures above 100 K (Aga et al., 2000). This progress triggered a renewed interest in research of physical properties of the Hg-HTS films and application of these materials in electronic applications, such as passive microwave devices, as well as electric power-related applications, such as transmission cables, generators, motors, etc.

This chapter intends to review the recent progress in epitaxy of Hg-HTS thin and thick films with emphasis on various growth techniques developed, their advantages, and their technical limitations (Section 18.2). The physical properties of these films are discussed in Section 18.2 with an update of film quality using several routine characterization techniques, such as scanning electron microscopy (SEM), x-ray diffraction (XRD), magnetic, and electric transport measurements. Since applications are the major driving force for epitaxy of Hg-HTS films, a review of various efforts in applications (Section 18.4) of Hg-HTS thin and thick films will also be included. In Section 18.5, we discuss the remaining challenges and future research topics associated to the Hg-HTS films. Due to rapid advancing of technologies and a limited time available for writing this chapter, we apologize for many excellent works we may not be aware of or unable to include in this chapter.

18.2 FABRICATION OF Hg-HTS FILMS

Most Hg-HTS films are produced in *ex situ* processes. Although differing in details, all the *ex situ* processes developed so far for Hg-HTS films consist of two steps: deposition of precursor films and annealing in vapors of Hg and oxygen at high temperatures. There are two different types of processes: one use amorphous precursor films that are mixtures of simple oxides of Hg, Ba, Ca, and Cu, and the other, epitaxial precursor films of similar structures and composition to that of targeted Hg-HTS's. These two processes have completely different growth mechanisms while both produce good-quality Hg-HTS films. When the amorphous precursor films are used, we call the process "*conventional*" since it adopted the route typical to the conventional *ex situ* processes, in which a targeted material is formed during post annealing according to the required phase equilibrium. Section 18.2.1 will discuss the technical details in the conventional process, as well as some of its variations, that were developed for growth of Hg-HTS films. When an epitaxial precursor film is employed, the post annealing is used for quite a different purpose: to replace certain cations with others, which may or may not cause any changes in the films crystalline structure during the annealing. This so-called "*cation exchange*" process adopts a growth mechanism differing dramatically from that of the conventional process. Section 18.2.2 will discuss this cation exchange process.

The Hg-vapor annealing is carried out at pressures of several atmospheres. The only exception is in Hg-1201 films due to a relatively lower partial pressure of Hg vapor required. Shortly after the Hg-HTS's were discovered, Adachi et al. reported fabrication of *c*-axis-oriented Hg-1201 films by rf sputtering from a $Hg_{2.5}Ba_2CuO_{5.5}$ target (Adachi et al., 1993a, 1993b) and a subsequent annealing in flowing mixed gas of nitrogen and oxygen. These films had T_c's up to 93 K and J_c's in the range of 10^4 A/cm^2 at 77 K and a zero magnetic field. Although Hg-1201 films are very interesting for many physical investigations, they have received much less attention than Hg-1212 and Hg-1223 films due to the lower T_c around 95 K for Hg-1201. Most efforts in development of Hg-HTS films have been focused on epitaxy of Hg-1212 and Hg-1223 films not only because they have high T_c's, but also because they are relatively easier to obtain than other Hg-HTS family members of high number of Cu–O planes. In the rest of this article, Hg-HTS's will mainly refer Hg-1212 and Hg-1223, if not indicated otherwise.

Since substrates provide the foundation for epitaxial nucleation of a film, undegraded film/substrate interfaces are crucial to high-quality epitaxy of Hg-HTS thin films. Unfortunately, Hg vapor attacks almost all oxides and metals at elevated temperature. Consequently, the film/substrate interfaces degrade seriously (Wu et al., 1997a), resulting in poor quality epitaxy for Hg-HTS films made in conventional process, in which a long processing time at elevated temperatures is employed. This excludes most technological compatible substrates for epitaxy of Hg-HTS films and most good quality films were reported only on few substrates, such as $SrTiO_3$ (STO) substrates, that have superior chemical stability. Even on STO, poor film/substrate interfaces were observed in Auger and Rutherford backscattering (RBS) studies of depth profiles of the cations (Wu et al., 1997a). Serious Ba diffusion into the substrates was found in the range of a few μms, which increases with the Hg-vapor annealing time and temperature.

By improving the processing condition, several other substrates have been recently used successfully for growth of Hg-HTS films. By reducing thermal budget, the film/substrate interface was shown to improve significantly and high T_c's and J_c's have also been achieved on $LaAlO_3$ (LAO) substrates (Yun et al., 1996; Wu et al., 1996a). Recently, Sun et al. also obtained epitaxial Hg-1212 films on several other substrates including $NdGaO_3$ (NGO) and YSZ with T_cs up to 117 K and <50 K, respectively. Inoue et al. reported very thin $(Re_{0.1}Hg_{0.9})$-212 films on LSAT substrates with high J_cs up to a few MA/cm^2 at 77 K and self-field.

Although this film/substrate interface problem remains in the conventional process, it is minimized in the cation exchange process as the precursor matrix serves as a diffusion barrier that prevents Hg from attacking the film/substrate interface. Using this cation-exchange process, high-quality Hg-HTS films have been obtained on several technologically important substrates including STO, LAO, MgO, LSAT, CeO_2, and $CeO_2/YSZ/CeO_2$ buffered Ni substrates (Xie et al., 2000). More details will be discussed later in this chapter.

18.2.1 Conventional Crucible Process

Many groups have reported successes in fabrication of *c*-axis-oriented Hg-1212 (Wang et al., 1993; Tsuei et al., 1994; Miyashita et al., 1994; Higuma et al., 1994; Wu et al., 1997a; Yu et al., 1997; Guo et al., 1997a, 1997b; Gasser et al., 1998; Sun et al., 1999) and Hg-1223 (Yun and Wu, 1996; Yun et al., 1996; Foong et al., 1996; Moriwaki et al., 1998; Kang, W.N. et al., 1998) films using conventional process. This *ex situ* crucible process was used for synthesis of Hg-HTS films shortly after Hg-HTS's were discovered. This process is similar to the closed-crucible method employed for epitaxy of Tl-based HTS films (see the review article by Siegal et al., 1997 and the references therein). It consists two steps: room-temperature deposition of precursor films of Hg–Ba–Ca–Cu is followed by annealing of the precursor films in vapors of Hg and O_2. The precursor films are amorphous and non-superconducting. The superconducting $HgBa_2Ca_{n-1}Cu_nO_{2n+2}$ phases are formed through a solid/vapor reaction as Hg diffuses into precursor film from the film surface during the Hg-vapor annealing at temperatures typically between 750–950°C. This annealing step is hence critical to achieving high-quality epiatxial Hg-HTS thin films. Since most thus-made Hg-HTS films are slightly oxygen deficient, post oxygen anneal at low temperatures in the range of 300–400°C is sometime necessary to optimize the content of oxygen in Hg-HTS films.

Deposition of Precursor Films. The precursor films may be laid down using almost any thin or thick film coating techniques, such as pulsed laser deposition (PLD), magnetron sputtering, electron-beam or thermal evaporation, chemical vapor deposition (CVD), Sol-Gel coating, etc. As in the case of Tl-based HTS films, the method adopted for fabricating the precursor films is not particularly important as long as it provides a precise and uniform cation composition in the film. A stringent requirement is, however, that the precursor films should not be exposed to air so that the detrimental effect of moisture and carbonates in air can be minimized. In fact, sensitivity of the precursors to air has been a major problem in fabrication of Hg-based HTS bulks as well as thin films, resulting in extremely poor sample quality and reproducibility. The problem stems from the presence of various simple metal oxides in the precursor films, particularly CaO and BaO as the starting materials for Ca or Ba-sources, since they are chemically reactive even at the room temperature. For example, CaO, can easily change to $Ca(OH)_2$ or $CaCO_3$ through reactions with H_2O and CO_2 in air:

$$CaO + H_2O \rightarrow Ca(OH)_2$$
$$Ca(OH)_2 + CO_2 \rightarrow CaCO_3$$

The air detrimental effect affects the quality of the Hg-HTS films via several ways. First, the air contamination of the original CaO and BaO powders can change the stoichiometry of the target used, for example, for PLD or magnetron sputtering, which in turn results in deviation of the precursor film composition from the desired value. Typically, those precursor films are Ca or Ba deficient. Consequently, Cu or Hg-rich impurity phases are formed after the annealing and degrade the superconducting properties of the Hg-HTS films. Secondly, air contamination of CaO and BaO in the precursor films also result in the similar non-superconducting phases since the $Ca(OH)_2$, $Ba(OH)_2$, $CaCO_3$ and $BaCO_3$ phases are unlikely to decompose at the annealing temperature, typically in the neighborhood of ~800°C, for formation of Hg-based HTSs. Thirdly as we will discuss in detail later, the Hg-source pellet used to control the phase equilibrium during the Hg-vapor annealing could be easily contaminated, resulting in deviation from the required phase equilibria for Hg-HTS's and hence formation of impurity phases. Since a short-term air exposure of few seconds can result in severely degraded sample quality, most groups reported careful dry-box processing in the early stage. Nevertheless, the reproducibility of the Hg-based HTS films was extremely poor.

To eliminate the air detrimental effect on precursor films, modified precursor films were experimented. The IBM group included Hg into the precursor films through a layer-by-layer mixing of HgO and Ba–Ca–Cu–O at an atomic scale (Tsuei et al., 1994). In addition, a thin protective cap layer of either MgO or HgO was deposited on top of the precursor film. Significantly improved film quality, in terms of T_c and J_c, was reported. For example, a zero-resistance T_c was as high as 124 K and J_c was close to 1 MA/cm^2 at 100 K (Krusin-Elbaum et al., 1995). Although it was believed that the atomic-scale mixing of HgO and Ba–Ca–Cu–O improves the uniformity of Hg across the film thickness, it was later found that mixing Tl into the Ba–Ca–Cu–O precursor films by magnetron sputtering from Tl-HTS targets serves the same purpose and results in even better quality Hg-HTS films (Xie et al., 1999). This suggests that mixing Tl or Hg into the Ba–Ca–Cu–O may chemically stabilize the simple oxides in the precursors from reacting with moisture or carbonates in the air.

Improvement of sample quality and reproducibility were also reported using chemical-doping-assisted growth, including Re (Gasser et al., 1998; Moriwaki et al., 1998; Kang, W.N. et al., 1998), Tl (Brazdeikis et al., 1996; Xie et al., 1999), Pb (Higuma et al., 1994; Yu et al., 1997), Bi (Guo et al., 1997b), alkali (Li and Na)-doping (Wu et al., 1998; Gapud et al., 1998) in the Ba–Ca–Cu–O precursor films, or by partial substitution of HgO with Hg-halides (Kang, B.W. et al., 1998) or Tl_2O_3 (Foong et al., 1996) in the Hg-source pellet. Chemical doping/substitution in precursor films may provide another solution to the problem of air contamination. For example, by partially replace Hg with other rare earths, such as Re, Pb, etc., the air contamination problem seems much less severe in the form of $Hg_{1-x}R_x$–Ba–Ca–Cu–O precursors, where x is typically in the range of 0 to 0.2. Among others, Re-substituted precursors are particularly tough in air, although the mechanism is yet fully understood. One may speculate that the same mechanism as in the case of mixing Hg or Tl in the precursor films may apply to the cases of rare earth substitution. Several groups have reported high-quality films of $Hg_{1-x}R_x$-1223 and $Hg_{1-x}R_x$-1212 (Gasser et al., 1998; Moriwaki et al., 1998; Kang, W.N. et al., 1998). Some even obtained high quality Hg-HTS's without using dry boxes, suggesting that either Re stabilizes BaO and CaO in the amorphous precursor films, or promotes decomposition of contaminated Ba or Ca precursors. A related effort is to partially substitute HgO with $HgCl_2$ or HgF_2 in the Hg-source pellet by up to 25% (Kang, B.W. et al., 1998). It was argued that Cl^-/F^- ions react with the contaminated Ba or Ca precursors so as to recover these precursors. Besides the improved sample quality, the sample reproducibility was reportedly improved too.

Post Annealing in Hg and Oxygen Vapors. Since Hg-HTS's are formed during the post annealing, there is no doubt that this step is crucial to achieving high-quality Hg-HTS films. Hg-HTSs are formed through solid/vapor reaction during the Hg-vapor annealing, processing parameters, such as Hg-vapor partial pressure (P_{Hg}) and annealing temperature (T_{Hg}), need to be controlled precisely in order to maintain the required phase equilibrium for Hg-HTSs (Xue et al., 1996). Although the growth mechanism of Hg-HTS films was believed quite similar to that of Tl-HTS films, it turned out to be much more difficult to grow Hg-HTSs in the close-crucible process due to the high volatility of the Hg-related compounds. Unlike Tl-HTS's that can be formed at a pressure $\sim$1 atm, most Hg-HTS's require a relatively higher pressure, typically in the range of 5–10 atm. The only exception is Hg-1201 that has been synthesized at an atmospheric pressure. It is, however, the higher-T_c family members, such as Hg-1212 and Hg-1223 that seem more appealing for applications and have attracted much attention and efforts in epitaxy of the corresponding films. To reach the high pressure required in a controllable way, a torch-sealed quartz crucible was adopted to encapsulate the precursor films with a Hg-vapor source.

It is difficulty to control processing parameters, such as Hg vapor pressure, during the post annealing because most Hg sources vaporize in an uncontrollable fashion when the sample assembly is heated. This leads to an over pressure of the Hg vapor at the beginning of the annealing and Hg deficiency before the annealing is completed. The key to solve this problem is to control release of Hg on one hand and absorption of Hg on the other hand so as to reach a well-balanced Hg-vapor pressure during the whole period of the post anneal. One way to control release of Hg from the source is to densify the Hg source pellet so that Hg is released slowly from the source. No simple Hg-related compounds, such as HgO, HgF_2, $HgCl_2$, etc. have been used successfully so far as the source because most Hg atoms are released too fast even in the densified form. Based on a similar idea working for bulk Hg-HTS's (Meng et al., 1993), a more or less stoichiometric pellet composed of a powder mixture of HgO and pre-calcined Ba–Ca–Cu–O precursor powders of the same (or similar) nominal composition as the precursor films was then adopted for Hg-HTS film fabrication (Wang et al., 1993). This pellet is compressed at relatively high pressure, typically above 100 ton/in^2, to slow down release of Hg. The reaction between Hg and the Ba–Ca–Cu–O precursor in the pellet locks some Hg so that Hg could be released slowly during the annealing period of a few hours.

Besides this Hg-containing pellet, another pellet of pre-calcined Ba–Ca–Cu–O precursor powders was used as an absorber of Hg vapor to further control available Hg in the vapor form. Wang et al. first used two pellets: one with nominal composition of $Ba_2CaCu_2O_x$ and the other $HgBa_2CaCu_2O_x$. The two pellets and the Ba–Ca–Cu–O precursor films were encapsulated in a quartz ampoule with the precursor film to control the P_{Hg} for growth of Hg-1212 films. The total pressure in the quartz crucible is controlled, under selected annealing temperature, through the ratio between the mass of the Hg source and the volume of the crucible. The idea is that when a precursor film with same composition as that of the absorber pellet is brought to the proximity of the two pellets, Hg-HTS phases should be favorably formed at the desired phase equilibrium maintained by the two pellets. Although their films had a diamagnetic transition at around 120 K, the resistive transition is incomplete due to dominance of impurity phase induced possibly by air detrimental effect on the precursor film (Wang et al., 1993).

In the conventional process, slow heating/cooling is usually preferred in synthesis of volatile materials, such as Hg-HTS's, in order to maintain the phase equilibrium. Problems arise as the commonly used Hg source–HgO decomposes at ~500°C, much lower than that for formation of Hg-HTS's. Premature chemical reaction of Hg vapor with the precursor film became a major problem, since impurities, such as $HgCaO_2$ phase, are formed during the heating period from 500–800°C. If a long sample-heating period of a few hours is adopted, this premature reaction can result in an extremely high Ca deficiency in the precursor films at the end of the heating period. Since $HgCaO_2$ phase is fairly stable at the typical processing conditions for Hg-HTS films, little superconducting phase can be possibly obtained as no Ca is left in the precursor films.

Reducing the sample-heating period, such as fast heating/cooling, may provide a solution to the problem. The question is whether the phase equilibrium within the crucible could be maintained when the temperature is varied rapidly. This actually depends on two time scales: the heating time constant, that is defined as the inverse of the heating rate, and the transport time constant that measures the time needed to transport the Hg source to the sample. As long as the latter is much shorter than the former, the phase equilibrium is reached at each instant no matter how fast the temperature is changed. A simple way to minimize the transport time constant is to minimize the distance between the sample and the source by placing the sample in close proximity of the source. This idea has been proved experimentally working effectively well in minimizing impurity formation for Hg-1212 and Hg-1223 films. In this so-called fast temperature ramping annealing (FTRA) process, the heating period is reduced from a few hours to at most few tens of minutes (Yun et al., 1996; Wu et al., 1996a). This significantly suppresses the premature chemical reaction, resulting in high purity Hg-1212 and Hg-1223 films.

18.2.2 Epitaxy of Hg-HTS Films in Cation Exchange Process

A recently developed cation-exchange process not only provided an alternative process for epitaxy of Hg-HTS films, but also offers a promise to circumvent major problems encountered in conventional crucible process (Wu et al., 1999). This process employs an epitaxial precursor matrix, instead of an amorphous precursor film, and achieves epitaxial Hg-HTS films via replacement of certain cations in the precursor matrix. In Figure 18.1, a comparison of the conventional thermal process (Figure 18.1, panel A) and the cation-exchange process (Figure 18.1, panel B) is schematically described. In the cation-exchange process, the precursor matrices are chosen to have a similar structure and composition to that of the target material and to have at least one weakly bonded cation (cation "b") to be replaced by the desired cation (cation "a") to form the target material. When the cation "b" is perturbed using various different methods such as thermal heating or light/particle-beam irradiation, it will vibrate around the equilibrium site where the Gibbs free energy is minimized. The spatial deflection of this cation is proportional to the energy of perturbation. When the threshold perturbation energy (U_{th}) is reached, at which the deflection of cation "b" is comparable to the lattice constant, the precursor matrix may collapse due to quick escape of many cation "b". In the cation-exchange process, however, the perturbation energy is maintained to be close to but below U_{th} so that the precursor matrix is well kept while the cation "b" is slowly escaping. If a vapor of cation "a" is provided simultaneously, the overwhelming population of cation "a" may induce the replacement of cation "b" with cation "a" and the target material is formed.

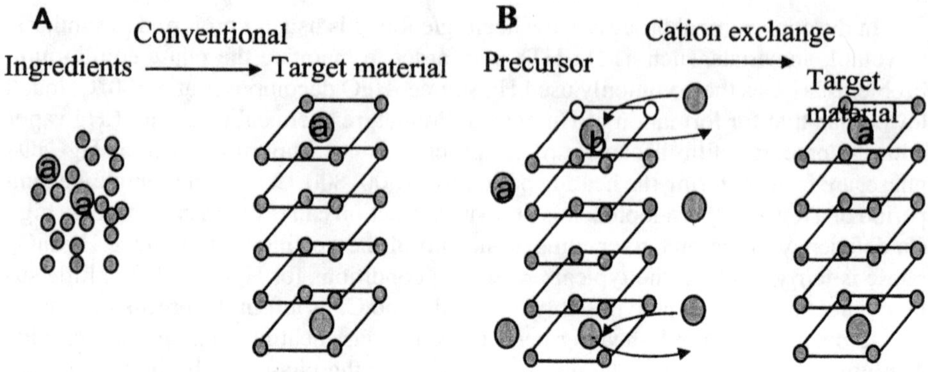

Figure 18.1. Schematic description of the conventional (panel A) and the cation-exchange (panel B) processes.

Selection of Precursor Matrices. Although many others may serve as the precursor matrices for Hg-HTS's, Tl-based high-T_c superconductors (Tl-HTS's) were selected in our experiments considering their structural similarity to that of Hg-HTS's. Generally speaking, Tl-HTS's are much less volatile, insensitive to air, and easy to be synthesized into forms of bulks or epitaxial films on many single-crystal substrates (Siegal et al., 1997 and references therein). Tl-HTS's have two series: one contains a single Tl–O plane [$TlBa_2Ca_{n-1}Cu_nO_{2(n+1)+1}$, $n = 1, 2, 3, \ldots$] and the other, double Tl–O planes [$Tl_2Ba_2Ca_{n-1}Cu_nO_{2(n+2)}$, $n = 1, 2, 3, \ldots$] in a unit cell (see, for example, Tl-2212, Tl-2223, Tl-1212 and Tl-1223 in Figure 18.2). The members of the former have nearly the same structures as that of their Hg-HTS's counterparts and therefore are ideal precursor matrices for Hg-HTS's. For example, to make Hg-1223, we may choose Tl-1223 as the precursor matrix and to obtain Hg-1223 by replacing Tl cations with Hg cations while leaving the lattice unchanged. $Tl_2Ba_2Ca_{n-1}Cu_nO_{2(n+2)}$ may also be employed as the precursor matrices for the Hg-HTS's with two possible routes. In one case, the two-Tl–O planes may collapse into one Hg–O plane to form $HgBa_2Ca_{n-1}Cu_nO_{2(n+1)}$, $n = 1, 2, 3, \ldots$, while in the other, the two Tl–O planes may be transferred to two Hg–O planes to form, presumably, $Hg_2Ba_2Ca_{n-1}Cu_nO_{2(n+2)+1}$, $n = 1, 2, 3, \ldots$. It should be pointed out that the Hg-HTS's containing two Hg–O planes in a unit cell have to be synthesized at high pressures with partial replacement of Ba by other cations such as Sr. Moreover, these superconducting phases are not stable in atmospheric environments. Based on these arguments the most probable phases to result from a Tl-HTS's precursor of two Tl–O planes are Hg-HTS's of one Hg–O plane, unless the processing conditions for cation exchange are specially manipulated. This speculation has been confirmed experimentally as we will describe in the following. Table 18.1 lists several possible Tl-HTS's precursor matrices for Hg-HTS's and some of them have been tested experimentally.

Converting Tl-HTS Films to Hg-HTS Films. The cation-exchange process employs a quite different growth mechanism. It does not require a stringent phase equilibrium and allows epitaxy of Hg-HTSs through a diffusion process so that it can be carried out effectively in a large processing window for conversion from Tl-2212 and Tl-1212 to Hg-1212 (Wu et al., 1999). Figure 2A of this article shows that three Hg-1212 films processed at dramatically different P_{Hg} of $1P_0$, $0.75P_0$ and $0.5P_0$ have nearly identical T_cs and J_cs. Figure 2B of the same article depicts results for two

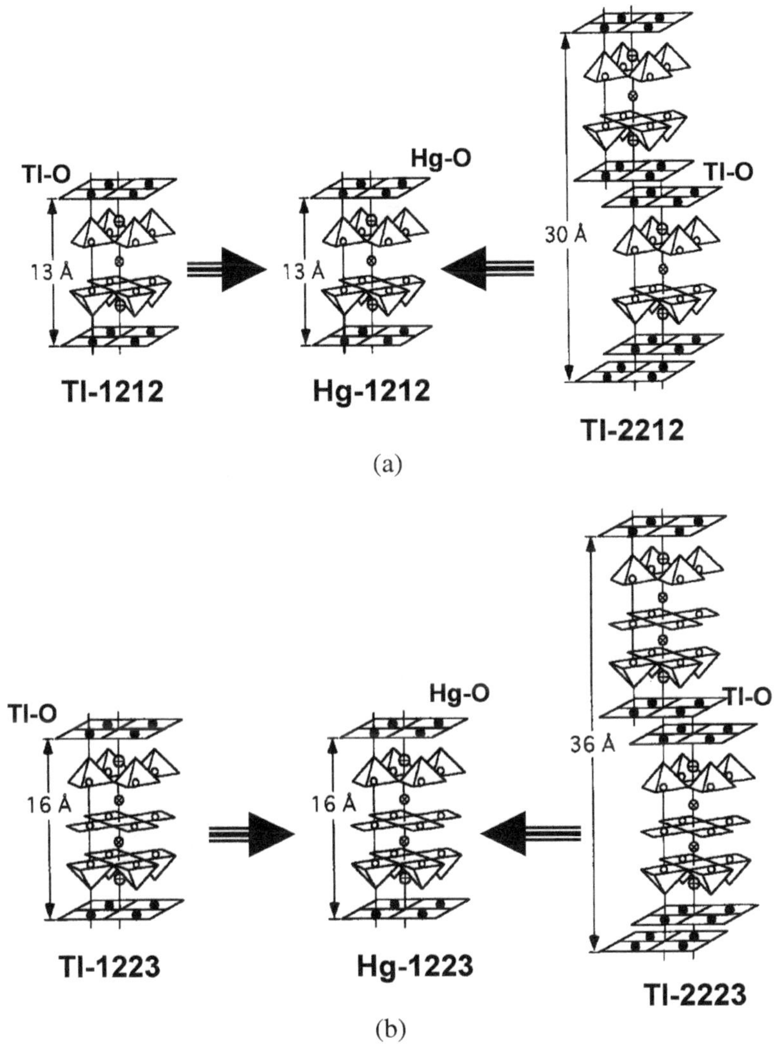

Figure 18.2. Structural variations of samples during cation exchange. Panel (a) illustrates the structural change from Tl-1212 to Hg-1212 and Tl-2212 to Hg-1212; and panel (b), that from Tl-1223 to Hg-1223 and Tl-2223 to Hg-1223.

Hg-1212 films with one annealed at 700°C for 12 hours and the other at 780°C for 3 hours in the cation-exchange process. Again, both T_cs and J_cs are nearly the same in these films while the processing conditions are far from the required phase equilibrium, $P_{Hg} \approx 1 P_0$ and $T_{Hg} \approx 800$°C, for Hg-1212. Hg-1212 and Hg-1223 thin films processed via cation exchange inherit high-quality epitaxy and smooth surface morphology from their precursor matrices (Yan et al., 1998, 2000; Fang et al., 2000). Consequently, the quality of the film made in cation-exchange process is superior to that made in the conventional process. For example, their J_cs close to 1 MA/cm^2 at 110 K are nearly an order of magnitude higher than the best value reported previously on Hg-HTS thin film (Yun and Wu, 1996a, 1996b; Krusin-Elbaum et al., 1995) and can be further improved with optimization of processing conditions.

Table 18.1. Precursors for Hg-based high temperature superconductors

Target material	Possible precursor matrix 1	Possible precursor matrix 2	Possible precursor matrix 3
Hg-1201 $a = 0.3875$ nm, $c = 0.951$ nm, $T_c \sim 85$–90 K	$TlBa_2CuO_5$(Tl-1201) $a = 0.377$ nm, $c = 0.89$ nm, $T_c \sim 15$ K	$Tl_2Ba_2CuO_6$(Tl-2201)* $a = 0.387$ nm, $c = 2.31$ nm, $T_c \sim 85$–90 K	$Bi_2Sr_2CuO_6$(Bi-2201) $a = 0.387$ nm, $c = 2.31$ nm, $T_c \sim 85$–90 K
Hg-1212 $a = 0.386$ nm, $c = 1.271$ nm, $T_c \sim 125$–127 K	$TlBa_2CaCu_2O_7$* $a = 0.385$ nm, $c = 1.277$ nm, $T_c \sim 90$–103 K	$Tl_2Ba_2CaCu_2O_8$* $a = 0.387$ nm, $c = 2.93$ nm, $T_c \sim 105$–110 K	$Bi_2Sr_2CaCu_2O_8$(Bi-2212) $a = 0.387$ nm, $c = 2.93$ nm, $T_c \sim 105$–110 K
Hg-1223 $a = 0.385$ nm, $c = 1.578$ nm, $T_c \sim 133$–135 K	$TlBa_2Ca_2Cu_2O_9$* $a = 0.385$ nm, $c = 1.595$ nm, $T_c \sim 115$–120 K	$Tl_2Ba_2Ca_2Cu_3O_{10}$* $a = 0.387$ nm, $c = 3.56$ nm, $T_c \sim 125$–127 K	$Bi_2Sr_2Ca_2Cu_3O_{10}$(Bi-2223) $a = 0.387$ nm, $c = 3.56$ nm, $T_c \sim 125$–127 K
Hg-1234 $a = 0.385$ nm, $c = 1.897$ nm, $T_c \sim 130$ K	$TlBa_2Ca_3Cu_4O_{11}$ $a = 0.385$ nm, $c = 1.915$ nm, $T_c \sim 120$ K	$Tl_2Ba_2Ca_3Cu_4O_{12}$ $a = 0.387$ nm, $c = 4.20$ nm, $T_c \sim 110$–115 K	$Bi_2Sr_2Ca_3Cu_4O_{12}$(Bi-2234) $a = 0.387$ nm, $c = 4.20$ nm, $T_c \sim 110$–115 K

A partial list of possible precursor matrices for Hg-1201, Hg-1212, Hg-1223 and Hg-1234 superconductors is included in Table 18.1. Those precursor matrices labeled with "*" have been tested experimentally. It should be noted that only the original forms of the single- and double-layers Tl-HTS's and double-layer Bi-HTS's are listed in the table. In fact for all the three families, there are numerous ways of cation replacement, which generates similar materials with different chemical compositions. For example, Tl can be replaced partially by Bi or Pb in the Tl-HTS's. One can imagine that all these variations can also be chosen as precursor matrices for corresponding Hg-HTS's

The conversion from Tl-2212 to Hg-1212 is evidenced first in the shortening of the c-axis lattice constant as shown in the X-ray diffraction (XRD) θ–2θ patters of a precursor Tl-2212 film (see Figure 1(a) of Yan et al., 2000) and that of the Hg-1212 film processed at 700°C for 12 h (Figure 1(b) of Yan et al., 2000). The formed Hg-1212 film is also *c*-axis-oriented and the *c*-axis lattice constant for Hg-1212 is ~1.27 nm. Since the *c*-axis of Hg-1212 is shorter than that of Tl-2212, the film experiences a volume reduction along the *c*-axis after the conversion. This implies reduction in film thickness by approximately 14% after a complete conversion, as confirmed in the film thickness measurement using the Rutherford backscattering (RBS), in which a 300 nm thick Tl-2212 film, for example, was found to be 260 nm thick after the conversion. The XRD pole figures of the Tl-2212 films and Hg-1212 film indicate the film grows epitaxially with its (100) axis aligned with the (100) axis of the $LaAlO_3$ substrates (see Figure 18.3). This observation suggests an interesting procedure during the Hg–Tl cation exchange: while the Tl-2212 lattice is collapsing along the *c*-axis, the in-plane alignment is well kept, allowing Hg-1212 films to inherit epitaxy from their Tl-2212 precursor films. For the conversion from Tl-1212 to Hg-1212 no structural variation has been detected experimentally since the two have the identical structures (Wu, 2000; Fang et al., 2000). Interestingly, the full-width-half-maximum of the (103) poles for Hg-1212 films, made either from Tl-1212 or Tl-2212 precursor films, are similar (see Figure 18.3) and comparable to that of their precursor films. Nevertheless, slight increase in the RBS/channeling χ_{min} from typically less than 10 on Tl-1212 films to 12–20 on Hg-1212 films after the conversion. This indicates increase of lattice disordering, most probably via generation of points defects such as interstitials and vacancies, and after the cation exchange.

Conversion from Tl-2223 and Tl-1223 to Hg-1223 is relatively more difficult, but has also been demonstrated recently (Xie et al., 1999; Yan et al., 2000). The bulk conversion is straightforward and T_c's up to 135 K has been obtained (Xie et al., 1999). This is not surprising as Hg can diffuse into (and Tl diffuse out of) Tl-HTS bulks fairly easily through grain boundaries in bulk Tl-HTS samples. Hg-1223 films have also been obtained in cation-exchange process, but the quality of the film is not as good as expected. The major difficulty is in epitaxy of high-quality, pure-phase Tl-2223 or Tl-1223 precursor films. Using a relatively pure Tl-2223 film (up to 90% phase purity) of T_c in the range of 110 to 118 K, Hg-1223 films can be obtained via cation exchange with T_cs up to 130 K. On the other hand, if Tl-1223 films are used as precursor matrices, the T_cs can be increased from ~100–110 K to 120–126 K after cation exchange. Although this T_c increase is large, the J_cs of Hg-1223 films made in cation exchange process are not as good as that of Hg-1212 films made in the same process. This was attributed to the poorer epitaxy and phase purity in the Tl-HTS precursor films as Hg-1223 films made in cation exchange process are expected to have structures and phase purity no better than their Tl-HTS precursors.

18.3 PHYSICAL PROPERTIES OF Hg-1212 AND Hg-1223 FILMS

18.3.1 Surface Morphology

Hg-HTS films made in the conventional process have typically rough surface due to presence of massive impurity phases as indicated by scanning electron microscopy (SEM) studies. Most common impurities include $HgCaO_2$, and several Ba–Cu–O phases that are non-superconducting. Misaligned grains visible on the film surface

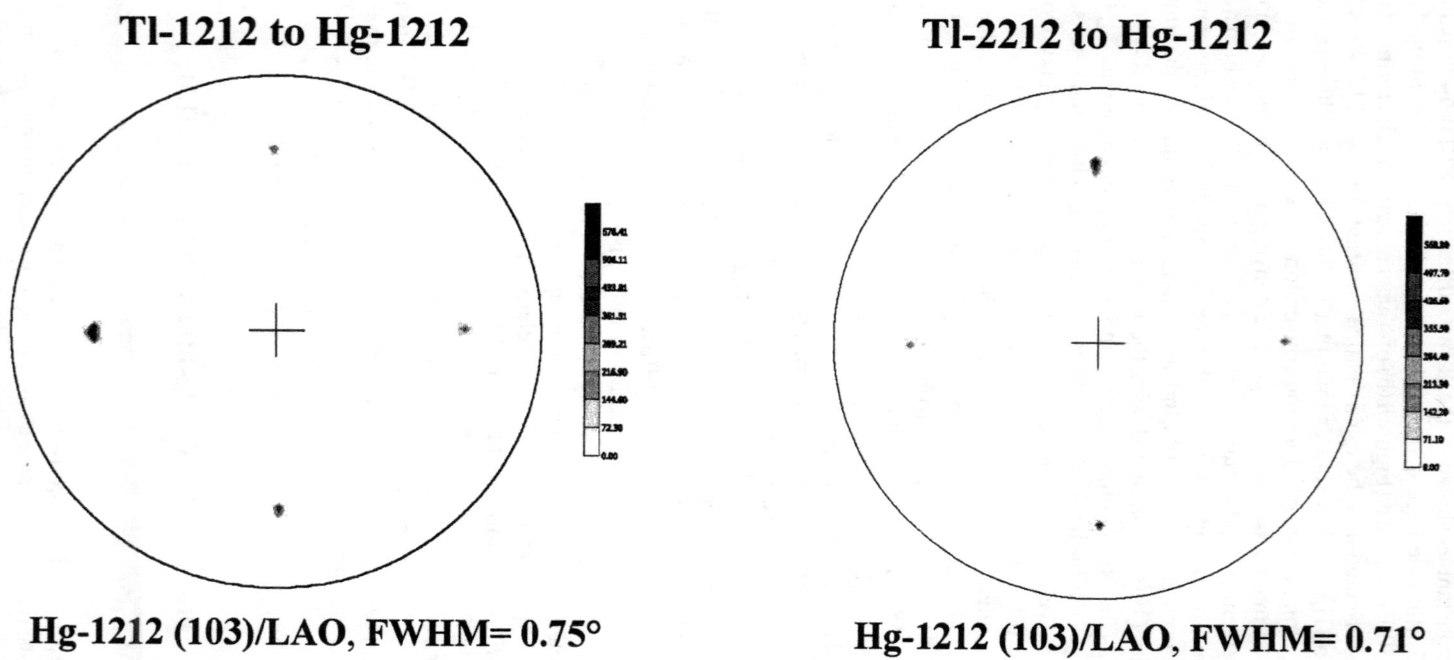

Figure 18.3. XRD (103) pole figures of Hg-1212 films converted from Tl-1212 (left panel) and Tl-2212 (right panel) thin films. The thickness of the film is in the range of 200–250 nm and the substrates for both films are (100) single crystal $LaAlO_3$.

are another reason responsible for the rough film surface. Wu et al. reported improved surface morphology on films made using FTRA and attributed it to the reduced impurities as well as better aligned grains (Wu et al., 1996b). However, outgrowths and pinholes are popular with *ex situ* processed films. This rough surface morphology results in high microwave surface resistance (R_S) in these Hg-HTS films, despite their high T_cs and J_cs (see Section 18.3.5 for more details of R_S measurements). The best R_S observed on Hg-HTS films made in conventional crucible process is at least one order of magnitudes higher than the typical values of YBCO or Tl-HTS films at the same condition (see the inset of Figure 18.8). It is therefore necessary to optimize the film growth process to achieve lower R_S. Recently Sun et al. reported mirror-like surfaces on their Hg-1212 films on NGO substrates (Figure 2A in Sun et al., 1999). Inoue et al. observed flat surface that is free of pinholes and outgrowths on ($Re_{0.1}Hg_{0.9}$)-212 films on LSAT substrates and argued that when the film thickness is small, the surface morphology is significantly improved (Inoue et al., 1999). It would be certainly interesting to know the microwave R_S on these films.

The surface morphology was considerably improved on the Hg-HTS films made in the cation exchange process. Basically, it looks similar to that of their precursor matrices, which is advantageous as Tl-2212 and Tl-1212 films have much smoother surface morphology. As shown in Figure 18.4, Hg-1212 films converted from Tl-1212 look almost identical before and after the cation exchange, which is not unexpected as no structural variation occurs during the cation exchange. In the case of Tl-2212 as the precursor, voids are visible on the Hg-1212 films while the overall surface morphology remains more or less the same (see also Figure 3 of Yan et al., 2000). A chemical mapping of the Hg and Tl elements on samples quenched at different stages of cation exchange process shows more Hg distributed near the voids at earlier stage of the process. Since no voids were observed on Hg-1212 films converted from Tl-1212 films, one may speculate that the extra Tl cations, by making a way out, are the major reason for the formation of the voids. Although a uniform Hg distribution was reached on Hg-1212 films after the cation exchange is completed, Cu-deficiency was detected near the voids, suggesting that the voids were most probably formed on grain boundaries in the Tl-2212 precursor films. Nevertheless, Hg-1212 films made in cation-exchange process, whether from Tl-1212 or from Tl-2212 films, have much improved surface morphology than that of the same films made in conventional process. One of the remarkable features associated with the improved surface morphology is the low microwave surface resistance (Aga et al., 2000) on the cation-exchange processed Hg-1212 films. The R_S (see Figure 18.8) was found comparable to that of other high-quality HTS films at 77 K and remains nearly unchanged at temperatures near their T_cs (details are given in Section 18.3.5).

18.3.2 High T_cs

The T_cs of Hg-1212 and Hg-1223 films, made either in conventional or cation exchange processes, are generally comparable. As shown in Table 18.2, T_cs for Hg-1212 films are up to 125 K, the same as the value for Hg-1212 bulk samples, while that of Hg-1223, though in exceeding 130 K, are a few degrees lower than its bulk value of 135 K. This lower T_cs may be attributed to slight oxygen deficiency of the film. It should be realized that the most Hg-HTS films are oxygen deficient after the Hg-vapor anneal and a post anneal at low temperatures around 350°C is necessary to bring the oxygen content to the optimum level. In the case of Hg-1212 film, however, the post oxygen annealing in the temperature range of 290°C to 360°C, does not seem adequate

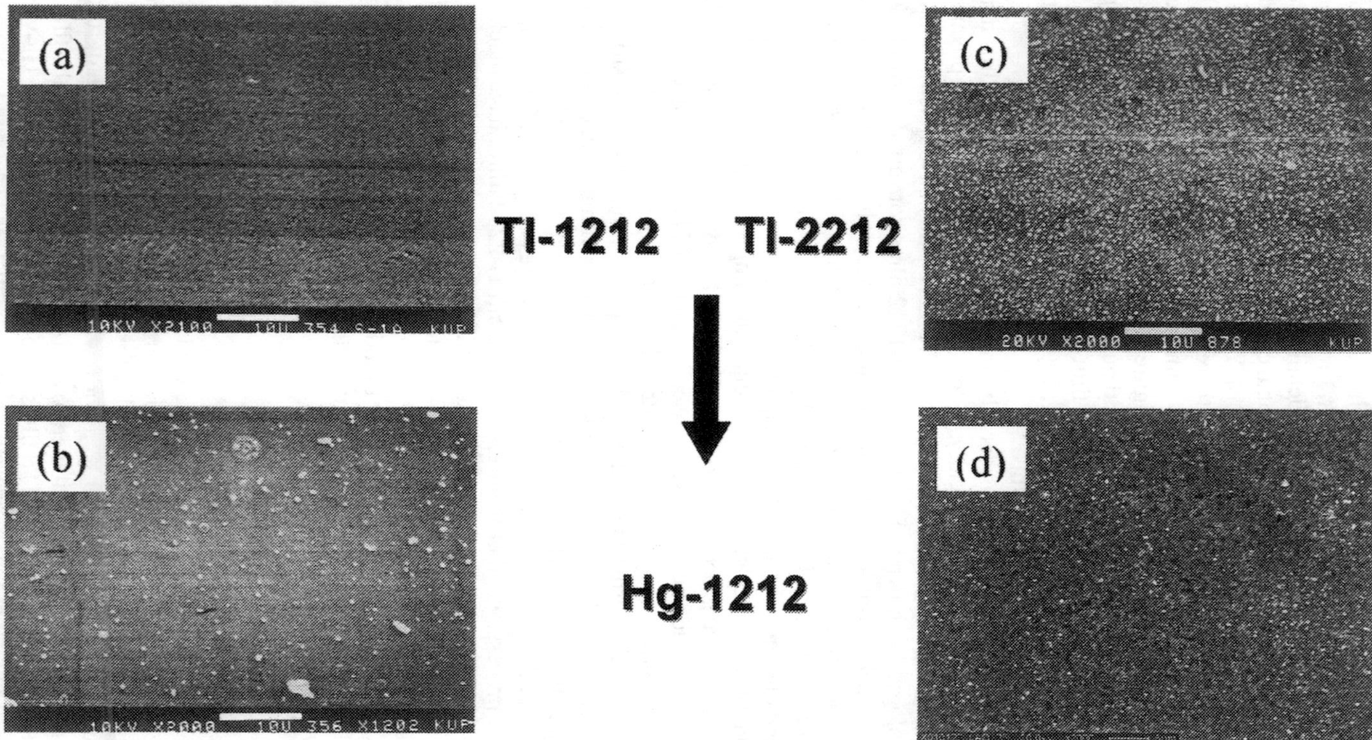

Figure 18.4. Scanning electron micrographs (SEM) of Hg-1212 films converted from Tl-1212 (b) and Tl-2212 (d) thin films. The SEM's of Tl-1212 (panel (a)) and Tl-2212 (panel (c)) precursor films are also included. The thickness of the film is in the range of 200–250 nm and the substrates for are (100) single crystal $LaAlO_3$. The magnification for all four panels were 2000 and the scale shown is 10 micrometers.

Table 18.2. Superconducting properties of Hg-HTS thin films

Film/substrates	Method	Phase-Purity/Thickness	T_{c0} (K)	J_c (5 K)/J_c (77 K)/J_c (100 K), $H = 0$ (MA/cm^2)	H_{irr} (77 K)/H_{irr} (100 K) (H//c-axis)
Hg-1201/$SrTiO_3$	Conventional	Pure/200-300 nm	93 K	NA	~0.5 T/NA
Hg-1212/$SrTiO_3$	Conventional	Pure/200-300 nm	120–124 K	10/1/0.1	~0.5 T/0.1 T
Hg-1212/$LaAlO_3$	Conventional	Pure/200-300 nm	116–124 K	20/3/1.5	~0.5 T/0.1 T
(Hg,Re)-1223/$SrTiO_3$	Conventional	Predomin/200-300 nm	120–131 K	20/4/1.6	~2 T/1 T
Hg-1223/$LaAlO_3$	Conventional	Predomin/200-300 nm	120–128 K	13/1.4/1.5	~2.2 T/0.8 T
Hg-1212/$LaAlO_3$	Cation exchange	Pure/200-300 nm	120–125 K	20–40/2–5/0.5–2	2–3 T/0.8–1 T
Hg-1223/$LaAlO_3$	Cation exchange	Pure/300-500 nm	125–130 K	10–20/1–1.5/0.1–0.5	2–3 T/0.3–0.5 T
Hg-1212/$LaAlO_3$	Cation exchange	Pure/50-80 nm	110–118 K	5–10/0.7–1.1/0.1–0.4	1–2 T/0.3 T
Hg-1212/Ni	Cation exchange	Predomin./300-800nm	123–125 K	5–10/1–2/0.5–0.7	2–3 T/0.8–1 T

Table 18.2 summarizes some superconducting properties, such as T_cs, J_cs, and H_{irr}s of Hg-1212 and Hg-1223 films made in conventional and cation-exchange processes.The Ni substrates were provided by the Oak Ridge National lab using the Rolling assisted bi-axially textured substrates (RABiTS) process. These substrates were buffered with a standard CeO_2/YSZ/CeO_2 tri-layers.

to raise the T_c to 135 K (Kang et al., 1999). Another point worth mentioning is that some of Hg-1212 films cation-exchange processed in the mixture vapor of Hg, oxygen and Fluorine are optimally doped before the post oxygen anneal and could be over-doped with oxygen after the post oxygen anneal (Wu and Xie, 2001). This suggests that the presence of Fluorine may facilitate diffusion of oxygen into the Hg-1212 films and it would be interesting to see whether the T_cs of Hg-1223 films could be brought up to the bulk value of 135 K when Fluorine is included as part of vapor during the Hg-vapor annealing.

18.3.3 High J_cs

J_c's of Hg-HTS films were characterized using both electrical transport or magnetic induction methods. In the former, a standard four-probe configuration was employed and J_cs were determined from the $I-V$ curves by a criterion of 1 μV/cm between the two voltages leads. Most samples used for transport measurements were patterned into narrow bridges. J_cs can also be derived from magnetic induction using the Bean's model, although the J_c's values calculated this way are slightly lower than that obtained from the transport measurement. The J_c's listed in Table 18.2 are based on both transport and magnetic measurements.

The left panel of Figure 18.5 shows comparison of zero-field J_cs (T) of several HTS thin films including YBCO, Tl-2212, Tl-1212, Tl-1223, Hg-1212 and Hg-1223. The film thickness is in the range of 200–250 nm and all the films were regarded as standard good quality. Below 77 K, the YBCO film has the highest J_c in the group while above 77 K, Hg-1212 and Hg-1223 films show substantially higher J_cs, especially above 90 K when YBCO becomes normal. It has been noticed that the J_c's of Hg-1212 films made in cation exchange process are higher than that of Hg-1223 films made in the same process, although the T_c of the Hg-1223 films is higher. The major reason is in the poorer quality of epitaxy in Hg-1223 films obtained thus far and the problem stems from the difficulties in epitaxy of high-quality pure phase Tl-2223 or Tl-1223 films. As we discussed earlier, the Hg-HTS films made in the cation exchange process inherit structures and surface morphologies from their Tl-HTS precursor films. In order to improve the quality of Hg-1223 films, other precursor films, such as doped Tl-1223 or Tl-2223, may be employed.

The left panel of Figure 18.5 depicted transport J_cs as function of temperature in magnetic fields up to 5 Tesla applied along the normal direction of the film (Gapud et al., 1999b). Films used for the transport measurement were patterned into thin lines of typical line width of 50–100 μm and a threshold of 1 μV/cm was applied to determine J_cs from the $I-V$ curves measured at different temperatures. Only high temperature transport J_cs were measured due to limitation of our current source. The zero-field J_c vs. temperature curve obtained in magnetic measurement was also included in this figure for comparison. It is interesting to see that the zero field J_cs, measured either magnetically or in transport, are quite consistent, but the former are generally lower than the latter when the magnetic field is applied, due to possibly more sensitive threshold used in magnetic measurement for J_c determination. Consequently, the irreversible fields (H_{irr}) determined from transport measurement are higher than that from magnetic measurement (see Figure 18.6).

When a magnetic field is applied, J_c's decrease faster with increasing field in Hg-HTS films than in YBCO films at fixed temperatures near or below 77 K. This is not surprising as YBCO is more isotropic than Hg-HTS (Welp et al., 1993; Huang et al., 1994), which implies higher irreversibility fields (H_{irr}) at the same reduced temperatures. At the absolute temperature scale, the H_{irr} of Hg-1212 or Hg-1223

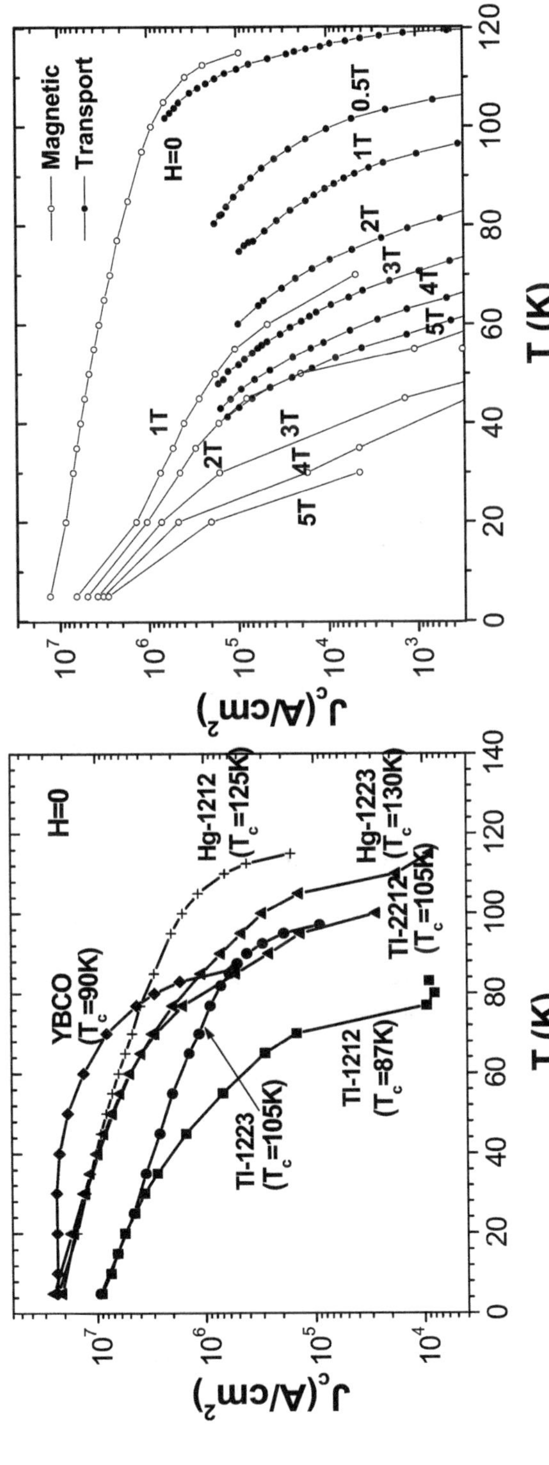

Figure 18.5. Left panel: comparison of zero-field J_Cs in thin films of several high temperature superconductors of standard quality; right panel: J_C as a function of temperature in magnetic fields up to 5 Tesla applied along perpendicular to the film (Gapud et al., 1999b).

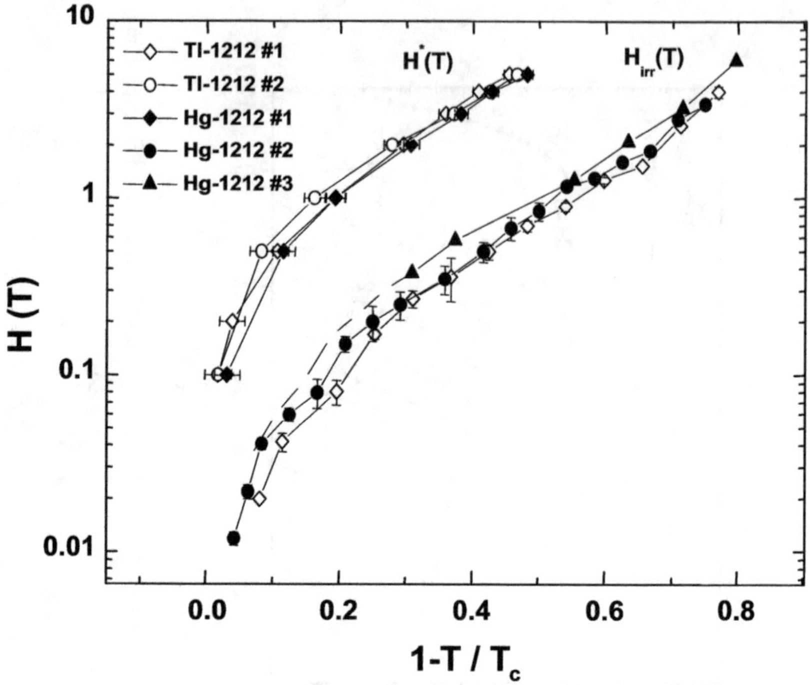

Figure 18.6. Irreversibility fields for Hg-1212 and Tl-1212 films are depicted as functions of temperature. The upper group of curves were obtained from resistive measurement and the lower group of curves, from magnetic measurement. The magnetic field was applied along the normal direction of the films (Gapud et al., 1999a). The Tl-1212 films used for this experiment were provided by M. Siegal at the Sandia National Laboratory, while the Hg-1212 films, by our group using the cation exchange process.

crosses over that of YBCO at ~77–80 K. While that of YBCO decreases quickly to zero as its $T_c \sim 90$ K is approached, the H_{irr} of Hg-1212 or Hg-1223 remains considerably high up to 110–115 K. For example, the H_{irr} for Hg-1212 films is up to 0.8–1 T at 100 K. The $H_{irr} \sim 2$–3 T of Hg-HTS films at 77 K can be improved via introducing defects (Krusin-Elbaum et al., 1997a, 1997b). A H_{irr} of 4–5 T at 77 K has been reported on Hg-1223 powder samples after ion-beam irradiation. It should be especially mentioned that using the cation exchange process, we have successfully coated Hg-1212 on in-plane textured Ni substrates and achieved J_c's comparable to that of high-quality Hg-1212 films on single-crystal substrate (Xie et al., 2000b). Such high J_c's are more than an order of magnitude higher than the best value reported previously on Hg-HTS's coated conductors (Xie et al., 2000a) and other Hg-HTS films on metal substrates (Schwartz and Sastry, 2001).

18.3.4 Irreversibility

Irreversibility sets the upper boundary for practical applications of superconductors in the temperature and field phase diagram. The H_{irr} (T) for a superconducting thin film can be characterized either magnetically from the close point (usually $\Delta M = M^+ - M^- < 10^{-5}$ emu/cm^3) in H vs. M (magnetization) hysteretic loops at a fixed temperature, or in electrical transport measurements, as the temperature J_cs drop below a certain threshold value (usually $\sim 10^3$ A/cm^2). One need to be careful when compare the H_{irr} (T) for a thin film with that for bulks as the latter regards the intra-grain properties and while in thin films, weak links across the grain boundaries may dramatically affect the irreversibility behavior. Therefore the bulk H_{irr} (T) usually represents the upper limit for H_{irr} (T) of a thin film since the latter includes both intra-grain and inter-grain effects.

Several groups have studied the H_{irr} (T) in Hg-1212 and Hg-1223 bulks (Welp et al., 1993; Huang et al., 1994). The H_{irr} at 77 K and 100 K, for example, are around 2–3 T and 0.2–0.5 T, respectively. These values can be significantly increased after the linear defects are introduced via high-energy ion beam irradiation. Krusin-Elbaum et al. reported a H_{irr} above 4 T at 77 K and 0.8 T at 100 K for Hg-1223 bulks irradiated with splayed GeV proton beams that generates nuclear fission tracks (or linear defects) in the materials (Krusin-Elbaum et al., 1997a, 1997b).

Gapud et al. did a careful study of H_{irr} (T) in the "1212 system" that includes both Hg-1212 and Tl-1212 films (Gapud et al., 1999). As shown in Figure 18.6, the H_{irr} defined via a transport measurement is up to 3 T at 77 K and 1 T at 100 K, respectively, approaching the intra-grain H_{irr} (T). Moreover, by plotting the H_{irr} (T) for the "1212 system" at the reduced temperature scale, it was found that H_{irr} (T) for Hg-1212 and Tl-1212 films coincide nicely, indicating not only similar anisotropy but also similar inter-grain connectivity in these two type of films. Since the Hg-1212 films used for this study were obtained from Tl-1212 precursor films in cation exchange processes, the similar grain connectivity confirms again that the microstructure of "1212" film remains almost the same during the cation exchange. The H_{irr} (T) for Hg-1223 films reported so far is lower than that of Hg-1212 films.

The H_{irr} of Hg-HTS thin films can be significantly improved when linear defects are added to the films using high-energy heavy ion-beam irradiation, resulting in "optimized" H_{irr}, or magnetic pinning strength, in these films. Figure 18.7 depicts the "optimized" H_{irr} (T) curves for Hg-1212 and several other HTS films including YBCO, Bi-2212, Bi-2223, and Tl-1223. For Hg-1212, linear defects of 2-Tesla-equivalent density were generated with 4 GeV Pb ions. Although this density may be further tuned to yield the best H_{irr}, the H_{irr} for Hg-1212 films is the highest among the HTS films studied in this figure in at 77 K and higher temperatures. This certainly provides headroom

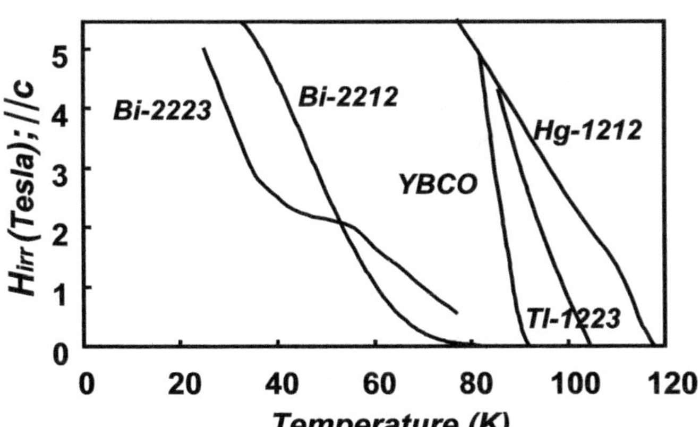

Figure 18.7. Irreversibility fields, optimized by GeV heavy ion-beam irradiation, are depicted as functions of temperature for thin films of several high-temperature superconductors (by Christen et al., private communication).

for applications in both electronics as well as electrical components/systems using Hg-HTS thin and thick films. This result is consistent with our earlier study of irradiation effect on Hg-1223 films using splayed GeV proton beams that generates nuclear fission tracks (Thompson et al., 1997).

18.3.5 Microwave Surface Resistance

HTSs are very attractive for microwave applications due to their low-loss performance in this frequency range. Generally, a microwave loss is characterized with the surface resistance (R_S) of a superconductor defined as dissipation per unit area at a given power level. Results of R_S measurements have been reported on many HTS thin films (see a review article by Gallop, 1997). Low R_Ss in the range of 0.1 to 0.5 mΩ at 77 K and 10 GHz have been observed on several HTSs including YBCO, Tl-2212, and Tl-2223 (Gallop, 1997; Lancaster, 1997; Holstein et al., 1992). This R_S value is nearly an order of magnitude lower than that of the copper ($R_S \sim 0.87$ mΩ, see Lancaster, 1997) at the same temperature and frequency. Microwave passive devices fabricated on these films show very encouraging results for practical applications. For example, planar filters with up to 11 poles have been produced with an insertion loss lower than 0.05 dB at 2 GHz and 77 K. Operation of these devices at temperatures ~77 K has been demonstrated by many groups.

The higher T_Cs of Hg-HTS's makes it possible to push the microwave device operation temperature to even higher values. Most Hg-HTs films made in conventional process, however, suffer from high R_S that is at least an order of magnitude higher than that of YBCO at 77 K and 10 GHz (Aga et al., 2000a). This has been shown in the inset of Figure 18.8. This high R_S was attributed to the existence of impurity phases, rough surfaces and poor-quality epitaxy of the film as much lower R_S was observed on cation-exchange processed Hg-1212 films when these properties are greatly improved. As shown in the left panel of Figure 18.8 (Aga et al., 2000a), the R_S (10 GHz) of Hg-1212 films made in cation exchange process is as low as ~0.2 mΩ at 77 K and 0.3 mΩ at about 120 K. Shown in the same figure is one of the best (in terms of R_S) Hg-1212 films made in the conventional process (inset of the same figure). At 77 K, its R_S is still up to five times higher than that of the cation-exchange processed Hg-1212 film although the two films have similar T_Cs. The R_S of Tl-2212 precursor film is also included showing comparable values to that of the cation-exchange processed Hg-1212 at lower temperatures. Since the R_S's of all the films remain nearly constant in a wide temperature range below their T_Cs, an increase of 20 K in the operation temperature is possible in Hg-1212 microwave devices over their Tl-2212 counterpart.

The right panel of Figure 18.8 shows the power handling capability of Hg-1212 microstrip resonators (Aga et al., 2000c). The results of YBCO and Tl-2212 resonators in the same configuration are also included for comparison. Although the power handling capability of Hg-1212 resonators is higher than that of YBCO and Tl-2212 in the high temperature range, the critical power P_c vs. T curves for the three systems follows the same behavior indicating the mechanism for the high-power loss is the same for these resonators. It is interesting to note that the P_c vs. T for Hg-1212 is lower than that of YBCO (or Tl-2212) at the reduced temperature scale, suggesting higher power handling capability might be achievable on Hg-1212 microwave devices.

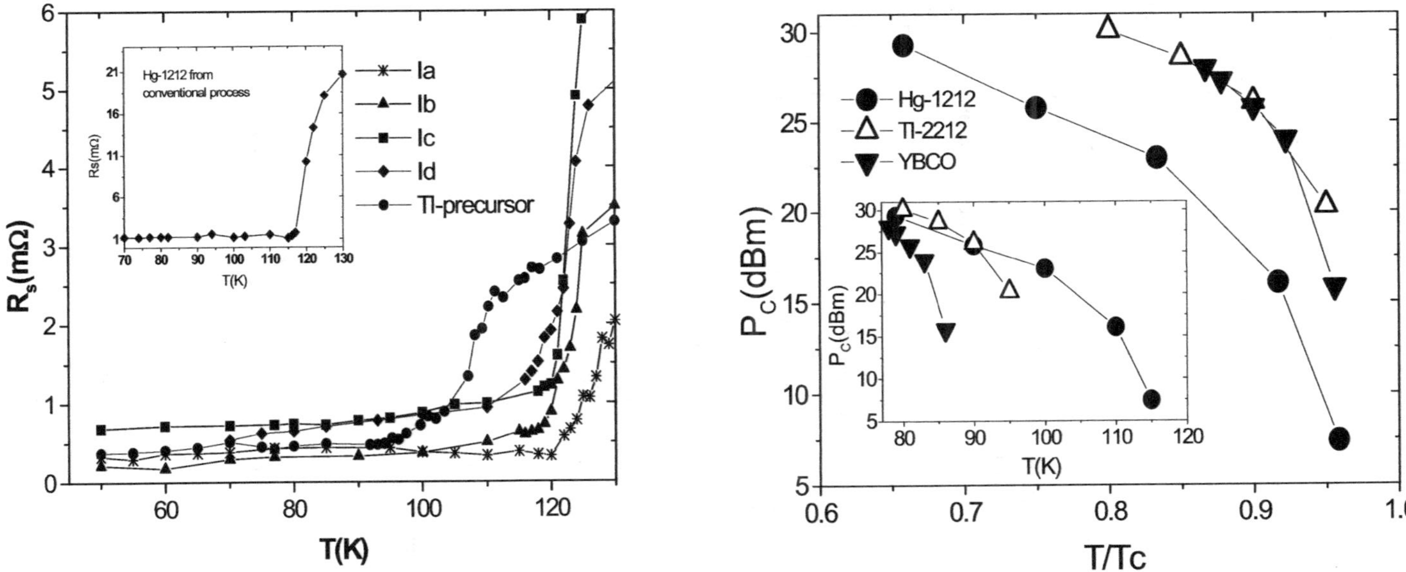

Figure 18.8. Left panel: microwave surface resistance (R_S) of Hg-1212 films made in cation exchange process. The inset shows the R_S for a Hg-1212 film made in conventional process. The microwave frequency is 10 GHz. Right panel: a comparison of power handling capability of YBCO, Tl-2212 and Hg-1212 microstrip resonators at 1 GHz at a reduced temperature scale.

18.4 APPLICATIONS OF Hg-HTS THIN FILMS

18.4.1 Bi-Crystal Grain Boundary Junctions and SQUIDs

Josephson devices are one of the important applications for HTS's since they can be operated at 77 K or higher temperatures. Hg-HTS's are particularly interesting because of their T_cs above 120 K that may bring the device operation temperature above 100 K. Several groups reported fabrication of grain-boundary Hg-1212 junctions (Hg-1212 GBJ) and SQUIDs (superconducting quantum interference devices) on commercial bi-crystal STO substrates (Gupta et al., 1994; Yu et al., 1999; Tsukamoto et al., 1998). These Hg-1212 GBJ's behave typically as resistive shunted junctions. Gupta et al. first obtained Hg-1212 GBJ's and SQUIDs on 36.8° STO bi-crystal substrates. They used laser ablation for patterning of the devices and the typical junction width is about 10 μm. The $I_c R_n$ products of their junctions are in the range of 60–120 μV at 77 K, comparable to that in other HTS GBJ's. The Hg-1212 SQUIDs made of two GBJ's can be operated at temperatures above 100 K. For example, at 107 K, a flux responsivity $dV/d\Phi$ is $\sim 8\ \mu V/\Phi_0$, but the flux noise at 77 K (for example, S_Φ, of $\sim 10^{-6}\Phi_0^2$/Hz) is about an order of magnitude higher than that of YBCO and Tl-2212 SQUIDs. Higher $I_c R_n$ products above 200 μV at 77 K were later reported on Hg-1212 GBJ's grown on 24° STO bi-crystal substrates by Hitachi/ISTEC and Kansas groups and the highest is around 450–460 μV, which in fact is the highest reported so far on HTS junctions. Although the SQUIDs on 24° STO bi-crystal substrates show similar flux responsivity at 110 K, their flux responsivity at 77 K is much higher than those on 36.8° STO bi-crystal substrates.

18.4.2 Hg-HTSs Tapes and Coated Conductors

The high T_c's and J_c's make Hg-HTS's desirable materials for high-current-carrying superconducting wire/tape applications at 77 K or higher. Several groups have reported experimental results on fabrication of Hg-HTS thick films on oxides and metal substrates. First of all, the selection of metal substrates for Hg-HTS's is restricted to few, such as Ni, due to amalgamation of most metals with Hg, which results in subsequent melting of most metals under the processing conditions for Hg-HTS's. For example, Amm et al. (1997) found that platinum and palladium react with Hg-based compounds. Silver and Gold were reported compatible with Hg-HTs's while still absorb significant amount of Hg (Lechter et al., 1995; Schwartz et al., 1996). The Ag/Hg amalgam, however, decomposes at high temperatures to release Hg (Meng et al., 1996). Nevertheless, high volume portion of Hg-1201 phase were obtained with Ag as the sheath material (Schwartz et al., 1996; Peacock et al., 1997) using the powder-in-tube (PIT) that has been successfully applied for fabrication of Bi-HTS tapes (Heine et al., 1989; Malozemoff, 1996). A technical problem associated with these Hg-1201 tapes is the porosity that results in poor inter-grain connectivity and therefore poor J_c's. Since the Ag/Hg reaction is promoted at higher processing temperatures, reducing the temperature may suppress the reaction and lead to better inter-grain connectivity. Recently, Sastry et al. (2000) and Su et al. (2000) reported fabrication of Pb-doped Hg-1223 on Ag at a much lower processing temperature ~780°C. They observed large regions of aligned grains of $(Hg_{1-x}Pb_x)$-223 on which T_c is up to ~133 K. Gold seems to be a better choice and little Au/Hg reaction was observed when gold foils were used as encapsulation in hot isostatic pressing (HIP) process. Predominant Hg-1223 phase was obtained with zero-resistance

T_cs ~ 127 K and magnetic J_cs ~ 1.8 kA/cm^2 at 77 K and zero field (Lechter et al., 1995).

Among all the metals, Ni may be the best choices due to its extremely low solubility in Hg. For example, the solubility of Ni in Hg is 8.5×10^{-3} wt% at 550°C and no mechanical change was observed on Ni after the Hg-HTS's processing. Meng et al. (1996) first reported coating of Ni substrates with $(Hg_{1-x}Re_x)$-223 and obtained zero-resistance T_cs up to 117 K and transport J_cs ~ 2.5×10^4 A/cm^2 at 77 K and zero field. To minimize the reaction between Ni and Hg, they buffered the Ni with 30–50 nm thick Cr, that is about 100 times less soluble in Hg, and toped with 5 nm thick Ag to promote melting so as to lower the processing temperature.

Most of these Hg-HTS tapes are, however, polycrystalline or untextured in the plane of the substrate and therefore suffer low J_cs due to a dramatic reduction of J_c across large-angle grain boundaries. To eliminate these large-angle grain boundaries, a coated-conductor technology that employs biaxially textured metal substrates, such as Ni or Ni-alloys, becomes the most promising approach (Iijima et al., 1992, 1993; Reade et al., 1993; Wu et al., 1995; Goyal et al., 1996; Norton et al., 1996). These metals are generally buffered and these buffer layers serves as a chemical diffusion barrier between the metal and the HTS film, on one hand, and allows HTS films to replicate the biaxial texture from the metal substrate. A major problem encountered in Hg-HTS's coated conductors was the degradation of the interface between the Hg-HTS's films and metal substrates due to the highly volatility of Hg and Hg-related compounds. Consequently, Hg-HTS's films grown on these biaxially textured substrates, including using either the conventional or the cation exchange processes, show poor-quality epitaxy and low J_c values typically below 10^5 A/cm^2 at 77 K in zero applied field (Xie et al., 2000).

One possible solution for this interface problem is to search for suitable buffer layers, but could be time consuming. For example, STO shows superior chemical compatibility and good lattice match with Hg-HTS's, but is difficult to grow epitaxially on Ni substrates due to the three-component complexity, high growth temperature, and high oxygen partial pressure required. Xie at al. (2000b) took a different route by modifying the original cation-exchange process in two aspects: (1) Oxygen, instead of argon, was employed during epitaxy of Tl-2212 precursor films to stabilize the CeO_2/YSZ/CeO_2 buffers on Ni substrates, and (2) FTRA was applied to minimize the unnecessary high-temperature processing time. The quality of Hg-1212 coated conductors has been significantly improved using this modified process in terms of epitaxy, surface (see the right panel of Figure 18.9) and film/substrate interface morphology. High J_c's close to that on single-crystal Hg-1212 films have been obtained. For example, the J_c for a 600 nm-thick Hg-1212 on CeO_2/YSZ/CeO_2/Ni is up to 7×10^5 A/cm^2 at 100 K and 2.4×10^5 A/cm^2 at 110 K. At 77 K, J_c increases to 2.2×10^6 A/cm^2 in zero field, comparable to the best J_c on YBCO coated conductors. Irreversibility fields of ~2.4 T and ~0.8 T were observed at 77 K and 100 K, respectively. A comparison of J_cs in Hg-1212 and YBCO coated conductors and that in Bi-2223 wires made with powder-in-tube technique is shown in the right panel of Figure 18.9. It is exciting to see that Hg-1212 coated conductors can still carry J_cs > 1 MA/cm^2 at temperatures YBCO turns normal, indicating Hg-1212 is promising for applications at 77 K and higher temperatures.

18.4.3 Microwave Devices

Applications of Hg-HTS films in microwave devices are very minimal mainly due to difficulties in fabrication of large-area Hg-HTS films. Up to now, almost all good

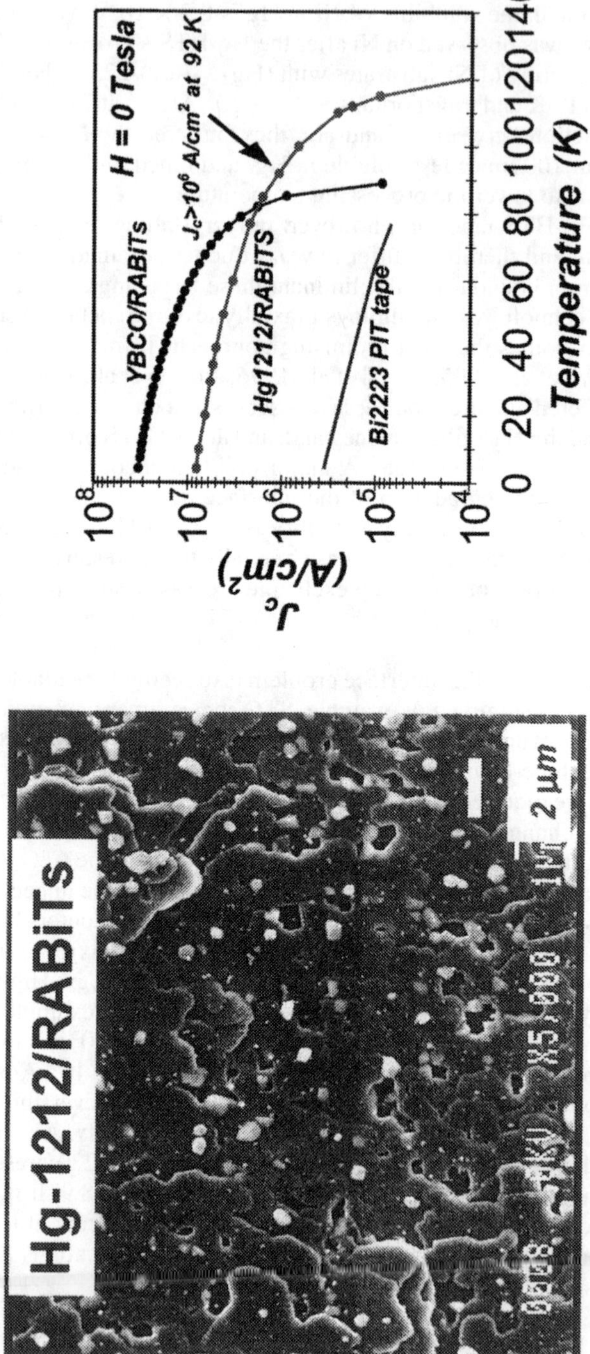

Figure 18.9. Left panel: SEM picture of a Hg-1212 film/RABiTs made in cation exchange process. Right panel: a comparison of transport J_Cs (in unit of A/cm^2) of YBCO/RABiTs, Hg-1212/RABiTs and Bi-2223 wire made in powder-in-tube process (Christen et al., unpublished).

results for Hg-HTS films were reported on small samples of dimensions of only few millimeters. In contrast, most practical applications require samples of larger dimensions. Using cation exchange process, Xie et al. (2000c) obtained Hg-1212 films of $12 \times 12\ \mathrm{mm}^2$ of uniform superconducting properties, such as T_cs and J_cs. It is difficult to go further as torch sealing the quartz tube would be technically impractical when the diameter of the tube is bigger. New techniques need to be developed for scale up the Hg-HTS films to larger area or longer length.

Nevertheless, some simple microwave devices, such as micro-strip resonators, have been fabricated and characterized recently (Aga et al., 2000b, 2000c). It should also be mentioned that these devices were fabricated first on Tl-2212 films using standard photolithography and converted to Hg-1212 devices in cation-exchange process (Xie et al., 2000d). This, on one hand, minimizes the exposure of Hg-1212 films to various chemicals, water-based solutions/gases that may lead to degradation of Hg-1212 films due to existence of Ba- and Cu-based impurity phases (Tolga et al., 1998, 1999) and, on the other hand, provide a shortcut for fabrication of Hg-HTS devices by taking advantages of the more matured Tl-HTS device technology.

Measurements of power handling capability on these Hg-1212 micro-strip transmission lines showed that a stable output power up to 19 dBm can be attained at 110 K (Aga et al., 2000c). When the input power was further increased, the output started to decrease continuously and self heating was observed. At 112 K, the critical operating power level (P_c) at 1 GHz drops moderately from 19 dBm to 16 dBm. Such results make Hg-1212 films very appealing for microwave applications at 77 K and higher temperatures. To make a comparison with the Tl-2212 film, the power handling capability is depicted at reduced temperature scale and shown for both Hg-1212 and Tl-2212 films. It is interesting to see that Tl-2212 has better temperature performance at the reduced temperature scale. It is not clear whether this is due to a difference in the intrinsic properties of the two films. If it is not, this result suggests that the quality of Hg-1212 films can be further improved.

18.5 REMAINING CHALLENGES

Despite the exciting progress achieved in epitaxy of Hg-HTS thin and thick films during the past few years, challenges remain in development of these materials for practical applications. First of all, it is necessary to scale up the existing film fabrication processes to either large area of the order of several inches in diameter or long length in the range of several kilometers. The former is required by many microelectronic device applications, such as Josephson junctions and microwave components/systems, and the latter is needed for coated conductors used for power transmission cables, superconducting magnets, motors/generators, etc. The major obstacle in scale-up Hg-HTS's is the high Hg-vapor pressure requirement. Up to now, almost all Hg-HTS films were made under high Hg-vapor pressures in the range of a few atmospheres. Although these high pressures can be achieved in a small size torch-sealed quartz ampoule of few millimeter in diameter using several solid pellets as the vapor source, it is not hard to imagine that extending such a configuration to large size will become incredibly difficult, if at all not impossible. It should be realized that it is unlikely possible to reduce the Hg-vapor pressure in the conventional process according to the required chemical phase equilibrium. Novel processes that allow epitaxy of Hg-HTS films at low Hg-vapor pressure, preferably close or even lower than the atmospheric pressure need to be developed in order to facilitate the system design for large-area or long-length

Hg-HTS films. Encouraging results have been obtained in our recent experiments in fabrication of Hg-1212 films at reduced Hg-vapor pressures using the cation-exchange process. We have found that the quality of the film remains the same although the Hg-vapor pressure was reduced to 1/4 of the original pressure and degrades slightly at 1/8 of the original pressure. This pressure may be further reduced. This means that it might be possible to demonstrate epitaxy Hg-1212 films at an atmospheric pressure, which facilitates design of new systems for scale-up of Hg-HTS films.

The second challenge in the research of Hg-HTS films is to develop multilayered structures. Although single-layered Hg-HTS films can be used directly for passive microwave devices and high-current-carrying superconducting cables, multilayered structures that comprise Hg-HTSs and other materials, such as insulators and metals, are desirable for many other applications including Josephson junction based electronic circuits. On the other hand, coated conductors may be benefited from superconductor/insulators multilayered structures to maximize the J_c. This seems impossible, at least at this stage, unless an *in situ* process could be applied for growth of Hg-HTS films.

ACKNOWLEDGMENTS

It is a pleasure to acknowledge my students and colleagues at the University of Kansas with whom I have collaborated with over the past few years on these studies. They include S.L. Yan, Y.Y. Xie, L. Fang, S.H. Yun, B.W. Kang, A. Gapud, T. Aytug, R. Aga, Jr., and S.Y. Han. I am very grateful to S.C. Tidrow, M.P. Siegal, and D.K. Christen, for valuable discussions and advice. This work has been supported in part by AFOSR, NSF, BMDO, DOE through ORNL and NREL contracts, NSF EPSCoR, DEPSCoR, and the University of Kansas new faculty startup and GRF funds.

REFERENCES

Adachi, H., Satoh, T., and Setsune, K., 1993a, Highly oriented Hg–Ba–Ca–Cu–O superconducting thin films, *Appl. Phys. Lett.*, 63:3628.

Adachi, H. et al., 1993b, *Jpn. J. Appl. Phys.*, 32:L1798.

Aga, R.S., Yan, S.L., Xie, Y.Y., Han, S.Y., and Wu, J.Z., 2000a, Microwave surface resistance of Hg-1212 thin films, *Appl. Phys. Lett.*, 76:1606.

Aga, R.S., Xie, Y.Y., Wu, J.Z., and Han, S., 2000b, Microwave characterization of $HgBa_2CaCu_2O_{6+\delta}$ thin films, *Physica C*, 341–348:2721.

Aga, R.S., Xie, Y.Y., Wu, J.Z., and Han, S., 2000c, Microwave power handling capability of Hg-1212 microstrip resonators, *Appl. Phys. Lett.* (submitted).

Amm, K.M., Wolters, Ch., Knoll, D.C., Peterson, S.C., and Schwartz, J., 1997, Growth of $HgBa_2Ca_2Cu_3O_{8+\delta}$ on a metallic substrate, *IEEE Trans. Appl. Supercond.*, 7:1973.

Antipov, E.V., Loureiro, S.M., Chaillout, C., Capponi, J.J., Bordet, P., Tholence, J.L., Putilin, S.N., and Marezio, M., 1993, The synthesis and characterization of the $HgBa_2Ca_2Cu_3O_{8+\delta}$ and $HgBa_2Ca_3Cu_4O_{10+\delta}$ phases, *Physica C*, 215:1.

Aytug, T., Kang, B.W., Yan, S.L., Xie, Y.Y., and Wu, J.Z., 1998, Stability of Hg-based superconducting thin films, *Physica C*, 307:117.

Aytug, T., Gapud, A.A., Yoo, S.W., Kang, B.W., and Wu, J.Z., 1999a, Effect of Na-doping on oxygen distribution of Hg-1223 superconductors, *Physica C*, 312.

Aytug, T., Yan, S.L., Xie, Y.Y., and Wu, J.Z., 1999b, Response of superconducting characteristics of Hg-1212 thin films to photolithographic processes, *Physica C*, 325:56.

Capponi, J.J., Kopnin, E.M., Loureiro, S.M., Antipov, E.V., Gautier, E., Chaillout, C., Souletie, B., Brunner, M., Tholence, J.L., and Marezio, M., 1996, High-pressure synthesis and heat treatments of the $HgBa_2Ca_4Cu_5O_{12+\delta}$ and $HgBa_2Ca_5Cu_6O_{14+\delta}$ phases, *Physica C*, 256:1.

Chu, C.W., Gao, L., Chen, F., Huang, Z.H., Meng, R.L., and Xue, Y.Y., 1993, Superconductivity above 150 K in $HgBa_2Ca_2Cu_3O_{8+\delta}$ at high pressure, *Nature*, 365:323.

Fang, L., Yan, S.L., Gapud, A.A., Aytug, T., Kang, B.W., Xie, Y.Y., Wu, J.Z., Tidrow, S.C., Liu, J.Z., and Chu, W.K., 2000, Epitaxial growth of very-thin Hg-1212 films, *Physica C* (to appear).

Foong, F., Bedard, B., Xu, Q.L., and Liou, S.H., 1996, C-oriented (Hg,Tl)-based superconducting films with $T_c > 125$ K, *Appl. Phys. Lett.*, 68:1153.

Gallop, J., *Supercond. Sci. Technol.*, 10:A120.

Gasser, G., Moriwaki, Y., Sugano, T., Nakanishi, K., Wu, X.J., Adachi, S., and Tanabe, K., 1998, Orientation control of *ex situ* $(Hg_{1-x}Re_x)Ba_2CaCu_2O_y$ $(x \approx 0.1)$ thin films on $LaAlO_3$, *Appl. Phys. Lett.*, 72:972.

Goyal, A., Norton, D.P., Budai, J.D., Paranthaman, M., Specht, E.D., Kroeger, D.M., Christen, D.K., He, Q., Saffian, B., List, F.A., Lee, D.F., Martin, P.M., Klabunde, C.E., Hartfield, E., and Sikka, V.K., 1996, High critical current density superconducting tapes by epitaxial deposition of YBCO thick films on biaxially textured metals, *Appl. Phys. Lett.*, 69:1795.

Gupta, A., Sun, J.Z., and Tsuei, C.C., 1994, Mercury-based cuprate high-transition temperature grain-boundary junctions and SQUIDs operating above 110 K, *Science*, 265:1075.

Gapud, A.A., Aytug, T., Xie, Y.Y., Yoo, S.H., Kang, B.W., Gapud, S.D., Wu, J.Z., Wu, S.M., Liang, W.Y., Cui, X.T., Liu, J.R., and Chu, W.K., 1998, Li-doping-assisted growth of Hg-1223 superconducting phase in bulks and thin films, *Physica C*, 308:264.

Gapud, A.A., Kang, B.W., Wu, J.Z., Yan, S.L., Xie, Y.Y., and Siegal, M.P., 1999a, Nature of giant T_c shift in '1212' superconductors due to Hg/Tl exchange, *Phys. Rev. B*, 59:203.

Gapud, A.A., Wu, J.Z., Kang, B.W. et al., 1999b, Critical current densities of Hg-1212 and Tl-1212 thin films, *Appl. Phys. Lett.*, 9:2387.

Goyal, A., Norton, D.P., Budai, J.D. et al., 1996, *Appl. Phys. Lett.*, 69:1795.

Guo, J.D., Xiong, G.C., Yu, D.P., Feng, Q.R., Xu, X.L., Lian, G.L., and Hu, Z.H., 1997a, Preparation of superconducting $HgBa_2CaCu_2O_y$ films by means of annealing of mercury-free precursor films, *Physica C*, 276:277.

Guo, J.D., Xiong, G.C., Yu, D.P., Feng, Q.R., Xu, X.L., Lian, G.L., Xiu, K., and Hu, Z.H., 1997b, Preparation of superconducting $HgBa_2CaCu_2O_y$ and $Hg_{0.8}Bi_{0.2}Ba_2CaCu_2O_y$ films with a zero-resistance transition temperature of 121 K, *Physica C*, 282–287:645.

Heine, Tenbrink, J., and Thoner, M., *Appl. Phys. Lett.*, 55:2441.

Holstein, W.L., Parisi, L.A., Wilker, C., and Flippen, R.B., 1992, *Appl. Phys. Lett.*, 60:2014-2016.

Higuma, H., Miyashita, S., and Uchikawa, F., 1994, Synthesis of superconducting Pb-doped $HgBa_2CaCu_2O_y$ films by laser ablation and post annealing, *Appl. Phys. Lett.*, 65:743.

Huang, Z.J., Xue, Y.Y., Meng, R.L., and Chu, C.W., 1994, Irreversibility line of the $HgBa_2CaCu_2O_{6+\delta}$ high-temperature superconductors, *Phys. Rev. B*, 49:4218.

Iijima, Y., Tanabe, N., Kohno, O., and Ikeno, Y., 1992, *Appl. Phys. Lett.*, 60:769.

Iijima, Y., Onabe, K., Futaki, N., Sadakata, N., and Kohno, O., 1993, *J. Appl. Phys.*, 74:1905.

Ivanov, Z. and Johansson, L.-G., Thallium- and mercury containing cuprates in electronic devices, *Supercond. Sci. Technol.*, 10:896.

Lancaster, M.J., *Passive Microwave Device Applications of HTS*, Cambridge University Press, Cambridge, p. 315.

Kang, B.W., Gapud, A.A., Fei, X., Aytug, T., and Wu, J.Z., 1998, Minimization of detrimental effect of air in $HgBa_2CaCu_2O_{6+\delta}$ thin film processing, *Appl. Phys. Lett.*, 72:1766.

Kang, W.N., Meng, R.L., and Chu, C.W., 1998, Growth of $HgBa_2Ca_2Cu_3O_{8+\delta}$ thin films using stable $Re_{0.1}Ba_2Ca_2Cu_3O_x$ precursor by pulsed laser deposition, *Appl. Phys. Lett.*, 73:381.

Kang, W.N., Lee, Sung-lk, and Chu, C.W., 1999, Oxygen annealing and superconductivity of $HgBa_2Ca_2Cu_3O_{8+\delta}$ thin films, *Physica C*, 315:223.

Krusin-Elbaum, L., Tsuei, C.C., and Gupta, A., 1995, High current densities above 100 K in the high-temperature superconductor $HgBa_2CaCu_2O_{6+\delta}$, *Nature*, 373:679.

Krusin-Elbaum, L., Thompson, J.R., Wheeler, R., Ullmann, J., Chu, C.W., and Lin, Q.M., 1997a, *Nature*, 389:243.

Krusin-Elbaum, L. et al., 1997b, in: *Proceedings of the 8th US–Japan Workshop on HTS's*, p. 158.

Lechter, W., Toth, L., Osofsky, M., Skelton, E., Soulen, R.J., Jr., Qadri, S., Schwartz, J., Wolters, Ch., and Kessler, J., 1995, One step reaction and consolidation of Hg based high temperature superconductors by hot isostatic pressing, *Physica C*, 249:213.

Locquet, L.-P., Perret, J., Fompeyrine, J., Machler, E., Seo, J.W., and Van Tendeloo, G., 1998, Doubling the critical temperature of $La_{1.9}Sr_{0.1}CuO_4$ using epitaxial strain, *Nature*, 394:453.

Malozemoff, A.P., 1996, in: *Proceedings of 10th Anniversary HTS Workshop*, Houston, TX, March 12–16.

Meng, R.L., Hickey, B.R., Wang, Y.Q., Sun, Y.Y., Gao, L., Xue, Y.Y., and Chu, C.W., 1996, Processing of highly oriented $(Hg_{1-x}Re_x)Ba_2Cu_3O_{8+\delta}$ tape with $x \sim 0.1$, *Appl. Phys. Lett.*, 68:3177.

Miyashita, S., Higuma, H., and Uchikawa, F., 1994, Structure and superconducting properties of $HgBa_2CaCu_2O_y$ films prepared by laser ablation, *Jpn. J. Appl. Phys.*, 33(7A):L931.

Moriwaki, Y., Sugano, T., Tsukamoto, A., Gasser, C., Nakanishi, K., Adachi, S., and Tanabe, K., 1998, Fabrication and properties of *c*-axis Hg-1223 superconducting thin films, *Physica C*, 303:65.

Norton, D.P. et al., 1996, *Science*, 274:755.

Peacock, G.B., Gameson, I., Edwards, P.P., Khaliq, M., Yang, G., Shields, T.C., and Abell, J.S., 1997, Fabrication of high-temperature superconducting $HgBa_2CuO_{4+\delta}$ within silver-sheathed tapes, *Physica C*, 273:193.

Putilin, S.N., Antipov, E.V., Chmaissem, O., and Marezio, M., 1993a, Superconductivity at 94 K in $HgBa_2CuO_{4+\delta}$, *Nature*, 362:226.

Putilin, S.N., Antipov, E.V., and Marezio, M., 1993b, Superconductivity above 120 K in $HgBa_2CaCu_2O_{6+\delta}$, *Physica C*, 212:266.

Reade, R.P., Burdahl, P., Russo, R.E., and Garrison, S.M., 1993, *Appl. Phys. Lett.*, 61:2231.

Sastry, P.V.O.S.S., Li, Y., Su, J.H., and Schwartz, J., 2000, Appempts to fabricate thick HgPb1223 superconducting films on silver, *Physica C*, 335:112.

Schwartz, J., Amm, K.M., Sun, Y.R., and Wolters, Ch., 1996, HgBaCaCuO superconductors: Processing, properties, and potential, *Physica B*, 216:261.

Schwartz, J. and Sastry, P.V.O.S.S., 2001, Emerging materials, in: *Handbook of Superconducting Materials* (in press).

Schilling, A.M., Cantoni, J.D., Guo, H., and Ott, R., 1993, Superconductivity above 130 K in the Hg–Ba–Ca–Cu–O system, *Nature*, 363:56.

Siegal, M.P., Venturini, E.L., and Aselage, T.L., 1997, Synthesis and properties of Tl–Ba–Ca–Cu–O superconductors, *J. Mat. Res.*, 12:2825.

Su, Jianhua, Sastry, P.V.P.S.S., and Schwartz, J., 2000, Growth of $Hg_{0.8}Pb_{0.2}Ba_2CaCu_2O_{6+\delta}$ thick films on Ag using a modified process route, in: *Proceeding of Appl. Supercond. Conf.*

Sun, Y., Guo, J.D., Xu, X.L., Lian, G.J., Wanh, Y.Z., and Xiong, G.C., 1999, Superconducting $HgBa_2CaCu_2O_y$ thin films grown on $NdGaO_3$, $SrTiO_3$, $LaAlO_3$, and Y-ZrO_2 substrates, *Physica C*, 312:197.

Thompson, J.R., Krusin-Elbaum, L., Christen, D.K., Ullman, J.L., Wu, J.Z., Ren, Z.F., Tkaczyk, J.E., and Deluca, J.A., 1997, Generic enhancements of the highest-T_c superconductors with GeV protons, *Appl. Phys. Lett.*, 71:536.

Tsuei, C.C., Gupta, A., Trafas, G., and Mitzi, D., 1994, Superconducting mercury-based cuprate films with a zero-resistance transition temperature of 124 Kelvin, *Science*, 263:1259.

Tsukamoto, A., Takagi, K., Moriwaki, Y., Sugano, T., Adachi, S., and Tanabe, K., 1998, High-performance $(Hg,Re)Ba_2CaCu_2O_y$ grain-boundary Josephson junctions and dc superconducting quantum interference devices, *Appl. Phys. Lett.*, 73:990.

Welp, U., Crabtree, G.W., Wagner, J.L., and Hinks, D.G., 1993, Flux pinning and the irreversibility lines in the $HgBa_2CuO_{4+\delta}$, $HgBa_2CaCu_2O_{6+\delta}$ and $HgBa_2Ca_2Cu_3O_{8+\delta}$ compounds, *Physica C*, 218:373.

Wang, Y.Q., Meng, R.L., Sun, Y.Y., Huang, Z.J., Ross, K., and Chu, C.W., 1993, Synthesis of preferred-oriented $HgBa_2CaCu_2O_{6+\delta}$ thin films, *Appl. Phys. Lett.*, 63:3084.

Wu, J.Z., 2000, Cation exchange: Synthesis and epitaxy of Hg-based superconductin thin films via atomic perturbations, in: *Proceeding of SPIE Aerosense Conference*, Orlando, FL, April 24–28.

Wu, J.Z., Yun, S.H., Kang, W.N., Kang, B.W., Gapud, A.A., Tidrow, S.C., and Eckart, D., 1996a, Fabrication of high quality Hg-1212 and Hg-1223 thin films, in: *Proceedings of 10th Anniversary HTS Workshop on Physics, Materials and Applications*, World Scientific, Singapore, p. 89.

Wu, J.Z., Eckart, D., Yun, S.H., and Tidrow, S., 1996b, Microstructures of mercury-based cuprate thin films, *Physica C*, 271:1.

Wu, J.Z., Yun, S.H., Gapud, A., Kang, B.W., Kang, W.N., Tidrow, S.C., Monahan, T.P., Cui, X.T., and Chu, W.K., 1997a, Epitaxial growth of $HgBa_2CaCu_2O_{6+\delta}$ thin films on $SrTiO_3$ substrates, *Physica C*, 277:219.

Wu, J.Z., Yun, S.H., Gapud, A., Kang, B.W., Kang, W.N., Tidrow, S.C., Eckart, D., Cui, X.T., and Chu, W.K., 1997b, Fast temperature ramping Hg-vapor annealing technique for growth of Hg-based cuprate thin films, *IEEE Trans. Appl. Supercond.*, 7:1907.

Wu, J.Z., Yoo, S.W., Aytug, T., Gapud, A., Kang, B.W., Wu, S., and Zhou, W., 1998, Superconductivity in sodium and lithium doped mercury-based cuprates, *J. Superconductivity*, 11:169.

Wu, J.Z., Yan, S.L., and Xie, Y.Y., 1999, Cation exchange: A new scheme for synthesis of high-quality epitaxial Hg-based superconducting thin films, *Appl. Phys. Lett.*, 74:1469.

Wu, J.Z. and Tidrow, S.N., 1999, Recent progress in high-T_c superconducting heterostructures, in: *Thin Films: Heteroepitaxial Systems*, W.K. Liu and M.B. Santos, eds., World Scientific, Singapore, p. 267.

Wu, J.Z. and Xie, Y.Y., 2001, Over-doping Hg-1212 thin films, Preprint.

Wu, X.D., Foltyn, S.R., Arendt, P.N. et al., 1995, *Appl. Phys. Lett.*, 67:2397.

Xie, Y.Y., Wu, J.Z., Gapud, A., Yu, Y., and Xin, Y., 1999a, Synthesis of Hg-1223 superconductors using a cation-exchange process, *Physica C*, 322:19–24.

Xie, Y.Y., Wu, J.Z., Yan, S.L., Yu, Y., Aytug, T., and Fang, L., 1999b, Elimination of air detrimental effect using Tl-assisted growth process for Hg-1212 thin films, *Physica C*, 328:241.

Xie, Y.Y., Aytug, T., Wu, J.Z., and Christen, D.K., 2000a, Growth and physical properties of Hg-1212 HTSC tapes on buffered metal substrates, *Physica C* (to appear).

Xie, Y.Y., Aytug, T., Wu, J.Z., Verebelyi, D.T., Paranthaman, M., Goyal, A., and Christen, D.K., 2000b, Epitaxy of Hg-1212 superconducting films on textured Ni substrates, *Appl. Phys. Lett.* (to appear).

Xie, Y.Y., Wu, J.Z., Aytug, T., Gapud, A.A., Christen, D.K., Verebelyi, D.T., and Song, K., 2000c, Uniformity of the physical properties of large-area Hg-1212 thin films, *Supercond. Sci. Technol.*, 13:225.

Xue, Y.Y., Meng, R.L., Lin, Q.M., Hickey, B., Sun, Y.Y., and Chu, C.W., 1996, Phase stability and defects of $HgBa_2Ca_2Cu_3O_{8+\delta}$, in: *Proceedings of the 10th Anniversary HTS Workshop, Houston*, March 12–16, p. 93.

Yan, S.L., Xie, Y.Y., Wu, J.Z., Aytug, T., Gapud, A.A., Kang, B.W., Fang, L., He, M., Tidrow, S.C., Kirchner, K.W., Liu, J.R., and Chu, W.K., 1998, High critical current density in epitaxial Hg-1212 thin films, *Appl. Phys. Lett.*, 73:2989.

Yan, S.L., Wu, J.Z., Fang, L., Xie, Y.Y., Aytug, T., Gapud, A.A., and Kang, B.W., 2000, Converting Tl-2212 thin films to Hg-1212 thin films using cation-exchange process, *J. Appl. Phys.* (submitted).

Yan, S.L., Fang, L., and Wu, J.Z., 2000, Epitaxy of Hg-1223 films using cation exchange process, Preprint.

Yu, Y., Shao, H.M., Zheng, Z.Y., Sun, A.M., Qin, M.J., Xu, X.N., Ding, S.Y., Jin, X., Yan, X.X., Zhou, J., Ji, Z.M., Yang, S.Z., and Zhang, W.L., 1997, $HgBa_2CaCu_2O_y$ superconducting thin films prepared by laser ablation, *Physica C*, 289:199.

Yu, Yang, Yan, S.L., Fang, L., Xie, Y.Y., Wu, J.Z., Han, Siyuan, Shimakage, H., and Wang, Zhen, 1999, Fabrication of $HgBa_2CaCu_2O_y$ grain boundary junctions using the cation exchange method, *Supercond. Sci. Technol.*, 12:1020–1022.

Yun, S.H. and Wu, J.Z., 1996, Superconductivity above 130 K in high-quality mercury-based cuprate thin films, *Appl. Phys. Lett.*, 68:862.

Yun, S.H., Wu, J.Z., Tidrow, S.C., and Eckart, D.W., 1996b, Growth of $HgBa_2Ca_2Cu_3O_{8+\delta}$ thin films on $LaAlO_3$ substrates using fast temperature ramping Hg-vapor annealing, *Appl. Phys. Lett.*, 68:2565.

SUBJECT INDEX

www.ingramcontent.com/pod-product-compliance
Lightning Source LLC
LaVergne TN
LVHW020018181026
850497LV00003B/110
* 9 7 8 1 4 0 2 0 8 1 1 7 0 *